Advances in Intelligent Systems and Computing

Volume 932

The series "Advances in Intelligent Systems and Computing" contains publications on theory, applications, and design methods of Intelligent Systems and Intelligent Computing. Virtually all disciplines such as engineering, natural sciences, computer and information science, ICT, economics, business, e-commerce, environment, healthcare, life science are covered. The list of topics spans all the areas of modern intelligent systems and computing such as: computational intelligence, soft computing including neural networks, fuzzy systems, evolutionary computing and the fusion of these paradigms, social intelligence, ambient intelligence, computational neuroscience, artificial life, virtual worlds and society, cognitive science and systems, Perception and Vision, DNA and immune based systems, self-organizing and adaptive systems, e-Learning and teaching, human-centered and human-centric computing, recommender systems, intelligent control, robotics and mechatronics including human-machine teaming, knowledge-based paradigms, learning paradigms, machine ethics, intelligent data analysis, knowledge management, intelligent agents, intelligent decision making and support, intelligent network security, trust management, interactive entertainment, Web intelligence and multimedia.

The publications within "Advances in Intelligent Systems and Computing" are primarily proceedings of important conferences, symposia and congresses. They cover significant recent developments in the field, both of a foundational and applicable character. An important characteristic feature of the series is the short publication time and world-wide distribution. This permits a rapid and broad dissemination of research results.

** Indexing: The books of this series are submitted to ISI Proceedings, EI-Compendex, DBLP, SCOPUS, Google Scholar and Springerlink **

More information about this series at http://www.springer.com/series/11156

Álvaro Rocha · Hojjat Adeli ·
Luís Paulo Reis · Sandra Costanzo
Editors

New Knowledge
in Information Systems
and Technologies

Volume 3

 Springer

Editors
Álvaro Rocha
Departamento de Engenharia Informática
Universidade de Coimbra
Coimbra, Portugal

Luís Paulo Reis
Faculdade de Engenharia/LIACC
Universidade do Porto
Porto, Portugal

Hojjat Adeli
The Ohio State University
Columbus, OH, USA

Sandra Costanzo
DIMES
Università della Calabria
Arcavacata di Rende, Italy

ISSN 2194-5357 ISSN 2194-5365 (electronic)
Advances in Intelligent Systems and Computing
ISBN 978-3-030-16186-6 ISBN 978-3-030-16187-3 (eBook)
https://doi.org/10.1007/978-3-030-16187-3

Library of Congress Control Number: 2019934961

This Springer imprint is published by the registered company Springer Nature Switzerland AG
The registered company address is: Gewerbestrasse 11, 6330 Cham, Switzerland

Preface

This book contains a selection of papers accepted for presentation and discussion at The 2019 World Conference on Information Systems and Technologies (WorldCIST'19). This Conference had the support of IEEE SMC (IEEE Systems, Man, and Cybernetics Society), AISTI (Iberian Association for Information Systems and Technologies/Associação Ibérica de Sistemas e Tecnologias de Informação), GIIM (Global Institute for IT Management), and University of Vigo. It took place at La Toja, Galicia, Spain, April 16–19, 2019.

The World Conference on Information Systems and Technologies (WorldCIST) is a global forum for researchers and practitioners to present and discuss recent results and innovations, current trends, professional experiences and challenges of modern Information Systems and Technologies research, technological development and applications. One of its main aims is to strengthen the drive toward a holistic symbiosis between academy, society, and industry. WorldCIST'19 built on the successes of WorldCIST'13 held at Olhão, Algarve, Portugal; WorldCIST'14 held at Funchal, Madeira, Portugal; WorldCIST'15 held at São Miguel, Azores, Portugal; WorldCIST'16 held at Recife, Pernambuco, Brazil; WorldCIST'17 held at Porto Santo, Madeira, Portugal; and WorldCIST'18 took place at Naples, Italy.

The Program Committee of WorldCIST'19 was composed of a multidisciplinary group of more than 200 experts and those who are intimately concerned with Information Systems and Technologies. They have had the responsibility for evaluating, in a 'blind review' process, the papers received for each of the main themes proposed for the Conference: (A) Information and Knowledge Management; (B) Organizational Models and Information Systems; (C) Software and Systems Modeling; (D) Software Systems, Architectures, Applications and Tools; (E) Multimedia Systems and Applications; (F) Computer Networks, Mobility and Pervasive Systems; (G) Intelligent and Decision Support Systems; (H) Big Data Analytics and Applications; (I) Human Computer Interaction; (J) Ethics, Computers and Security; (K) Health Informatics; (L) Information Technologies in Education; (M) Information Technologies in Radiocommunications; and (N) Technologies for Biomedical Applications.

The Conference also included workshop sessions taking place in parallel with the conference ones. Workshop sessions covered themes such as: (i) Air Quality and Open Data: Challenges for Data Science, HCI and AI; (ii) Digital Transformation; (iii) Empirical Studies in the Domain of Social Network Computing; (iv) Health Technology Innovation: Emerging Trends and Future Challenges; (v) Healthcare Information Systems Interoperability, Security and Efficiency; (vi) New Pedagogical Approaches with Technologies; (vii) Pervasive Information Systems.

WorldCIST'19 received about 400 contributions from 61 countries around the world. The papers accepted for presentation and discussion at the Conference are published by Springer (this book) in three volumes and will be submitted for indexing by ISI, Ei Compendex, Scopus, DBLP, and/or Google Scholar, among others. Extended versions of selected best papers will be published in special or regular issues of relevant journals, mainly SCI/SSCI and Scopus/Ei Compendex indexed journals.

We acknowledge all of those that contributed to the staging of WorldCIST'19 (authors, committees, workshop organizers, and sponsors). We deeply appreciate their involvement and support that was crucial for the success of WorldCIST'19.

April 2019

Álvaro Rocha
Hojjat Adeli
Luís Paulo Reis
Sandra Costanzo

Organization

Conference

General Chair

Álvaro Rocha — University of Coimbra, Portugal

Co-chairs

Hojjat Adeli — The Ohio State University, USA
Luis Paulo Reis — University of Porto, Portugal
Sandra Costanzo — University of Calabria, Italy

Local Chair

Manuel Pérez Cota — University of Vigo, Spain

Advisory Committee

Ana Maria Correia (Chair) — University of Sheffield, UK
Andrew W. H. Ip — Hong Kong Polytechnic University, China
Cihan Cobanoglu — University of South Florida, USA
Chris Kimble — KEDGE Business School and MRM, UM2, Montpellier, France
Erik Bohlin — Chalmers University of Technology, Sweden
Eva Onaindia — Universidad Politecnica de Valencia, Spain
Eugene H. Spafford — Purdue University, USA
Gintautas Dzemyda — Vilnius University, Lithuania
Gregory Kersten — Concordia University, Canada
Janusz Kacprzyk — Polish Academy of Sciences, Poland

João Tavares	University of Porto, Portugal
Jon Hall	The Open University, UK
Karl Stroetmann	Empirica Communication & Technology Research, Germany
Kathleen Carley	Carnegie Mellon University, USA
Keng Siau	Missouri University of Science and Technology, USA
Salim Hariri	University of Arizona, USA
Marjan Mernik	University of Maribor, Slovenia
Michael Koenig	Long Island University, USA
Miguel-Angel Sicilia	Alcalá University, Spain
Peter Sloot	University of Amsterdam, the Netherlands
Reza Langari	Texas A&M University, USA
Robert J. Kauffman	Singapore Management University, Singapore
Wim Van Grembergen	University of Antwerp, Belgium

Program Committee

Abdul Rauf	RISE SICS, Sweden
Adnan Mahmood	Waterford Institute of Technology, Ireland
Adriana Peña Pérez Negrón	Universidad de Guadalajara, Mexico
Adriani Besimi	South East European University, Macedonia
Agostinho Sousa Pinto	Polytecnic of Porto, Portugal
Ahmed El Oualkadi	Abdelmalek Essaadi University, Morocco
Alan Ramirez-Noriega	Universidad Autónoma de Sinaloa, Mexico
Alberto Freitas	FMUP, University of Porto, Portugal
Aleksandra Labus	University of Belgrade, Serbia
Alexandru Vulpe	Politehnica University of Bucharest, Romania
Ali Alsoufi	University of Bahrain, Bahrain
Ali Idri	ENSIAS, Mohammed V University, Morocco
Almir Souza Silva Neto	IFMA, Brazil
Amit Shelef	Sapir Academic College, Israel
Ana Isabel Martins	University of Aveiro, Portugal
Ana Luis	University of Coimbra, Portugal
Anabela Tereso	University of Minho, Portugal
Anacleto Correia	CINAV, Portugal
Anca Alexandra Purcarea	Politehnica University of Bucharest, Romania
André Marcos Silva	Centro Universitário Adventista de São Paulo (UNASP), Brazil
Aneta Poniszewska-Maranda	Lodz University of Technology, Poland
Angeles Quezada	Instituto Tecnologico de Tijuana, Mexico
Ankur Singh Bist	KIET, India
Antoni Oliver	University of the Balearic Islands, Spain
Antonio Borgia	University of Calabria, Italy

Francesco Bianconi   Università degli Studi di Perugia, Italy
Francisco García-Peñalvo   University of Salamanca, Spain
Francisco Valverde   Universidad Central del Ecuador, Ecuador
Frederico Branco   University of Trás-os-Montes e Alto Douro,
                                Portugal
Gabriel Pestana   Universidade Europeia, Portugal
Galim Vakhitov   Kazan Federal University, Russia
George Suciu   BEIA, Romania
Ghani Albaali   Princess Sumaya University for Technology,
                                Jordan
Gian Piero Zarri   University Paris-Sorbonne, France
Giuseppe Di Massa   University of Calabria, Italy
Gonçalo Paiva Dias   University of Aveiro, Portugal
Goreti Marreiros   ISEP/GECAD, Portugal
Graciela Lara López   University of Guadalajara, Mexico
Habiba Drias   University of Science and Technology Houari
                                Boumediene, Algeria
Hafed Zarzour   University of Souk Ahras, Algeria
Hamid Alasadi   University of Basra, Iraq
Hatem Ben Sta   University of Tunis at El Manar, Tunisia
Hector Fernando   Universidad Tecnica de Ambato, Ecuador
  Gomez Alvarado
Hélder Gomes   University of Aveiro, Portugal
Helia Guerra   University of the Azores, Portugal
Henrique da Mota Silveira   University of Campinas (UNICAMP), Brazil
Henrique S. Mamede   University Aberta, Portugal
Hing Kai Chan   University of Nottingham Ningbo China, China
Hugo Paredes   INESC TEC and Universidade de
                                Trás-os-Montes e Alto Douro, Portugal
Ibtissam Abnane   Mohammed V University in Rabat, Morocco
Imen Ben Said   Université de Sfax, Tunisia
Ina Schiering   Ostfalia University of Applied Sciences,
                                Germany
Inês Domingues   University of Coimbra, Portugal
Isabel Lopes   Instituto Politécnico de Bragança, Portugal
Isabel Pedrosa   Coimbra Business School ISCAC, Portugal
Isaías Martins   University of Leon, Spain
Ivan Lukovic   University of Novi Sad, Serbia
Jan Kubicek   Technical University of Ostrava, Czech Republic
Jean Robert Kala Kamdjoug   Catholic University of Central Africa, Cameroon
Jesús Gallardo Casero   University of Zaragoza, Spain
Jezreel Mejia   CIMAT Unidad Zacatecas, Mexico
Jikai Li   The College of New Jersey, USA
Jinzhi Lu   KTH Royal Institute of Technology, Sweden
Joao Carlos Silva   IPCA, Portugal

João Manuel R. S. Tavares	University of Porto, FEUP, Portugal
João Reis	University of Lisbon, Portugal
João Rodrigues	University of Algarve, Portugal
Jorge Barbosa	Polytecnic Institute of Coimbra, Portugal
Jorge Buele	Technical University of Ambato, Ecuador
Jorge Esparteiro Garcia	Polytechnic Institute of Viana do Castelo, Portugal
Jorge Gomes	University of Lisbon, Portugal
Jorge Oliveira e Sá	University of Minho, Portugal
José Álvarez-García	University of Extremadura, Spain
José Braga de Vasconcelos	Universidade New Atlântica, Portugal
Jose Luis Herrero Agustin	University of Extremadura, Spain
José Luís Reis	ISMAI, Portugal
Jose Luis Sierra	Complutense University of Madrid, Spain
Jose M. Parente de Oliveira	Aeronautics Institute of Technology, Brazil
José Machado	University of Minho, Portugal
José Martins	Universidade de Trás-os-Montes e Alto Douro, Portugal
Jose Torres	University Fernando Pessoa, Portugal
José-Luís Pereira	Universidade do Minho, Portugal
Juan Jesus Ojeda-Castelo	University of Almeria, Spain
Juan M. Santos	University of Vigo, Spain
Juan Pablo Damato	UNCPBA-CONICET, Argentina
Juncal Gutiérrez-Artacho	University of Granada, Spain
Justyna Trojanowska	Poznan University of Technology, Poland
Katsuyuki Umezawa	Shonan Institute of Technology, Japan
Khalid Benali	LORIA, University of Lorraine, France
Korhan Gunel	Adnan Menderes University, Turkey
Krzysztof Wolk	Polish-Japanese Academy of Information Technology, Poland
Kuan Yew Wong	Universiti Teknologi Malaysia (UTM), Malaysia
Laila Cheikhi	Mohammed V University, Rabat, Morocco
Laura Varela-Candamio	Universidade da Coruña, Spain
Laurentiu Boicescu	E.T.T.I. U.P.B., Romania
Leonardo Botega	University Centre Eurípides of Marília (UNIVEM), Brazil
Leonid Leonidovich Khoroshko	Moscow Aviation Institute (National Research University), Russia
Letícia Helena Januário	Universidade Federal de São João del-Rei, Brazil
Lila Rao-Graham	University of the West Indies, Jamaica
Luis Alvarez Sabucedo	University of Vigo, Spain
Luis Mendes Gomes	University of the Azores, Portugal
Luiz Rafael Andrade	Tiradentes University, Brazil
Luis Silva Rodrigues	Polytencic of Porto, Portugal

Luz Sussy Bayona Oré	Universidad Nacional Mayor de San Marcos, Peru
Magdalena Diering	Poznan University of Technology, Poland
Manuel Antonio Fernández-Villacañas Marín	Technical University of Madrid, Spain
Manuel Pérez Cota	University of Vigo, Spain
Manuel Silva	Polytechnic of Porto and INESC TEC, Portugal
Manuel Tupia	Pontifical Catholic University of Peru, Peru
Marco Ronchetti	Università di Trento, Italy
Mareca María PIlar	Universidad Politécnica de Madrid, Spain
Marek Kvet	Zilinska Univerzita v Ziline, Slovakia
María de la Cruz del Río-Rama	University of Vigo, Spain
Maria João Ferreira	Universidade Portucalense, Portugal
Maria João Varanda Pereira	Polytechnic Institute of Bragança, Portugal
Maria José Sousa	University of Coimbra, Portugal
María Teresa García-Álvarez	University of A Coruna, Spain
Marijana Despotovic-Zrakic	Faculty Organizational Science, Serbia
Mário Antunes	Polytecnic of Leiria and CRACS INESC TEC, Portugal
Marisa Maximiano	Polytechnic of Leiria, Portugal
Marisol Garcia-Valls	Universidad Carlos III de Madrid, Spain
Maristela Holanda	University of Brasilia, Brazil
Marius Vochin	E.T.T.I. U.P.B., Romania
Marlene Goncalves da Silva	Universidad Simón Bolívar, Venezuela
Maroi Agrebi	University of Polytechnique Hauts-de-France, France
Martin Henkel	Stockholm University, Sweden
Martín López Nores	University of Vigo, Spain
Martin Zelm	INTEROP-VLab, Belgium
Mawloud Mosbah	University 20 Août 1955 of Skikda, Algeria
Michal Adamczak	Poznan School of Logistics, Poland
Michal Kvet	University of Zilina, Slovakia
Miguel António Sovierzoski	Federal University of Technology - Paraná, Brazil
Mihai Lungu	Craiova University, Romania
Milton Miranda	Federal University of Uberlândia, Brazil
Mircea Georgescu	Al. I. Cuza University of Iasi, Romania
Mirna Muñoz	Centro de Investigación en Matemáticas A.C., Mexico
Mohamed Hosni	ENSIAS, Morocco
Mokhtar Amami	Royal Military College of Canada, Canada
Monica Leba	University of Petrosani, Romania

Muhammad Nawaz	Institute of Management Sciences, Peshawar, Pakistan
Mu-Song Chen	Dayeh University, China
Nastaran Hajiheydari	York St John University, UK
Natalia Grafeeva	Saint Petersburg State University, Russia
Natalia Miloslavskaya	National Research Nuclear University MEPhI, Russia
Naveed Ahmed	University of Sharjah, United Arab Emirates
Nelson Rocha	University of Aveiro, Portugal
Nelson Salgado	Pontifical Catholic University of Ecuador, Ecuador
Nikolai Prokopyev	Kazan Federal University, Russia
Niranjan S. K.	JSS Science and Technology University, India
Noemi Emanuela Cazzaniga	Politecnico di Milano, Italy
Noor Ahmed	AFRL/RI, USA
Noureddine Kerzazi	Polytechnique Montréal, Canada
Nuno Melão	Polytechnic of Viseu, Portugal
Nuno Octávio Fernandes	Polytechnic Institute of Castelo Branco, Portugal
Paôla Souza	Aeronautics Institute of Technology, Brazil
Patricia Zachman	Universidad Nacional del Chaco Austral, Argentina
Paula Alexandra Rego	Polytechnic Institute of Viana do Castelo and LIACC, Portugal
Paula Viana	Polytechnic of Porto and INESC TEC, Portugal
Paulo Maio	Polytechnic of Porto, ISEP, Portugal
Paulo Novais	University of Minho, Portugal
Paweł Karczmarek	The John Paul II Catholic University of Lublin, Poland
Pedro Henriques Abreu	University of Coimbra, Portugal
Pedro Rangel Henriques	University of Minho, Portugal
Pedro Sobral	University Fernando Pessoa, Portugal
Pedro Sousa	University of Minho, Portugal
Philipp Brune	University of Applied Sciences Neu-Ulm, Germany
Piotr Kulczycki	Systems Research Institute, Polish Academy of Sciences, Poland
Prabhat Mahanti	University of New Brunswick, Canada
Radu-Emil Precup	Politehnica University of Timisoara, Romania
Rafael M. Luque Baena	University of Malaga, Spain
Rahim Rahmani	Stockholm University, Sweden
Raiani Ali	Bournemouth University, UK
Ramayah T.	Universiti Sains Malaysia, Malaysia
Ramiro Delgado	Universidad de las Fuerzas Armadas ESPE, Ecuador

Ramiro Gonçalves	University of Trás-os-Montes e Alto Douro and INESC TEC, Portugal
Ramon Alcarria	Universidad Politécnica de Madrid, Spain
Ramon Fabregat Gesa	University of Girona, Spain
Reyes Juárez Ramírez	Universidad Autonoma de Baja California, Mexico
Rui Jose	University of Minho, Portugal
Rui Pitarma	Polytechnic Institute of Guarda, Portugal
Rui S. Moreira	UFP & INESC TEC & LIACC, Portugal
Rustam Burnashev	Kazan Federal University, Russia
Saeed Salah	Al-Quds University, Palestine
Said Achchab	Mohammed V University in Rabat, Morocco
Sajid Anwar	Institute of Management Sciences, Peshawar, Pakistan
Salama Mostafa	Universiti Tun Hussein Onn Malaysia, Malaysia
Sami Habib	Kuwait University, Kuwait
Samuel Fosso Wamba	Toulouse Business School, France
Sanaz Kavianpour	University of Technology, Malaysia
Sandra Costanzo	University of Calabria, Italy
Sandra Patricia Cano Mazuera	University of San Buenaventura Cali, Colombia
Sergio Albiol-Pérez	University of Zaragoza, Spain
Shahnawaz Talpur	Mehran University of Engineering and Technology, Jamshoro, Pakistan
Silviu Vert	Politehnica University of Timisoara, Romania
Simona Mirela Riurean	University of Petrosani, Romania
Slawomir Zolkiewski	Silesian University of Technology, Poland
Solange N. Alves-Souza	University of São Paulo, Brazil
Solange Rito Lima	University of Minho, Portugal
Sorin Zoican	Polytechnica University of Bucharest, Romania
Souraya Hamida	University of Batna 2, Algeria
Stefan Pickl	UBw München COMTESSA, Germany
Sümeyya Ilkin	Kocaeli University, Turkey
Syed Asim Ali	University of Karachi, Pakistan
Taoufik Rachad	Mohammed V University, Morocco
Tatiana Antipova	Institute of certified Specialists, Russia
The Thanh Van	HCMC University of Food Industry, Vietnam
Thomas Weber	EPFL, Switzerland
Timothy Asiedu	TIM Technology Services Ltd., Ghana
Tom Sander	New College of Humanities, Germany
Tomaž Klobučar	Jozef Stefan Institute, Slovenia
Toshihiko Kato	University of Electro-Communications, Japan
Tzung-Pei Hong	National University of Kaohsiung, Taiwan

Valentina Colla	Scuola Superiore Sant'Anna, Italy
Veronica Segarra Faggioni	Private Technical University of Loja, Ecuador
Victor Alves	University of Minho, Portugal
Victor Georgiev	Kazan Federal University, Russia
Victor Hugo Medina Garcia	Universidad Distrital Francisco José de Caldas, Colombia
Vincenza Carchiolo	University of Catania, Italy
Vitalyi Igorevich Talanin	Zaporozhye Institute of Economics and Information Technologies, Ukraine
Wolf Zimmermann	Martin Luther University Halle-Wittenberg, Germany
Yadira Quiñonez	Autonomous University of Sinaloa, Mexico
Yair Wiseman	Bar-Ilan University, Israel
Yuhua Li	Cardiff University, UK
Yuwei Lin	University of Roehampton, UK
Yves Rybarczyk	Universidad de Las Américas, Ecuador
Zorica Bogdanovic	University of Belgrade, Serbia

Workshops

First Workshop on Air Quality and Open Data: Challenges for Data Science, HCI and AI

Organizing Committee

Kai v. Luck	Creative Space for Technical Innovation, HAW Hamburg, Germany
Susanne Draheim	Creative Space for Technical Innovation, HAW Hamburg, Germany
Jessica Broscheit	Artist, Hamburg, Germany
Martin Kohler	HafenCity University Hamburg, Germany

Program Committee

Ingo Börsch	Technische Hochschule Brandenburg, Brandenburg University of Applied Sciences, Germany
Susanne Draheim	Hamburg University of Applied Sciences, Germany
Stefan Wölwer	HAWK University of Applied Sciences and Arts Hildesheim/Holzminden/Goettingen, Germany
Kai v. Luck	Creative Space for Technical Innovation, HAW Hamburg, Germany

Tim Tiedemann   Hamburg University of Applied Sciences,
                                    Germany
Marcelo Tramontano   University of São Paulo, Brazil

Second Workshop on Digital Transformation

Organizing Committee

Fernando Moreira   Universidade Portucalense, Portugal
Ramiro Gonçalves   Universidade de Trás-os-Montes e Alto Douro,
                                    Portugal
Manuel Au-Yong Oliveira   Universidade de Aveiro, Portugal
José Martins   Universidade de Trás-os-Montes e Alto Douro,
                                    Portugal
Frederico Branco   Universidade de Trás-os-Montes e Alto Douro,
                                    Portugal

Program Committee

Alex Sandro Gomes   Universidade Federal de Pernambuco, Brazil
Arnaldo Martins   Universidade de Aveiro, Portugal
César Collazos   Universidad del Cauca, Colombia
Jezreel Mejia   Centro de Investigación en Matemáticas A.C.,
                                    Mexico
Jörg Thomaschewski   University of Applied Sciences, Germany
Lorna Uden   Staffordshire University, UK
Manuel Ortega   Universidad de Castilla–La Mancha, Spain
Manuel Peréz Cota   Universidade de Vigo, Spain
Martin Schrepp   SAP SE, Germany
Philippe Palanque   Université Toulouse III, France
Rosa Vicardi   Universidade Federal do Rio Grande do Sul,
                                    Brazil
Vitor Santos   NOVA IMS Information Management School,
                                    Portugal

First Workshop on Empirical Studies in the Domain of Social Network Computing

Organizing Committee

Shahid Hussain   COMSATS Institute of Information Technology,
                                    Islamabad, Pakistan
Arif Ali Khan   Nanjing University of Aeronautics
                                    and Astronautics, China
Nafees Ur Rehman   University of Konstanz, Germany

Program Committee

Abdul Mateen	Federal Urdu University of Arts, Science & Technology, Islamabad, Pakistan
Aoutif Amine	ENSA, Ibn Tofail University, Morocco
Gwanggil Jeon	Incheon National University, Korea
Hanna Hachimi	ENSA of Kenitra, Ibn Tofail University, Morocco
Jacky Keung	City University of Hong Kong, Hong Kong
Kifayat Alizai	National University of Computer and Emerging Sciences (FAST-NUCES), Islamabad, Pakistan
Kwabena Bennin Ebo	City University of Hong Kong, Hong Kong
Mansoor Ahmad	COMSATS University Islamabad, Pakistan
Manzoor Ilahi	COMSATS University Islamabad, Pakistan
Mariam Akbar	COMSATS University Islamabad, Pakistan
Muhammad Khalid Sohail	COMSATS University Islamabad, Pakistan
Muhammad Shahid	Gomal University, DIK, Pakistan
Salima Banqdara	University of Benghazi, Libya
Siti Salwa Salim	University of Malaya, Malaysia
Wiem Khlif	University of Sfax, Tunisia

First Workshop on Health Technology Innovation: Emerging Trends and Future Challenges

Organizing Committee

Eliana Silva	University of Minho & Optimizer, Portugal
Joyce Aguiar	University of Minho & Optimizer, Portugal
Victor Carvalho	Optimizer, Portugal
Joaquim Gonçalves	Instituto Politécnico do Cávado e do Ave & Optimizer, Portugal

Program Committee

Eliana Silva	University of Minho & Optimizer, Portugal
Joyce Aguiar	University of Minho & Optimizer, Portugal
Victor Carvalho	Optimizer, Portugal
Joaquim Gonçalves	Instituto Politécnico do Cávado e do Ave & Optimizer, Portugal

Fifth Workshop on Healthcare Information Systems Interoperability, Security and Efficiency

Organizing Committee

José Machado	University of Minho, Portugal
António Abelha	University of Minho, Portugal

| Luis Mendes Gomes | University of Azores, Portugal |
| Anastasius Mooumtzoglou | European Society for Quality in Healthcare, Greece |

Program Committee

Alberto Freitas	University of Porto, Portugal
Ana Azevedo	ISCAP/IPP, Portugal
Ângelo Costa	University of Minho, Portugal
Armando B. Mendes	University of Azores, Portugal
Cesar Analide	University of Minho, Portugal
Davide Carneiro	University of Minho, Portugal
Filipe Portela	University of Minho, Portugal
Goreti Marreiros	Polytechnic Institute of Porto, Portugal
Helia Guerra	University of Azores, Portugal
Henrique Vicente	University of Évora, Portugal
Hugo Peixoto	University of Minho, Portugal
Jason Jung	Chung-Ang University, Korea
Joao Ramos	University of Minho, Portugal
José Martins	UTAD, Portugal
Jose Neves	University of Minho, Portugal
Júlio Duarte	University of Minho, Portugal
Luis Mendes Gomes	University of Azores, Portugal
Manuel Filipe Santos	University of Minho, Portugal
Paulo Moura Oliveira	UTAD, Portugal
Paulo Novais	University of Minho, Portugal
Teresa Guarda	Universidad Estatal da Península de Santa Elena, Ecuador
Victor Alves	University of Minho, Portugal

Fourth Workshop on New Pedagogical Approaches with Technologies

Organizing Committee

Anabela Mesquita	ISCAP/P.Porto and Algoritmi Centre, Portugal
Paula Peres	ISCAP/P.Porto and Unit for e-Learning and Pedagogical Innovation, Portugal
Fernando Moreira	IJP and REMIT – Univ Portucalense & IEETA – Univ Aveiro, Portugal

Program Committee

| Alex Gomes | Universidade Federal de Pernambuco, Brazil |
| Ana R. Luís | Universidade de Coimbra, Portugal |

Armando Silva	ESE/IPP, Portugal
César Collazos	Universidad del Cauca, Colombia
Chia-Wen Tsai	Ming Chuan University, Taiwan
João Batista	CICE/ISCA, UA, Portugal
Lino Oliveira	ESMAD/IPP, Portugal
Luisa M. Romero Moreno	Universidade de Sevilha, Espanha
Manuel Pérez Cota	Universidade de Vigo, Espanha
Paulino Silva	CICE & CECEJ-ISCAP/IPP, Portugal
Ramiro Gonçalves	UTAD, Vila Real, Portugal
Rosa Vicari	Universidade de Rio Grande do Sul, Porto Alegre, Brazil
Stefania Manca	Instituto per le Tecnologie Didattiche, Italy

Fifth Workshop on Pervasive Information Systems

Organizing Committee

Carlos Filipe Portela	Department of Information Systems, University of Minho, Portugal
Manuel Filipe Santos	Department of Information Systems, University of Minho, Portugal
Kostas Kolomvatsos	Department of Informatics and Telecommunications, National and Kapodistrian University of Athens, Greece

Program Committee

Andre Aquino	Federal University of Alagoas, Brazil
Carlo Giannelli	University of Ferrara, Italy
Cristina Alcaraz	University of Malaga, Spain
Daniele Riboni	University of Milan, Italy
Fabio A. Schreiber	Politecnico Milano, Italy
Filipe Mota Pinto	Polytechnic of Leiria, Portugal
Hugo Peixoto	University of Minho, Portugal
Gabriel Pedraza Ferreira	Universidad Industrial de Santander, Colombia
Jarosław Jankowski	West Pomeranian University of Technology, Szczecin, Poland
José Machado	University of Minho, Portugal
Juan-Carlos Cano	Universitat Politècnica de València, Spain
Karolina Baras	University of Madeira, Portugal
Muhammad Younas	Oxford Brookes University, UK
Nuno Marques	New University of Lisboa, Portugal
Rajeev Kumar Kanth	Turku Centre for Computer Science, University of Turku, Finland

Ricardo Queirós   ESMAD- P.PORTO & CRACS - INESC TEC,
                             Portugal
Sergio Ilarri   University of Zaragoza, Spain
Spyros Panagiotakis   Technological Educational Institute of Crete,
                             Greece

Contents

Pervasive Information Systems

Health Informatics

Blood4Life: A Mobile Solution to Recruit and Retain Blood Donors Through Gamification and Trans-Theoretical Model

Lamyae Sardi[1], Manal Kharbouch[1], Taoufik Rachad[1], Ali Idri[1(✉)],
Juan Manuel Carillo de Gea[2], and José Luis Fernández-Alemán[2]

[1] Software Project Management Research Team, ENSIAS,
Mohammed V University, Rabat, Morocco
lamyasardi@gmail.com, kharbouch.manal16@gmail.com,
rachad.taoufik@gmail.com, ali.idri@um5.ac.ma
[2] Department of Informatics and Systems, Faculty of Computer Science,
University of Murcia, Murcia, Spain
{jmcdgl,aleman}@um.es

Abstract. The worldwide demand for blood and its components is critically growing owing to the rise in target diseases, accidents and surgeries. As blood supplies are considerably outstripped by the immediate and crucial need of blood transfusions, the recruitment and retention of voluntary qualified donors pose an acute challenge for blood centers. In this regard, digital technology has proven to be effective in the optimization of blood donation in many ways. Based on that fact, it was designed a solution consisting of a hybrid mobile application named 'Blood4Life' which employs gamification techniques and integrates the principles of Trans-Theoretical Model (TTM) of behavior change. This paper presents, therefore, the requirements and characteristics of 'Blood4Life' which aims at targeting all types of users in terms of their stage of change which defines their readiness and willingness to donate blood. By means of a variety of gamification elements and according to the initial stage of change of the user, the mobile application is likely to trigger the processes of change that are assumed to encourage users to progress towards later stages.

Future work includes the development of a facet for blood centers to manage donors' appointments. By then, an empirical evaluation of Blood4Life solution will be required to thoroughly examine the effects of combining gamification and behavior change theory in achieving sustained engagement of blood donors.

Keywords: Blood donation · Mobile application · Gamification ·
Trans-Theoretical Model

1 Introduction

Blood donation is traditionally described as a remarkably noble and altruistic act of service to humanity [1]. Due to the frequency of transplants and surgeries, there is a worldwide rising demand for blood transfusions whilst a critical scarcity is occurring across the blood supplies because of the steady shrinkage of blood donations.

© Springer Nature Switzerland AG 2019
Á. Rocha et al. (Eds.): WorldCIST'19 2019, AISC 932, pp. 3–12, 2019.
https://doi.org/10.1007/978-3-030-16187-3_1

Several factors come into play to threaten the blood supplies such as the overly strict eligibility criteria, the ageing populations and the decline in altruism [2]. Blood services face, therefore, a huge challenge to make available a sufficient and safe blood supply through voluntary and non-remunerated blood donations.

To reach self-sufficiency in blood products, efforts must focus on recruiting and retaining more regular, volunteer blood donors. Accordingly, it is important to gain insight into characteristics of donors, motives and potential deterrents to voluntary blood donation. In this regard, a good number of studies have thoroughly explored factors influencing the willingness to donate blood. Given the prosocial nature of blood donation, empathy and altruism are by far the most commonly reported motives for giving blood [2, 3]. Moreover, many studies have emphasized that social influence, personal norms, awareness of the need of blood and perceived psychological and health benefits, each play a prominent role in motivating individuals to donate blood [4, 5]. Yet, there is a disparity among the motivators that stand out across the four donors' profiles (first-time donor, repeat, lapsed and eligible non-donor) [6]. It stands to reason that understanding the factors that influence donors to donate blood can be of particular benefit to blood collection agencies [5]. Besides, mobile technology can be tremendously effective in terms of donors' recruitment and engagement. Indeed, blood donation apps are considered a promising approach for promoting donors' behavior and enhancing their motivation to donate [7, 8]. Moreover, the features of health mobile applications are now strengthened by means of gamification techniques [9] that nurture users' loyalty and commitment to the foreseen health behavior change while ensuring an entertaining experience through game mechanics.

In the realm of blood donation, there is a paucity of mobile solutions that incorporate gamification techniques. According to a recent review [10], only ten gamified blood donation apps were found in the four app repositories. In the same context, Domingos et al. [11] designed an application that combines gamification techniques and social networking elements in order to make possible the interaction and information exchange between donors and blood centers. Another blood donation system proposed by Fotopoulos et al. [12], combines cloud-computing and mobile technology to empower blood agencies to effectively recruit and retain a healthy donors pool. The system is also expected to take advantages of gamification to bolster donors' motivation and minimize volunteer relapses. Nonetheless, many researchers reported that the positive effect of gamification is appeared to fade over time [13, 14]. In fact, most of gamified systems do not adopt a user-centered approach as they fail to cater different users' needs. Thus, it is important to understand the users and grasp the determinants of their behavior change for better implementation of gamification techniques. Accordingly, this study seeks to propose the design of a gamified mobile solution for blood donation that can be tailored to any donor profile through the integration of the Trans-Theoretical Model (TTM) of behavior change.

The remainder of this paper is structured as follows: Sect. 2 describes gamification and the TTM. Section 3 details the purpose and the requirements of Blood4Life solution. The implementation of Blood4Life solution is presented in Sect. 4. Finally, Sect. 5 provides some final conclusions and directions for future work.

2 Theoretical Aspects

This section provides a thorough description of the main theoretical elements of the proposed solution, namely the concept of gamification and the TTM.

2.1 Gamification

Gamification is a growing phenomenon that lies in the application of game theory concepts and techniques to non-game activities [15]. This process of integrating game elements into an existing system is intended to motivate users' engagement and loyalty whilst promoting a compelling and an entertaining game experience. Given that motivation is perceived as one of the major propellants for human activity, gamification can potentially be employed to induce positive behavioral change. Nowadays, gamification has been widely applied to various areas including education, software industry, health and business [16]. The game essence in gamification can be brought through a combination of elements which include points, badges, levels, challenges, leaderboards, among others [17]. Research suggests that for gamification to yield positive outcomes, it is important to combine multiple game elements. Nonetheless, it is important to ensure an appropriate balance between these elements to foster both intrinsic and extrinsic motivations.

In the domain of healthcare, gamification provides promising directions for designing persuasive interventions leveraging the capabilities of mobile technology and ubiquitous computing that promote positive health behavior change [18]. Current health gamification applications cover all major domains including chronic disease rehabilitation, mental well-being, physical activity and diet management [9]. Blood donation is also considered as a potential application area of gamification. The constant need for blood transfusions is ineluctably leading researchers to direct efforts to designing and developing digital solutions that support blood donation.

In this respect, building gamified mobile applications is one of the feasible approaches to address blood donation shortage. However, the number of existing gamified blood donation apps is very scarce owing to several reasons ranging from security issues to globalization obstacles [11]. Moreover, most of these apps lack an understanding of the different donor profiles and the motivating factors for each particular type of donor, hence failing in targeting groups of potential donors [19].

2.2 The Trans-Theoretical Model

The TTM is one of the most useful models for understanding health-related behaviors and conducting efforts to promote health. Originally developed by Prochaska and DiClemente in the early 1980s [20], this model deals with behavior change and operates on the assumption that habitual and sustainable change in behavior occurs through a cyclic progress. In fact, individuals move through the following five stages of changes: Pre-contemplation (Not making any effort to change behavior), Contemplation (thinking about changing behavior in the next six months), Preparation (prepared to change behavior in the next month), Action (actively involved in the route of change) and Maintenance (maintaining the behavior change for a period of six months by avoiding any temptations).

The TTM theorizes that people engage in overt and covert activities to move between progressive stages of change. These activities constitute another major dimension of the TTM, commonly known as Processes of Change. There are ten stage-related processes of change divided into experiential and behavioral patterns.

On the one hand, the experiential processes of change include Consciousness raising (increasing awareness about the problem behavior), Dramatic relief (emotional arousal about the problem behavior), Environmental re-evaluation (social assessment on how the problem behavior affects others), Self-re-evaluation (self-reappraisal with respect to the problem behavior) and Social liberation (notice public support through alternatives for problem behavior).

On the other hand, the behavioral processes of change consist of Counter conditioning (substituting healthy behavior for the problem behavior), Helping relationships (finding supportive relationships that help attain to desired behavior change), Reinforcement management (rewarding oneself for healthy behavior), Self-liberation (belief and commitment to change the problem behavior) and Stimulus control (re-engineering the environment to remove cues for the problem behavior and add prompts for healthier alternatives).

3 Blood4Life Solution

This section presents the purpose, major features, quality characteristics and gamification incentives that constitute the core components of Blood4Life solution.

3.1 Purpose

The main objective of Blood4Life solution resides in promoting blood donation through raising awareness about its importance and increasing the recruitment and retention of prospective donors. Given that Blood4Life solution consists of a mobile application that integrates game mechanics, it is expected to interest technology enthusiasts. This will likely ensure the acquisition of young and motivated donors who are predisposed to potentially long careers of blood donation. Besides, the use of gamification techniques and social media sharing will render the blood donation experience more enjoyable, inspiring and rewarding, and will immensely help to maintain the pool of existing donors and motivate them to inherently recruit new ones.

Another aspect of the key objective of Blood4Life solution lies in integrating principles of behavior change theory to target all types of blood donors ranging from novice donors to loyal ones. In this respect, the TTM constructs and gamification elements were unified to design a tailored solution that provides users with a unique experience according to their stage of change. The initial stage of change of users will be determined, at first use, through users' answers to a short quiz based on a staging algorithm inspired from a study by Burditt et al. [4]. This quiz is composed of four questions and it basically assesses the users' eligibility to donate blood, their history of donations and their willingness to donate blood in the future. Blood4Life solution is, by then, intended to emphasize the processes of change that are assumed to positively influence the individuals' behavior considering the inter-stages transition they are about to perform, by means of a variety of gamification elements.

Blood4Life solution is conceptually dedicated to donors and blood centers, however, in this paper, only the characteristics of the facet for donors will be presented.

3.2 Requirements Specification

A Software Requirements Specification covering functional and non-functional requirements for Blood4Life solution, has been elaborated according to the IEEE 29148 standard [21]. Initially, the following functional requirements were implemented:

- **User registration:** The user should be able to register using Google or Facebook credentials or by filling in the registration form including email, password, full name, phone number, age, gender and blood type.
- **Login:** A registered user should be able to log in to the mobile application using login credentials (email and password). The login information will be stored on the phone and in the future, the user should be logged in automatically.
- **Retrieve password:** A user should be able to retrieve his/her password by email.
- **Visualize and edit profile:** A user should be able to edit his/her profile information including email, password and phone number.
- **Take TTM quiz:** Once a user gets registered, he/she should take TTM quiz to obtain his/her stage of change upon which the app will be adapted.
- **Find nearby blood centers:** Given that a user is logged in to the mobile app, he/she should be able to search for nearby blood centers. The search results should be viewed on a map. The closest blood centers according to the users' location are displayed using specific pins.
- **Schedule appointments for blood donation:** The user should be able to schedule an appointment for donating blood (when eligible) at the selected blood center.
- **View donation history:** The user should be able to visualize the list of his/her past donations including date, center and type of blood donation.
- **Share donations on social media:** The user should be able to share his/her blood donations on Facebook and Twitter.
- **Receive eligibility notifications:** The user should be notified about their eligibility to donate blood based on his/her history of donations.
- **Receive appointment reminders:** Given that a user has scheduled an appointment, he/she should receive a push notification to remind him/her of the upcoming blood donation appointment. The user shall set the frequency by which he/she wants to receive reminders.
- **Receive alerts during blood shortages:** The user should be able to receive notifications when his/her own blood type is needed.
- **Change notifications settings:** The user should be able to choose how and when he/she would like to receive notifications.
- **Create/Join teams:** The user should be able to create and/or join teams of blood donors and to visualize the prevalence of different donors' teams across the city.

In line with previous research studies on the application of ISO/IEC 25010 standard [22] to health-related software products including blood donation mobile applications and mobile health records [23–25], a set of non-functional properties has been determined to improve the product quality of Blood4Life solution. The core quality requirements that were considered are the following:

- **Functional suitability:** Blood4Life solution should meet users' needs through well integrated functions and appropriate content.
- **Performance efficiency:** Blood4Life solution should have a short response time to enhance user experience (UX).
- **Reliability:** Blood4Life solution should remain operational and accessible in a specific manner under the possible circumstances (background/foreground, with/without internet connection).
- **Operability:** Blood4Life solution should be scalable, it should be able to handle a large number of users or quantities of data.
- **Security:** Blood4Life solution should ensure encrypted communication, protection and security of users' accounts and sensitive information.
- **Compatibility:** Blood4Life solution should work well on different mobile devices with various features and appliances.
- **Maintainability:** Blood4Life solution should have a readable and extendible code to easily implement new functions and to avoid increasing maintenance cost.
- **Transferability:** Blood4Life should support the common mobile platforms.

3.3   Integration of TTM Principles and Gamification Techniques

To effectively help individuals progress throughout the TTM stages, a broad range of gamification elements has been implemented to trigger the TTM processes of change specific to each transition. These gamification elements comprise the following:

Status. Status determines the relative position of the user in relation to others and is considered one of the most desired and sticky potential prize to win as it incorporates pride and motivation dynamics. Four statuses are to be obtained according to the stage of change of the user. The status should be systematically attributed to the user after obtaining TTM quiz result and should be updated during inter-stages progression. If the quiz result is 'Contemplation', the user should be attributed 'Blood Noob' status. The user should be granted the 'Good Samaritan' status if his/her quiz result is 'Preparation'. The quiz results 'Action' and 'Maintenance' are associated with 'Red Ninja' and 'Red Blooded Hero' status, respectively. Whereas, no status should be attributed to the user if her/his quiz result is 'Precontemplation'.

In-app Point System. Being one of the most used gamification elements, point system is a powerful and important gamification element since it gives the users the extra nudge they need to get actively involved. Nonetheless, rewarding points may be wrongly utilized if they don't align with the desired behavior. Blood4Life solution primarily aims to increase blood donation, three point-based rewards are therefore implemented. The user's score should be incremented by a total of 100 Donation Points (DPs) upon making a blood donation appointment and by 150 DPs if the blood donation appointment is scheduled when blood stocks are running low. In contrast, the user's score should be decremented by 50 DPs after cancelling the blood donation appointment.

Badges. Badges are a visual and a collectible reward that marks tasks completion. Six badges are implemented in Blood4Life solution.

- The user should earn 'Welcome' badge after signing-up.
- The user should earn 'Let them know' badge upon sharing blood donation appointments on social networks.
- The user should earn 'Be a member of' badge for joining a team of donors.
- The user should earn 'Spread the good will' badge for referring three friends within a week.
- The user should earn 'Red hat-trick' badge for taking the third blood donation appointment within one year.
- The user should earn 'Be there when needed' badge for donating when blood stocks are running low.

Trophies. Trophies are recognition items that are commonly used in games owing to their versatility. Three types of trophies can be obtained in Blood4Life solution. The 'bronze trophy' should be awarded to the user upon unlocking three badges within a month. The 'silver' and 'the gold' trophies should be awarded to the user upon collecting a total of 500 DPs and 1000 DPs, respectively.

Progress Bar. People are inherently driven to have goals and then accomplish them. Progress bar is therefore an effective visual element that allows users track their progress towards goal attainment. In Blood4Life solution, the progress bar was conceived in a way that displays the milestones (Stages of Change) reached by the user.

Leaderboard. The purpose of leaderboards is to show the ranking of users. In order to render it more social yet competitive, the leaderboard in the Blood4Life solution will display the ranking of users in each team based on their total earned DPs.

Given the characteristics of each stage of change, the inter-stage progression will be possible upon the fulfilment of a condition specific to each transition as depicted in Fig. 1. The specified gamification aspects align well with the definition of the processes of change that are appropriate to each stage-transition. On the whole, the progression from Pre-contemplation to Contemplation primarily requires enhancing knowledge and awareness about blood donation process and its importance. Whilst, both transitions Contemplation-Preparation and Preparation-Action entail regular rewarding and increasing social influence whilst the progression towards maintenance stage implies substantial recognition through obtaining real-world items.

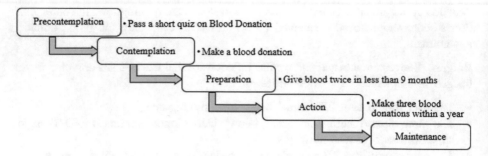

Fig. 1. Conditions to transition between the five stages of change

4   Implementation

Blood4Life solution consists of a cross-platform mobile application being currently developed using Angular and Ionic framework (version CLI 4.2.1). Hybrid development allows to develop mobile applications that are consistent across different mobile platforms where only a single codebase is used. Several advantages come along hybrid app development such as cost-effectiveness, easy scalability and maintenance.

In the actual development phase, this application is solely dedicated to blood donors. Figure 2 in Appendix shows a few snapshots of user interfaces for the mobile application. The user will be firstly asked to register and answer a short quiz to determine her/his stage of change. At this point, the user might be ineligible to donate blood, hence, he/she can access the application in Guest Mode where only very limited functionalities are available such as consulting nearby blood centers and referring new friends to use the application.

Once registered, the user will be able to schedule a blood donation appointment in the blood center he/she prefers from the list of the available blood centers. Besides accessing his/her blood donation history, the user will be able to create or join a team to build a pervasive and competitive ambiance to further promote blood donation. All the obtained virtual rewards will be systematically updated and displayed to the user. Moreover, the user will be able to switch on/off blood shortage alerts and eligibility notifications along with precising the frequency of which he/she would prefer to receive appointment reminders. The Appendix can be found at the following link: https://www.um.es/giisw/Blood4Life/Appendix.pdf.

5   Conclusion and Future Perspectives

Seeking to aid in the recruitment and the retention of motivated young blood donors, it was designed a solution consisting of a hybrid mobile application named 'Blood4Life'. Taking advantage of gamification techniques and behavior change theory principles, Blood4Life solution represents a novel approach in the development of tailored blood donation mobile applications. It offers a unique experience to users based on their initial stage of change by triggering the processes of changes that are assumed to help them transition to the next stages through gamification elements and social influence. Blood4Life solution is most likely useful and efficient in promoting blood donation. Researchers and developers can therefore benefit from this study by reusing the specified functional requirements, gamification and TTM aspects to further tailor their solutions to a larger community of blood donors.

For future work, it is intended, prior to actual implementation, to develop the facet for blood centers in the mobile application to efficiently track eligibility and manage blood appointments of prospective donors. A further perspective worth mentioning is that of performing a longitudinal empirical evaluation with real participants to assess the effectiveness of the application in blood donation behavior change and measure its potency in the optimization of donors' recruitment and retention.

Acknowledgments. This work was conducted within the research project PEER, 7-246 supported by the US Agency for International Development. The authors would like to thank the NAS and USAID for their support.

References

1. Evans, R., Ferguson, E.: Defining and measuring blood donor altruism: a theoretical approach from biology, economics and psychology. Vox Sang. **106**, 118–126 (2014)
2. Sojka, B.N., Sojka, P.: The blood donation experience: self-reported motives and obstacles for donating blood. Vox Sang. **94**, 56–63 (2008)
3. Kasraian, L., Maghsudlu, M.: Blood donors' attitudes towards incentives: influence on motivation to donate. Blood Transfus. **10**, 186–190 (2012)
4. Burditt, C., Robbins, M.L., Paiva, A., Velicer, W.F., Koblin, B., Kessler, D.: Motivation for blood donation among African Americans: developing measures for stage of change, decisional balance, and self-efficacy constructs. J. Behav. Med. **32**, 429–442 (2009)
5. Buciuniene, I., Stonienė, L., Blazeviciene, A., Kazlauskaite, R., Skudiene, V.: Blood donors' motivation and attitude to non-remunerated blood donation in Lithuania. BMC Public Health. **6**, 166 (2006)
6. Bednall, T.C., Bove, L.L.: Donating blood: a meta-analytic review of self-reported motivators and deterrents. Transfus. Med. Rev. **25**, 317–334 (2011)
7. Yuan, S., Chang, S., Uyeno, K., Almquist, G., Wang, S.: Blood donation mobile applications: are donors ready? Transfusion **56**, 614–621 (2016)
8. Ouhbi, S., Fernández-Alemán, J.L., Toval, A., Idri, A., Rivera Pozo, J.: Free blood donation mobile applications. J. Med. Syst. **39**, 1–20 (2015)
9. Sardi, L., Idri, A., Fernández-alemán, J.L.: A systematic review of gamification in e-Health. J. Biomed. Inform. **71**, 31–48 (2017)
10. Sardi, L., Idri, A., Fernández-alemán, J.L.: Gamified mobile blood donation applications. In: 5th International Work-Conference on Bioinformatics and Biomedical Engineering, pp. 165–176 (2017)
11. Domingos, D.C.L., Lima, L.F.S.G., Messias, T.F., Feijó, J.V.L., Anthony, A.R., Soares, H. B.: Blood Hero : an application for encouraging the blood donation by applying gamification. In: IEEE 38th Annual International Conference of Engineering in Medicine and Biology Society (EMBC), pp. 5624–5627 (2016)

12. Fotopoulos, I., Palaiologou, R., Kouris, I., Koutsouris, D.: Cloud-based information system for blood donation. In: XIV Mediterranean Conference on Medical and Biological Engineering and Computing 2016, pp. 796–801 (2016)
13. Looyestyn, J., Kernot, J., Boshoff, K., Ryan, J., Edney, S., Maher, C.: Does gamification increase engagement with online programs? Syst. Rev. PLoS One **12**, 1–19 (2017)
14. Zuckerman, O., Gal-Oz, A.: Deconstructing gamification: evaluating the effectiveness of continuous measurement, virtual rewards, and social comparison for promoting physical activity. Pers. Ubiquitous Comput. **18**, 1705–1719 (2014)
15. Deterding, S., Dixon, D., Khaled, R., Nacke, L.: From game design elements to gamefulness: Defining "Gamification." In: Proceedings of the 15th International Academic MindTrek Conference on Envisioning Future Media Environments - MindTrek 2011, pp. 9–15 (2011)
16. Mora, A., Riera, D., Gonzalez, C., Arnedo-Moreno, J.: A literature review of gamification design frameworks. In: VS-Games 2015, 7th International Conference on Games and Virtual Worlds for Serious Applications, pp. 1–8. IEEE (2015)
17. Gabe, Z., Cunningham, C.: Gamification by design: Implementing game mechanics in web and mobile apps. O'Reilly Media Inc., Newton (2011)
18. Johnson, D., Deterding, S., Kuhn, K.-A., Staneva, A., Stoyanov, S., Hides, L.: Gamification for health and wellbeing: a systematic review of the literature. Internet Interv. **6**, 89–106 (2016)
19. Sabani, A.C., Manuaba, I.B.K., Adi, E.: Gamification: blood donor apps for iOS devices. J. Game, Game Art Gamification, **1**, 14–26 (2016)
20. Prochaska, J.O., Diclemente, C.C.: Stages and processes of self- change of smoking - toward an integrative model of change. J. Consult. Clin. Psychol. **51**, 390–395 (1983)
21. IEEE 29148 Standard: Systems and software engineering—Life cycle processes—Requirements Engineering (2011)
22. ISO/IEC-25010: Systems and software engineering—Systems and software Quality Requirements and Evaluation (SQuaRE)—System and software quality models (2011)
23. Ouhbi, S., Idri, A., Fern, L.: Applying ISO/IEC 25010 on Mobile Personal Health Records. In: 8th International Conference on Health Informatics, pp. 405–412 (2015)
24. Idri, A., Bachiri, M., Fernández-alemán, J.L., Toval, A.: ISO/ IEC 25010 based evaluation of free mobile personal health records for pregnancy monitoring. In: IEEE 41st Annual Computer Software and Applications Conference, pp. 262–267 (2017)
25. Idri, A., Sardi, L., Fernández-alemán, J.L.: Quality evaluation of gamified blood donation apps using ISO/IEC 25010 Standard. In: 12th International Conference on Health Informatics, pp. 607–614 (2018)

Breast Cancer Classification
with Missing Data Imputation

Imane Chlioui[1], Ali Idri[1(✉)], Ibtissam Abnane[1],
Juan Manuel Carillo de Gea[2], and Jose Luis Fernández-Alemán[2]

[1] Software Project Management Research Team, ENSIAS,
University Mohammed V of Rabat, Rabat, Morocco
ali.idri@um5.ac.ma
[2] Department of Informatics and Systems, Faculty of Computer Science,
University of Murcia, Murcia, Spain

Abstract. Missing Data (MD) is a common drawback when applying Data Mining on breast cancer datasets since it affects the ability of the Data mining classifier. This study evaluates the influence of MD on three classifiers: Decision tree C4.5, Support vector machine (SVM), and Multi-Layer Perceptron (MLP). For this purpose, 162 experiments were conducted using KNN imputation with three missingness mechanisms (MCAR, MAR and NMAR), and nine percentages (form 10% to 90%) applied on two Wisconsin breast cancer datasets. The MD percentage affects negatively the classifier performance. MLP achieved the lowest accuracy rates regardless the MD mechanism/percentage.

Keywords: KNN imputation · Data mining · Breast cancer

1 Introduction

In the past twenty years, the number of patients with breast cancer (BC) continues to rise, and it became the second leading cause of death among women [1]. It is a malignant tumor that has developed from cells in the breast [2]. The key to an effective treatment is early diagnosis: the earlier the disease is diagnosed the less it progresses. Nowadays, the amount of data is increasing constantly in all fields such us: education, agriculture and medicine [3]. Which the necessity of the use of data mining (DM) techniques to analyze the huge amount of available data and extract knowledge [1]. According to Idri et al. [4] the use of DM techniques has increased lately, and become a powerful tool to help radiologists and practitioners to deal with BC challenges.

To successful the use of DM techniques, the preprocessing step is recommended to avoid biased and deceptive results. Cleaning, transformation, reduction, and integration are subfields of preprocessing. Handling missing data (MD) as a part of the cleaning process is a major problem facing the use of DM tools [5]. Thus, several MD techniques have been proposed and experimented; they can be grouped in three categories [6, 7]: (1) toleration technique which consists on ignoring the MD, (2) deletion technique which consists on deleting the MD, and (3) imputation techniques which consist on filling in the MD with appropriate values.

© Springer Nature Switzerland AG 2019
Á. Rocha et al. (Eds.): WorldCIST'19 2019, AISC 932, pp. 13–23, 2019.
https://doi.org/10.1007/978-3-030-16187-3_2

Therefore, this paper analyses and discusses the impact of the use of KNN-imputation on the accuracy of three classifiers: decision tree C4.5, support vector machine (SVM) and multi-layer perceptron (MLP) over two datasets: Wisconsin breast cancer original and Wisconsin breast cancer prognosis. Moreover, the empirical evaluations used three MD missingness mechanisms (MCAR, MAR, NMAR), nine MD percentages (from 10% to 90%), and were performed using the experimental process proposed by Idri et al. [7]. To the best of our knowledge, no existing study that analyzes the impact of KNN imputation using different MD mechanisms (MCAR, MAR, and NMAR) with nine percentages (from 10% to 90%) on the performance of classification techniques in breast cancer, which motivates this study.

This paper is structured as follows. Section 2 introduces the different MD mechanisms and MD imputation techniques. Related work dealing with missing values in breast cancer is presented in Sect. 3. Section 4 introduces the datasets as well as the classification techniques used. The experimental design followed in this study is detailed in Sect. 5, while the Sect. 6 presents the results and discuss the findings. Threats to validity are presented in Sect. 7. Section 8 concludes the paper and proposes further research lines.

2   Missing Data Concepts

The MD mechanisms and technique used in this study are presented in the section follow.

2.1   Missing Data Types

The missingness mechanism indicates the reason of data missingness and could help to choose the suitable MD technique. Three MD mechanisms where defined by Rubin [8]:

Missing Completely at Random (MCAR): MD are independent of other variables and there is no specific reason of missingness [9].

Missing at Random (MAR): MD is not related to the missing values themselves, but is related to other observed variables. This mechanism can bias the results and may cause unbalanced data [10].

Not Missing at Random (NMAR): MD are dependent to the MD themselves. This can happen when the variable is not observed or it takes a value out of its representation range. This MD type may give highly biased estimation results [10].

2.2   Missing Data Imputation Techniques

In contrast to deletion technique that permits to discard instances that contain missing values, imputation techniques replace missing items with plausible values [11]. KNN imputation has been widely used by several researchers, and it is performed by considering the K closest instances to the incomplete instance according to a given distance metric. Several distance measures were proposed in literature such as: Euclidian distance, Manhattan distance and Hamming distance [12]. This study employs the Euclidean distance which assesses the distance between instances x_i and y_i by the Eq. (1).

$$D_E(x_i, y_i) = \sqrt{\sum_{i=1}^{n}(x_i - y_i)^2} \qquad (1)$$

Where n is the number of attributes describing the instances x_i and y_i.

3 Related Work

This section presents a summary of two papers selected from the systematic mapping study of [4].

García-Laencina et al. [13] applied several MD techniques on a breast cancer dataset with a high percentage of MD in order to predict breast cancer survivability. The dataset used in this study is collected from the Institute Portuguese of Oncology of Porto (IPO). They found that KNN imputation was the best in terms of accuracy, while the Mode imputation was the worst.

Jerez et al. [14] Compared statistical/machine learning imputation techniques with deletion to predict the breast cancer prognosis. The study proved that all imputation techniques except Hot-deck could improve the accuracy of an artificial neural network (ANN) based prediction. Moreover, ML imputation techniques outperformed statistical ones and led to statistically significant improvements in prediction accuracy. The best predictions were obtained using the KNN imputation with an improvement of 2.71% over deletion.

4 Datasets and Classifiers Description

This section presents a brief description of the datasets used and the classification techniques investigated.

4.1 Datasets Description

The datasets used in this study were collected at the University of Wisconsin–Madison Hospital [25]. The first one is the Wisconsin breast cancer original dataset, most commonly employed by researchers investigating machine learning techniques for breast cancer. It contains 699 instances described by 10 numerical attributes. The second one is the Wisconsin breast cancer prognosis dataset; it contains 198 records descried by 35 numerical attributes.

All cases containing missing data were discarded, which reduces the size of each dataset: 683 in Wisconsin original and 194 in Wisconsin prognosis. Moreover, the attributes of Wisconsin breast cancer prognosis dataset were normalized within the interval [0–1] in order to avoid bias of attributes' ranges. Note that the attribute values of the Wisconsin breast cancer original dataset were already normalized within the interval [1–10]. Table 1 presents datasets information, along with the number of instances and attributes. All the attributes used in this study are numerical.

Table 1. Datasets description

Database	Instances	Attributes	Attributes type	Source
Wisconsin breast cancer original	194	35	Numeric	[15]
Wisconsin breast cancer prognosis	683	10	Numeric	[16]

4.2 Classification Techniques

Hereafter, we describe the three classifiers used in this study.

C4.5. A supervised learning classification algorithm used to construct decision trees from the data developed by Quinlan [17].

SVM. A group of supervised learning methods developed by Vapnik in the 90's [18]. It is used to model data not linearly separable [19, 20].

MLP. A type of artificial neural network (ANN) that can represent complex input-output relationships [21]. MLP consists of neurons organized in three layers: input, hidden, and output layers which each one performs simple task of information processing by converting received inputs into processed output.

5 Empirical Design

Figure 1 presents the empirical process we used in this study. We followed the same empirical process used by [7] to handle missing data in software development effort estimation. This process consists of four main phases: data removal, complete dataset generation using imputation techniques, application of classification techniques, and accuracy evaluation. Each step of this process is detailed in the following subsections.

5.1 Data Removal

The datasets should be complete to work with. For this reason, the two datasets were cleaned by deleting all the cases with MD. Thereafter, the MD were generated artificially using the three missing data mechanisms:

- The MCAR mechanism relies on the randomization; the MD was induced completely at random for each variable.
- The MAR mechanism was simulated relying on a single attribute of each dataset: cell_shape_uniformity for Wisconsin original dataset and lymph_node_status for Wisconsin prognosis dataset. First, the instances were sorted in an ascending order of the selected attribute. Thereafter, the data were split into three equal subsets. The MD were distributed as follows: (1) 60% * p assigned randomly to the first subset, (2) 40% * p assigned to the second subset, and (3) 0% to the third subset; p

is the percentage of MD. The MD were induced to all attributes with bias related to the two selected attributes.

- The NMAR mechanism is similar to the MAR, except that the Wisconsin original dataset was sorted according to cell_size_uniformity. The only difference between NMAR and MAR mechanisms is that for NMAR, MD were not induced to all attributes but only to the attribute to which the datasets were sorted (i.e. cell_size_uniformity and lymph_node_status).

For each missingness mechanism, 9 percentages (10%, 20%, 30%, 40%, 50%, 60%, 70%, 80% and 90%) were simulated, which gave us a total of 54 incomplete datasets (54 = 3 MD mechanisms * 9 percentages * 2 datasets).

5.2   Complete Dataset Generation

In this step, KNN imputation was applied on the datasets resulted from the data removal step. The Euclidean distance was used to evaluate the similarity between instances since all the attributes are numerical. Thereafter, the median of the closest instances was used to fulfill the missing values. At the end of this step, 54 complete datasets were willing to be used for the classification task (54 = 54 incomplete datasets * 1 MD technique).

5.3   Generating Classifiers

For the classification task, three classifiers were applied on the complete datasets: C4.5, SVM, and MLP. To fulfill this task, the datasets were divided into training and testing sets using the 10-fold cross validation. After applying the three classifiers to the 54 data sets of the complete dataset generation step, we obtained 162 classification experiments (162 = 54*3). For each classifier, the grid search method was used to vary the classifiers parameters.

5.4   Performance Evaluation

To evaluate the performance of the three classifiers, the accuracy measure was used; it represents the probability of correctly predicting the class of an instance [22]. The accuracy rate was evaluated using the Eq. (2). TP or true positives is the number of positives cases correctly classified, TN or true negatives is the number of negatives cases correctly classified, FP or false positives, the number of positives cases classified as negatives, and FN or false negatives, the number of negatives cases classified as positives [23].

$$\text{Accuracy rate} = (TP + TN) / (TP + TN + FP + FN) \tag{2}$$

6   Results and Discussion

This section presents and discusses the empirical results when applying the three classifiers using KNN imputation, three MD mechanisms, and nine percentages over the two Wisconsin datasets. Therefore, we analyze the impact of each MD mechanism on the classifiers accuracy under different MD percentages. All the empirical evaluations were implemented using the WEKA (3.8.0) tool [26].

In order to determine the best performance of each classifier, the grid search method was used to set up the ranges of parameters and test each combination to find the best one [27]. The optimal configuration (i.e., that which attained the best performance results) of each classifier was then used in the subsequent experiment. Table 2 presents the parameter ranges of each classifier and the optimal configuration of each classifier as well.

6.1   Evaluation of Classifiers Using MD Mechanisms

This section presents and discusses the influence of the three missingness mechanisms on the mean accuracy rates of the three classifiers using KNN imputation and nine percentages.

- For the SVM classifier using KNN imputation, Fig. 2a shows that the mean accuracy rates obtained under MCAR were better than under MAR and NMAR regardless the MD percentages (at 10% of MD the mean accuracy rates were: under MCAR 90.10%, under MAR 88.26% and under NMAR 88.30%). Moreover, under NMAR the mean accuracy rates were higher than those obtained under MAR regardless the MD percentages (at 90% of MD the mean accuracy rates were: under NMAR 88.16% and under MAR 87.31%.

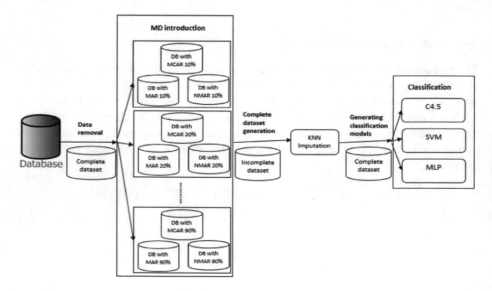

Fig. 1. Experimental design

Table 2. Parameters ranges and optimal configuration of each classifier

Algorithm	Parameters ranges	Optimal configuration
C4.5	C = {0.1 –> 5, increment = 0.1}; M = {10 –> 100, increment = 10};	C = 0.25 M = 2
SVM	Kernel = RBFKernel; C = {100 –> 200, increment = 10}; G = {0.01 –> 0.1, increment = 0.01}	Kernel = RBFKernel C = 1 G = 0.01
MLP	L = {0.1 –> 1, increment = 0.1} M = {0.1–> 1, increment = 0.1}	L = 0.3 M = 0.2
KNN imputation	K = {2 –> 7, increment = 1}	K = 5

- For the MLP classifier using KNN imputation, according to Fig. 2b the mean accuracy rates achieved under MAR were better than under MCAR and NMAR regardless the MD percentages (at 10% of MD the mean accuracy rates were: under MAR 87.82%, under MCAR 86.53% and under NMAR 84.77%). Furthermore, under MCAR the mean accuracy rates were higher than those obtained under NMAR (at 90% of MD the mean accuracy rates were: under MCAR 86.02%, and under NMAR 81.19%).
- For the C4.5 classifier using KNN imputation, according to Fig. 2c the mean accuracy rates realized under MAR were better than under MCAR and NMAR regardless the MD percentages (at 10% of MD the mean accuracy rates were: under MAR 89.80%, under NMAR 88.11% and under MCAR 88.08%). Furthermore, under NMAR the mean accuracy rates were higher than those obtained under MCAR (at 90% of MD the mean accuracy rates were: under NMAR 88.04%, and under MCAR 85.61%).

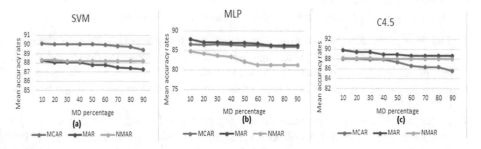

Fig. 2. Mean accuracy rates of SVM, MLP and C4.5 using KNN imputation with three MD mechanisms and nine MD percentages.

6.2   Comparison of Classifiers Using MD Mechanisms

This section compares the mean accuracy rates of the three classifiers C4.5, SVM and MLP, using KNN imputation, three MD mechanisms, and nine MD percentage. Figure 3a–c show the mean accuracy rates of each classifier using KNN imputation, with three MD mechanisms and nine MD percentages.

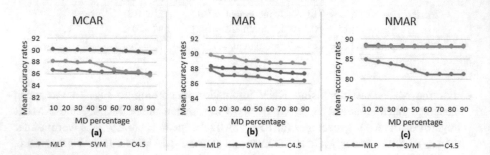

Fig. 3. Comparison of three classifiers using KNN imputation with three MD mechanisms and nine MD percentages

As can be seen in Fig. 3a–c, SVM achieved the highest accuracy rates followed by C4.5 under MCAR and NMAR regardless the MD percentage (for example at 10% of MD the accuracy rates obtained under MCAR are: 90.10% for SVM, 88.08% for C4.5 and 86.53% for MLP). While under MAR C4.5 achieved better results than SVM and MLP regardless the MD percentage (for example at 10% of MD the accuracy rates obtained under MAR are: 89.80% for C4.5, 88.26% for SVM, and 87.88% for MLP).

It's noteworthy that MLP achieved the lowest accuracy rates regardless the MD mechanisms and percentage.

According to Figs. 2a–c and 3a–c, we summarize the findings:

1. The MD percentage has an important influence on the accuracy rates of the classifiers, as long as the MD percentage increases the accuracy rate decreases; which can be explained by the fact that imputing 10% of MD is more reliable due to the remaining large sample of instances, unlike imputing 90% of MD that can bias the dataset [24].
2. SVM showed the better accuracy rates comparing to C4.5 and MLP, and this can be explained by the fact that SVM proved helpful in breast cancer diagnosis [22]. The imputation can perform better for SVM while C4.5 is more missing data resistant, and can handle missing data [23].
3. Although ANNs are often considered an advanced machine learning and very sophisticated, MLP showed the lowest results in our study regardless the MD mechanism/percentage.

7 Threats to Validity

Internal validity: Internal threats of this paper are in general corresponding to the evaluation of the classifiers. The first one is related to the evaluation metric used; the results of this study were based on the accuracy rate metric because it is widely used to evaluate the classification performance. The second threat is related to the evaluation method, 10 cross-validation model was adopted because it is considered as a standard for performance estimation and technique selection [24].

External validity: this study used only numerical attributes. Further investigations on other datasets are required to discuss categorical attributes. In this empirical evaluation, only three classifiers were applied to investigate the impacts of MD mechanisms. Thus, more classifiers should be evaluated in order to assess the impact of MD. Moreover, there is several imputation techniques used to handle MD; in this work only KNN imputation was used due to its popularity, yet other imputation techniques can achieve better results.

8 Conclusion and Future Work

In this study, the impacts of MD on the three classifiers: C4.5, SVM and MLP were evaluated over two datasets from the UCI repository, the Wisconsin breast cancer original and prognosis datasets. The performance of each classifier was evaluated using KNN imputation with three MD mechanisms (MCAR, MAR and NMAR) and nine percentages (from 10% to 90). According to the obtained results, the missingness mechanism influences the accuracy rates of the classifiers, and varies for different classifiers. On the other hand, the MD percentage influences negatively the classifiers accuracy, while the MD percentage increases the accuracy rate decreases regardless the MD mechanism and technique. SVM yelled to better results under MCAR and NMAR regardless the MD percentage, while MLP achieved the lowest accuracy rates.

Ongoing research intends to carry out more empirical evaluations of the impact of MD on the performance of breast cancer classification in order to refute or confirm the findings of the present study. Moreover, we intend to compare the impact of KNN imputation technique and deletion on more classifiers such us Random Forest (RF) and Case-based reasoning (CBR). Since the present study only deals with numerical attributes, it would be of great interest to deal with missing categorical data.

References

1. Oskouei, R.J., Kor, N.M., Maleki, S.A.: Data mining and medical world: breast cancers' diagnosis, treatment, prognosis and challenges. Am. J. Cancer Res. (2017)
2. Akay, M.F.: Support vector machines combined with feature selection for breast cancer diagnosis. Expert Syst. Appl. **36**, 3240–3247 (2009). https://doi.org/10.1016/j.eswa.2008.01.009

3. Esfandiari, N., Babavalian, M.R., Moghadam, A.M.E., Tabar, V.K.: Knowledge discovery in medicine: current issue and future trend. Expert Syst. Appl. (2014). https://doi.org/10.1016/j.eswa.2014.01.011

4. Idri, A., Chlioui, I., Ouassif, B.E.: A systematic map of data analytics in breast cancer. In: ACSW 2018 Proceedings pf Australasian Computer Science Week Multiconference, Brisband, pp. 26:1–26:10 (2018). https://doi.org/10.1145/3167918.3167930

5. Cismondi, F., Fialho, A.S., Vieira, S.M., Reti, S.R., Sousa, J.M.C., Finkelstein, S.N.: Missing data in medical databases: impute, delete or classify? Artif. Intell. Med. (2013). https://doi.org/10.1016/j.artmed.2013.01.003

6. Idri, A., Benhar, H., Fernández-Alemán, J.L., Kadi, I.: A systematic map of medical data preprocessing in knowledge discovery. Comput. Methods Programs Biomed. **162**, 69–85 (2018). https://doi.org/10.1016/j.cmpb.2018.05.007

7. Idri, A., Abnane, I., Abran, A.: Missing data techniques in analogy-based software development effort estimation. J. Syst. Softw. **117**, 595–611 (2016). https://doi.org/10.1016/j.jss.2016.04.058

8. Rubin, D.B.: Inference and missing data (with discussion). Biometrika **63**, 581–592 (1976)

9. Garciarena, U., Santana, R.: An extensive analysis of the interaction between missing data types, imputation methods, and supervised classifiers. Expert Syst. Appl. **89**, 52–65 (2017). https://doi.org/10.1016/j.eswa.2017.07.026

10. Curley, C., Krause, R.M., Feiock, R., Hawkins, C.V.: Dealing with missing data : a comparative exploration of approaches using the integrated city sustainability database (2017). https://doi.org/10.1177/1078087417726394

11. Schafer, J.L., Graham, J.W.: Missing data: our view of the state of the art. Psychol. Methods **7**, 147–177 (2002). https://doi.org/10.1037/1082-989X.7.2.147

12. Yenduri, S.: An empirical study of imputation techniques for software data sets (2005)

13. García-Laencina, P.J., Abreu, P.H., Abreu, M.H., Afonoso, N.: Missing data imputation on the 5-year survival prediction of breast cancer patients with unknown discrete values. Comput. Biol. Med. **59**, 125–133 (2015). https://doi.org/10.1016/j.compbiomed.2015.02.006

14. Jerez, J.M., Molina, I., García-Laencina, P.J., Alba, E., Ribelles, N., Martín, M., Franco, L.: Missing data imputation using statistical and machine learning methods in a real breast cancer problem. Artif. Intell. Med. **50**, 105–115 (2010). https://doi.org/10.1016/j.artmed.2010.05.002

15. Index of /ml/machine-learning-databases/breast-cancer-Wisconsin (2017). Archive.ics.uci.edu. https://ww.archive.ics.uci.edu/ml/datasets/breast+cancer+wisconsin+(original). Accessed 20 Jul 2003

16. Index of /ml/machine-learning-databases/breast-cancer-wisconsin (2017). Archive.ics.uci.edu. https://archive.ics.uci.edu/ml/datasets/breast+cancer+wisconsin+(Prognostic). Accessed 20 Jul 2003

17. Song, Q., Shepperd, M., Chen, X., Liu, J.: Can k-NN imputation improve the performance of C4.5 with small software project data sets? a comparative evaluation. J. Syst. Softw. (2008). https://doi.org/10.1016/j.jss.2008.05.008

18. Hall, M., Witten, I., Frank, E.: Data Mining, 4th Edn., Elsevier (2011)

19. Alpaydın, E.: Introduction to Machine Learning, 2nd Edn., The MIT Press, London (2014). https://doi.org/10.1007/978-1-62703-748-8-7

20. Cristianini, N., Shawe-Taylor, J.: An Introduction to Support Vector Machines and other kernel based learning methods. Cambridge University Press, Cambridge (2000). citeulike-article-id:114719

21. Ghosh, S., Mondal, S., Ghosh, B.: A comparative study of breast cancer detection based on SVM and MLP BPN classifier. In: 2014 First International Conference on Automation, Control, Energy and System, pp. 1–4 (2014). https://doi.org/10.1109/aces.2014.6808002
22. Baldi, P., Brunak, S., Chauvin, Y., Andersen, C.A., Nielsen, H.: Assessing the accuracy of prediction algorithms for classification: an overview. Bioinformatics (2000). https://doi.org/10.1093/bioinformatics/16.5.412
23. Fawcett, T.: An introduction to ROC analysis. Pattern Recognit. Lett. (2006). https://doi.org/10.1016/j.patrec.2005.10.010
24. Salzberg, S.L.: On comparing classifiers: Pitfalls to avoid and a recommended approach. Data Min. Knowl. Discov. (1997). https://doi.org/10.1023/a:1009752403260
25. Jhajharia, S., Varshney, H.K., Verma, S., Kumar, R.: A neural network based breast cancer prognosis model with PCA processed features. In: 2016 International Conference on Advances in Computing, Communications and Informatics, ICACCI 2016, pp. 1896–1901 (2016). https://doi.org/10.1109/ICACCI.2016.7732327
26. The university of Waikato, Weka the university of Waikato, (n.d.). https://www.cs.waikato.ac.nz/ml/weka/
27. Ma, X., Zhang, Y., Wang, Y.: Performance evaluation of kernel functions based on grid search for support vector regression. In: 2015 IEEE 7th International Conference on Cybernetics and Intelligent Systems and IEEE Conference on Robotics, Automation and Mechatronics, pp. 283–288 (2015). https://doi.org/10.1109/ICCIS.2015.7274635

COSMIC Functional Size Measurement of Mobile Personal Health Records for Pregnancy Monitoring

Mariam Bachiri[1], Ali Idri[1(✉)], Leanne Redman[2], Alain Abran[3],
Juan Manuel Carrillo de Gea[4], and José Luis Fernández-Alemán[4]

[1] Software Project Management Research Team, ENSIAS,
Mohammed V University in Rabat, Rabat, Morocco
{mariam_bachiri, ali.idri}@um5.ac.ma

[2] Reproductive Endocrinology and Women's Health Laboratory,
Pennington Biomedical Research Center, Louisiana State University,
Baton Rouge, USA
leanne.redman@pbrc.edu

[3] Department of Software Engineering, École de Technologie Supérieure,
Montréal, Canada
alain.abran@etsmtl.ca

[4] Faculty of Computer Science, Department of Computing,
University of Murcia, Murcia, Spain
{jmcdgl, aleman}@um.es

Abstract. An empirical evaluation of the Common Software Measurement International Consortium (COSMIC) method has been conducted in this study, by measuring the functional size of 17 mobile Personal Health Records (mPHRs) for pregnancy monitoring. The aim of this evaluation is to compare the functional size of each app measured using the COSMIC method to the score of the app obtained in a previous evaluation that relied on functions extraction using a quality assessment questionnaire. The shift between the rankings of both evaluations was small for the majority of the selected apps. Therefore the findings of this study support the use of the COSMIC method for these apps in regards to measuring the functional size for further updates or improvements. Moreover, the use of COSMIC is more effective since it covers the full features and functionalities of mPHRs for pregnancy monitoring.

Keywords: COSMIC · Pregnancy monitoring ·
Mobile Personal Health Records · mhealth · Functional size measurement

1 Introduction

Mobile applications are software applications designed to run on smartphones, tablets or other mobile devices [1, 2]. According to statistics performed by March 2017, the number of apps available for download in leading app stores as of that month is: 2.8 million apps in Google Play and 2.2 million apps in Apple's App Store [3]. Moreover,

© Springer Nature Switzerland AG 2019
Á. Rocha et al. (Eds.): WorldCIST'19 2019, AISC 932, pp. 24–33, 2019.
https://doi.org/10.1007/978-3-030-16187-3_3

the development of mobile apps is considered relatively easier than web or desktop apps. In addition, their availability and lower price encourage usage, thus this industry experiences considerable growth in app production each year.

Mobile health (mhealth) apps are applications that refer to medicine and health services through mobile devices. The health, fitness and medical industries have been named as the top three fields to accelerate the growth of mobile devices [4, 23]. Moreover, according to recent statistics, more than 84,000 mobile apps were released for the medical and health & fitness markets in 2017 [5].

Mobile Personal Health Records (mPHRs) for pregnancy monitoring are mhealth apps dedicated for pregnant women in order to track pregnancy [6]. In previous work [7], a review of the mPHRs for pregnancy monitoring was conducted by applying the Systematic Literature Review (SLR) protocol in the analysis process. The method followed to perform this analysis included developing a quality assessment question-naire, based on a rigorous review of scientific literature on pregnancy and mobile applications available on the market. The quality assessment questionnaire was then applied to a set of mPHRs for pregnancy monitoring (available on iOS and Android app stores), in order to analyze the features and functionalities of these apps that are specific to pregnancy monitoring. Instead of using such a classical and qualitative method for assessing the functionality of pregnancy monitoring mPHRs, Functional Size Measurement (FSM) methods offer an objective quantification of the functional size of such applications. Initially, FSM methods measured the software size in terms of functions required by the users, rather than how software is implemented. For example, it provides software size as an input to a number of the software effort estimation techniques and tools [8].

The Common Software Measurement International Consortium, called COSMIC – ISO 19761, is the second generation of FSM methods and is based on fundamental principles of software engineering and metrology. COSMIC is applicable to business, real-time and infrastructure software, complying with the standard ISO/IEC14143/1 [8]. The COSMIC method has several assets, such as being publicly available and applicable to a wide range of software and underlying concepts that are compatible with modern concepts of software engineering [9].

The need for efficient and standardized development processes for the pregnancy monitoring mPHRs requires an evaluation of their size, which can be realized by using a functional size measurement method. Therefore, reports utilization of the COSMIC method to measure the functional size of 17 free pregnancy monitoring mPHRs available in the Google Play and Apple App Stores. Thereafter, the COSMIC func-tional size values are compared with those obtained using a qualitative questionnaire [7]. The aim of this comparison is to empirically apply the COSMIC method on mPHRs for pregnancy monitoring, in order to identify its accuracy as regards mea-suring the functional size of these apps compared to the qualitative method that relies on objectivity using a quality assessment questionnaire. To the best of our knowledge, no research work has previously applied the COSMIC method on mhealth apps, in particular on mPHRs for pregnancy monitoring.

The remainder of this paper is structured as follows: Sect. 2 introduces the concepts and methods used in the present study, in particular mPHRs applications and FSM methods: COSMIC. Section 3 introduces the related work on the application of

COSMIC method on mobile applications. A description of the method used in this study is presented in Sect. 4. Section 5 presents and discusses the findings of this study. Finally, the conclusions and perspectives are summarized in Sect. 6.

2 Background

This section provides an overview of the mPHRs for pregnancy monitoring, along with an introduction to the FSM method COSMIC.

2.1 MPHRs for Pregnancy Monitoring

MPHRs have been considered as a point of interest for several research studies. MPHRs are available for chronic diseases such as cardiovascular diseases [2], diabetes [10] or obesity [11]. They are also used for specific health conditions such as pregnancy, which requires specific monitoring during the 40 weeks. MPHRs for pregnancy monitoring provide several functionalities to help the pregnant woman keep track of her health status and the health of her baby. A previous study has analyzed the features and functionalities of pregnancy monitoring mPHRs [7].

2.2 Functional Size Measurement: COSMIC

According to the standard ISO/IEC 14143/1:2011 [8], a functional size is defined as "a size of software derived by quantifying the functional user requirements". Functional size measurement is, therefore, aimed to quantify the effort and duration of software projects and the evaluation of productivity in the software development process as stated in ISO/IEC 14143 [8]. Five ISO-based measurement methods have been established in the software industry [9]: (i) ISO 19761-COSMIC; (ii) ISO 20926-IFPUG; (iii) ISO 20968-MKII, (iv) ISO 24570- NESMA; and (v) ISO 29881-FISMA.

The FSM method used in this study is ISO 19761-COSMIC. The COSMIC method is designed to measure the functionality of: business application software, real-time software, infrastructure software and some types of scientific and engineering software [12]. The COSMIC method has two models based on the fundamental software engineering principles: the 'Software Context Model', which enables to define the software to be measured and the size measurement, and the 'Generic Software Model', which defines how the Functional User Requirements (FURs) of the software to be measured are modeled so that they can be measured [12]. The COSMIC method consists of 3 phases: (1) The Measurement Strategy Phase: it is the preliminary phase in which the key parameters of the measurement are defined carefully [12]. (2) The Mapping Phase. In this phase, the functional processes are identified from the available FURs, in addition to the object of interest, the data groups and the data movements as described in [12]. (3) The Measurement Phase: The data movements are counted in this phase, by associating 1 Cosmic Function Point (CFP), which is the measurement unit, to each data movement. Thus, they are summed to represent the functional value of the measurement. The measurement process is more detailed in [12].

At the moment of writing this paper, the COSMIC Measurement manual version 4.0.2 [12] is the latest version released, which is applied in this study.

3   Related Work

This section introduces the previous studies that have been conducted as regards the analysis of functionalities for Personal Health Records (PHRs), in addition to studies about the use of the COSMIC method to measure the functional size of mobile apps.

3.1   PHRs: Analysis of Functionalities

Previous studies have been conducted in order to analyze the features and functionalities of PHRs.

In [13], the functionalities of Web-based PHRs were analyzed and assessed according to health information, user actions and connection with other tools. An SLR was conducted to select 19 free Web-based PHRs from the 47 PHRs identified. The findings of this study showed that none of the PHRs selected met all of the functions analyzed in the study. Moreover, the content, functions, security and marketing characteristics of mPHRs were previously evaluated for iOS, Blackberry and Android [14]. Nineteen mPHRs were selected and evaluated. The evaluation of the mPHRs covered the product characteristics, data elements, application features and marketing tactics. The study found none of the mPHRs included all the evaluated aspects. Furthermore, in our previous study [7], we analyzed the features and functionalities of 33 pregnancy monitoring mPHRs for iOS and Android, which were selected from Apple App store and Google Play store, respectively. The evaluation was conducted according to 9 data items: calendars, pregnancy information, health habits, counters, diaries, mobile features, security, backup, configuration and architectural design, in addition to 35 questions included in a quality assessment questionnaire. We found none of the mPHRs selected met 100% of the functionalities analyzed. Moreover, the highest score achieved was 77%, while the lowest was 17% [7].

3.2   COSMIC Functional Size Measurement for Mobile Applications

Measuring the functional size of mobile applications is a new branch for applying the COSMIC method, which has been investigated in several previous studies [15–20]:

An empirical study has been carried out to measure the functional size of Android mobile applications using COSMIC to estimate the amount of needed memory, in addition to identify some possible recurrent patterns for the measurement [15]. The results of this evaluation showed that it is possible to accurately predict the needed memory for mobile applications. A case study of the mobile game application Angry Bird aimed to demonstrate how UML (Unified Modeling Language) representations improve the use of COSMIC measurement by mapping the UML context with COSMIC FSM rules and measurement [16]. In addition, the case study showed that contributing to build UML models according to the COSMIC measurement rules, eases measurement procedures. A COSMIC FSM based approximative method has been introduced in order to quickly and accurately measure the functional size of mobile applications [17]. The aim of this approximative method is the quick and easy use of COSMIC in the case of mobile apps. The COSMIC method was applied on a mobile application that provides a basic course management [18]. The study concluded that COSMIC is suitable to measure the functional size of a mobile application [18]. However, some characteristics of the mobile application development, such as non-functional requirements and small teams and projects might represent serious limitations for these apps. Moreover, a Use Case based measurement method was proposed [19], to estimate the functional size of mobile and web applications using the COSMIC method. The proposed method is based on a set of measurement formulae evaluated through a case study of a restaurant management system. Moreover, the proposed method will help to apply COSMIC and avoid measurement errors. Furthermore, in a review of the measurement process and rules used by FSM for mobile applications and UML modelling [20], it is stated that most of the literature adapted FSM in the estimation of the mobile development effort. However, some literature used FSM with UML modelling, since UML model can represent the functional requirement of a mobile application.

4   Method

MPHRs for pregnancy monitoring belong to business applications, which are characterized by managing large amount of data about events and objects in the real world related to business administrations [21]. Therefore, COSMIC can be applied to measure the functional size of these apps.

This study targets 17 mPHRs for pregnancy monitoring running under Android and iOS platforms, since they are currently the leading mobile operating systems and the preferred for mhealth apps development [22]. These apps were the result of a selection process that was conducted in our previous study [7], and revised by excluding the apps that were no longer available in the apps repositories or became outdated. The previous review of the mPHRs for pregnancy monitoring [7] targeted Apple App store and Google Play store as the sources for the selection of the mPHRs for pregnancy monitoring, since they are the official apps repositories for both iOS and Android respectively. The terms and keywords have been chosen to fulfill the search in these sources. Moreover, four inclusion criteria and two exclusion criteria were used in order to select a set of apps to be evaluated. Thereafter, a list of data items to be extracted from each app was defined in order to establish a quality assessment questionnaire to evaluate the apps selected. These data items focused only on pregnancy monitoring.

The method used in the present study is thereafter described and consists of the three phases of COSMIC:

a. Measurement Strategy Phase
The main parameters of this phase are identified as follows:

- The purpose: Measurement of the size of FURs of the selected mPHRs for pregnancy monitoring, to be compared with the functional scores of these apps, which were obtained in a previous study [7] based on a regular method relying on functions extraction. These scores represent the number of features and function-alities analyzed that are present in these apps, according to the data items used.

- The overall scope: The measurement of all the FURs within the selected apps only. The FURs executed in external apps are excluded from this measurement. There-fore, only data movements between the external apps and the measured apps are considered.
- Functional users: The functional users in this case are human users, which are pregnant women, that use the mPHRs for pregnancy monitoring.
- Layer: A mobile application is considered as an application layer, which is developed on the top of other data layers [17].
- Level of granularity: The measures will be conducted at the level of the screens of each mPHR for pregnancy monitoring.

In order to visualize the interaction between the defined key parameters at this stage, Fig. 1 presents the context diagram of the measurement that will be conducted in this study. The context diagrams are used to show the measured software, along with its functional users, the boundary, the persistent storage and the movements of data between them [12].

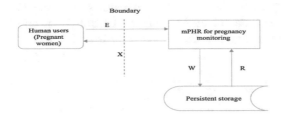

Fig. 1. Context diagram. Acronyms: Entry (E), Exit (X), Write (W) and Read (R)

b. Mapping Phase

A functional user starts a functional process in response to a triggering event. In the case of mPHRs for pregnancy monitoring, the triggering events are the users' inputs. Moreover, FURs can be mapped into unique functional processes, which are identified in this phase, for each selected mPHR for pregnancy monitoring, along with the objects of interest, data groups and data movements. The functional processes can be extracted based on each screen of the mobile apps [18]. This method was adopted in this study by the authors, by installing each mPHR for pregnancy monitoring, then browsing the features and functionalities listed in each screen, in order to extract the functional processes. Hence, the types of the data movements (Entry (E), Exit (X), Read (R) or Write (W)) were defined.

In this study, the mPHRs for pregnancy monitoring were considered as being used for the first time for the extraction of the functional processes. The use of the mPHRs for pregnancy monitoring allows the progression of the processes be followed while using the apps, and to extract some processes that are generally executed only while using the app for the first time, such as creating a profile or setting the due date.

c. Measurement Phase

In this phase, the data movements were counted based on the measurements that have been conducted according to some common cases [15] such as create, select, delete, add, share and display, which can be repeatedly executed by the user, which has facilitated the measurement process. The measurement was conducted independently by the first two authors. Any measurement discrepancies were resolved through discussions.

5   Results and Discussion

The aim of this study was to evaluate the functional size of 17 mPHRs for pregnancy monitoring using the COSMIC method. Thereafter, we compared the COSMIC functional size values with the functionality scores of these apps obtained from our previous work [7]. A ranking of the 17 apps was made based on both criteria: COSMIC functional size and the scores of [7].

The functionality scores of the 17 apps in ascending order is presented in Fig. 2. The app Pregnancy & Baby App – Nurture achieved the highest score of provided functionalities (27), while the app My pregnancy guider achieved the lowest score of 11 [7]. Figure 3 presents the COSMIC functional size of the 17 apps in ascending order. The app Gestavida obtained the highest COSMIC functional size of 400 CFP, while the app Pregnancy mode free obtained the lowest functional size of 33 CFP.

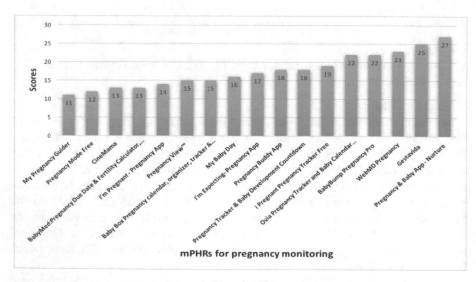

Fig. 2. Scores of the 17 selected mPHRs for pregnancy monitoring from our previous study [7]

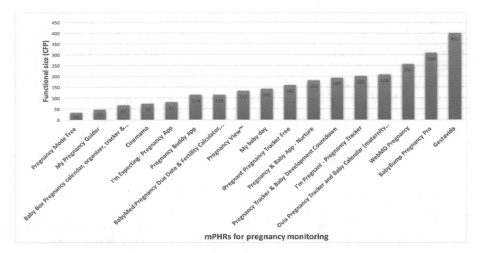

Fig. 3. COSMIC functional size of the 17 selected mPHRs for pregnancy monitoring

The differences noticed between the functional sizes may be due to the number of functions for each purpose provided by the apps, since the apps may have multiple purposes and each purpose provides a different number of functionalities, depending on itd importance.

However, the overall difference between the scores is considered relatively small. Furthermore, there is considerable variability in the COSMIC functional sizes of the measured apps.

Regarding congruence in terms of comparative ranking between scores and the COSMIC functional size: some apps got only one shift between the two rankings, such as: Gestavida, Ovia Pregnancy Tracker and Baby Calendar (maternity app), Pregnancy Tracker & Baby Development Countdown and My Baby Day. This difference indicates that each criteria used (scores of [7] and CFP) ranked these apps similarly. Moreover, some apps got a significant ranking variation, such as for I'm Pregnant - Pregnancy App and Pregnancy & Baby App – Nurture. For the remaining apps, the shift between the rankings ranged between two and three positions, which is considered as a slight difference. Furthermore, the only pregnancy monitoring mPHR that got the same ranking in both cases is WebMD Pregnancy.

From these results we conclude that there was no important shift between the rankings of the COSMIC functional sizes and the scores of the majority of apps (15 of 17), except for two: I'm Pregnant-Pregnancy App and Pregnancy & Baby App-Nurture.

The functional size has been calculated according to the COSMIC rules and guidelines and it has covered the features and functionalities within the 17 apps. However, the scores obtained in the prior study [7] have been calculated according to an evaluation based on a quality assessment questionnaire that has been focusing only on pregnancy oriented aspects. Hence, the difference noticed by using these two evaluations could be due to the fact that the scores are mainly related to pregnancy monitoring. Thus, they did not cover some complementary features and functionalities provided by these mPHRs.

It is deduced that the COSMIC method is flexible and suitable for the measurement of the functional size of mobile apps [18]. Moreover, COSMIC, as an FSM method, is more accurate and practical to measure the functional size of mPHRs for pregnancy monitoring, since it is based on a rigorous basis of the software engineering principles, and it is independent of the technologies used or the development methods. COSMIC can target a specific set of functionalities as well, in order to measure their functional size, instead of measuring the functional size of the entire mobile app.

6 Conclusion and Future Work

We conducted an empirical evaluation of the COSMIC method on 17 mPHRs for pregnancy monitoring, to compare their CFP values with the scores obtained in a previous evaluation [7]. Although the number of the evaluated apps is few, we noted that the shift between the rankings of both the functional size evaluations was small for the majority of the selected apps. Thus, COSMIC is a preferred method to measure the functional size of such apps, as it covers all features and functionalities, and gives an overview about the richness of the app contents according to the main purpose, which is tracking the health of the pregnant woman and her baby during pregnancy. As for the developers of the mPHRs for pregnancy monitoring, using COSMIC will provide valuable results that can be exploited to evaluate and compare the available apps on the market.

Acknowledgments. This work was conducted within the research project PEER, 7-246 supported by the US Agency for International Development. The authors would like to thank the NAS and USAID for their support.

References

1. Sardi, L., Idri, A., Fernández-Alemán, J.L.: Gamified mobile blood donation applications. In: International Conference on Bioinformatics and Biomedical Engineering, pp. 165–176. Springer, Cham (2017)
2. Ouhbi, S., Idri, A., Fernandez-Aleman, J.L., Toval, A.: Mobile personal health records for cardiovascular patients. In: Third World Conference on Complex Systems, pp. 1–6 (2015)
3. Statista.com: Number of apps available in leading app stores as of March 2017. https://www.statista.com/statistics/276623/number-of-apps-available-in-leading-app-stores/. Accessed 18 Apr 2018
4. Statista.com: mHealth - Statistics & Facts. https://www.statista.com/topics/2263/mhealth/. Accessed 18 Apr 2018
5. 84,000 health app publishers in 2017 – Newcomers differ in their go-to-market approach. https://research2guidance.com/84000-health-app-publishers-in-2017/. Accessed 29 May 2018
6. Idri, A., Bachiri, M., Fernández-Alemán, J.L.: A framework for evaluating the software product quality of pregnancy monitoring mobile personal health records. J. Med. Syst. **40**, 50 (2016)

7. Bachiri, M., Idri, A., Fernández-Alemán, J.L., Toval, A.: Mobile personal health records for pregnancy monitoring functionalities: analysis and potential. Comput. Methods Programs Biomed. **134**, 121–135 (2016)
8. ISO/IEC 14143-1:2007: Information technology - software measurement - functional size measurement - Part 1: definition of concepts, International Organization for Standardization, Geneva (2007)
9. Abran, A.: Software Metrics and Software Metrology. Wiley, Hoboken (2010)
10. El-Gayar, O., Timsina, P., Nawar, N., Eid, W.: Mobile applications for diabetes self-management: status and potential. J. Diabetes Sci. Technol. **7**, 247–262 (2013)
11. Lopes, I.M., Silva, B.M., Rodrigues, J.J.P.C., Lloret, J., Proenca, M.L.: A mobile health monitoring solution for weight control. In: International Conference on Wireless Communications and Signal Processing, pp. 1–5 (2011)
12. Common Software Measurement International Consortium (COSMIC): Measurement Manual v4.0.1 (2015)
13. Fernández-Alemán, J.L., Seva-Llor, C.L., Toval, A., Ouhbi, S., Fernández-Luque, L.: Free web-based personal health records: an analysis of functionality. J. Med. Syst. **37**, 9990 (2013)
14. Kharrazi, H., Chisholm, R., VanNasdale, D., Thompson, B.: Mobile personal health records: an evaluation of features and functionality. Int. J. Med. Inform. **81**, 579–593 (2012)
15. D'Avanzo, L., Ferrucci, F., Gravino, C., Salza, P.: COSMIC functional measurement of mobile applications and code size estimation. In: Proceedings of the 30th Annual ACM Symposium on Applied Computing - SAC 2015, pp. 1631–1636 (2015)
16. Abdullah, N.A.S., Rusli, N.I.A., Ibrahim, M.F.: A case study in COSMIC functional size measurement: angry bird mobile application. In: IEEE Conference on Open Systems, ICOS 2013, pp. 139–144 (2013)
17. Van Heeringen, H., Van Gorp, E.: Measure the functional size of a mobile app: using the cosmic functional size measurement method. In: Proceedings of the Joint Conference of the International Workshop on Software Measurement, IWSM 2014, the International Conference on Software Process and Product Measurement, pp. 11–16 (2014)
18. Nitze, A.: Measuring Mobile Application Size Using COSMIC FP. MetriKon, pp. 101–114 (2013)
19. Haoues, M., Sellami, A., Ben-Abdallah, H.: A rapid measurement procedure for sizing web and mobile applications based on COSMIC FSM method. In: Proceedings of the 27th International Workshop on Software Measurement and 12th International Conference on Software Process and Product Measurement 2003, pp. 129–137 (2017)
20. Abdullah, N.A.S., Rusli, N.I.A.: Reviews on functional size measurement in mobile application and UML model. In: Proceedings of the 5th International Conference on Computing and Informatics, ICOCI, pp. 353–358 (2015)
21. Baklizky, D.: The COSMIC Functional Size Measurement Method Versions 4.0.1/4.0.2 Guideline for Sizing Business Application Software (2017)
22. Healthcare Mobile App Development and mHealth Apps in 2017. https://medium.com/@Adoriasoft_Com/healthcare-mobile-app-development-and-mhealth-apps-in-2017-eb307d4cad36. Accessed 24 Sept 2018
23. Vázquez, M.Y., Sexto, C.F., Rocha, Á., Aguilera, A.: Mobile phones and psychosocial therapies with vulnerable people: a first state of the art. J. Med. Syst. **40**(6), 157 (2016)

Environmental Quality Monitoring System Based on Internet of Things for Laboratory Conditions Supervision

Gonçalo Marques and Rui Pitarma[✉]

Polytechnic Institute of Guarda – Unit for Inland Development,
Av. Dr. Francisco Sá Carneiro, no 50, 6300-559 Guarda, Portugal
goncalosantosmarques@gmail.com, rpitarma@ipg.pt

Abstract. Indoor environment quality (IEQ) has a significant impact on health and in all human activities. On the one hand, laboratories are spaces characterised by numerous pollution sources that can lead to relevant unhealthy indoor environments. On the other hand, experimental testing requires the supervision of several indoor parameters, such as the case of thermography experiments. With the proliferation of the Internet of Things (IoT) devices and Ambient Assisted Living (AAL) technologies, there is a great potential to create automatic solutions for IEQ supervision. In fact, due to people spend about 90% of our lives indoors, it is crucial to monitor the IEQ in real-time to detect unhealthy conditions to quickly take interventions in the building for enhanced living environments and occupational health. Buildings are responsible for about 40% of the global energy consumption, and over 30% of the CO_2 emissions; also a considerable proportion of this energy is used for thermal comfort. This paper aims to present a solution for real-time supervision of laboratory environmental conditions based on IoT architecture named *iLabM*. The solution is composed by a hardware prototype for ambient data collection and smartphone compatibility for data consulting. The *iLabM* provides temperature, relative humidity, barometric pressure and qualitative air quality data supervision in real-time. The smartphone application can be used to access the collected data in real time but also provides the history of the IEQ data. The results obtained are promising, representing a significant contribution to IEQ monitoring systems based on IoT.

Keywords: Indoor air quality · Indoor environmental quality ·
Internet of Things · Smart cities · Laboratory environmental conditions

1 Introduction

The basic idea of the Internet of Things (IoT) is the pervasive presence of a variety of objects with interaction and cooperation capabilities among them to reach a common objective [1]. Is expected that the IoT will have a high impact on several aspects of everyday life and this concept will be used in several applications such as domotics, assisted living, e-health and is also an ideal emerging technology to provide new evolving data and computational resources for creating revolutionary software applications [2].

Á. Rocha et al. (Eds.): WorldCIST'19 2019, AISC 932, pp. 34–44, 2019.
https://doi.org/10.1007/978-3-030-16187-3_4

Ambient Assisted Living (AAL) is an emerging multi-disciplinary field aiming at providing an ecosystem of different types of sensors, computers, mobile devices, wireless networks and software applications for personal healthcare monitoring and telehealth systems [3]. Currently, there are different AAL having as basis several sensors for measuring weight, blood pressure, glucose, oxygen, temperature, location and position and which are usually applied wireless technologies such as ZigBee, Bluetooth, Ethernet and Wi-Fi.

At 2050 20% of the world population will be age 60 or older [4] that will result in an increase of diseases, health care costs, shortage of caregivers, dependency and brutal social impact. It is a fact that 87% of people prefer to stay in their homes with and support the amazing cost of nursing care [5].

There is a lot of challenges in designing and implementation of an effective AAL system such as information architecture, interaction design, human-computer interaction, ergonomics, usability and accessibility [6]. There are also social and ethical problems like the acceptance by the older adults and the privacy and confidentiality that should be a requirement of all AAL devices. In fact, it is also essential to ensure that technology does not replace human care and should be an amazing compliment.

IoT and AAL technologies provide many benefits to the healthcare domain in activities such as tracking of objects, patients and staff, identification and authentication of people, automatic data collection and sensing [7].

Indoor environmental quality (IEQ) in buildings includes indoor air quality (IAQ), acoustics, thermal comfort, and lighting [8]. It is well known that poor IEQ has a negative impact on occupational health, particularly on children and old people.

The problem of poor IAQ is of utmost importance affecting especially severe form the poorest people in the world who are most vulnerable presenting itself as a severe problem for world health such as tobacco use or the issue of sexually transmitted diseases [9]. In the USA, indoor and outdoor air quality is regulated by the Environmental Protection Agency (EPA). EPA considers that indoor levels of pollutants may be up to 100 times higher than outdoor pollutant level and ranked poor air quality as one of the top 5 environmental risks to the public health [10].

High-quality research should continue to focus on the quality problems of IAQ to adopt legislation, inspection and creating mechanisms that act in real time to improve public health, both in public places such as schools and hospitals and private places and further increase the rigorousness of the building's construction rules. In the significant cases, simple interventions provided by home-owners and building operators can produce significant positive impacts on IAQ such as the avoidance of smoking indoors and the use of natural ventilation are important behaviours that should be taught to children through educational programs that address the IAQ [11].

Thermal comfort is ranked by building occupants to be of greater importance compared with visual and acoustic comfort and IAQ [12]. The comfort temperature might be as low as 17 °C and as high as 30 °C. Thermal comfort is influenced by six factors: air temperature, radiant temperature, air velocity, humidity, clothing insulation and metabolic heat. The first four factors can be measured, and the last two are personalised factors [13].

Buildings represent about 40% of the global energy utilisation and are responsible for over 30% of the CO2 emissions. A large proportion of this energy is related to thermal comfort in buildings. The use of personalised conditioning systems is apparently the most ideal approach to increase user acceptability with the thermal environment as thermal comfort is a complex subject and we are far from knowing all its related viewpoints [14].

Currently, IEQ in buildings is based on random sampling. However, these procedures are providing only information relating to a specific sampling and are devoid of information of spatio-temporal variations, particularly in laboratory tests such as thermography experiments.

The concept of the "smart city" has recently been introduced as a strategic device to encompass modern urban production factors in a common framework and, in particular, to highlight the importance of Information and Communication Technologies (ICTs) in the last 20 years for enhancing the competitive profile of a city as proposed by [15]. The smart city is directly related to an emerging strategy to mitigate the problems generated by the urban population growth and rapid urbanisation [16]. The most relevant issue in smart cities is the no interoperability of the heterogeneous technologies; the IoT can provide the interoperability to build a unified urban-scale ICT platform for smart cities [17]. IoT has an incredible potential for creating new real-life applications and services for the smart city context [18], particularly for environmental quality supervision.

This paper aims to present a Laboratory Environmental Conditions (LEC) real-time monitoring solution based on IoT architecture named *iLabM*. To create a cost-effective system, only one sensor was chosen. The Bosch BME680 was selected since it is an integrated environmental sensor developed specifically for mobile applications and wearables where size and low power consumption are key requirements. This sensor provides temperature, relative humidity, barometric pressure and qualitative air quality. The solution is composed by a hardware prototype for ambient data collection and a smartphone compatibility for data consulting. The *iLabM* is based on open-source technologies and is a totality Wi-Fi system, with several advantages compared to existing systems, such as its modularity, scalability, low-cost and easy installation. The data is uploading to a SQL SERVER database and the can be accessed using a smartphone. This system is based on an ESP8266 microcontroller with built-in Wi-Fi communication technology and has been tested in the context of infrared thermography experiments.

The paper is structured as follows: besides de introduction (Sect. 1), Sect. 2 presents the related work and Sect. 3 is concerned to the methods and materials used in the implementation of the sensor system; Sect. 4 demonstrates the system operation and experimental results, and the conclusion is presented in Sect. 5.

2   Related Work

Several examples of projects on environmental quality monitoring are available in the literature. In this section, the most outstanding solutions are the use of open-source, low cost and mobility technologies.

A Wireless sensor network (WSN) low-cost application for precision agriculture deployed in a pepper vegetable greenhouse that takes an appropriate measure such as remote control for drip irrigation and facilities with the guidance of the expert system, is proposed by [19].

A real-time system for environmental monitoring based on WSN capable of monitoring ambient temperature, relative humidity, acoustic levels and concentration of dust particles in the air for smart cities is proposed by [20].

A strategy of using WSN to monitor the temperature distribution in a large-scale indoor space, which aims to improve the quality of the measurements transmitted by wireless signals, identify the temperature distribution pattern of the large-scale space, and optimize the allocation of supply air rated flow rate to multiple supply air terminals that serve the space according to the identified temperature distribution pattern is proposed by [21].

Several IoT architectures for IAQ monitoring that incorporates open-source technologies for processing and data transmission and microsensors for data acquisition but also allows access to data collected from different places simultaneously through web access and through mobile applications in real time are proposed by [22–30].

The *iLabM* system aims to provide a useful tool for management enhanced living environments of smart cities. The benefits for health, comfort and productivity of good IAQ conditions can be improved by decreasing the pollution load while the ventilation remained unchanged [31]. The authors develop a completely wireless solution using the ESP8266 module which implements the IEEE 802.11 b/g/n networking protocol.

3 Materials and Methods

The *iLabM* solution has been created as a low-cost, reliable system that can be easily configured and installed by the average user without supporting the cost of an installation done by certified professionals.

The main objective is to provide accurate supervision of temperature, humidity and barometric pressure and a qualitative air quality index. Therefore, the authors selected a cost-effective BOSCH BME680 sensor a 4-in-1 multi-functional micro-electromechanical system (MEMS) environmental sensor which integrates Volatile Organic Compounds (VOC) sensor, temperature sensor, humidity sensor and barometer. The DFRobot Gravity BME680 environmental sensor was used as it provides a Gravity I2C connector, plug & play, easy to connect.

The *iLabM* incorporate a microcontroller with native Wi-Fi support, a FireBeetle ESP8266 (*DFRobot*). Figure 1 represents the prototype used by the authors. In this section will be discussed in detail the hardware and software used for the system development.

A brief introduction of each component used is shown below

- FireBeetle ESP8266 – is a development board integrated with IoT WiFi, TCP/IP, 32-bit MCU, 10-bit ADC and multiple interfaces such as HSPI, UART, PWM, I2C and I2S. In DTIM10, the full power consumption to maintain WiFi connection reached to 1.2mW. Equipped with 16 MB outer SPI flash memory, ESP8266 is available for programs and firmware storage.

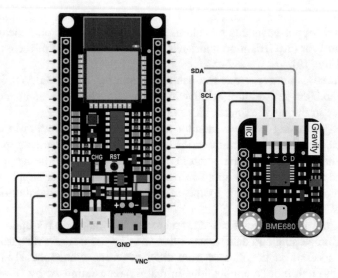

Fig. 1. *iLabM* prototype.

- DFRobot Gravity BME680 – is an I2C environmental VOC sensor, temperature sensor, humidity sensor and barometer. It supports an input voltage of 3.3 V–5.0 V, the operating current consumption is 5 mA without air quality sensing and 25 mA with air quality features. This sensor module size is 30 × 22 (mm)/1.18 × 0.87 (in.). The temperature measurement range is from –40 °C to +85 °C with a precision of ±1.0 °C (0–65 °C). The humidity measurement range is from 0 to 100% r.H with a precision of a ±3% r.H. (20–80% r.H., 25 °C). The atmospheric pressure measurement range is from 300 to 1100 hPa with a precision of ±0.6 hPa (300–1100 hPa, 0–65 °C). The IAQ qualitative range is from 0 to 500 (the larger, the worse). Figure 2 illustrates the IAQ index of the selected sensor.

IAQ Index	Air Quality
0 – 50	good[10]
51 – 100	average
101 – 150	little bad
151 – 200	bad
201 – 300	worse[2]
301 – 500	very bad

Fig. 2. IAQ index sheet.

The ESP8266 microcontroller with built-in Wi-Fi capabilities is used both as processing and communication unit. The monitored data is stored in a SQL SERVER database using web services. For data consulting this solution uses a mobile phone application developed in SWIFT for the iOS operating system (Fig. 3).

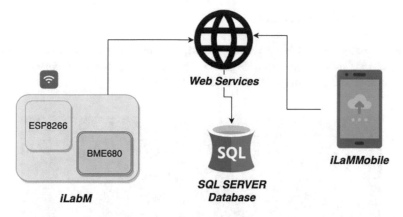

Fig. 3. *iLabM* architecture.

The iOS application is denominated by *iLabMMobile* was developed with SWIFT programming language in XCode IDE (Integrated Development Environment), and it is compatible with iOS 12 and above. This app has two important features as it permits not only real-time consulting of the last data collected and also to receive real-time notifications to advise the user when the air quality is defective (Fig. 4).

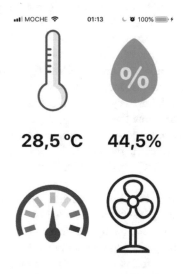

Fig. 4. *iLabMMobile* app.

4   Results and Discussion

For testing purposes, a laboratory of a Portuguese university was on-site monitored, and one *iLabM* module was used. Figure 5 represents the laboratory experiments done by the authors and supported by the *iLabM* system.

Fig. 5.   Thermography experiments supported by *iLabM*.

As in most buildings, the space monitored is naturally ventilated, without any dedicated ventilation slots on the facades.

The module is powered by the power grid using a 230 V–5 V AC-DC 2 A power supply. Environmental data were collected for two months which showed that under certain conditions indoor conditions are significantly lower than those considered healthy for standards. The tests conducted show the system capability to analyse in real-time monitoring of the laboratory. The collected data was correlated with the thermography experiments to know the impact of the environmental conditions on the thermography tests. The *iLabM* allows history data consulting using the smartphone app. A sample of the data collected by *iLabM* is shown in Fig. 6 that represents the temperature monitoring data (°C) and in Fig. 7 that represents relative humidity monitoring data (%).

The smartphone software also allows the user to access the data anytime and anywhere, allowing a precise analysis of the detailed temporal evolution. Therefore, the system is a powerful tool for analysing environmental data parameters to detect unhealthy conditions but also to support experimental tests. Compared to other systems, the *iLabM* system has the following advantages: modularity, small size, low-cost construction and easy installation.

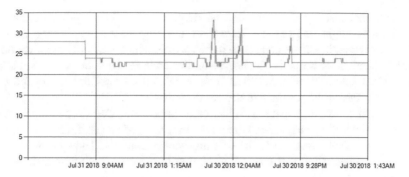

Fig. 6. Temperature monitoring data (°C).

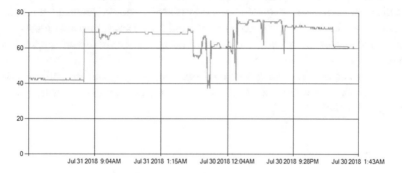

Fig. 7. Relative humidity monitoring data (%).

5 Conclusion

This paper had presented an IoT architecture for LEC real-time monitoring composed by a hardware prototype for ambient data collection and smartphone compatibility for data consulting.

IEQ includes IAQ, acoustics, thermal comfort, and lighting conditions. Poor IEQ has a negative impact on occupational health, particularly on children and old people. Particularly, thermal comfort is ranked by occupants to be of greater importance. Buildings are responsible for about 40% of the global energy consumption, and over 30% of the CO_2 emissions; also a considerable proportion of this energy is used for thermal comfort.

With the proliferation of IoT devices and AAL technologies, there is great potential to create automatic solutions for IEQ supervision. A real-time IEQ supervision system makes possible to measure the IEQ levels in different places and provide data to detect unhealthy indoor conditions automatically.

The results obtained are auspicious, representing a significant contribution to IEQ monitoring systems based on IoT. In the one hand, the monitored data can be particularly valuable to analyse the indoors parameters and its evolution throughout the day to detect unhealthy situations for planning interventions to improve occupational health.

These data can also be used for medical diagnosis by clinical professionals as the medical team might analyse the history of the IEQ parameters of the ecosystem everywhere the patient lives and relate these records with his health complications. On the other hand, it is possible to supervisor the ambient conditions of places where technological activities are conducted such as laboratory tests to support the correct correlation between the results and the environment.

Compared to existing systems, it has great importance due to the combination of sensitivity, flexibility, easy installation and accuracy of measurement in real time, allowing significant evolution in the existing quality controls. The system has advantages both in installation and configuration, due to the use of wireless technology for communications.

As future work, the main goal is to make technical improvements, including not only an integrated web application but also the development of important alerts and notifications to notify the user when the environmental conditions reached parametrised setpoints. The authors also planned software and hardware improvements to adapt the system to specific laboratory tests, such as thermographic experiments applied to wood and trees.

In spite of the influence of indoor environments in daily human activities, systems like this will contribute not only to enhanced living environments but also to support technological activities.

References

1. Giusto, D. (ed.): The Internet of Things: 20th Tyrrhenian Workshop on Digital Communications. Springer, New York (2010)
2. Gubbi, J., Buyya, R., Marusic, S., Palaniswami, M.: Internet of Things (IoT): a vision, architectural elements, and future directions. Future Gener. Comput. Syst. **29**(7), 1645–1660 (2013)
3. Marques, G., Pitarma, R.: An indoor monitoring system for ambient assisted living based on Internet of Things architecture. Int. J. Environ. Res. Public. Health **13**(11), 1152 (2016)
4. UN: World population ageing: 1950–2050, pp. 11–13 (2001)
5. Centers for Disease Control and Prevention: 'The state of aging and health in America 2007,' N. A. on an Aging Society (2007). https://www.cdc.gov/aging/pdf/saha_2007.pdf
6. Koleva, P., Tonchev, K., Balabanov, G., Manolova, A., Poulkov, V.: Challenges in designing and implementation of an effective ambient assisted living system. In: 12th International Conference on Telecommunication in Modern Satellite, Cable and Broadcasting Services (TELSIKS), pp. 305–308 (2015)
7. Atzori, L., Iera, A., Morabito, G.: The Internet of Things: a survey. Comput. Netw. **54**(15), 2787–2805 (2010)
8. Vilcekova, S., Meciarova, L., Burdova, E.K., Katunska, J., Kosicanova, D., Doroudiani, S.: Indoor environmental quality of classrooms and occupants' comfort in a special education school in Slovak Republic. Build. Environ. **120**, 29–40 (2017)
9. Bruce, N., Perez-Padilla, R., Albalak, R.: Indoor air pollution in developing countries: a major environmental and public health challenge. Bull. World Health Organ. **78**(9), 1078–1092 (2000)

10. Seguel, J.M., Merrill, R., Seguel, D., Campagna, A.C.: Indoor air quality. Am. J. Lifestyle Med. (2016). https://doi.org/10.1177/1559827616653343
11. Jones, A.P.: Indoor air quality and health. Atmos. Environ. **33**(28), 4535–4564 (1999)
12. Yang, L., Yan, H., Lam, J.C.: Thermal comfort and building energy consumption implications – a review. Appl. Energy **115**, 164–173 (2014)
13. Havenith, G., Holmér, I., Parsons, K.: Personal factors in thermal comfort assessment: clothing properties and metabolic heat production. Energy Build. **34**(6), 581–591 (2002)
14. Rupp, R.F., Vásquez, N.G., Lamberts, R.: A review of human thermal comfort in the built environment. Energy Build. **105**, 178–205 (2015)
15. Caragliu, A., Del Bo, C., Nijkamp, P.: Smart cities in Europe. J. Urban Technol. **18**(2), 65–82 (2011)
16. Chourabi, H., et al.: Understanding smart cities: an integrative framework, pp. 2289–2297 (2012)
17. Zanella, A., Bui, N., Castellani, A., Vangelista, L., Zorzi, M.: Internet of Things for smart cities. IEEE Internet Things J. **1**(1), 22–32 (2014)
18. Hernández-Muñoz, J.M., et al.: Smart cities at the forefront of the future internet. In: Domingue, J., Galis, A., Gavras, A., Zahariadis, T., Lambert, D., Cleary, F., Daras, P., Krco, S., Müller, H., Li, M.-S., Schaffers, H., Lotz, V., Alvarez, F., Stiller, B., Karnouskos, S., Avessta, S., Nilsson, M. (eds.) the future internet, vol. 6656, pp. 447–462. Springer, Heidelberg (2011)
19. Srbinovska, M., Gavrovski, C., Dimcev, V., Krkoleva, A., Borozan, V.: Environmental parameters monitoring in precision agriculture using wireless sensor networks. J. Clean. Prod. **88**, 297–307 (2015)
20. Sanchez-Rosario, F., et al.: A low consumption real time environmental monitoring system for smart cities based on ZigBee wireless sensor network. In: International Wireless Communications and Mobile Computing Conference (IWCMC), Dubrovnik, pp. 702–707 (2015)
21. Zhou, P., Huang, G., Zhang, L., Tsang, K.-F.: Wireless sensor network based monitoring system for a large-scale indoor space: data process and supply air allocation optimization. Energy Build. **103**, 365–374 (2015)
22. Marques, G., Pitarma, R.: IAQ evaluation using an IoT CO2 monitoring system for enhanced living environments. In: Rocha, Á., Adeli, H., Reis, L.P., Costanzo, S. (eds.) Trends and Advances in Information Systems and Technologies, vol. 746, pp. 1169–1177. Springer, Cham (2018)
23. Pitarma, R., Marques, G., Ferreira, B.R.: Monitoring indoor air quality for enhanced occupational health. J. Med. Syst. **41**, 23 (2017)
24. Pitarma, R., Marques, G., Caetano, F.: Monitoring indoor air quality to improve occupational health. In: Rocha, Á., Correia, A.M., Adeli, H., Reis, L.P., Mendonça Teixeira, M. (eds.) New Advances in Information Systems and Technologies, vol. 445, pp. 13–21. Springer, Cham (2016)
25. Marques, G., Pitarma, R.: Smartwatch-based application for enhanced healthy lifestyle in indoor environments. In: Omar, S., Haji Suhaili, W.S., Phon-Amnuaisuk, S. (eds.) Computational Intelligence in Information Systems, vol. 888, pp. 168–177. Springer, Cham (2019)
26. Marques, G., Pitarma, R.: Using IoT and social networks for enhanced healthy practices in buildings. In: Rocha, Á., Serrhini, M. (eds.) Information Systems and Technologies to Support Learning, vol. 111, pp. 424–432. Springer, Cham (2019)

27. Marques, G., Pitarma, R.: Monitoring health factors in indoor living environments using Internet of Things. In: Rocha, Á., Correia, A.M., Adeli, H., Reis, L.P., Costanzo, S. (eds.) Recent Advances in Information Systems and Technologies, vol. 570, pp. 785–794. Springer, Cham (2017)

28. Marques, G., Roque Ferreira, C., Pitarma, R.: A system based on the Internet of Things for real-time particle monitoring in buildings. Int. J. Environ. Res. Public. Health **15**(4), 821 (2018)

29. Salamone, F., Belussi, L., Danza, L., Galanos, T., Ghellere, M., Meroni, I.: Design and development of a nearable wireless system to control indoor air quality and indoor lighting quality. Sensors **17**(5), 1021 (2017)

30. Akkaya, K., Guvenc, I., Aygun, R., Pala, N., Kadri, A.: IoT-based occupancy monitoring techniques for energy-efficient smart buildings. In: IEEE Wireless Communications and Networking Conference Workshops (WCNCW), New Orleans, pp. 58–63 (2015)

31. Wargocki, P., Wyon, D.P., Sundell, J., Clausen, G., Fanger, P.O.: The effects of outdoor air supply rate in an office on perceived air quality, Sick Building Syndrome (SBS) symptoms and productivity. Indoor Air **10**(4), 222–236 (2000)

Noise Monitoring for Enhanced Living Environments Based on Internet of Things

Gonçalo Marques and Rui Pitarma[✉]

Polytechnic Institute of Guarda – Unit for Inland Development,
Av. Dr. Francisco Sá Carneiro, nº 50, 6300–559 Guarda, Portugal
goncalosantosmarques@gmail.com, rpitarma@ipg.pt

Abstract. Environmental noise has a direct influence on human health and also on the quality of life. The environmental noise effects on health are not only related to annoyance, sleep and cognitive performance but can also be linked with raised blood pressure. Therefore, noise pollution must be seen as severe world public health challenge and should be monitored not only inside buildings, as people spend about 90% of our lives indoors, for enhanced occupational health but also in outside for enhanced living environments in smart cities. Noise real-time monitoring allows the detection of unhealthy situations and to notify the building or the city managers to take interventions to decrease the sound levels quickly. Considering the proliferation of Internet of Things (IoT) devices and technologies, the *iSound*, a solution for real-time noise monitoring based on IoT has been developed. This solution is composed by a hardware prototype for ambient data collection and web portal for data consulting. The *iSound* is based on open-source technologies and is a totality Wi-Fi system, with several advantages compared to existing systems, such as its modularity, scalability, low-cost and easy installation.

Keywords: Ambient assisted living · Enhanced livings environments · Monitoring · Environmental noise · Internet of Things · Smart cities · Smart home

1 Introduction

The 'environmental noise' can be seen as an unwanted sound produced by human activities that are considered harmful or detrimental to human health and quality of life, while 'noise' was identified as being sound that is 'out of place' or as a form of acoustic pollution as much as carbon oxide (CO) is for air pollution [1].

Air pollution and environmental noise have a direct influence on human health. Although as people typically spend more than 90% of their time in indoor environments, the building attributes are closely related to the air-noise interaction as they affect noise more than air pollution. Buildings act as a shield for the dispersion of pollution, but the decreasing effect is much larger for noise than for air pollution [2].

The noise effects on health are not only related to annoyance, sleep and cognitive performance for both adults and children but can also be associated with raised blood pressure [3]. Environmental noise pollution may be a novel risk factor for

© Springer Nature Switzerland AG 2019
Á. Rocha et al. (Eds.): WorldCIST'19 2019, AISC 932, pp. 45–54, 2019.
https://doi.org/10.1007/978-3-030-16187-3_5

pregnancy-related hypertension, particularly more severe variants of preeclampsia [4]. Long-term exposure to railway and road noise, especially at night, may affect arterial stiffness, a major determinant of cardiovascular disease. Therefore the noise monitoring can be significant to the enhanced understanding of noise-related health symptoms and effects [5]. Poor sleep causes endocrine and metabolic disturbances, several cardio-metabolic and psychiatric problems and anti-social behaviour, both in children and adults. The duration and quality of sleep are risk factors significantly affected by the environment but amenable to modification through awareness and counselling and measures of public health [6]. Pregnancy and childhood exposure to road traffic noise can be associated with a higher risk for childhood overweight has been concluded by [7]. The World Health Organization (WHO) has recently acknowledged that contrary to the trend for other environmental stressors, noise exposure is increasing in Europe [8]. Therefore, the majority of the developed countries support laws noise regulation at specific hours [9].

Environmental noise must be assumed as a serious public health issue throughout the world. Overall, the evidence suggests that environmental noise should be placed at the forefront of national and international health policies to prevent unnecessary adverse health impacts on the general population [10].

Several public and private entities associate exposure to environmental pollution (noise and air) with health. In fact, there is good evidence from large population studies that environmental noise, regardless of association with air pollution, has harmful effects on health. Therefore, environmental planning and policy should take both exposures into account when assessing environmental impacts [11].

The variation in noise levels in cities is determined by the cumulative effect of unfavourable or thoughtful city design elements at several scales of a city's general and neighbourhood layout. This is related with the transportation system, the buildings structures, population density, the design of street and building facades, the amount of green space, and the quality of the dwellings concerning sound and vibration features inherent to each city [12]. In the same city, we can find high sound levels at some locations when compared with other places of the same town that are quieter. This is many times related to the city design, especially in cities created a long time ago, not planned and even entirely away from the current mobility needs of the citizens.

Considering the health damage caused by noise pollution, over the years, the occupants have been demanding from the authorities' mechanisms of evaluation and control [13]. Therefore, is necessary to ensure the supervision of the noise pollution levels to provide a healthy and safe environment.

In spite of the rapid increase in infrastructure and industrial plants, environmental issues have influenced the need for real-time monitoring systems [14]. Among the panoply of applications enabled by the Internet of Things (IoT), smart and connected healthcare is a particularly important one [15]. With the proliferation of IoT devices and technologies, there is a great potential to exploit their communications for occupancy monitoring [16]. Currently, smartphones can provide noise measurement data through mobile/cell phones in the form of 'citizen science' [17].

City Sensing is a new approach to supervisor the region and the environment using miniaturised cost-effective sensors to provide real-time monitoring together with interaction with citizens for enhanced smart cities [18]. The concept of the "smart city" has recently been introduced as a strategic device to encompass modern urban production factors in a common framework and, in particular, to highlight the importance of Information and Communication Technologies (ICTs) in the last 20 years for enhancing the competitive profile of a city as proposed by [19]. Nowadays cities face interesting challenges and problems to meet socio-economic development and quality of life objectives and the concept of "smart cities" correspond to answer to these challenges [20]. The smart city is directly related to an emerging strategy to mitigate the problems generated by the urban population growth and rapid urbanisation [21]. The most relevant issue in smart cities is the no interoperability of the heterogeneous technologies; the IoT can provide the interoperability to build a unified urban-scale ICT platform for smart cities [9].

A real-time IoT noise monitoring system not only can measure the noise levels in different places and provide data to the municipal authorities to plan interventions to reduce environmental noise pollution but also offer a space-time map of noise pollution in the area and contribute for public security [22]. Therefore, the noise monitoring service can also be used to enforce public safety, using sound detection.

Ambient Assisted Living (AAL) is an emerging multi-disciplinary field aiming at providing an ecosystem of different types of sensors, computers, mobile devices, wireless networks and software applications for personal healthcare monitoring and telehealth systems [23]. Due to the well-studied health effects of environmental noise an IoT noise monitoring system can be used to provide real-time supervision of sound levels inside buildings for enhanced occupational health. In significant cases, simple interventions provided by home-owners, building operators or municipal authorities can produce significant positive impacts to decrease the noise pollution.

This paper aims to present a cost-effective IoT real-time noise monitoring system for enhanced ambient assisted living and occupational health. To provide accurate data sensing a calibrated sound sensor was chosen. The solution is composed by a hardware prototype for ambient data collection and a web portal for data consulting. The *iSound* is based on open-source technologies and is a totality Wi-Fi system, with several advantages compared to existing systems, such as its modularity, scalability, low-cost and easy installation. This system is based on an ESP8266 microcontroller with built-in Wi-Fi technology as communication and processing unit and incorporates a sound sensor as sensing unit.

The paper is structured as follows: besides de introduction (Sect. 1), Sect. 2 presents the related work and Sect. 3 is concerned to the methods and materials used in the implementation of the sensor system; Sect. 4 demonstrates the system operation and experimental results, and the conclusion is presented in Sect. 5.

2 Related Work

Currently, several healthcare projects for enhanced living environments and occupational health are available in the literature. In this section, several solutions that incorporate open-source, low cost and mobility technologies are presented.

Numerous IoT systems for real-time ambient quality supervision that incorporates open-source technologies for data collection, processing and transmission from different places simultaneously and mobile computing technologies for data consulting are proposed by [14, 24–26].

A new approach for the assessment of noise pollution involving the general public to turn GPS-equipped mobile phones into noise sensors that enable citizens to measure their personal exposure to noise in their everyday environment has been proposed by [27].

A method of compressing the noise events of an entire day (24 h) into a one-minute summary that can be sent over SMS text messages, and is simple enough to run on an inexpensive microprocessor was proposed by [28].

A MCS (Mobile Crowd Sensing)-based platform for gathering noise measurements by using mobile-embedded microphones that perform opportunistic/participatory measurements, a data warehouse system for managing data (storage, aggregation and filtering) and a web app providing city managers with multiple views about collected data has been proposed by [29].

The *iSound* system aims to provide a useful tool for noise pollution monitoring for enhanced living environments in smart cities. The authors develop a completely wireless solution using the ESP8266 module which implements the IEEE 802.11 b/g/n networking protocol. This microcontroller with built-in Wi-Fi capabilities is used both as processing and communication unit. The monitored data is stored in a SQL SERVER database using web services. For data consulting a web portal has been created by the authors using ASP.NET C# (Fig. 1).

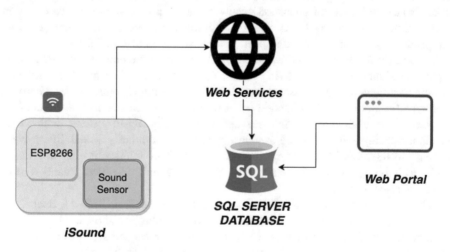

Fig. 1. *iSound* architecture

3   Materials and Methods

The objective of the authors is to develop a cost-effective and accurate system that can be easily configured and installed by the average user. Therefore, we use a cost-effective but very reliable sound sensor and a microcontroller with native Wi-Fi support. This system consists of two components, a LOLin Node Mcu V3 microcontroller developed by Wemos and an analog sound level meter developed by Gravity (Fig. 2).

Fig. 2. *iSound* prototype

A brief introduction of each component used is shown below:

- LOLin Node Mcu V3 – is a Wi-Fi chip with integrated antenna switches, RF balun, power amplifier, low noise receive amplifier, filters and power management modules. It supports 802.11 b/g/n protocols, Wi-Fi 2.4 GHz, support WPA/WPA2, has a standby power consumption of <1.0mW (DTIM3) and can operate at temperature range –40C–125C. This microcontroller is one of the most used ESP8266 development boards support 4 MB of flash memory, 11 GPIO pins, and one analogue-to-digital converter (ADC) pin with 10-bit resolution. It also has a built-in voltage regulator, and you can power up the module using the mini USB socket or the Vin pin.
- Gravity Sound level meter - is a noise sensor that can accurately measure the sound level of the surrounding environment. This sensor incorporates an instrument circuit, low noise microphone, which makes it highly precious. It supports 3.3–5.0 V wide input voltage and provides and a 0.6–2.6 V voltage output. This sensor has a measurement range of 30dBA–130dBA with a measurement error of ±1.5 dB. The frequency response is 31.5 Hz–8.5 kHz and the response time is 125 ms. The current consumption is 22 mA at 3.3 V or 14 mA at 5.0 V. The sensor size is 60 mm * 43 mm.

The web portal is denominated by *iSoundWeb* and was developed with ASP.NET C#. This app has two important features as it permits not only real-time consulting of the last data collected and also provide a list of anomalies.

For testing purposes, a laboratory of a Portuguese university was on-site monitored, and one *iSound* module was used (Fig. 3). The module is powered by the power grid using a 230 V-5 V AC-DC 2A power supply. Noise monitoring data were collected for two months which showed that under certain conditions sound values are significantly higher than those considered healthy for standards.

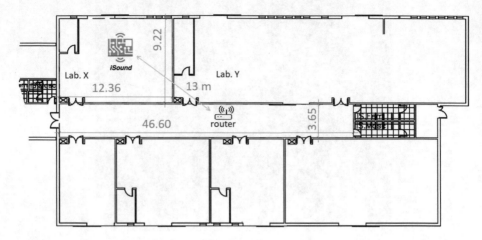

Fig. 3. *iSound* installation schema.

4   Results and Discussion

The *iSound* provides data consulting in as graphical and numerical form. A sample of the data collected by *iSound* is shown in Fig. 4 that represents the sound level data measured in dBA.

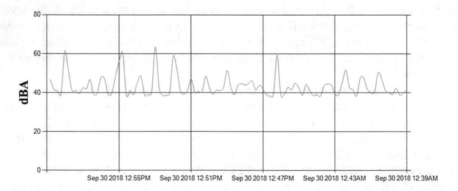

Fig. 4. Noise monitoring data (dBA)

The graphics display of the noise level data allows a higher perception of the behaviour of the monitored parameter than the numerical display format. On the other hand, the web portal also allows the user to access the data easily to provide a more precise analysis of the detailed temporal evolution. Thus, the system is a powerful tool for noise analysis.

Is extremely important to provide noise monitoring at hospitals as the noise pollution can be a source of stress for the patients. In 1977, the concerning about noise levels in hospitals was already studied to encourage the establishment of a set of specifications which the vendors would agree to abide by when placing equipment in hospitals. Right to be told, with the introduction of noise-making devices such as computers, it is almost easy to become acclimated to this intrusion of privacy [30].

A study has been carried at Zeynep Kamil Hospital. In this test, the highest average noise level measured is 81.25 ± 3.21 dB, and the lowest noise level is 52.51 ± 2.37 dB which are much higher than the internationally recommended noise levels [31].

In another study about the people's willingness to pay (WTP) for noise reduction measures concludes that 80% of respondents were unwilling to pay anything for a noise-control policy [32]. Therefore, cost-effective solutions should be developed in order to allow real-time noise monitoring.

Is imperative to control noise pollution effectively and the authors believe that the first step is to provide real-time supervision to perceive its variation in real-time and detect non-healthy situations for enhanced living environments.

Compared to other systems, the *iSound* system has the following advantages: modularity, small size, low-cost construction and easy installation.

Noise monitoring is relevant not only to support planned interventions for enhanced acoustic conditions but also to support medical diagnostics.

As future work, the main goal is to make technical improvements, including the development of important alerts and notifications to notify the building manager or the city authorities when the sound levels are above the allowed levels considering the hour of the day. Improvements to the system hardware and software are planned to make it much more appropriate for specific purposes not only as thermographic experiments applied to wood and trees but also for hospitals and schools.

5 Conclusion

Environmental noise has a direct influence on human health and WHO recognised that noise exposure is increasing in Europe. The noise effects on health are not only related to annoyance, sleep and cognitive performance but can also be linked with raised blood pressure. Therefore, noise pollution must be seen as severe world public health challenge.

With the increase of IoT devices and technological enhancements, is now possible to provide real-time noise supervision to perceive its variation in real-time and detect non-healthy situations for enhanced living environments. Consequently, this paper had presented an IoT architecture for noise pollution real-time monitoring composed by a hardware prototype for ambient data collection and a web portal for data consulting.

The results obtained are promising, representing a significant contribution to noise monitoring systems based on IoT. In the one hand, the monitored data inside buildings can be particularly valuable to offer support to a medical diagnosis by clinical professionals as the medical team might analyse the history of noise pollution parameters of the ecosystem everywhere the patient lives and relate these records with his health complications. On the other hand, it is possible to supervisor the cities noise pollution in real time and plans interventions and control the sound level for enhanced smart cities. The system has advantages both in installation and configuration, due to the use of wireless technology for communications, but also because it was developed for to be compatible with all domestic house devices and not only for smart houses or high-tech houses.

Future work includes the development of important alerts and notifications to notify the building manager or the city authorities when the sound levels for enhanced living environments. The authors planned software and hardware improvements to adapt the system to specific laboratory tests, such as thermographic experiments applied to wood and trees.

We believe that in the future, systems like this will contribute to enhanced living environments but also be an integral part of smart cities.

References

1. Murphy, E., King, E.A.: Principles of environmental noise. In: Environmental Noise Pollution, Elsevier, pp. 9–49 (2014)
2. Khan, J., Ketzel, M., Kakosimos, K., Sørensen, M., Jensen, S.S.: Road traffic air and noise pollution exposure assessment–a review of tools and techniques. Sci. Total Environ. **634**, 661–676 (2018)
3. Stansfeld, S.A., Matheson, M.P.: Noise pollution: non-auditory effects on health. Br. Med. Bull. **68**(1), 243–257 (2003)
4. Auger, N., Duplaix, M., Bilodeau-Bertrand, M., Lo, E., Smargiassi, A.: Environmental noise pollution and risk of preeclampsia. Environ. Pollut. **239**, 599–606 (2018)
5. Foraster, M., et al.: Exposure to road, railway, and aircraft noise and arterial stiffness in the SAPALDIA study: annual average noise levels and temporal noise characteristics. Environ. Health Perspect. **125**(9), 097004 (2017)
6. Gupta, A., Gupta, A., Jain, K., Gupta, S.: Noise pollution and impact on children health. Indian J. Pediatr. **85**(4), 300–306 (2018)
7. Christensen, J.S., Hjortebjerg, D., Raaschou-Nielsen, O., Ketzel, M., Sørensen, T.I.A., Sørensen, M.: Pregnancy and childhood exposure to residential traffic noise and overweight at 7 years of age. Environ. Int. **94**, 170–176 (2016)
8. Murphy, E., King, E.A.: An assessment of residential exposure to environmental noise at a shipping port. Environ. Int. **63**, 207–215 (2014)
9. Zanella, A., Bui, N., Castellani, A., Vangelista, L., Zorzi, M.: Internet of Things for smart cities. IEEE Internet Things J. **1**(1), 22–32 (2014)
10. Murphy, E., King, E.A.: Environmental noise and health. In: Environmental Noise Pollution, pp. 51–80. Elsevier (2014)
11. Stansfeld, S.: Noise effects on health in the context of air pollution exposure. Int. J. Environ. Res. Public Health **12**(10), 12735–12760 (2015)
12. Lercher, P.: Noise in cities: urban and transport planning determinants and health in cities. In: Nieuwenhuijsen, M., Khreis, H. (eds.) Integrating Human Health into Urban and Transport Planning, pp. 443–481. Springer, Cham (2019)
13. Morillas, J.M.B., Gozalo, G.R., González, D.M., Moraga, P.A., Vílchez-Gómez, R.: Noise pollution and urban planning. Curr. Pollut. Rep. **4**(3), 208–219 (2018)

14. Saha, A.K., et al.: A raspberry Pi controlled cloud based air and sound pollution monitoring system with temperature and humidity sensing. In: 2018 IEEE 8th Annual Computing and Communication Workshop and Conference (CCWC), Las Vegas, NV, pp. 607–611 (2018)
15. Hassanalieragh, M., et al.: Health monitoring and management using Internet-of-Things (IoT) sensing with cloud-based processing: opportunities and challenges. In: 2015 IEEE International Conference on Services Computing, New York City, NY, USA, pp. 285–292 (2015)
16. Akkaya, K., Guvenc, I., Aygun, R., Pala, N., Kadri, A.: IoT-based occupancy monitoring techniques for energy-efficient smart buildings. In: 2015 IEEE Wireless Communications and Networking Conference Workshops (WCNCW), New Orleans, LA, USA, pp. 58–63 (2015)
17. Murphy, E., King, E.A.: Conclusions and future directions. In: Environmental Noise Pollution, pp. 247–260. Elsevier (2014)
18. Camporese, R., Borga, G., Iandelli, N., Ragnoli, A.: New technologies and statistics: partners for environmental monitoring and city sensing. In: Crescenzi, F., Mignani, S. (eds.) Statistical Methods and Applications from a Historical Perspective, pp. 347–358. Springer, Cham (2014)
19. Caragliu, A., Del Bo, C., Nijkamp, P.: Smart cities in Europe. J. Urban Technol. 18(2), 65–82 (2011)
20. Schaffers, H., Komninos, N., Pallot, M., Trousse, B., Nilsson, M., Oliveira, A.: Smart cities and the future internet: towards cooperation frameworks for open innovation. In: Domingue, J., Galis, A., Gavras, A., Zahariadis, T., Lambert, D., Cleary, F., Daras, P., Krco, S., Müller, H., Li, M.-S., Schaffers, H., Lotz, V., Alvarez, F., Stiller, B., Karnouskos, S., Avessta, S., Nilsson, M. (eds.) The Future Internet, vol. 6656, pp. 431–446. Springer, Heidelberg (2011)
21. Chourabi, H., et al.: Understanding Smart Cities: An Integrative Framework, pp. 2289–2297 (2012)
22. Talari, S., Shafie-khah, M., Siano, P., Loia, V., Tommasetti, A., Catalão, J.: A review of smart cities based on the Internet of Things concept. Energies 10(4), 421 (2017)
23. Universal Open Platform and Reference Specification for Ambient Assisted Living. http://www.universaal.org/
24. Pitarma, R., Marques, G., Ferreira, B.R.: Monitoring indoor air quality for enhanced occupational health. J. Med. Syst. 41(2) (2017)
25. Marques, G., Pitarma, R.: An indoor monitoring system for ambient assisted living based on Internet of Things architecture. Int. J. Environ. Res. Public Health 13(11), 1152 (2016)
26. Marques, G., Roque Ferreira, C., Pitarma, R.: A system based on the Internet of Things for real-time particle monitoring in buildings. Int. J. Environ. Res. Public Health 15(4), 821 (2018)
27. Maisonneuve, N., Stevens, M., Niessen, M.E., Steels, L.: NoiseTube: measuring and mapping noise pollution with mobile phones. In: Athanasiadis, I.N., Rizzoli, A.E., Mitkas, P. A., Gómez, J.M. (eds.) Information Technologies in Environmental Engineering, pp. 215–228. Springer, Heidelberg (2009)
28. Zimmerman, T., Robson, C.: Monitoring residential noise for prospective home owners and renters. In: Lyons, K., Hightower, J., Huang, E.M. (eds.) Pervasive Computing, vol. 6696, pp. 34–49. Springer, Heidelberg (2011)
29. Zappatore, M., Longo, A., Bochicchio, M.A., Zappatore, D., Morrone, A.A., De Mitri, G.: Improving urban noise monitoring opportunities via mobile crowd-sensing. In: Leon-Garcia, A., Lenort, R., Holman, D., Staš, D., Krutilova, V., Wicher, P., Cagáňová, D., Špirková, D., Golej, J., Nguyen, K. (eds.) Smart City 360°, vol. 166, pp. 885–897. Springer, Cham (2016)
30. Sellers, D.E., Grams, R.R., Horty, T.: Documentation of hospital communication noise levels. J. Med. Syst. 1(1), 87–97 (1977)

31. Yarar, O., Temizsoy, E., Günay, O.: Noise pollution level in a pediatric hospital. Int. J. Environ. Sci. Technol. (2018)
32. Huh, S.-Y., Shin, S.-Y.: Economic valuation of noise pollution control policy: does the type of noise matter? Environ. Sci. Pollut. Res. (2018)

Exploring the Potential of IT-Enabled Healthcare in Rural Regions in Cameroon: A Case Study

Doriane Micaela Andeme Bikoro[1]([✉]), Samuel Fosso Wamba[2],
and Jean Robert Kala Kamdjoug[1]

[1] GRIAGES, Catholic University of Central Africa, Yaounde, Cameroon
dorianebikoro@gmail.com, jrkala@gmail.com
[2] Toulouse Business School, Toulouse, France
s.fosso-wamba@tbs-education.fr

Abstract. In industrialized countries, the rate at which technological tools is use in the present century is exponential. This is felt in all sectors of society and in this case the field of health. In Africa in hole and Cameroon in particular, the speed is slow to emerge because of the huge digital gap between urban and rural areas. In this paper, it is a question of exploring the technological potential to see informatics of rural medical institutions; the case of three health centers and rural hospitals in the Southern part of the country. The main objective of this qualitative study is to make an inventory and propose solutions to overcome the problems detected. The results obtained really show technological shortcomings in the hospital management of these establishments and in the administration of health care to the patients. This led to some proposals at the institutional, technological and research level. All this is going in the direction to improve the health sector in Cameroon.

Keywords: E-health · Potential · Rural regions · Qualitative method · Cameroon

1 Introduction

The technological discoveries of the present century have greatly contributed to the renewal and restructuring of modern medicine [28]. One of these innovations is the implementation and use of information technologies in the health sector [4]. Indeed, these information and communication technologies (ICT) are perceived as "catalysts of development"; as if to say that ICTs are actually tools to better manage information [32]. Technology and its equipment play an important role in health services. The role of technology management must be understood in order to establish clear communication with health professionals; because it directly contributes to improving patient's health outcomes [28]. This was quickly considered in most hospitals in industrialized countries that have introduced and recognized the technological component as an integral part of hospital management. [28]. This is why predictive research has suggested that information and communication technologies (ICTs) in health have the potential to transform health services [18]. In addition, information and communication

© Springer Nature Switzerland AG 2019
Á. Rocha et al. (Eds.): WorldCIST'19 2019, AISC 932, pp. 55–62, 2019.
https://doi.org/10.1007/978-3-030-16187-3_6

technologies (ICTs) are revolutionizing the lives of human beings every day and their modes of interaction. The application of ICTs in health is referred to as eHealth, which includes "telemedicine, electronic medical records and health information systems with decision support tools, mobile health and e-learning" [33]. E-Health has come to improve the health care delivery system and as a direct result improve overall health and universal health coverage. All this by creating access and improving the quality of health services by filling the scarcity of health human resources [11, 33]. In a nut shell, information and communication technologies have come to improve the functioning of health systems [2, 30].

In poor and less poor income countries, research in the health sector plays a key role in the development process. Several initiatives and publications have helped to strengthen the capacity of this sector of activity in Sub-Saharan Africa particularly [7, 22, 35]. Nowadays, thank to technological innovations in health research, many health professionals are now communicating more quickly with their colleagues and their patients [15, 33]. This is the case in sub-Saharan Africa with mobile health or "m-Health", millions of people who have never had to use traditional fixed telephone lines now use their mobile phone. M-Health, refers to the use of short messaging service (SMS) applications and wireless data transmission primarily [2, 3, 6].

M-Health has reoriented daily tasks including "Electronic Health Records" (EHR) [8, 27]. Information technology has huge potential for Africa. They reduce costs by improving access to care, standardize operational efficiencies in the health system, enable clinical automation, improve process management, make available information on the most effective care and enhance safety (patient confidentiality and medical secrets) [7, 19]. A recent study in South Africa presented ICT as a means of bridging the digital divide between urban and rural areas to address the shortcomings of the rural health sector [31] and as a means of decreasing medication errors [13]. More particularly in Cameroon, the government system wants to achieve broader goals in health and development [14].

This shows that telemedicine and some e-Health applications are seen as tools to facilitate the development of health services. Mobile health strategies should be considered as integral systems that should be integrated into existing systems [12, 33]. Because of their growing importance, health information systems are spreading around the world, improving human health and prosperity [11, 26].

After presenting the importance of technologies in health in general, in Africa and Cameroon in particular; in this study, we will take stock of the challenges of health in Cameroon. Then to present explicitly some examples of ICT use in health. To close we will present the results of our qualitative survey followed by the proposal of some tracks to remedy the problems detected.

2   Cameroon and Key Healthcare Challenges

Cameroon, a country in Central Africa, located at the bottom of the Gulf of Guinea with a population of more than 20 million, of which nearly 52% live in urban areas. In the health sector, the epidemiological profile of the country is dominated by infections and parasitic diseases and a tendency to increase the prevalence of cardiovascular diseases, diabetes and cancers. The Cameroonian health system is focused on three sectors, including a public sub-sector consisting of public hospitals and health structures under the supervision of ministerial departments. A private non-profit sub-sector and for-profit sub-sector. A subsector of traditional medicine which is an important component of the system [23, 24]. While malaria is considered to be the leading cause of mortality and morbidity, respiratory infections and gastrointestinal diseases are also causes of death, especially in children under the age of 5. Less than 10% of the population has health insurance and households contribute about 83% to health financing. Life expectancy at birth is about 57 years old. The infant mortality rate experienced a significant reduction of 28% between 2004 and 2014 [34] and is estimated at 53.63 deaths/100,000 live births [5, 16, 17]. The hospital-specific malaria mortality rate from 43% to 22.4% between 2008 and 2013. The prevalence of HIV/AIDS increased from 5.5% to 4.3% between 2004 and 2011 [20].

From 2001 to 2015, the country's health policy was recorded and framed in the Health Sector Strategy (SSS), the guiding framework for government action in the health sector. Having reached maturity, an evaluation of the content and implementation of this strategy has led to the development of a new one that will cover the period 2016–2027 [23, 34]. This new strategy aligns with the "Strategy Paper for Growth and Employment" and the Sustainable Development Goals (SDGs). A country where universal access to quality health services is ensured for all social strata by 2035, with the full participation of communities; this is the overall vision of Cameroon devoted in this strategy from 2016 to 2027 [10, 23]. It should be emphasized that the budget allocation to Cameroon for health is 8% and the Abuja Declaration advocates the allocation of a proportion of 15% of the budget of the State to this sector [34].

Among the challenges of the health sector in Cameroon, we must note the lack of essential infrastructure such as drinking water supplies [10, 23]. Several articles present the bad maintenance of the Cameroonian medical structures, with only 20 to 25% of the operational medical equipment and a rate of attendance of the hospitals of 30% [16]. It is therefore clear that the problem of infrastructures and medical equipment is precisely focused on their availability and optimal functioning [9]. This low rate of hospital use finds the problem of the cost of health care on average high and not affordable for the average citizen (peasants, unemployed etc.) [9, 10]. In addition, about 40% of local doctors work in the central region, where only 18% of the population lives; which shows an inequitable distribution of doctors [34]. Most of these medical staff are concentrated in the metropolis, and their living conditions are rather precarious, which can sometimes be at the root of the lack of ethics and deontology observed in the practice of their professions [9]. We must not forget the essential challenge of the Cameroonian sanitary system today, which is to continue to modernize its medical equipment and equip itself with more and more tools at the cutting edge of technology [23]. As in many countries, the Information and Communications Technologies (ICT) revolution has created new opportunities and challenges for developing countries in their efforts to strengthen their health information systems [25, 30, 33].

3   IT-Enabled Healthcare: A Survey

With the advent of Information and Communication Technologies (ICT), many e-health applications have emerged and are now used in many health systems [33]. In industrialized countries like Australia, the "HealthConnect" online health application provides patients with direct and rapid access to general health information, reduces duplication of services, and enables better portability of medical records. Telemedicine is very well developed in Australia [12]. Another mobile health initiative is the "ubiquitous program" in South Korea that uses sensors to monitor patient health from a distance [8].

In both these countries and in the developing countries, with telemedicine, Information and Communication Technologies play an important role in the health sector. Advanced ICTs are used for diagnostics, to conduct research, transfer patient data and improve disease management [29]. In South Africa and Rwanda, e-health has improved the surveillance and treatment of Tuberculosis and HIV AIDS [33]. In Cameroon, the startup "Himore Medical" which was created by the Cameroonian engineer Arthur Zang to market the first African medical tablet, the "Cardio pad". It is a solution that helps heart medical professionals to follow patients living in the field by allowing them to conduct remote cardiac exams and pass the results to specialists for interpretation [21]. This is consistent with activities in Botswana to improve the provision and management of information in rural health care [1, 4, 30]. In Senegal, an application to locate health structures around oneself has been set up; it is the "SenGeoSanté" application, it works with or without an Internet connection. In the same vein, in Uganda the "Matibabu" was set up by young researchers to detect by infrared rays the malaria virus [1].

4   Methodology and Survey Results of Three Rural Health Facilities

The methodology used in this study is essentially qualitative. In fact, we conducted three direct observation sessions, one for each health facility and three semi-structured interviews with the heads of health establishments mentioned below. To this end, we developed an observation grid before descending into the field for direct observation. This observation guide consisted of five main items that we fixed ourselves to observe in view of our problematic. These are the patient's circuit, the technological health infrastructures, the material infrastructures, the post-medical follow-up of the patients and the hospital administration. The interview guide resulted from the first remarks made at the end of the observation to have details on the details observed. It must be emphasized here that all the field trips were recorded in a document we call "the Field Journal"; it was he who made the restitution of this collection of data. This survey was conducted in three health facilities in the Southern Region, "Vallée du Ntem" Division and "Olamze" sub-division. The observation and interviews were spread over a week.

The first institution in which this survey was conducted is the Integrated Health Center of "Mekomengona". Indeed, created since 1959 this health center now has two buildings. This including an old nurse's office compound, waiting room for patients, the consultation room, a treatment room, a hospital ward with two beds and a pharmacy. The other new building has two delivery rooms. This integrated health center is managed by two employees, they are: a Nurse Center Manager in service since 2014 and a pharmacy clerk in service since 2008.

In terms of technical equipment, the integrated health center has many medical equipment. But technologically this institution does not have computers. Patient records are listed in drawers as pointed out by the head nurse.

Regarding the circuit of the patient, it is as follows: upon arrival the patient settles in the waiting room if he finds that the nurse is busy with other patients. Then the patient is received and consulted; after consulting the doctor the latter is either hospitalized or he returns with a medical prescription and will pay the bill of care and consultation and possibly buy the drugs from the pharmacy clerk. The one who stays in hospital will follow care until his recovery and before leaving the hospital settles the bill with the pharmacy clerk.

After the "Mekomengona" Health Center, we moved closer to the "Olamze" District Hospital. The latter, until inauguration of the new building is still housed in an old building. There is a guard room, a medical laboratory, a pharmacy, two doctor's offices, an antenatal consultation room, the office of the bursar, a delivery room, a men's hospital ward with 07 beds, a hospitalization room for women with 07 beds. The staff consists of two doctors, one thrift, 03 laboratory technicians, 04 nurses and 01 pharmacy clerks.

Technologically, no presence of computers despite the many requests to those who are entitled as pointed out by one of the doctors. Regarding the circuit of the patient, when the patient arrives in the waiting room nurses receive them and take certain medical parameters before sending them to the doctor with these parameters. The doctor then takes care of the consultation, then the patient is sent back to the nurses for specific care or for hospitalization when the doctor discovers the importance. Patient records are not listed in storage cabinets. All payments are made at the thrift except the payment of products that is done at the pharmacy.

The third institution where this survey was conducted is the Integrated Health Center of "Meyo-Biboulou", the premises of which is a new building built in the shape of "H". This building includes a maternity ward with a delivery room, 03 hospital rooms, a non-operational laboratory room, a treatment room, a consultation room, a meeting room and three modernized toilets. The staff consists of two nurses, one of whom is under house arrest and the other is not stable.

Similarly, in terms of technology, there are no computers for the health center. Only for personal use as pointed out by one of the nurses we met. The patients are not listed; this is done by recognition as emphasized by this nurse.

The circuit of the patient here is presented in this way: on arrival the patient is directly received in the consultation room when the nurses are not busy otherwise he waits in the premises of the center. When it is received, the patient is consulted and given the necessary care. The patient then pays the bill to the head nurse before leaving if he or she is not hospitalized. When the head nurse is absent the bill is settled with his deputy who resides on the spot.

5 Conclusion

It has been found in previous research and throughout this study that Information and Communication Technologies are now indispensable in the health sector [2, 30]. The advent of these technological tools has opened new perspectives and many health challenges around the world, particularly in Africa. While the majority of the major powers have taken the bandwagon on medical technology advances, some developing countries are still making their first steps in the field. These innovations often face many local structural constraints; this is the case of Cameroon. This study has shown that in Cameroon, access to quality health care is often restricted to geographical areas, as is the case in large metropolises where there is an overabundance of doctors and therefore medical care. In addition, it is in these hospitals in large cities that we find the majority of hospitals with technological tools. What has been demonstrated with this case study in a rural area in the south of the country where access to medical care and hospital management face many structural and institutional constraints. It shows that the technological potential is rare or almost non-existent. It is therefore appropriate at the end of this study to propose some avenues that will serve both health professionals and decision-makers as well as future researchers.

- It would be necessary by the measure of things that the digital divide be improved beforehand. This will be beneficial on the one hand for the decision-makers in the direction of improving the living conditions of the populations, the improvement of the telecommunication infrastructures and especially to ensure the continuity of the public and medical service. On the other hand it will be a gain for medical staff who can at any time have information on medical news.
- To sensitize the medical staff on technological awakening. This will help them both in hospital management (e.g. keeping patient records that may be more confidential in computers via applications or database), in medication errors and in improving the services delivered. Thereafter it would be wise for the state to provide these health centers and district hospitals with computer tools to break the barriers of the theory.
- Implementation of the "Computerized Patient Record" concept would improve disease monitoring and management. Indeed, after their first visit to the health center, the patients will have personal files that will be well kept in a database. This will allow a good capitalization of the disease experiences on the one hand and on the other hand to avoid or reduce the errors of medication.
- In addition, we should think about setting up an IT master plan; important to manage the digitization of an institution. All this takes into account the mapping of processes and the implementation of a security policy for their information system.
- Once again at the institutional level, the financing and support of medical projects and start-ups must be done more efficiently. This is to encourage research and innovation in health information systems.
- Finally, for future researchers, it is advisable to develop strategies and tools (medical devices, applications, etc.) enabling these rural health establishments to offer better services in order to satisfy local populations.

References

1. Afrique-Economique: Cameroun: les secteur de l'e-santé décolle (2016). http://societe. economie-afrique.com/societe/cameroun-le-secteur-de-le-sante-decolle/. Accessed 23 March 2016
2. Ammenwerth, E., et al.: Evaluation of health information systems—problems and challenges. Int. J. Med. Inform. **71**(2–3), 125–135 (2003)
3. Avison, D., Young, T.: Time to rethink health care and ICT? Commun. ACM **50**(6), 69–74 (2007)
4. Bastos, P., et al.: The importance of technology for achieving superior outcomes from intensive care. Intensive Care Med. **22**(7), 664–669 (1996)
5. Bawack, R.E., Kamdjoug, J.R.K.: Adequacy of UTAUT in clinician adoption of health information systems in developing countries: the case of Cameroon. Int. J. Med. Inform. **109**, 15–22 (2018)
6. Betjeman, T.J., et al.: mHealth in sub-Saharan Africa. Int. J. Telemed. Appl. **2013**, 6 (2013)
7. Blaya, J.A., et al.: E-health technologies show promise in developing countries. Health Aff. **29**(2), 244–251 (2010)
8. Broens, T., et al.: Towards an application framework for context-aware m-health applications. Int. J. Internet Protoc. Technol. **2**(2), 109–116 (2007)
9. Camer.be: Proposition de reforme du systeme de sante du camerou: Cameroon (2016). http://www.camer.be/51719/30:27/proposition-de-reforme-du-systeme-de-sante-du-cameroun-cameroon.html%29. Accessed 9 May 2016
10. Cameroon-tribune: Un meilleur accès à la santé pour tous les Camerounais (2018). https://www.cameroon-tribune.cm/article.html/21999/fr.html/-lacces-populations-leau-potable-reste. Accessed 5 Nov 2018
11. Castillejo, P., et al.: Integration of wearable devices in a wireless sensor network for an E-health application. IEEE Wirel. Commun. **20**(4), 38–49 (2013)
12. Celler, B.G., et al.: Using information technology to improve the management of chronic disease. Med. J. Aust. **179**(5), 242–246 (2003)
13. Chaudhry, B., et al.: Systematic review: impact of health information technology on quality, efficiency, and costs of medical care. Ann. Intern. Med. **144**(10), 742–752 (2006)
14. Chetley, A., et al.: Improving health connecting people: the role of ICTs in the health sector of developing countries (2006)
15. Dünnebeil, S., et al.: Determinants of physicians' technology acceptance for e-health in ambulatory care. Int. J. Med. Inform. **81**(11), 746–760 (2012)
16. Editions2015: La santé au Cameroun (2015). https://www.editions2015.com/cameroun/index.php/le-cameroun-mis-a-nu/la-sante/
17. Fomulu, F.J., et al.: Mortalité maternelle à la Maternité du Centre Hospitalier et Universitaire de Yaoundé, Cameroun: étude rétrospective de 5 ans (2002 à 2006). Health Sciences and Diseases 10(1) (2013)
18. Goldzweig, C.L., et al.: Costs and benefits of health information technology: new trends from the literature. Health Aff. **28**(2), w282–w293 (2009)
19. Hesse, B.W., et al.: Trust and sources of health information: the impact of the Internet and its implications for health care providers: findings from the first Health Information National Trends Survey. Arch. Intern. Med. **165**(22), 2618–2624 (2005)
20. Keugoung, B., et al.: Trente ans de lutte antituberculeuse au Cameroun: une alternance entre systemes d'offre de soins de santé «vertical» et «horizontal». Revue d'Épidémiologie et de Santé Publique **61**(2), 129–138 (2013)

21. lebledparle: 5 startups du secteur de la sante qui changent la donne en Afrique (2016). https://www.lebledparle.com/high-tech/1101220-5-startups-du-secteur-de-la-sante-qui-changent-la-do. Accessed 9 Aug 2016
22. Mair, F.S., et al.: Factors that promote or inhibit the implementation of e-health systems: an explanatory systematic review. Bull. World Health Organ. **90**, 357–364 (2012)
23. MinSanté: Strategie Sectorielle de Santé 2016–2027, pp. 7–17 (2016)
24. Mpondo, M., et al.: État actuel de la médecine traditionnelle dans le système de santé des populations rurales et urbaines de Douala (Cameroun). J. Appl. Biosci. **55**, 4036–4045 (2012)
25. Murray, E., et al.: Why is it difficult to implement e-health initiatives? A qualitative study. Implement. Sci. **6**(1), 6 (2011)
26. Oak, M.: A review on barriers to implementing health informatics in developing countries. J. Health Inform. Dev. Ctries. **1**(1), 19–22 (2007)
27. Odekunle Florence, F.: Current roles and applications of electronic health record in the healthcare system. Health Sci. **5**(12), 48–51 (2016)
28. Ogembo-Kachieng'a, M., Ogara, W.: Strategic management of technology in public health sector in Kenya and South Africa. East Afr. Med. J. **81**(6), 279–286 (2004)
29. Omona, W., Ikoja-Odongo, R.: Application of information and communication technology (ICT) in health information access and dissemination in Uganda. J. Libr. Inf. Sci. **38**(1), 45–55 (2006)
30. Panir, M.J.H.: Role of ICTs in the health sector in developing countries: a critical review of literature. J. Health Inform. Dev. Ctries. **5**(1) (2011)
31. Ruxwana, N.L., et al.: ICT applications as e-health solutions in rural healthcare in the Eastern Cape Province of South Africa. Health Inf. Manag. J. **39**(1), 17–29 (2010)
32. Sein, M.K., Harindranath, G.: Conceptualizing the ICT artifact: toward understanding the role of ICT in national development. Inf. Soc. **20**(1), 15–24 (2004)
33. Shiferaw, F., Zolfo, M.: The role of information communication technology (ICT) towards universal health coverage: the first steps of a telemedicine project in Ethiopia. Glob. Health Action **5**(1), 15638 (2012)
34. Wathi: La situation sanitaire au Cameroun (2018). https://www.wathi.org/election-cameroun-2018/contexte-election-cameroun-2018/la-situation-sanitaire-au-cameroun/. Accessed 25 Sept 2018
35. Whitworth, J.A., et al.: Strengthening capacity for health research in Africa. Lancet **372** (9649), 1590–1593 (2008)

Indirect Measurement of Blood Pressure and Arm's Body Composition in Women: Identification of Rules and Patterns Using Statistics and Data Mining

Paôla de Oliveira Souza[1(⊠)], José Maria Parente de Oliveira[1], and Letícia Helena Januário[2]

[1] Instituto Tecnológico de Aeronáutica, São José dos Campos, SP, Brazil
paola.cefetmg@gmail.com, parente@ita.br
[2] Campus Centro-Oeste, Universidade Federal de São João del Rei, Divinópolis, MG, Brazil
leticiahj@ufsj.edu.br

Abstract. The objective of this paper is to analyze the relation of blood pressure values and the dimensions and different muscle and fat rates in healthy young women's arms by means of data mining techniques. Methodology: 341 women from 18 to 29 years old were appraised in Divinópolis, Brazil and data on anthropometric measurements and Blood Pressure was collected and developed in multiple linear regression models using data mining techniques. Results: the average was 105,55 mmHg for systolic blood pressure (SBP) and 64,56 mmHg for diastolic blood pressure (DBP). The right arm's SBP was higher when compared to the left arm (106,22 × 104,89) and DBP in the right arm was lower than it was in the left arm (63,94 × 65,19). The values of arm's length (AL), triceps skinfold (TS) and arm's muscle circumference (MC) correlates to SBP and DBP. The variables AL e MC can be considered forecasts of the increase of PAS and PAD's values. Higher levels of MC Values with TS higher than 21,05 was a relevant factor in SBP's increase. Conclusion: there are suggestions as to the dimensions and different fat and muscle rates being correlated with BP indirect measurement values: AL overestimates both SBP and DBP. TS and MC show distinct correlations between SBP and DBP according to specific intervals. Due to sex-related differences in the body composition of arms, can it be a measurement bias?

Keywords: Blood pressure · CART · Body composition

1 Introduction

The Arterial Hypertension (AH) is a global epidemic and the leading preventable risk factor for premature death and disability worldwide. It is a health problem directly associated with cardiovascular diseases that constitute a major cause of human mortality and morbidity [1, 2]. The AH affects approximately 25% of the population today, with projections of 40% in the year 2025 [3, 4].

© Springer Nature Switzerland AG 2019
Á. Rocha et al. (Eds.): WorldCIST'19 2019, AISC 932, pp. 63–71, 2019.
https://doi.org/10.1007/978-3-030-16187-3_7

Hypertension A is characterized by a persistent increase in blood pressure (BP) values to higher than normal levels, which is generally asymptomatic according to [5]. If there are no symptoms, it is through the systematic measurement of these values that it becomes possible to identify changes in the individual's blood pressure and making possible the introduction of earlier therapeutic measures.

Measurement of blood pressure can be performed by direct or indirect methods. The direct method is more precise, however it is highly invasive, complex and costly. It must be performed by qualified professionals. Indirect methods are simpler, non-invasive, and cheaper. Briefly, it consists of compression and decompression of the artery, correlating the pulse with the unit in millimeter of mercury (mmHg) of the chosen manometer.

Considering that in the indirect method, the reading of pressure values results from the compression and decompression of the artery, it is inferred that the arterial pressure is not measured but rather that the pressure necessary to be applied on the corporal structures that surround the artery until occlusion. If the measurement occurs in the arm, as it often does, the cuff compresses the skin and its attachments in addition to the adipose tissue, muscles, nerves, blood vessels and bone to occlude the brachial artery. These structure's levels of density and thickness vary among individuals. According to [6], women show higher percentage of triciptal and bicipital fat and lower body dimensions when compared to men. Considering pressure as quantity given by the ratio of a perpendicularly applied force over a surface and its area [7], it is possible that the indirect measure of women's AP underestimates hypertension and slows an eventually necessary treatment.

In the standardization of the BP measurement technique, it is recommended that the selection of cuffs be proportional to the circumference of the arm. However, in the same circumferential value there are different compositions such as more adipose tissue than muscle or more muscle than fat. [8], evaluated the relation amongst arm circumference, arterial pressure and total body composition of 569 adults. The authors concluded that the highest values of arterial pressure found in persons with higher Body Mass Index may be related to sphygmomanometer cuffs that are unfit for these individuals brachial circumference. The authors see the correlation of arterial pressure and obesity found in epidemiologic studies as strongly influenced by the effects of the variation of arm circumference. The possible variation of pressure values according to sex and different percentages of fat and muscle in individuals with the same arm circumference was studied in [9].

The identification of rules and patterns that relates the results of pressure measurement with the dimensions and body composition of the arm may result in a more accurate estimation of blood pressure values by indirect methods. In this regard, data mining is a computational process of pattern discovery and extraction of huge amounts of data. Both the data mining and healthcare industry have published some reliable early detection systems and other various healthcare related systems from the clinical and diagnosis data [10]. Therefore, this paper's objective is to analyze the relation of blood pressure values with the size and different ratios of muscle and adipose in healthy young women's arms by means of data mining.

2   Methodology

This work deals with a cross-sectional descriptive study performed in the Federal University of São João del Rei (UFSJ) - Divinópolis, Brazil. Data was collected from a population of 489 healthy young adult university students (341 women and 148 men) aged between 18 and 29 years old, by means of questionnaires and blood pressure and anthropometric measurements. Opting for young adults benefits the population sample, due to the wide diversity of body composition and only one expected blood pressure interval. Participants that informed the presence of diseases or the use of drugs or medical substances with potential effect on blood pressure.

The equipment used was duly validated and calibrated for anthropometric measurements of height, weight, perimeters, circumferences and skinfolds according to NHANES [11] (Lufkin anthropometric tape, scientific adipometer harpenden skinfold caliper). Anthropometric measurements were triple-checked and averaged. The variables derived from the anthropometric measurements were estimated according to [9].

BP was measured on both upper arms simultaneously with properly calibrated electronic devices and a cuff suitable for upper arm circumference. Three measurements were taken and the average of the last two measurements was used. The technique used to measure BP is in accordance with international guidelines [12–15].

The collected and estimated data of women was submitted to statistical analysis and data mining. Multiple linear regression and classification and regression tree (CART) models were performed using R software.

3   Results and Discussion

The number of female participants in this study was 341 (variable n). Their blood pressure values were normal and the overall average was 105.55 mmHg for systolic blood pressure (SBP) and 64.56 mmHg for diastolic blood pressure (DBP). The average pressure in the right upper arm was SBP: 106.22 mmHg and DBP: 63.94 mmHg. In the left upper arm SBP average was 104.89 mmHg and DBP was 65.19 mmHg (Table 1). The confidence interval was 95%. The average age of participating women was 21.57.

According to Table 1, average values of SBP, DBP, triceps skinfold (TS) - measured in millimeters (mm), arm length (AL) - measured in centimeters (cm), muscular circumference (MC), upper arm muscle index (AMI) and upper arm fat index (AFI), were different between upper arms, with statistically significant values ($p < 0.05$). However, publications generally related to body composition use TS values, precluding muscle fractions. Significant differences were not observed in relation to upper arm circumference (AC).

From these results, assuming the similarity of AC and the difference between the values of BP, TS, MC, AL, AFI and AMI, we chose to use n of 682 [sum of n of both upper arms (341)]. This is justified by measures that statistically show that the right and left upper arms are symmetrical in circumference, but asymmetrical in pressure values and adipose and muscle ratio values.

The average values of women's blood pressure values were different inter-arms. SBP and DBP was higher in the right upper arm than in the left upper arm. This

Table 1. Comparison of the average of variables by hemi-corps

Variables	Overall average (n = 682)	Hemi-corps mean (n = 341)		p
		Right	Left	
SBP	105,55	106,22	104,89	0,000
DBP	64,56	63,94	65,19	0,000
AL	32,8	32,13	32,03	0,000
AC	26,15	26,17	26,13	0,243
TS	20,15	20,77	19,52	0,000
MC	19,82	19,65	20,00	0,000
AFI	0,41	0,43	0,40	0,000
AMI	0,46	0,45	0,47	0,000

asymmetry of pressure values has been widely discussed in literature but there is still no consensus among publications. Differences in inter-arm BP which was greater in the right upper arm was also found, but without statistically significant values [16, 17]. Another study aimed at clarifying if both upper arms are equally good for assessing BP in the general population found only a small difference in BP between the upper arms in a healthy population of 484 participants aged between 25 and 74 years old [18].

A review of meta-analysis was performed with 22 articles that related the difference in blood pressure values between upper arms, vascular diseases and increased cardiovascular mortality and all-cause mortality. According to the authors, a difference of 15 mmHg or higher may be a useful indicator of risk of vascular disease and death by [19]. An increased inter-arms SBP difference (≥ 6 mmHg) is associated with the burden of atherosclerotic coronary disease [20, 21]. However, the consistency determining the difference in inter-arms blood pressure values lacks concordance, for which the authors suggest further studies [9].

The AL, TS, MC values correlated with SBP. The DBP values were correlated with age, AL, TS and MC. The variables that had the strongest correlations (AL and MC) with both BP were included in multiple linear regression models to verify the relationship of SBP and DBP with upper arm dimension and upper arm body local composition (estimated by TS and AC). The analysis resulted in statistically significant models for SBP and DBP. The SBP's model - which AL and MC explained 26,3% of the increase in SBP. AL and MC also composed the best model to predict DBP, explaining the increase of 22,8%. Mathematical Eqs. 1 and 2 describe these relationships.

$$SBP = 7,761 + 0,056AL + 0,401MC \tag{1}$$

$$DBP = 44,382 + 0,311AL + 0,467MC \tag{2}$$

Tables 2 and 3 show the coefficients used to interpret the influence of each independent variable on the model.

Table 2. Table of coefficients for SBP

SBP	ß	T	p
AL	0.227	5.905	0.000
MC	0.069	1.799	0.006

Table 3. Table of coefficients for DBP

DBP	ß	T	p
AL	0.143	3.676	0.000
MC	0.111	2.848	0.005

According to Eqs. 1 and 2, and Tables 2 and 3, AL and MC variables can be considered as predictors of increase in SBP and DBP in the indirect measurement of BP. The MC results from the relationship between TS and AC, leading to variation of fat and muscle ratios, consequently, variation in arm density. The density of human tissues is different - specifically of muscle and fat. The tissue density of mammalian skeletal muscle is estimated to be 1.06 kg/l and adipose tissue (fat) is 0.92 kg/l [22, 23]. The higher density in the area of BP measurement may require more force for brachial artery occlusion, and therefore, overestimating blood pressure values in indirect measurements.

This result is similar to the study by [24]. The authors found a positive association between lean mass determined by MC and BP in university students. However, this correlation has not been explored [24]. Frequently, studies explore the relationship between BP and body composition estimated by body mass index or the sum of skinfolds, others with AC, and still others with TS.

BP has frequently been correlated positively with body fat in cross-sectional epidemiological studies. For the authors [8] this may be a bias: the use of only one cuff size for adults, disregarding the differences between upper arm circumferences. The correlations between BP and obesity found in epidemiological studies may have been significantly influenced by the effects of AC variation [8].

The arm muscle circumference has been widely discussed in recent scientific literature in function of the relation of muscle mass diminution and of fat free mass with higher risk of human mortality by all causes between middle-aged adults and elderly [25].

Classification and Regression Trees (CART) methodology was also used to develop models that can predict SBP and DBP values. CART analysis is a tree-building technique in which several "predictor" variables are tested to determine how they impact the "outcome" variable. The algorithm selects the predictor that provides the best or "optimal" split, such that each of the two subgroups is more homogeneous with respect to outcome. Each subgroup is further dichotomized into smaller and more homogeneous groups by choosing the variable that best splits the subgroup [1]. Thus, the CART method is able to determine the complex interactions among variables in the final tree. In studies which the distributions of variables are not well known, CARTs provide a model void of any assumptions about the distribution of the variables, preventing model misspecifications.

The variables included in the CART analysis were AL, AC, TS and MC. Using the CART hierarchically, AL and TS emerged as predictors of SBP and DBP. The trees resulted from CART are shown in Fig. 1 for SBP and Fig. 2 for DBP

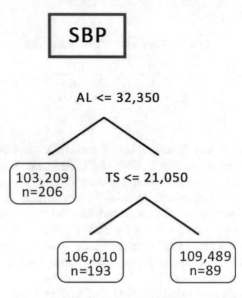

Fig. 1. CART tree SBP

The mean values of SBP are lower in women with AL equal to or less than 32, 35 and greater when AL is greater than 32,35. The interaction of longer arm length with TS values higher than 21,05 is a relevant factor in SBP increase. The observed SBP value increase in described interaction may be related to bigger area, explained by longer arm's length associated to larger quantity of adipose tissue.

The mean values of DBP are lower in women with AL equal to or less than 32, 25 and greater when the AL is above 32, 25. However it can be observed that interactions with TS have happened. In the first situation in which TS is higher than 24, 10 the values of DBP increased considerably. In the second situation in which TS is equal to or lower than 24, 10, lower DBP values are presented. In this case, the probable explanation consists in the reduction of area density characterized by the distribution of a lower quantity of adipose tissue in a longer arm length. The values grouped by CART for DBP are close to the average of AL and above average if TS: 32, 80 and 20, 15 respectively.

The relation of arm length with triceps skinfold can be a variable that offers data for a better understanding of BP values' behavior in indirect measurement. However, only TS distribution patterns in function of sex and age was found. As for AL, most publications found aimed at subsidizing ergonomic and clothing industries.

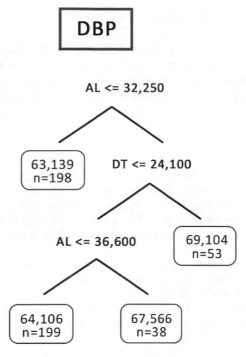

Fig. 2. CART tree DBP

4 Conclusion

The mean values of BP in healthy young women were different between hemi-bodies: higher SBP in the right upper arm and greater DBP in the left upper arm. Results suggest the dimensions and different fat and muscle ratios in the arm are correlated with BP values in indirect measurement: AL overestimates both SBP and DBP. TS and MC show distinct correlations between SBP and DBP according to specific intervals.

Women have lower BP values than men in their childhood, adolescence and - is suggested - as young adults. In light of this, the following issues are incited: can it be a physiological pattern? In this case, will it be necessary to review values to classify arterial hypertension by sex? Or, due to differences in the body composition of arms relative to sex, can it be a measurement bias? In this case, will it be necessary to add equations to the algorithms of electronic devices to correct pressure values? Further research is suggested on the topic of the relationship between the measurement of BP and different body compositions in the upper arm, as well as investigations concerning possible relations of arm's length with triceps skinfold. Moreover, data analysis on the identification of rules that relate anthropometric variables with blood pressure measure using other data mining techniques

5 Theoretical and Practical Contributions

The research contributes with the discussion concerning blood pressure values and proposes the inclusion of variables related to arm's body composition in the evaluation of results obtained through indirect measurement. It also contributes raising questions on the standardization of pressure values between men and women.

References

1. Mills, K.T., et al.: Global disparities of hypertension prevalence and control: a systematic analysis of population-based studies from 90 countries. Circulation **134**(6), 441–450 (2016)
2. Brook, R.D., et al.: Guideline for the prevention, detection, evaluation, and management of high blood pressure in adults. a report of the American college of cardiology/American heart association task force on clinical practice guidelines. J. Am. Soc. Hypertens. **12**(3), 238 (2017)
3. Teo, K., et al.: Prevalence of a healthy lifestyle among individuals with cardiovascular disease in high-, middle- and low-income countries the prospective urban rural epidemiology (PURE) study. JAMA **309**(15), 1613–1621 (2013)
4. Moreira, J.P.L., Moraes, J.R., Luiz, R.R.: Prevalence of self-reported systemic arterial hypertension in urban and rural environments in Brazil: a population-based study. Cad. Saúde Pública **29**(1), 62–72 (2013)
5. Zhou, B., et al.: Worldwide trends in blood pressure from 1975 to 2015: a pooled analysis of 1479 population-based measurement studies with 19_1 million participants. Lancet **389**(10064), 37–55 (2017)
6. Frisancho, A.R.: Anthropometric Standards for the Assessment of Growth and Nutritional Status. The University of Michigan Press, Ann Arbor (1990)
7. David, H., Robert, R., Jearl, W.: Fundamentals of Physics Extended, 10th edn. Editora Wiley, USA (2013)
8. Uluaszek, S.J., Hennberg, M.: Results of epidemiological studies of blood pressure are biased by continuous variation in arm size related to body mass. Hum. Biol. **84**(4), 437–444 (2012)
9. Januário, L.H., et al.: Relationship between upper arm muscle index and upper arm dimensions in blood pressure measurement in symmetrical upper arms: statistical and classification and regression tree analysis. In: Rocha, Á., Adeli, H., Reis, L., Costanzo, S. (eds.) Trends and Advances in Information Systems and Technologies. WorldCIST2018 Advances in Intelligent Systems and Computing, vol 746. Springer, Cham (2018)
10. Neesha, J., Nur, A.A.R., Wahidah, H.: Data mining in healthcare – a review. Procedia Comput. Sci. **72**, 306–313 (2015)
11. NHANES – National Health and Nutrition Examination Survey. Anthropometry Procedures Manual, http://www.cdc.gov/nchs/data/nhanes/nhanes07-08/manual_an.pdf. Accessed 17 Nov 2017
12. James, P.A. et al.: Evidence-based guideline for the management of high blood pressure in adults report from the panel members appointed to the Eighth Joint National Committee (JNC 8). JAMA **311**(5), 507–520 (2014). Erratum in JAMA **311**(17), 1809 (2014)
13. Weber, M.A., et al.: Clinical practice guidelines for the management of hypertension in the community: a statement by the American society of hypertension and the international society of hypertension. J. Hypertens. **32**(1), 03–15 (2014)

14. Sociedade Brasileira de Cardiologia: Departamento de Hipertensão Arterial. VII Diretrizes brasileiras de hipertensão. Arq Bras Cardiol. **107**(3), 01–83 (2016)
15. Leung, A.A., et al.: Hypertension Canada's 2017 guidelines for diagnosis, risk assessment, prevention, and treatment of hypertension in adults. Can. J. Cardiol. **33**(5), 557–576 (2017)
16. Kim, K.B., et al.: Inter-arm differences in simultaneous blood pressure measurements in ambulatory patients without cardiovascular diseases. Korean J. Fam. Med. **34**(2), 98–106 (2013)
17. Fonseca-Reyes, S., Forsyth-MacQuarrie, A.M., García de Alba-García, J.E.: Simultaneous blood pressure measurement in both arms in hypertensive and nonhypertensive adult patients. Blood Press Monit. **17**(4), 149–154 (2012)
18. Johansson, J.K., Puukka, P.J., Jula, A.M.: Interarm blood pressure difference and target organ damage in the general population. J. Hypertens. **32**(2), 260–266 (2014)
19. Song, X., et al.: Association of simultaneously measured four-limb blood pressures with cardiovascular function: a cross-sectional study. Biomed. Eng. Online **15**(2), 147, 247–260 (2016)
20. Her, A.Y., et al.: Association of inter-arm systolic blood pressure difference with coronary atherosclerotic disease burden using calcium scoring. Yonsei Med. J. **58**(5), 954–958 (2017)
21. Hirono, A., et al.: Development and validation of optimal cut-off value in inter-arm systolic. blood pressure difference for prediction of cardiovascular events. J. Cardiol. **71**(1), 24–30 (2017)
22. Farvid, M.S., et al.: Association of adiponectin and resistin with adipose tissue compartments, insulin resistance and dislipidaemia. Diabetes Obes. Metab. **7**(4), 406–413 (2005)
23. Mendez, J., Keys, A.: Density and composition of mammalian muscle. Metabolism **9**(2), 184–188 (1960)
24. Vaziri, Y., et al.: Lean body mass as a predictive value of hypertension in young adults, in Ankara, Turkey, Iran. J. Publ. Health **44**(12), 1643–1654 (2015)
25. Wu, L.-W., et al.: Mid-arm muscle circumference as a significant predictor of all-cause mortality in male individuals. PLoS One **12**(2), 01–11 (2017)

Problems and Barriers in the Transition to ICD-10-CM/PCS: A Qualitative Study of Medical Coders' Perceptions

Vera Alonso[1,2(✉)], João Vasco Santos[1,2,3], Marta Pinto[2,4,5],
Joana Ferreira[1,2], Isabel Lema[1,2], Fernando Lopes[1,2],
and Alberto Freitas[1,2]

[1] Department of Community Medicine, Information and Health Decision
Sciences (MEDCIDS), Faculty of Medicine, University of Porto, Porto, Portugal
vera.alonsop@gmail.com
[2] CINTESIS – Center for Health Technology and Services Research,
Porto, Portugal
[3] Public Health Unit, ACeS Grande Porto VIII Espinho-Gaia,
Vila Nova de Gaia, Portugal
[4] Faculty of Psychology and Education Science,
University of Porto, Porto, Portugal
[5] Subgroup of Terrorism and Security of the Crime and Justice Group
of Campbell Collaboration, Brisbane, Australia

Abstract. Background. Clinical coding is the process of transforming the information about diseases or procedures recorded in the health records, into numeric or alphanumeric codes, using an international classification system. Recently, a new version of the classification system (ICD-10-CM/PCS) started being used in Portugal.

Objective. To explore the perceptions of medical coders (medical doctors) regarding problems related to the transition from ICD-9-CM to ICD-10-CM/PCS that may affect the quality of coded data.

Methods. A qualitative study using four focus groups sessions with ten medical coders was performed between October and November 2017. The convenience sample was obtained from four public hospitals in Portugal. Questions related to problems with the coding process were developed by reviewing literature and with authors' expertise. The focus groups sessions were taped, transcribed and analyzed to elicit themes.

Results. Several problems related to the classification system transition were identified by the medical coders: more time spent in coding, decreased productivity, lack of clinical coding audits and lack of consensus and guidelines.

Conclusion. The availability of more tools to help medical coders, the establishment of guidelines and consensus, and clinical coding audits could increase the quality of coded data, contributing to achieve the expected improvements with the introduction of ICD-10-CM/PCS, namely in the purposes it is collected for.

Keywords: Clinical coding · Data quality ·
International Classification of Diseases · Health information management ·
Diagnosis Related Groups · Qualitative research

© Springer Nature Switzerland AG 2019
Á. Rocha et al. (Eds.): WorldCIST'19 2019, AISC 932, pp. 72–82, 2019.
https://doi.org/10.1007/978-3-030-16187-3_8

1 Introduction

The International Classification of Diseases - Ninth Revision (ICD-9) was designed by the World Health Organization (WHO) in 1970 and contained about 6,882 codes [1–3]. It was updated to a Tenth version (ICD-10) in 1990 containing 12,420 codes [1–3].

The International Classification of Diseases - Ninth Revision - Clinical Modification (ICD-9-CM), created by the National Center for Health Statistics (NCHS), results from an ICD-9 adaptation with modifications that respond to country-specific needs [2]. In 1979, its used was introduced in the United States. In 1997, tests in a new version of this classification system started – ICD-10-CM/PCS – in order to encompass medical advances and to increase the specificity and accuracy of the classification system [4]. This transition is expected to result in major improvements in the quality and usability of data for various healthcare settings [4]. In Portugal, ICD-9-CM was used between 1989 and 2017, when ICD-10-CM/PCS replaced it.

Significant improvements in the format and content occurred with the transition to ICD-10-CM/PCS. ICD-9-CM codes have 3 to 5 digits while ICD-10-CM/PCS codes are alphanumeric and the number of characters varies between 3 and 7, which adds complexity to clinical coding [5–7].

Clinical coding is the basis of a national database with both administrative and clinical information from all public hospitals, grouping episodes of care into Diagnosis Related Groups (DRGs), with the main purpose of hospital reimbursement [8].

Due to the increase in the number of codes and characters and the improvements of the specificity in the classification system, more accurate payments are expected and it will be easier to detect frauds in them [1, 2, 9–13]. Another advantage of this change might be its increased reuse for research [10, 14].

However, during the transition, a decrease in the coded data quality is expected [15, 16]. Several difficulties have already been pointed out, such as adaptations in information systems, lack of productivity of physicians, coders and other staff, increased time for training and impact on the hospital funding [2, 9, 17].

This study aims at identifying the main problems of the transition from ICD-9-CM to ICD-10-CM/PCS according to the perspective of the coders, specifically problems associated with the design and complexity of the new classification system, with the adaptation of medical coders, with the impact in time spent, productivity and data quality, and with coding audits in ICD-10-CM/PCS. This study is paramount because the transition is recent in Portugal.

2 Material and Methods

We conducted four focus groups sessions in October and November 2017 with two groups of five medical coders each. Both groups participated in two focus groups sessions, but in one of the sessions of each group, there was a lack of one intervener.

Firstly, coder's contacts were obtained through participation in coding meetings and through researchers' contact lists. These provided contacts of other potential participants. The email addresses of 54 medical coders from four public hospitals were obtained. These four hospitals represent an accessible sample of the 105 existing hospitals, including public and public-private partnerships [18]. The only inclusion criterion for study participation was to have some experience in clinical coding.

Coders were invited by email to answer a short questionnaire about their availability to participate in a focus group, demographic data and experience in clinical coding (which could be important for participants' selection). Twenty-one replies were received and a new email was then sent in order to ascertain the most suitable date for conducting the focus groups.

Eleven medical coders, belonging to the four hospitals, showed interest in participating. A new email was sent to inform each participant about the selected date of the focus group session and, three days before the session, another reminder was sent by e-mail. One of the medical coders was not able to attend any of the sessions.

Most of the participants were female (8 out of 10), the median age was 55 years (SD = 12), the median coding experience was 9 years (Range = 0.5–28) and 7 out of 10 worked part-time.

The interview guide was developed based on problems already known in scientific literature and on the researchers' expertise. The focus group sessions were held in Porto, in the Faculty of Medicine of the University of Porto, Portugal and their duration ranged from 1h15 to 1h50.

Before starting the sessions, all participants received written and oral information about the study. They signed an informed consent document allowing the recording and use of data and were ensured the anonymity and confidentiality of the collected data.

Sessions were recorded on audio-files with a mobile phone, using the *Dictaphone* software. Microsoft Word was used to transcribe all the recordings. A clean transcript was produced: repetitions, false starts, and possible errors were removed for the text to be clearer and friendlier for the reader. Recordings were deleted after that process.

Through thematic content analysis, information was grouped by themes, categories and subcategories related to the aim of the study [19]. The analysis was conducted by the authors, one author grouped the information and afterwards grouping was validated by consensus from the other authors.

3   Results

Despite the negative points described in the next categories, all participants declared that they are having a good adaptation to the ICD-10-CM/PCS, and all of them preferred this new classification system:
"I think we are well and adapted." (P9)
"I only reason in ICD-10; I no longer remember ICD-9." (P8)
"I really like ICD-10. I think it allows more development, provides more working speed." (P2)
"ICD-10 responds to everything recorded in health records." (P3)

Four categories emerged from factors that medical coders pointed out as influencing coding in ICD-10-CM/PCS and, consequently, the quality of coded data: (1) ICD-9-CM vs ICD-10-PCS; (2) Transition to ICD-10-CM/PCS; (3) Coding audits in ICD-10-CM/PCS; and (4) Variation in coding due to lack of normalization/consensus. For each result, one or more examples from the focus groups are presented.

1. ICD-9-CM vs ICD-10-CM/PCS

Participants began by stating that coding in ICD-10-CM/PCS is difficult. The new classification system is more complex and coders cannot memorize the codes, making coding more laborious.
"Sincerely, with my long experience (…) I have never felt so unsure (…)." (P7)
"I have never worked so hard in ICD-9, but it is interesting (…) as we get more into the system, we have more doubts, and this is dramatic for me." (P7)
"(…) the main problem is that we cannot memorize the codes." (P2)

One participant talked about the possibility of coding with ICD-9-CM and then, through General Equivalence Mappings (GEMs), translating the codes into ICD-10-CM/PCS. P7 did not agree with the use of this tool.
"(…) we have the converser and we can use it, and then adapt to the ICD-10 (…)." (P6)
"(…) the converser is bad, sometimes it helps, but other times…" (P7)

1.1 Coding Procedures with ICD-10-PCS

Coding procedures arise uncertainty for participants, due to the lack of guidelines and consensus. They pointed out the updating of procedure codes as a pitfall.
"In ICD-10, especially in procedures, as we do not have guidelines for some situations, it sometimes allows variation." (P3)
"(…) on procedures with these multiple possibilities [of codes], you cannot be sure that a particular code is the best for a given procedure." (P4)
"Consensus of coding for some procedures is a problem that will be difficult to solve. These procedures need to be perfectly standardized at national level." (P7)
"The problem was the upgrade of 6000 procedures in a year." (P4)

After pointing out negative aspects, they indicated positive aspects in ICD-10-PCS, for example, the detail and coverage, and the structure that allows the possibility of building the code. Some participants stated their preference for coding pro-cedures other than diagnoses.
"In procedures, it is more specific. It allows us to code some techniques that could not be coded before." (P2)
"The way it is structured is very interesting." (P3)
"In almost 100% of cases, obeying the rules, coder will have the possibility of building the right code for the executed procedure." (P7)
"For me it is easier to code procedures than diagnoses." (P10)

Root operations were presented as a topic that causes many doubts; only two coders argued that its coding is easy. The existence of bilateral codes only in some root operations was pointed out as a problem. Furthermore, three participants stated that its coding is subjective, depending on the coders' interpretation, although another participant present did not agree.
"Root operations are the most difficult part, (…) interpretation and definitions vary widely (…) it is necessary to know them very well, to study, to read and to discuss them again, and often not even then." (P7)
"I think root operations are much simpler." (P5)

"Sometimes it is contradictory. Because in some root operations there are bilateral codes and in others there are not, which is not logic (…). There should be bilateral codes for all bilateral organs." (P2)

"It can also be a little subjective. I am talking about repair and others…" (P1)

"But it is not subjective. There are guidelines." (P2)

1.2. Specificity of ICD-10-CM/PCS, Increased Number of Codes and Characters

All participants agreed that specificity increased, recognizing that this is one of the objectives of the transition, which allows improvement in data quality. Coding finger deformities was presented as a clear example of this increase. However, one participant mentioned neoplasms classification as an example where the specificity is not enough; another participant, while agreeing, defended that specificity was increased in dermatological neoplasms in ICD-10-CM/PCS. Another intervener referred that Spain solved this problem by joining ICD-10-CM/PCS with ICD-O.

"The aim is that specificity increases. You are not dividing codes, or modifying procedures by chance. It is exactly for this, to answer health records." (P7)

"Quality has improved because of codes' specificity…" (P6)

"If I want to say "I have crooked fingers or crooked toes", whatever the finger is, I used to say "crooked finger". Now I can tell if is the finger on the right hand or on the left hand, and I can even tell you what finger it is." (P9)

"There are neoplasms that are neither explicitly malign nor benign. It is uncertain." (P10)

"It has improved a lot, as dermatological neoplasms are already very differentiated. In other neoplasms, it would be very important to have greater specificity." (P8)

"Spain solved this issue very well, they joined ICD-10 with ICD-O. They classify the neoplasm and then add the ICD-O morphology code." (P9)

Regarding the increment in the number of codes and characters, all interveners said it increases the coders' work. One participant agreed, but defended that what is important is the purpose of this increment:

"It does not facilitate the coding process, because it gives a lot more work." (P5)

"(…) it takes a long time to write all the codes, it is dramatic." (P2)

"The issue is the goal. What matters is to fulfill the objective. Will the coder get more work? Maybe they will." (P7)

Eight participants stated that the increment in injury codes difficult their coding, due to the specification of the conditions. One participant argued that the specification of the external causes will continue to not be coded because it is not documented in health records. The other two participants did not know if there was a variation in the number of injury codes between ICD-9-CM and ICD-10-CM/PCS.

"The falls in the hospital, that we coded "falling in hospital", now is: in the hospital, in the hospital ward, in the bathroom, in the hallway…" (P5)

"It specifies more, but it gives a lot of work." (P7)

"I do not know if there was any variation. I learned already with this detail… For me, there are many codes, I would probably feel the same with ICD-9." (P10)

2. Transition to ICD-10-CM/PCS

Concerning the transition, experienced ICD-9-CM coders stated that coding is easier those who began already with ICD-10-CM/PCS. The participant who worked only with ICD-10-CM/PCS said that he/she also has many difficulties.
"When we changed, we had to relearn how to select the documented information, while those who started already with ICD-10-CM/PCS were trained for it (…) we have to be aware of certain details, and sometimes we still fail." (P9)
"(…) but it is also hard for me. I have many difficulties." (P10) [intervener who started in ICD-10-CM/PCS]

In relation to the transition and data quality, only one participant stated that data quality did not change with the transition, while the remaining participants said that the quality will improve but during the transition some errors may occur:
"No, I think the quality was not altered." (P2)
"I think quality will improve. Of course, in the transition there can be many errors (…)." (P1)
"I think, in this moment, the quality may be worrying." (P9)

Productivity and the time spent in coding were affected by the transition. Participants argued that the time spent increased and, consequently, productivity decreased. One participant stated that in his/her hospital they kept the productivity, due to the hospital' support and to coders' effort.
"It takes longer, because writing the codes takes three times as long." (P2)
"[Time spent] has increased brutally." (P1)
"I do not code 1000 episodes in ICD-10, but a few months earlier I coded 1000 in ICD-9." (P5)
"But thanks to support, we kept [the productivity]." (P7)
"To keep my productivity, I spend some nights awake." (P9)

3. Coding Audits in ICD-10-CM/PCS

Participants stated that audits in ICD-10-CM/PCS do not exist and advanced some causes: (1) lack of auditors, (2) lack of time and (3) the impossibility to use the software previously developed in Portugal to perform audits as it available only for ICD-9-CM coded episodes. One participant did not agree with the last cause, arguing that another software created for ICD-10-CM/PCS by the _Serviços Partilhados do Ministério da Saúde_ (SPMS) – Ministry of Health Shared Services – can support audits:
"There are not auditors in ICD-10." (P5)
"The problem is that I do not have time to perform audits." (P2)
"We do not have a software; auditors do not have the tool that they had [in ICD-9-CM]. I think it is essential." (P4)
"We do not have this software, but we have the software created by the SPMS. (…) I audited all my coded episodes for six months, and I also audited coded episodes of three other coders." (P2)

External audits have not been performed since the transition to ICD-10-CM/PCS. These audits were performed by *Administração Central do Sistema de Saúde* **(ACSS) – Health System Central Administration - with the purpose of comparing coded data and the hospital reimbursement. Some participants stated that the performance of these audits will be difficult due to the lack of standardization. Two participants argued that external audits can only restart when standardization is achieved and that it should be based on the database that compiles coded data and DRGs.**

"Imagine that the external audits begin… At this moment, I would like to know how this work will be done, because it is not easy, each one codes in his/her own way." (P7)
"The authorized entities should look at the existing database, and see what is being done… If there is much variation in the coding of the same conditions, it is necessary to intervene and standardize." (P9)

4. Variation in Coding Due to Lack of Normalizations/Consensus

Lack of normalizations/consensus allows different ways of coding the same condition, due to different possible interpretations by the coders. In addition, participants argued that support books do not always provide clear coding resolutions:
"If someone told us the way ahead, we would be more confident." (P9)
"There is not a line of reasoning stipulated in the hospital. I talk to Doctor A and he/she codes in one way, when I talk to Doctor B he/she codes in another way, and I then have to follow the best line of reasoning." (P10)
"There are different forms of interpretation from coder to coder." (P1)
"For example, we do not have a consensus in how to code stress incontinence. In the latest updates from Nelly's books, she said "it can be this, it can be that", and afterwards we have a big list of codes and we feel lost." (P7)

One group stated that the creation of a new Portal da Codificação (platform to support the activity of medical coders in Portugal) would help to establish consensus/normalizations:
"For clearing doubts and standardizing conditions." (P3)
"If there was a new platform for ICD-10, as there was for ICD-9, it would be very good." (P3)

It is necessary, and urgent, to establish consensus. Coded data needs to achieve its primordial purpose, the hospital reimbursement:
"(…) there are concepts, standards and recommendations that need to be defined. It cannot be the coder, nor the auditor, nor the coordinator to decide (…)." (P8)
"There are small problems resulting from the lack of consensus, and these problems arise every day. Consensus are needed." (P7)
"But this variation must stop, for reimbursement purposes, for DRGs purposes, for equality throughout the hospitals (…)." (P7)

4 Discussion

Updates to classification systems are required to keep up with advances in medical knowledge and technology. In this study, some problems related to the transition from ICD-9-CM to ICD-10-CM/PCS were identified, such as increased time spent, decreased productivity and lack of audits, which may influence the quality of coded data. However, participants recognized the increase in specificity as a positive point.

With the transition, there was an increase in the difficulty of coding, partly due to a lack of consensus for coding specific conditions and due to the subjectivity of the guidelines. This could lead to multiple interpretations by coders, resulting in different codes for the same situation. These discrepancies may compromise the purposes of coded data leading to variations in hospitals reimbursement and unreliable research. Subjectivity in clinical coding with ICD-10-CM/PCS is an already known problem [20, 21].

The transition also led to an increase in time spent, with an associated decrease in productivity [2, 6, 9, 12, 15, 17, 22–24], which was partially corroborated in this study. Some participants stated that thanks to supports from the institution and to their own effort, they were able to maintain their productivity.

According to participants, difficulties associated with coding in ICD-10-CM/PCS vary according to their experience. Experienced ICD-9-CM coders feel they have more difficulties than coders that only worked with ICD-10-CM/PCS. They argued that they need to change the way how information is selected and become aware of details that in ICD-9-CM were not relevant, i.e., they have to revise the abstraction process. Root operations were presented as the most difficult part in the new classification system by some participants. Other studies already refer it as the most difficult part of the transition for experienced by ICD-9-CM coders [7, 25]. In addition, *Sand et al.* [7] also concluded that experienced ICD-9-CM coders had more difficulties transferring their knowledge to code procedures than diagnoses, since the major differences were in procedures. In this study, not all participants agreed that coding procedures with ICD-10-PCS is more difficult.

The increment in codes and characters was presented as another negative point, because coders cannot memorize the codes as in ICD-9-CM, a problem already presented in another study [26]. General Equivalence Mappings (GEMs) were suggested as a possible support for ICD-10-CM/PCS coding, but it was not accepted by all participants. Other authors presented GEMs as a useful and accurate tool [2, 27]. This tool may be used by coders, but the coder must verify if the resulting codes are correct.

Although the increment in codes and characters was presented as a difficulty, participants also defended that this brought about specificity to the coding, allowing a more accurate representation of the patients' condition. A lot of studies also presented the specificity increase as an advantage of the transition [1, 2, 24, 28–30]. In addition to the coding of finger deformities, other studies pointed out that obstetrics, diabetes, injuries and public health diseases are good examples of areas in which specificity increased [4, 31, 32].

Regarding the coding of injuries in ICD-10-CM/PCS, external causes were a topic addressed. In the participants' point of view, the specification of the causes of the injuries will continue to be not coded since it is not well documented in health records.

In order to surpass this problem, interaction between coders and physicians, and training for physicians, are paramount.

Audits are important for data quality improvements and for detecting overpayments [33–35]; lack of audits in ICD-10-CM/PCS may compromise not only the quality of coded data but also the purposes of this data.

Besides the negative points, and contrary to what was expected before this study, coders demonstrated that they had a general good adaptation and assumed their preference for ICD-10-CM/PCS. Furthermore, participants understand that the change to ICD-10-CM/PCS was done to improve the coding and the purposes of coded data.

4.1 Limitations

The sample of medical coders was assembled out of convenience, consisting of medical coders who participated voluntarily in the study and who worked in only four hospitals in the north of the country. Participants could not be selected randomly due to the low number of medical coders who showed willingness to participate, thereof a bias may have been produced. Moreover, we were unable to collect data until saturation.

Another limitation was the possible competitiveness between medical coders belonging to different hospitals, which may have compromised the exposure of the full experience in each institution. Some people are uncomfortable giving their opinion, a problem that can only be solved through individual interviews.

5   Conclusion

Consensus and more assertive guidelines are deemed to help the coders' work. Moreover, clinical coding audits seem necessary to assure not only the quality of the coded data, but also to identify areas in which further intervention is necessary.

Coded data quality can improve with the transition to ICD-10-CM/PCS, which would contribute to increase the accuracy of hospital reimbursements. Moreover, it would contribute to improve results in all the purposes of clinical coding, such as research. Epidemiological profiles may be more reliable and the assessment of the quality, safety and efficacy of health care can be more accurate, which would, in turn, increase their level of success. In order to achieve these improvements, it is necessary to tackle the identified problems arising from the transition.

Future research should be performed in order to gauge the degree of impact in hospitals' reimbursement and in clinical and health services research and policy.

Acknowledgments. The authors would like to thank the focus group participants, the MSc in Medical Informatics (http://mim.med.up.pt), and Project NanoSTIMA (NORTE-01-0145-FEDER-000016). Project NanoSTIMA is financed by the North Portugal Regional Operational Programme (NORTE 2020), under the PORTUGAL 2020 Partnership Agreement, and through the European Regional Development Fund (ERDF).

References

1. Topaz, M., Shafran-Topaz, L., Bowles, K.H.: ICD-9 to ICD-10: evolution, revolution, and current debates in the United States. Perspect. Health Inf. Manag. **10**(Spring), 1d (2013)
2. Manchikanti, L., Falco, F.J.E., Hirsch, J.A.: Necessity and implications of ICD-10: facts and fallacies. Pain Physician **14**(5), E405–E425 (2011)
3. Jetté, N., Quan, H., Hemmelgarn, B., Drosler, S., Maass, C., Moskal, L., et al.: The development, evolution, and modifications of ICD-10: challenges to the international comparability of morbidity data. Med. Care **48**(12), 1105–1110 (2010)
4. Hazlewood, A.: ICD-9 CM to ICD-10 CM: implementation issues and challenges. In: AHIMA's 75th Anniversary National Convention and Exhibit Proceedings (2003)
5. ACSS: Implementação do Sistema de Codificação Clínica ICD-10-CM/PCS em Portugal, em substituição da atual ICD-9-CM. http://www.acss.min-saude.pt//wp-content/uploads/2016/07/Perguntas-Frequentes_-ICD10CMPCS.pdf. Accessed 28 July 2017
6. Butz, J., Brick, D., Rinehart-Thompson, L.A., Brodnik, M., Agnew, A.M., Patterson, E.S.: Differences in coder and physician perspectives on the transition to ICD-10-CM/PCS: a survey study. Health Policy Technol. **5**(3), 251–259 (2016)
7. Sand, J.N., Elison-Bowers, P.: ICD-10-CM/PCS: transferring knowledge from ICD-9-CM. Perspect. Health Inf. Manag. **10**(Summer), 1g (2013)
8. ACSS: Grupos de Diagnósticos Homogéneos. http://www2.acss.min-saude.pt/Default.aspx?TabId=460&language=pt-PT. Accessed 13 Aug 2017
9. Watzlaf, V., Alkarwi, Z., Meyers, S., Sheridan, P.: Physicians' outlook on ICD-10-CM/PCS and its effect on their practice. Perspect. Health Inf. Manag. **12**, 1b (2015)
10. Rousse, J.T.: From novice to expert: problem solving in ICD-10-PCS procedural coding. Perspect. Health Inf. Manag. **10**(Summer), 1d (2013)
11. Wright, R.E.: Administrative simplification: change to the compliance date for the international classification of diseases, 10th revision (ICD–10–CM and ICD–10–PCS) medical data code sets. Fed. Reg. **79**(149), 45128–45134 (2014)
12. Sanders, T.B., Bowens, F.M., Pierce, W., Stasher-Booker, B., Thompson, E.Q., Jones, W.A.: The road to ICD-10-CM/PCS implementation: forecasting the transition for providers, payers, and other healthcare organizations. Perspect. Health Inf. Manag. **9**(Winter), 1f (2012)
13. Mills, R.E., Butler, R.R., McCullough, E.C., Bao, M.Z., Averill, R.F.: Impact of the transition to ICD-10 on Medicare Inpatient Hospital Payments. Medicare Medicaid Res. Rev. **1**(2) (2011)
14. Freitas, A., Lema, I., da Costa-Pereira, A.: Comorbidity coding trends in hospital administrative databases. In: Rocha, Á., Correia, A., Adeli, H., Reis, L., Mendonça Teixeira, M. (eds.) New Advances in Information Systems and Technologies. Advances in Intelligent Systems and Computing, vol. 445. Springer, Cham (2016)
15. Stanfill, M.H., Kang, L., Hsieh, K., Beal, R., Fenton, S.H.: Preparing for ICD-10-CM/PCS implementation: impact on productivity and quality. Perspect. Health Inf. Manag. **11**, 1f (2014)
16. Pongpirul, K., Walker, D.G., Rahman, H., Robinson, C.: DRG coding practice: a nationwide hospital survey in Thailand. BMC Health Serv. Res. **11**, 290 (2011)
17. Alakrawi, Z.M., Nemchik, S., Sheridan, P.T.: New study illuminates the ongoing road to ICD-10 productivity and optimization. J. AHIMA **88**(3), 40–45 (2017)
18. INE: Portal do Instituto Nacional de Estatística. https://www.ine.pt/xportal/xmain?xpid=INE&xpgid=ine_indicadores&indOcorrCod=0008121&contexto=bd&selTab=tab2. Accessed 4 Oct 2018
19. Bardin, L.: Análise de Conteúdo. 4 edicao. EDICOES, Lisboa, 70 (2011)

20. Ngene, N.C., Moodley, J.: Assigning appropriate and comprehensive diagnosis for scientific report. Med. Hypotheses **83**(6), 681–684 (2014)
21. Lucyk, K., Tang, K., Quan, H.: Barriers to data quality resulting from the process of coding health information to administrative data: a qualitative study. BMC Health Serv. Res. **17**(1), 766 (2017)
22. Houser, S.H., Hart-Hester, S.: Assessing the planning and implementation strategies for the ICD-10-CM/PCS coding transition in Alabama hospitals. Perspect. Health Inf. Manag. **10**, 1a (2013)
23. Manchikanti, L., Falco, F.J.E., Kaye, A.D., Hirsch, J.A.: The disastrous but preventable consequences of ICD-10. Pain Physician **17**(2), E111–E118 (2014)
24. Meyer, H.: Coding complexity: us health care gets ready for the coming of ICD-10. Health Aff. **30**(5), 968–974 (2011)
25. Ross-davis, S.V.: Preparing for ICD-10-CM/PCS: one payer's experience with General Equivalence Mappings (GEMs). Perspect. Health Inf. Manag. **9**(Winter), 1e (2012)
26. Libicki, M., Brahmakulam, I.: The Costs and Benefits of Moving to the ICD-10 Code Sets. RAND Corporation, Santa Monica (2004). https://www.rand.org/pubs/technical_reports/TR132.html
27. Office of the Secretary, HHS: HIPAA administrative simplification: modifications to medical data code set standards to adopt ICD-10-CM and ICD-10-PCS. Fed. Reg. **74**(11), 3328–3362 (2009)
28. DeAlmeida, D.R., Watzlaf, V.J., Anania-Firouzan, P., Salguero, O., Rubinstein, E., Abdelhak, M., et al.: Evaluation of inpatient clinical documentation readiness for ICD-10-CM. Perspect. Health Inf. Manag. **11**(Winter), 1h (2014)
29. Cartwright, D.J.: ICD-9-CM to ICD-10-CM codes: What? Why? How? Adv. Wound Care (New Rochelle) **2**(10), 588–592 (2013)
30. Kealey, B.Y.B., Howie, A.: ICD-10 is coming an update on medical diagnosis and inpatient procedure coding. Minn. Med. **96**(11), 48–50 (2013)
31. Barta, A., McNeill, G.C., Meli, P.L., Wall, K.E., Zeisset, A.M.: ICD-10-CM primer. J. AHIMA **79**(5), 64–66 (2008)
32. Chute, C.G., Huff, S.M., Ferguson, J.A., Walker, J.M., Halamka, J.D.: There are important reasons for delaying implementation of the new ICD-10 coding system. Health Aff. **31**(4), 836–842 (2012)
33. Carpentier, P.J.: The risk of getting paid: why ICD-10-CM may increase physician liability under the False Claims Act. Quinnipiac Health Law J. **16**, 117–148 (2013)
34. Tatham, A.: The increasing importance of clinical coding. Br. J. Hosp. Med. **69**(7), 372–373 (2008)
35. Naran, S., Hudovsky, A., Antscherl, J., Howells, S., Nouraei, S.A.R.: Audit of accuracy of clinical coding in oral surgery. Br. J. Oral Maxillofac. Surg. **52**(8), 735–739 (2014)

Logging Integrity with Blockchain Structures

Marco Rosa, João Paulo Barraca, and Nelson Pacheco Rocha

Universidade de Aveiro, Campus Universitário de Santiago, Aveiro, Portugal
{marcofrosa,jpbarraca,npr}@ua.pt

Abstract. In developed countries, it is frequent for family members do not have the time, knowledge, or live in a close distance of their senior loved ones, so that many institutions offer their services to provide a good quality of life of older adults. To enable distributed local support, there is the need of digital platforms to allow the exchange of information. These platforms need to create trustful environments and to guarantee the integrity of the information exchanged. In this paper, it is presented a solution for a Logging Service that was developed for the SOCIAL platform, based on FHIR, which aims to solve the interoperability and data integrity of the platform user's activity logs.

Keywords: Logging · Integrity · Auditing · Blockchain · FHIR

1 Introduction

Health care provision, once focused on the management of acute conditions, is increasingly shaped by the epidemiological transition towards Non-Communicable Diseases (NCD), which are related to the main causes of mortality [1]. Also, there is evidence related to disease burden and loss of economic output associated with NCD, mainly cardiovascular diseases, oncological diseases, chronic respiratory diseases, diabetes and neurodegenerative diseases [2].

Most of these NCD last for long periods of time, may have a chronic nature and progress slowly. Consequently, the fragmented nature of the health care systems, focused on the management of acute conditions, is unable to provide universal, equitable, high-quality and financially sustainable care [3]. This is an opportunity to move from organization-centered care to a paradigm focused on the needs of the care receivers [4]. Integrated care approaches are required, not solely focused on medical purposes, but also on a range of essential activities for the maintenance of the individual's quality of life, which increases the complexity of the care services, as a consequence of shared procedures between different organizations and a trend towards community care, including, in addition to health care provision, by medical and nursing staff, social care provision, by social care staff, or home assistance services [5].

The platform of services SOCIAL [6] aims to surpass the so-called health and social care divide [7]. It prides information services to support integrated care and assistance of community-dwelling older adults in scenarios of care provided by a distributed environment of cooperating entities, which is typical for (at least) many European regions.

In this respect, trust and the promotion of trust assume a paramount importance. To promote trustful relationships between all the stakeholders of the SOCIAL platform a

Á. Rocha et al. (Eds.): WorldCIST'19 2019, AISC 932, pp. 83–93, 2019.
https://doi.org/10.1007/978-3-030-16187-3_9

major issue is to ensure the integrity of the activity logs that should be traceable by auditing processes. The study reported by the present paper focused on the information audit trail and the mechanisms to promote its integrity in distributed environments.

In the following sections, the paper will introduce the related work that is relevant to understand the reasoning process behind the proposed Logging Service, a description of the SOCIAL platform and its logging requirements, the presentation of the proposed solution, an evaluation of its features and a conclusion.

2 Related Work

Because of the distributed nature and low connection between care providers in the community care context, the policies, rules and processes are frequently difficult to be fully implement or adopted. However, this doesn't alleviate the need for formal access control and the creation of audit trails to be validated by auditors.

An auditor can be someone from the inside or the outside of the auditing target and an audit should be divided into several stages, including planning or gathering of data and its analysis. Analyzing logs allows an auditor to see and evaluate the activities of the users for security purposes (e.g. to verify if someone tried to log in several times in short period of times or attempted to perform non-authorized information accesses). In the end, the results should indicate that everything is going according to the plan or lead to modifications to improve the efficiency of the security mechanisms.

When developing a Logging Service (i.e. a service to record and retrieve the activities of the different users), the simplest procedure is to store the logging messages in raw format inside a file. However, by only doing that, if there are different entities contributing with their logs (each one with a different structure), the log file will be disorganized, unnecessarily complex, and would make the tracking and analyzing process harder to execute. At the same time, obeying protection regulations (e.g. EU Directive 2016/679 on the protection of individuals) becomes more difficult.

At the application level, there are popular solutions, some adapted to the use case of electronic health records [8], which facilitate the production of effective logs by exploiting careful, sometimes automated, log placement [9]. These solutions make it possible to specify the service or component producing the log entry, the event time, and a message, while associating well known severity levels. The destination is frequently a log file, but some solutions adopt the use of document-based databases [10]. However, in either case, the stored information remains vulnerable to tampering.

Over time, there have been several applications that improve the generation and management of logging records. Retrace [11] stores all the log data on a centralized location and provides a range of tools (e.g. error tracking, performance analysis or detailed log searching). It does not solve the tampering problem, since all logging data is stored in the same place, and there is not a way to verify if someone modified an entry. Alternative log management tools suffer the same problems, such as LogRhytm [12] or Splunk [13], because their main concern is to store data and retrieve entry quickly and efficiently, without analyzing possible integrity problems.

A centralized logging service, located in an area of restricted access, is useful for the case of log integrity, as it can handle different system logs, from remote systems, exposing a read only access to the stream. While this solution, that is typical of many deployments, provides some assurance by means of barriers, it provides no cryptographic support for integrity.

There are some proposed solutions for logging services with integrity control, for different type of use cases. For example, approaches to guarantee integrity of service's logs include: adding markers to the log entries, such as sequence numbers and epoch, to detect reordering attacks [14]; adapting a Merkle-tree by using chameleon hash functions instead of regular hash functions [15]; using a key pair (secret and public) where the secret key is used to generate a signature for the log requests and the public key for auditors to validate the signature and integrity of the log data [16]; adapting the blokchain technology into a 2-layer blockchain architecture (one layer for fast operations and other for heavy operations) where a block's hash can prove the logs integrity [17]; using a domain-specific language (DSL) that can allow to structure a log and to select the security parameters that can guarantee its integrity [18]; or using a homomorphic encryption algorithm, where only the one that sends the encrypted logs knows how to decrypt them and understand their contents [19].

In recent years, the development of blockchain technology, mostly famous through its many digital currencies, provided the means to create distributed ledgers, that can validate the chronological consistent of a series of actions. Its inherent capability of validating the temporal order of events also makes the solution adequate for integrity control of log entries.

As its name indicates, blockchain is constructed by several data blocks that are chained, using a process named hash chain. Inside each block, there are several attributes, namely the data (or payload) itself, a timestamp, a hash (a unique identifier or a digital digest) and the hash from the previous block. These hashes are the key attributes to confirm the data integrity, because each block is dependent on the previous (and/or next) and if someone want to tamper one block it would need to tamper every single existent block. Because the chain exists in multiple systems, this would make the intrusion very easy to detect. The blockchain also differs from other data storage implementations by not having intermediaries able to handle all the data related to activity logs. Instead, there is a distributed network, in which every participant has access to the relevant data (shared ledger) and contributes to their validation, as each new block added frequently requires the validation of all previous chain. That means that every single block must go through various nodes before being inserted into the shared ledger or exist in multiple replicas [20].

By integrating the blockchain structure with the logging feature of a system, it can be guaranteed that events are not manipulated by an attacker, as the entities manipulating the chain will keep a distributed consensus, built on a trust relationship between the different parties that are involved with the logging. By requesting their validation, thus obtaining a proof that the integrity was verified by a specific entity at a certain time instant. This will avoid certain parties to deny their involvement in the validation process or to tamper the log entries because there is consensus about the log history. It should be considered that timestamping and validation can rely on metadata, as the timestamping of the content fingerprint, effectively timestamps the content itself. This bring the added benefit of not disclosing sensitive information (the log entry) to third parties.

Guaranteeing the integrity of logs is even more relevant in systems that handle health-related information. The blockchain technology has been applied to guarantee the integrity of electronic health records [21, 22]. There are also studies which main goal is integrity by adapting the blockchain technology, but they not consider the integration with interoperability standards such as Fast Healthcare Interoperability Resources (FHIR) [21] or are mainly focused on health care (and not social care) systems [22].

3 Proposed Solution

The SOCIAL platform [6] addresses the care provided in the community, namely supporting informal and formal caregivers who do not have access to structured information services regarding their care receivers, such as the social care and assistance providers, or informal care providers.

Several web and mobile applications are being developed to fulfil the needs of the stakeholders of the SOCIAL platform, including common applications (e.g. registration and self-registration on the platform, insertion and editing of registered data such as demographic data, or notifications and alerts of relevant information, such as tasks to be performed or enrolled activities), care receiver's applications (e.g. requests for assistance, individual care plans, consultation and enrolment in programs being proposed or integration of information of personal health devices), caregiver applications (e.g. management of care plans and tasks) or care management applications (e.g. administrative management of care receivers and care providers or allocation of human and material resources).

The FHIR was thought to be used to guarantee the internal and external interoperability since it is regarded as a next generation health information interoperability framework and it presents high flexibility, since it is compatible with web standards such as the REST architectural style or the JSON serialization format. Another supporting component of the SOCIAL platform is the Logging Service. To develop a Logging Service that can be integrated with the FHIR standard and able to manage logs from different applications and services, there are some requirements that need to be fulfilled:

Logging Service isolation - a vital requirement is that the execution of the Logging Service should suffer from minimal interference, which imposes that, from the SOCIAL platform point of view, it should be an independent service, accepting log entries from the remaining applications and services. The storage medium should also be isolated, both for security reasons and for exploiting the adequacy of a specific storage backend.

Seamless integration with FHIR services - the Logging Service should be a web service, so that other services can communicate with it and the support for logging doesn't introduce additional dependencies. Since the Logging Service should exchange information, the use of the REST architecture is recommended, due to its performance, scalability and uniform interface. Moreover, integrating the Logging Service with the FHIR framework will allow the processing of data without misinterpretations, and to communicate them in a format that is understood by different parties.

Multidimensional reconstruction of the event timeline - data from activity registries must be contextualized so that auditing processes will be able to reconstruct the timeline of any participating entity. This implies the need for timestamps in all log entries, but also implies the identification of the different entities. In the scope of the development of the SOCIAL platform, this results in the definition of several attributes (e.g. time of creation, time of insertion, service identity or user identity).

Resiliency to manipulation - log entries should be stored in a way that it should not be possible to manipulate them. Also, it should not be possible to alter the timeline by simply deleting entries from the database.

Resiliency to arbitrary losses - if any of the log entries are lost during an attack or other type of issues. These data should be available elsewhere and ready to be quickly retrieved. Proof that the new data are not manipulated is also required.

Non-repudiation - it should be possible to request a validation from the entities that provide their log entries to the Logging Service, so that there can be several integrity proofs accumulated, for future verification.

Privacy and data confidentiality - log entries should contain minimal data required to assert the execution of an action by a user or service, but personal information should not be directly exposed in the entries. The use of temporary identifiers, and other anonymization techniques is mandatory. At the same time, third parties should be able to validate the timeline, without having access to the content of the log. Full reconstruction and validation of the log content should be possible only to auditors.

In order to integrate a secure logging solution, relying on blockchain technology, to satisfy the defined requirements, the Logging Service architecture is based on a RESTful API (Fig. 1), allowing the communication of the different SOCIAL applications via HTTP POSTs. This service is complemented by a timestamping service (timer) that sends the events triggering the block's insertion process. A cryptographic secure blockchain was adopted, which stores the events and the blocks (of the blockchain) in a special purpose database of the SOCIAL platform. Moreover, audit processes will access the stored events and blocks to determine the relevant data. For that three separate processes were considered: log insertion, log consensus and log audit.

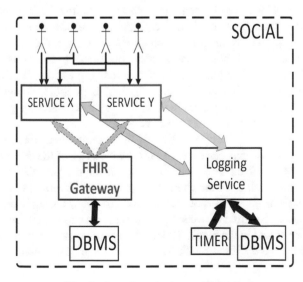

Fig. 1. Logging service architecture.

Log insertion is triggered when a service event occurs in the SOCIAL platform. The data describing the event is sent to the Logging Service, as an HTTP POST with the event inserted in its request body, as a JSON Object. The Logging Service processes its contents and categorizes the different elements involved in the event (e.g. date, user or application), so that it can construct the appropriate document to be persisted. These requests must also have a JSON Web Token (JWT), which is used to verify signatures (using an algorithm and a secret key previously established) and to validate requested events. The log integrity is validated locally upon block insertion, which happens at fixed intervals, as stated by the timer. Inserting a block requires the validation of the previous blocks in the chain (either by that service or offloaded to horizontally scalable instances), and the registration of the log entries that are pending to be inserted. This creates interdependencies between log entries from multiple sources, limiting single source spoofing by an adversary. Blocks can have no payload, effectively sealing the chain, proving that no activity was logged during that time interval. If there are data, a cryptographic hash is calculated (e.g. SHA-256), sorted by insertion order, as well as the hash of the previous block and other identification fields. The block is then inserted into a document-based storage, with multiple indexes, using the well-known ElasticSearch indexer. Since the SOCIAL platform main purpose is to coordinate supporting care services, there should be multiple streams in the blockchain (i.e. for each platform application or service and for each user). A global stream can also exist if an absolute timeline reference is desired, which is beneficial for the auditing process, in order to analyze the timeline of events of the users to verify their behaviors or to analyze the timeline of events of an application or service to verify what it is doing. The actual log data is not replicated, and this multi-stream approach is kept by linking digital hashes, and the block payload only contains indirect references. Privacy is inherent to the design as the one-way property of cryptographic digest functions prevents reversing the digest process.

Log consensus is achieved by sending data from the blocks inserted between two wider timeframes, to application or services belonging to the same ecosystem, which should reply with proofs of integrity. Each proof contains a digital signature, which may follow the standard X.509 timestamping practices. It can consist of full logs, allowing the effective replication of the entire event log and block chain in external services. However, only minimal indirect validation is possible, by synchronizing the block chain without actual log data, which is a tradeoff between performance and resilience to failures as well as audit effectiveness.

Log audit is done by direct observation of the event stream stored in the database. Existing tools such as Kibana can be used and the integrity of the data observed can be assured. The integrity of the chain can be verified anytime to check for any inconsistencies. Since every block of the chain has a clear identification, with fingerprints of the log data, as well as timestamps of the first and last event during that time period, the hash of all the events occurred between two checkpoints can be recalculated to check if it matches the one stored in the block. The next and previous hash fields on every block can also be used to verify if the block itself has not been tampered.

4   Evaluation

Besides the actual field validations, which are taking place as part of the planned development of the SOCIAL platform, to validate the proposed solution a simpler scenario was created with the following characteristics: 100 different users accessing four concurrent applications (i.e. an application to support city council's professionals, an application to support social care providers, an application to support assistance providers, and other application to support care receivers). The Logging Service goal is to handle all the events from all the four applications, while also guaranteeing the integrity of both the application's or service's and user's timelines.

For 60 min, it was planned that each application should continuously send events (up to 50), each one linked to a specific user, before stopping for a random period of time (i.e. 45 to 90 s for the application to support city council's professionals, 25 to 45 s for the application to support social care providers, 10 to 20 s for the application to support assistance providers and 40 to 60 s for the application to support care receiver's). These events were JSON Objects (from a standard FHIR integration), with different fields to identify the events, such as their users, time of occurrence and the respective applications or services. This was done to evaluate how the Logging Service can handle random situations, whether that be handling multiple events in the same time instant or none during certain time periods (inserting timestamping blocks).

The timer component triggers every 90 s, so that the Logging Service verifies if there were new events for each application or service and store the respective highest and lowest timestamps of that set of events. After the timer receives a completion response (showing that the Logging Service finished this stage), it waits a predefined amount of time (20 s in this case) and trigger the insertion by the Logging Service, so it creates, and inserts blocks into the respective blockchains (applications, services and users), updating the previous block values (the "nextBlockHash" value). After inserting every block, the Logging Service responds with a status message. After every successful block insertion process, the Logging Service also requests from all parties to validate the new blocks inserted, by sending their hash values to be signed. The signing process being used is the trusted timestamping, where each party concatenates the hash value with the current timestamp and signs it with its private key, and then sends that result, alongside with the timestamp used for the signing to the Logging Service, so it can be stored (Fig. 2).

Table 1 depicts the number of logs and blocks stored by the Logging Service for each application, and if the integrity of the application's timelines was verified. The user's blockchains (that varied between 16 and 19 blocks for each user) was also verified for their integrity. Moreover, it was assured the integrity of the activity logs of the 100 users that were considered for the test.

When an application accesses the auditing feature of the Logging Service, it should receive the events associated to a pair of user and application between two instants. From the Logging Service perspective, whenever it receives an audit request, first, the respective blockchain, with the timestamps between the two time periods requested is retrieved. Afterward, the integrity of those blocks is verified and, finally, if the integrity is validated, the requested events are sent to the platform. During the auditing tests (for application, service or user logs), the Logging Service always sent the requested events (in these tests, the auditor had privileges to access any event of the platform), which is possible if there were events between those two time periods and if their integrity was verified. The returned events, for every auditing test, are sorted by their timestamp, and it is possible to analyze the respective timeline of events and to carry out an audit with all the necessary data. Since the events follow the FHIR standard, the auditor could filter the events, by any of the fields available in them.

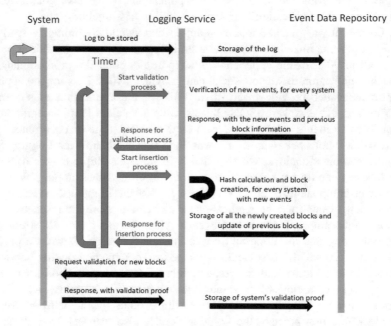

Fig. 2. Logging service workflow.

Table 1. Service events.

Service	Number of logs (total)	Number of blocks	Integrity verified
City council professionals	921	19	True
Social care providers	1974	19	True
Assistance providers	3751	19	True
Care receivers	1015	19	True

Table 2 shows, the time elapsed for seven different block insertion processes (this also includes the request of the events from the data storage to persist the new blocks and update the previous ones), the total number of events (from all of the different applications) used to calculate the hash and number of blocks created for the user's and application's blockchains. The first four results were obtained using the parameters established before, while the fifth and sixth result were obtained by increasing the number of events sent by the applications, between the respective two insertion processes, to verify if there were changes when the number of events used to calculate the hash deeply affected the elapsed time. We can see, for our scenario (four applications and 100 users), that the time the Logging Service took to retrieve events, create and update the respective blocks and insert them in the data storage, depends on the number of events, and it can take from 33 s to 45 s. Therefore, the platform needs to take in account the number of expected events that are going to be exchanged and respective number of users and applications, to determine the interval of time to create and insert new blocks into the blockchain.

Table 2. Block insertion elapsed times.

Number of events	Number of user's blocks	Number of application's blocks	Time elapsed (seconds)
143	81	4	37
417	97	4	33
315	97	4	38
310	97	4	45
2170	100	4	33
3977	100	4	39

The validation proofs (i.e. the trusted timestamping results) were also verified, by using the public key and the hash and timestamp which seals the signed values. Every validation was consistent, which guarantees the block's integrity. Moreover, the signed values cannot be repudiated by the entities that signed them, since the signatures are performed with private keys.

5 Conclusion

Considering the evaluation results, it is possible to conclude that even with large quantities of events in short periods of time, the Logging Service is able to create blocks and manage the respective blockchains for each user and application pair. Consequently, the Logging Service is able to guarantee the integrity of the generated timeline events. Therefore, with the proposed solution, tampering attempts can be efficiently detected in any timeline of events, which means that it is possible to create a trusted environment, since data confidentiality and log integrity are preserved.

The FHIR integration is also beneficial in promoting interoperability. Moreover, the same approach can be used for many other FHIR based instantiations, as it is based on readily available technology and has configurable levels of external validation and replication.

Acknowledgements. This work was supported by Sistema de Incentivos à Investigação e Desenvolvimento Tecnológico (SI I&DT) of the Programa Portugal 2020, through Programa Operacional Competitividade e Internacionalização and/or Programa Operacional do Centro do FEDER - Fundo Europeu de Desenvolvimento Regional, under Social Cooperation for Integrated Assisted Living (SOCIAL), project number 017861.

References

1. Chiarini, G., Ray, P., Akter, S., Masella, C., Ganz, A.: MHealth technologies for chronic diseases and elders: A systematic review. IEEE J. Sel. Areas Commun. **2**(9), 6–18 (2013)
2. Abegunde, D.O., Mathers, C.D., Adam, T., Ortegon, M., Strong, K.: The burden and costs of chronic diseases in low-income and middle-income countries. Lancet **370**(9603), 1929–1938 (2007)
3. WHO global strategy on people-centred and integrated health services. WHO (2015)
4. Blobel, B.: Co-production of health enabled by next generation personal health systems. Stud. Health Technol. Inform. **177**, 52–58 (2012)
5. Hägglund, M., Scandurra, I., Koch, S.: Studying intersection points - an analysis of information needs for shared homecare of elderly patients. J. Inf. Technol. Healthc. **7**(1), 23–42 (2009)
6. Sousa, M., Arieira, L., Queirós, A., Martins, A.I., Rocha, N.P., Augusto, F., Duarte, F., Neves, T., Damasceno, A.: SOCIAL platform. In: Advances in Intelligent Systems and Computing, vol. 746, pp. 1162–1168 (2018)
7. Rigby, M.: Integrating health and social care informatics to enable holistic health care. Stud. Health Technol. Inform. **177**, 41–51 (2012)
8. King, J., Williams, L.: Log your CRUD: design principles for software logging mechanisms. In: Proceedings of the 2014 Symposium and Bootcamp on the Science of Security (HotSoS 2014), pp. 5–15, ACM, Raleigh (2014)
9. Zhao, X., Rodrigues, K., Luo, Y., Stumm, M., Yuan, D., Zhou, Y.: The game of twenty questions: do you know where to log? In: Proceedings of the 16th Workshop on Hot Topics in Operating Systems (HotOS 2017), pp. 125–131. ACM, Whistler (2017)
10. Yang, K.: Aggregated containerized logging solution with Fluentd, Elasticsearch and Kibana. Int. J. Comput. Appl. **150**(3), 29–31 (2016)
11. Stackify, Retrace. https://stackify.com/retrace/. Accessed 22 Oct 2018
12. Sekharan, S.S., Kandasamy, K.: Profiling SIEM tools and correlation engines for security analytics. In: 2017 International Conference on Wireless Communications, Signal Processing and Networking (WiSPNET), pp. 717–721. IEEE, Chennai (2017)
13. Okumura, M., Fujimura, S.: Constructing a log collecting system using using splunk and its application for service support. In: Proceedings of the 2016 ACM on SIGUCCS Annual Conference (SIGUCCS 2016), pp. 103–106. ACM, Denver (2016)
14. Hartung, G.: Secure audit logs with verifiable excerpts. In: Topics in Cryptology - CT-RSA, pp. 183–199. Springer International Publishing (2016)
15. Ning, F., et al.: Efficient tamper-evident logging of distributed systems via concurrent authenticated tree. In: 36th IEEE International Performance Computing and Communications Conference, pp. 1–9. IEEE, San Diego (2017)
16. Lin, C.-Y., et al.: Secure logging framework integrating with cloud database. In: 2015 International Carnahan Conference on Security Technology, pp. 13–17. IEEE, Taipei (2015)

17. Aniello, L., Baldoni, R., Gaetani, E., Lombardi, F., Margheri, A., Sassone, V.: A prototype evaluation of a tamper-resistant high performance blockchain-based transaction log for a distributed database. In: 13th IEEE European Dependable Computing Conference, pp. 151–154. IEEE, Geneva (2017)

18. Zawoad, S., et al.: FAL: a forensics aware language for secure logging. In: Proceedings of the 2013 Federated Conference on Computer Science and Information Systems, Warsaw, Poland, pp. 1579–1586 (2013)

19. Rajalakshmi, J., et al.: Anonymizing log management process for secure logging in the cloud. In: 2014 International Conference on Circuit, Power and Computing Technologies, pp. 1559–1564. IEEE, Kanyakumari (2014)

20. Narayanan, A., Bonneau, J., Felten, E., Miller, A., Goldfeder, S.: Bitcoin and Cryptocurrency Technologies: A Comprehensive Introduction. Princeton University Press, Princeton (2016)

21. Magyar, G.: Blockchain: solving the privacy and research availability tradeoff for EHR data: a new disruptive technology in health data management. In: IEEE 30th Neumann Colloquium (NC), pp. 135–140. IEEE, Budapest (2017)

22. Zhang, P., et al.: FHIRChain: Applying Blockchain to Securely and Scalably Share Clinical Data. Comput. Struct. Biotechnol. J. **16**, 267–278 (2018)

Access Control for Social Care Platforms Using Fast Healthcare Interoperability Resources

Marco Rosa, João Paulo Barraca⍟, and Nelson Pacheco Rocha^(✉)

Universidade de Aveiro, Campus Universitário de Santiago, Aveiro, Portugal
{marcofrosa, jpbarraca, npr}@ua.pt

Abstract. The definition of authorization policies is essential to prevent information misuse and to guarantee that only authorized personnel can access specific information. Since not everyone is familiar with special purpose languages, an interpretation tool can allow the management of policies and rules using natural languages. This paper focuses on a parser developed as a component of a platform to support the care of community-dwelling older adults, the SOCIAL platform, allowing to create, read, update and delete authorization policies and rules, using natural languages.

Keywords: ABAC · XACML · Access control · Parser · Natural language · FHIR

1 Introduction

Health care and social care systems and platforms need to handle large quantities of care receiver's information. Since these systems and platforms manage sensitive personal information, including clinical parameters, a major concern is related to the authorization policies, especially because interoperability with other systems and platforms must be achieved.

Therefore, authorization mechanisms to decide if someone can access certain resources are crucial to promote privacy and confidentiality by preventing illegal accesses. However, the definition of authorization policies using special purpose policy languages may be hard to interpret and modify to those with a low technical knowledge whenever the access rules need to be updated. This means that specialized support are required to understand policy languages and to apply them, which implies additional resources.

The present paper purposes a natural language and policy language parser that can help anyone, even with no knowledge about specific policy languages, to create, read, update and delete rules related to authorization policies to be integrated with the Fast Healthcare Interoperability Resources (FHIR) [1], developed by Health Level Seven International (HL7), to promote interoperability across applications and systems.

The following sections will explain available solutions for access control, the target platform for the proposed solution, the SOCIAL platform [2], and why an access

Á. Rocha et al. (Eds.): WorldCIST'19 2019, AISC 932, pp. 94–104, 2019.
https://doi.org/10.1007/978-3-030-16187-3_10

control is important to fulfill its goals, the requirements for a natural language and policy language parser, its implementation and the evaluation to verify if the defined requirements are met.

2 Related Work

There are several open platforms [3], such as the Substitutable Medical Applications, Reusable Technologies (SMART) [4, 5] aiming to provide agile developments of new applications targeting electronic health records. Other platforms consider the specificities of the information required for the social care provision (i.e. electronic social records [6]), such as Senior Care Connect [7] and Ankira [8], aiming to deliver services to provide better quality of life for senior citizens, namely by promoting their empowerment and the empowerment of their formal and informal care, and assistance providers.

Both healthcare and social care platforms benefit with the integration with other systems and platforms, which means that interoperability standards assume a paramount importance to allow the exchange of the care receiver's information. In this respect, the FHIR has become increasingly important to provide longitudinal records of patient's health and healthcare [9] because it combines the best features of other known specifications, such as HL7 v2, HL7 v3, and Clinical Document. FHIR is being integrated in different systems and platforms, such as System C [10], SocialCare [11] or SOCIAL [2].

To avoid unauthorized accesses health care and social care platforms it is essential to develop access control mechanisms, such as Attribute-based Access Control (ABAC) or Role-based Access Control (RBAC).

The eXtensible Access Control Markup Language (XACML) [12], whose structure is a combination between Extensible Markup Language (XML) and Security Assertion Markup Language (SAML) [13], is a standard that allows handling requests, regarding data access, using an ABAC policy language. These requests related to the access of specific resources are be analyzed through a set of pre-established rules (of a certain policy) so that a decision can be achieved.

By using XACML, each rule can be divided into different categories, each one with different attributes, which can have several data types, namely double, boolean, string or date.

The decision for the access request is obtained after comparing all the values of the attributes of the respective categories, between the request and the previously defined rules. Depending on how a specific decision rule was defined, a request result could be a denial or a permit (the access). With XACML it is also possible to establish comparisons, either by verifying if two values are equal, greater or less or by using "and" or "or" operators between the different attributes of a rule, to reach a decision. Additionally, it is also possible to return more information than the decision result, such as the element "AdviceExpressions" that can specify additional restrictions regarding access decisions.

With XACML, policies and their respective rules can have different levels of detail, according to the different number of attributes needed before reaching a decision.

This feature makes this standard a perfect fit for systems that handle great quantities of information, with different types of data.

Several studies highlight how XACML can help a system with sensitive information (e.g. health-related information [14, 15]) and there are already examples of successful integration with existent systems [16], which proves its usefulness and flexibility. Because of this, tools were developed to help the management of access policies defined using XACML. Turner [17] developed a XACML parser and Stepien and colleagues [18] proposed a solution that can allow people with low technical understanding of XACML (e.g. clinicians working in the emergency, military or medical field) to create access rules.

However, those solutions are focused on healthcare, and do not consider the specificities of social care and informal care. Furthermore, these solutions don't take advantage of the "AdviceExpressions" XACML element, which is very dependent on the objectives and goals of the system being implemented. This turns even more important the development of custom XACML parsers, so that a system or platform can be able to adapt to any kind of scenario, regarding the access of its information.

Moreover, to the best of our knowledge, no previous studies have reported the development of a natural language and policy language parser to manage authorization policies considering the specificities of FHIR.

3 Profile Authorization for the SOCIAL Platform

The SOCIAL platform [2] aims to surpass the so-called health and social care divide [6] and to provide information services to support the care and assistance of community-dwelling older adults. The fundamental technological element is a platform of services, which include all the structural components that are required to support a range of coherent applications to ensure an integrated, consistent and cross-cutting view of the information of the care receivers, and efficient communication with and among caregivers [19].

In terms of architecture, the SOCIAL platform allows the integration of different applications by using FHIR. According to FHIR, data are represented as resources, which are JSON (or XML) objects with healthcare related fields and values (e.g. Patient, Practitioner, DiagnosticReport, Coverage, Questionnaire or Consent). In this respect, care receivers of the SOCIAL platform can be referenced as a Patient resource, where their demographic and other administrative information are structured as attributes (e.g. name or address). Other relevant FHIR resources that can be integrated with the SOCIAL platform are Practitioner (for the care or assistance provider), PractitionerRole (for the care or assistance provider role), Organization, CarePlan and Task.

The FHIR resources may have security labels for three different purposes: to indicate the permissions related to information operations, such as read or modify, to indicate what resources can be returned, and to indicate how specific information should be dealt with. Requests can also add their own security labels (e.g. break-the-glass protocol to allow a physician dealing with an emergency situation to access information of patient even when not having access permissions). The security labels can have different categorizations, such as confidentiality, sensitivity, compartment, integrity or handling.

Integrating ABAC mechanisms with this standard is possible, since the FHIR resources are also defined by their attributes.

3.1 Requirements

The main goal was to develop a parser for the SOCIAL platform compatible with FHIR resources. The parser should be versatile, so that any qualified user (whether it be an administrator, a case manager or a care provider) can manage the access rules of any exchanged information inside or outside the platform:

Flexibility in Policy Management - Anyone with the needed qualifications should be able to manage the access policies for their applications or services, regardless of their technical knowledge. A user of an integrated application or service should not need to know the policy language structure or ask for help to someone who knows, to create, read, update and delete access rules. This should be able to be done using a simple language that is understandable by most.

Moreover, the following requirements were also considered:

Attribute-Based Policies and Rules - Since the SOCIAL platform handles sensitive data, the access of these data should not be too generic. Using a single identifying attribute (e.g. the user's role) to decide if someone should access a resource can create scenarios of misuse of information. So, the SOCIAL platform access control should use and analyze several different attributes before reaching a decision, so that users with certain identifying attributes can access only the data to which they are authorized.

Simple Decision Responses, with Possible Additional Information - The responses regarding an access request should be simple enough, so that the requester can easily understand its meaning. However, in some cases, there can be some restrictions that must be enforced by the access control component or that should be notified to the requester, so that it can act accordingly.

Detailed Rules, for the Most Specific Use Cases - There can be several conditions for a specific access, and it should be possible to create a rule that can fulfill those requirements, regardless of their complexity. Those restrictions may be about the subject, the resource, the action, other categories, or even all of them. The access control should be able of handling any kind of situation.

FHIR Resources Integration - Since the FHIR resources have several attributes, these can be used to manage the policies and rules of a system, by using their respective values. Using these attributes plus the platform-specific attributes (e.g. IDs, roles or list of care receivers) allows the handling of different types of access scenarios, which can fulfill the requirements of the SOCIAL platform and respective applications and services.

3.2 XACML Parsing

XACML defines categories and attributes to distinguish the different entities involved in an access request. By default, there are four different categories: "access-subject" is the entity that is requesting the access, "resource" is what is going to be accessed, "action" is the type of access and "environment" is additional information regarding the access. Each one of these categories can have several attributes.

```
<Rule Effect="Permit" RuleId="2">
  <Description>{"Role":["ADMINISTRATIVE","CASE_MANAGER"]} RULES</Description>
  <Target/>
  <Condition>
    <Apply FunctionId="urn:oasis:names:tc:xacml:1.0:function:and">
      <Apply FunctionId="urn:oasis:names:tc:xacml:1.0:function:string-at-least-one-member-of">
        <Apply FunctionId="urn:oasis:names:tc:xacml:1.0:function:string-bag">
          <AttributeValue DataType="http://www.w3.org/2001/XMLSchema#string">ADMINISTRATIVE</AttributeValue>
          <AttributeValue DataType="http://www.w3.org/2001/XMLSchema#string">CASE_MANAGER</AttributeValue>
        </Apply>
        <AttributeDesignator AttributeId="urn:oasis:names:tc:xacml:1.0:subject:Role"
        Category="urn:oasis:names:tc:xacml:1.0:subject-category:access-subject"
        DataType="http://www.w3.org/2001/XMLSchema#string" MustBePresent="true"/>
      </Apply>
      <Apply FunctionId="urn:oasis:names:tc:xacml:1.0:function:string-at-least-one-member-of">
        <Apply FunctionId="urn:oasis:names:tc:xacml:1.0:function:string-bag">
          <AttributeValue DataType="http://www.w3.org/2001/XMLSchema#string">MENU_PATIENT</AttributeValue>
        </Apply>
        <AttributeDesignator AttributeId="urn:oasis:names:tc:xacml:1.0:resource:Other"
        Category="urn:oasis:names:tc:xacml:3.0:attribute-category:resource"
        DataType="http://www.w3.org/2001/XMLSchema#string" MustBePresent="true"/>
      </Apply>
    </Apply>
  </Condition>
</Rule>
```

Fig. 1. XACML structure example.

Considering the structure of XACML (see Fig. 1), its understanding can be difficult to most. However, managing XACML rules should not be dependent on the understanding of this language, and anyone with the proper authority should be able to create, read, update and delete policy rules. Developing a parser that can translate both from XACML to natural language and natural language to XACML can prove to be useful in maintaining this type of access control.

The rule (in natural language) might have the following structure:

> If a "<SUBJECT_ATTRIBUTE>" is "<SUBJECT_VALUE>",
> it can "<ACTION>" the "<ENVIRONMENT>"
> if the "<RESOURCE_ATTRIBUTE>" of
> the resource is "<RESOURCE_VALUE>"

An example of the application of this rule could be:

> If a "role" is "case manager",
> it can "read" the "profile"
> if the "role" of the resource" is "older adult"

Therefore, being established a structure for the definition of the rules, it is possible to parse the rules to their respective XACML format. Moreover, the inverse (i.e. the translation from XACML to natural language) is also possible. For instance, from the excerpt of XACML code presented in the Table 1 is possible to retrieve the following rule:

If a "role" is "administrative" or is "case manager",
it can "create" and "update" and "read" the
"address", the "name" and the "telecom"
if the "role" of the resource is "patient".
Advices: "SAME_ORGANIZATION"

Table 1. Category, AttributeId and respective value of XACML rule.

Category	AttributeId	Value
(…):xacml:1.0:subject-category: access-subject	(…):xacml:1.0:subject:role	"administrative" or "case manager"
(…):xacml:3.0:attribute-category:action	(…):xacml:1.0:action: action-id	"create" or "update" or "read"
(…):xacml:3.0:attribute-category:environment	(…):xacml:1.0: environment:type-data	"address" or "name" or "telecom"
(…):xacml:3.0:attribute-category:resource	(…):xacml:1.0:resource: role	"patient"

The access requests themselves can also be sent in a natural language structure, from the user's perspective and then be processed accordingly by the parser. For example:

Can someone with a "<SUBJECT_ATTRIBUTE>" as
"<SUBJECT_VALUE>",
perform the action "<ACTION>" over the
"<ENVIRONMENT>"
if the "<RESOURCE_ATTRIBUTE>"
of the resource is "<RESOURCE_VALUE>"?

The parser should then process this phrase and transform it into a JSON Object (see Fig. 2). Considering, the possibility to add the "AdviceExpressions" element using the parser, additional restrictions can be defined. For example, for scenarios where the access decision is "Permit", but only if a certain attribute value is the same between the

Can someone with a "Role" as "ADMINISTRATIVE" and someone with a "organization" as "IPN", perform the action "create" over the "DemographicData" if the "Role" of the resource is "Patient" and if the "organization" of the resource is "IPN"?

```
{
    "RESOURCES": {
        "Role": "Patient",
        "organization": "IPN"
    },
    "ACTION": {"action-id": "create"},
    "DATA": {"type-data": "DemographicData"},
    "SUBJECT": {
        "Role": "ADMINISTRATIVE",
        "organization": "IPN"
    }
}
```

Fig. 2. Parsing from natural language to XACML.

subject and resource, we can add, after an access statement, for instance to associated to a code, that is understood by the access control component:

> However, the "subject" "<SUBJECT_ATTRIBUTE>"
> must be the same as the "resource"
> "<RESOURCE_ATTRIBUTE>"

With the addition of "AdviceExpressions" several possibilities can be considered, such as restrictions to certain environment fields or accesses during specific periods of time. Therefore, additional steps must be considered before enforcing a final decision.

4   Evaluation

After the implementation of the parser using the JAVA language, an evaluation was performed. For the evaluation it was considered a scenario where it is required the possibility of sending access requests that should be verified if a set of rules was followed or not, by using both the SOCIAL platform identifying attributes (e.g. the "role" of the users in the platform) and the FHIR resources attributes (e.g. the attributes of the Patient FHIR Resource such as "name" or "address").

First, three distinct rules were created, using the natural language parser:

- Rule 1 - If a "role" is "case manager", it can "update" the "address" if the "role" of the resource is "patient".
- Rule 2 - If a "role" is "care receiver", it can "read", "update" and "create" the "photo" if the "role" of the resource is "patient". However, the "subject" "organization" must be the same as the "resource" "organization".
- Rule 3 - If an "organization" is "City Council A", it can "read" the "name" and the "address" if the "organization" of the resource is "City Council B".

Next, access requests were sent and it was verified if the rules were processed correctly (also using the natural language parser):

- Question 1 - Can someone with a "role" as "case manager", perform the action "delete" over the "address" if the "role" of the resource is "patient"?
- Question 2 - Can someone with a "role" as "case manager", perform the action "update" over the "address" if the "role" of the resource is "patient"?
- Question 3 - Can someone with a "role" as "care receiver" and with an "organization" as "Retirement Home A", perform the action "update" over the "photo" if the "role" of the resource is "patient" and if the "organization" of the resource is "Retirement Home A"?
- Question 4 - Can someone with a "role" as "care receiver" and with an "organization" as "Retirement Home B", perform the action "update" over the "photo" if the "role" of the resource is "patient" and if the "organization" of the resource is "Retirement Home A"?
- Question 5 - Can someone with a "role" as "care receiver" and with an "organization" as "Retirement Home A" perform the action "update" over the "photo" if the "role" of the resource is "patient" and if the "organization" of the resource is "Retirement Home B"?
- Question 6 - Can someone with an "organization" as "City Council A", perform the action "read" over the "address" if the "organization" of the resource is "City Council B"?
- Question 7 - Can someone with an "organization" as "City Council A", perform the action "read" over the "telecom" if the "organization" of the resource is "City Council B"?

Question 1 and 2 are for testing rule 1, while question 3, 4 and 5 are for testing rule 2 and finally, question 6 and 7 are for testing rule 3.

The seven questions are represented in Table 2 (for the attributes of the access-subject and resource category) and Table 3 (for the attributes of the action and environment category and access decision result).

As we can see in Table 3, the first question decision is denied, since the "case manager" can only "update" the "address" of the "patient". The fourth and fifth question also got a denial to its access request because only "care receiver" from "Retirement Home A" can "update" the "photo" of the "patient" from "Retirement Home A". The seventh question got a denial answer because the "City Council A" subjects can only "read" the "name" and "address" from resources from "City Council B".

Table 2. Access request tests (access-subject and resource).

Question	Access-subject	Resource
1	role:*case manager*	role:*patient*
2	role:*case manager*	role: *patient*
3	role:*care receiver*, organization:*Retirement Home A*	role: *patient*, organization:*Retirement Home A*
4	role:*care receiver*, organization:*Retirement Home B*	role:*patient*, organization:*Retirement Home A*
5	role:*care receiver*, organization:*Retirement Home A*	role:*patient*, organization:*Retirement Home B*
6	organization:*City Council A*	organization:*City Council B*
7	organization:*City Council A*	organization:*City Council B*

Table 3. Access request tests (action, environment and decision).

Question	Action	Environment	Decision
1	Delete	Address	Deny
2	Update	Address	Permit
3	Update	Photo	Permit
4	Update	Photo	Deny
5	Update	Photo	Deny
6	Read	Address	Permit
7	Read	Telecom	Deny

5   Conclusion

Using a natural language parser not only can help someone with low understanding of the XACML standard to be able to create and verify XACML rules. The study reported in the present paper shows that the parser do not disrupt the XACML structure and can be applied to FHIR resources (as proven by the decision answers obtained).

The parser can specify different attributes for any category and adapt them to create detailed access rules, which can be suitable to almost any kind of scenario. Moreover, by using the "AdviceExpressions" element, it can even add another layer of complexity over a certain rule and make the access over a certain resource even more restrictive.

Using XACML in a system's access control, that handles several distinct data, allows for a more secure system that can satisfy any requirement imposed by any institution or organization, regarding their data access. Therefore, developing a parser that can translate XACML to natural language, allowing anyone to directly manage their respective policies, might help an effective use of XACML.

Acknowledgements. This work was supported by Sistema de Incentivos à Investigação e Desenvolvimento Tecnológico (SI I&DT) of the Programa Portugal 2020, through Programa Operacional Competitividade e Internacionalização and/or Programa Operacional do Centro do

FEDER - Fundo Europeu de Desenvolvimento Regional, under Social Cooperation for Integrated Assisted Living (SOCIAL), project number 017861.

References

1. Baines, S., Hill, P., Garrety, K.: What happens when digital information systems are brought into health and social care? Comparing approaches to social policy in England and Australia. Soc. Policy Soc. **13**(4), 569–578 (2014)
2. Sousa, M., Arieira, L., Queirós, A., Martins, A.I., Rocha, N.P., Augusto, F., Duarte, F., Neves, T., Damasceno, A.: SOCIAL platform. In: Advances in Intelligent Systems and Computing, vol. 746, pp. 1162–1168 (2018)
3. Defining an Open Platform. Apperta Foundation (2017)
4. Mandl, K.D., Mandel, J.C., Murphy, S.N., Bernstam, E.V., Ramoni, R.L., Kreda, D.A., et al.: The SMART platform: early experience enabling substitutable applications for electronic health records. J. Am. Med. Inf. Assoc. **19**(4), 597–603 (2012)
5. Chaballout, B.H., Shaw, R.J., Reuter-Rice, K.: The SMART healthcare solution. Adv. Precis. Med. **2**(1), 1–3 (2017)
6. Rigby, M.: Integrating health and social care informatics to enable holistic health care. Stud. Health Technol. Inform. **177**, 41–51 (2012)
7. Kristal, L.: Senior Care Connect Inc. https://www.seniorcareconnect.co/. Accessed 22 Oct 2018
8. Metatheke Software: Ankira. https://ankira.pt/en/platform/. Accessed 22 Oct 2018
9. HL7, Fast Healthcare Interoperability Resources. https://www.hl7.org/fhir/. Accessed 22 Oct 2018
10. Hoeksma, J.: System C commits to 'full FHIR support' to drive interoperability, September 2018. https://www.digitalhealth.net/2018/09/system-c-commits-to-full-fhir-support-to-drive-interoperability/. Accessed 22 Oct 2018
11. Chu, D.: SocialCare. https://www.socialcare.com/. Accessed 22 Oct 2018
12. OASIS, XACML, January 2013. http://docs.oasis-open.org/xacml/3.0/xacml-3.0-core-spec-os-en.html. Accessed 22 Oct 2018
13. OASIS, Security Assertion Markup Language. https://wiki.oasis-open.org/security/FrontPage. Accessed 22 Oct 2018
14. Vora, J. et al.: Ensuring privacy and security in E-health records. In: International Conference on Computer, Information and Telecommunication Systems, pp. 1–5. IEEE, Colmar (2018)
15. Ray, I. et al.: Applying attribute based access control for privacy preserving health data disclosure. In: IEEE-EMBS International Conference on Biomedical and Health Informatics, pp. 1–4. IEEE, Las Vegas (2016)
16. Atiq, A.M., Alsulaiman, L.A.: Using XACML to enhance compliance with privacy regulations in health sector. In: 2016 World Symposium on Computer Applications & Research (WSCAR), pp. 53–58. IEEE, Cairo (2016)
17. Turner, R.C.: Proposed model for natural language ABAC authoring. In: Proceedings of the 2nd ACM Workshop on Attribute-Based Access Control - ABAC 2017, pp. 61–72. ACM, Scottsdale (2017)

18. Stepien, B. et al.: A non-technical XACML target editor for dynamic access control systems. In: 2014 International Conference on Collaboration Technologies and Systems, pp. 150–157. IEEE, Minneapolis (2014)
19. Santana, S., Dias, A., Souza, E., Rocha, N.: The domiciliary support service in Portugal and the change of paradigm in care provision. Int. J. Integr. Care **7**(1) (2007)

A Framework to Develop Gamified Health Applications to Aid the Treatment of Non-communicable Diseases

João Paulo Barros Borges, Eduardo Simões de Albuquerque[✉],
and Ana Paula Laboissière Ambrósio

Instituto de Informática, Universidade Federal de Goiás, Goiânia, Brazil
eduardo@inf.ufg.br

Abstract. Non-communicable chronic diseases represent one of the most serious health problems in the world. The development of *Gamified* applications for monitoring health treatments is one of the strategies that has been used to contribute to this problem. This work proposes a framework with *Gamification* elements to facilitate the development of applications focused on the treatment of chronic non-communicable diseases. Scenarios were used to evaluate the framework. The framework is targeted to developers providing the main structures used in the development of gamified applications aimed at non-communicable diseases patients. Tests show that the proposed framework is easily extensible and reduces development effort.

Keywords: Gamification · Non-communicable diseases

1 Introduction

Non-communicable chronic diseases (NCCDs) are lifelong diseases that demand prolonged treatment, examples are diabetes, obesity, cancer and cardiovascular diseases. They are the main cause of death worldwide, and create many limitations that impact on the quality of life of the affected individuals, generating numerous negative social and economic effects [22].

Research has shown that adherence to treatment of NCCDs is often difficult, and that there are many cases in which patients ignore (in whole or in part) medical recommendations, even when they are aware of the need for a routine change [3,18]. The researches conclude that information about diseases, medicines and a good health professional alone is not enough to guide the patient: adherence to treatment mainly involves motivational aspects related to economic needs, self perception of results and individual commitment.

Characteristics of motivation and behavioral changes are present in the electronic games, for example, when people dedicate many hours of a day to activities only for personal interests, for fun, even without external returns. Many publications, such as [14,15,20,21], began to look for game characteristics in which

© Springer Nature Switzerland AG 2019
Á. Rocha et al. (Eds.): WorldCIST'19 2019, AISC 932, pp. 105–114, 2019.
https://doi.org/10.1007/978-3-030-16187-3_11

they could be applied in other areas, with the aim of making activities typically monotonous in more fun, less tedious, or even participatory activities, which marked the beginning of gamification research.

Gamification has been defined as the use of game design elements in other contexts [2], such as to increase engagement and willingness of people to perform activities. In the last years many researches have been done involving the subject in the most diverse areas. The health and fitness applications industry, for example, has been a major player in the use of gamification, which is evident from the large number of applications containing game components today.

Studies have shown that many of these applications are successful, and that fitness and health applications represent a very promising and expanding market for disseminating behavioral change interventions in health. It has been observed, however, that many of these applications use a very small subset of gamification components, usually involving levels and a score for performed activities, which in itself does not generate a motivating environment and Challenge [7,12]. Works such as Lister et al. [7,11,12], also showed that there is a gap when it comes to developing applications that incorporate gamification elements. In other words, there is a lack of tools to aid the development of gamification applications, especially in the context of chronic diseases.

Considering the above aspects, this work proposes a framework with gamification elements to aid the development of applications aimed at the treatment of chronic diseases.

2   Related Work

Some works propose software structures aimed at the development of applications focused on NCCDs. *Octopus: A Gamification model to aid in ubiquitous care of chronic diseases* [12] proposes a software structure based on smartphones, which exploits the environment and recommends treatment of the patient. For example, nearby restaurants specializing in sugar-free foods are recommended, as well as places with access to sensors such as a glucose meter and scale. The model addresses gamification by generating a score for the patient. This score depends on acquiring or not acquiring the recommended feature, and it takes into account the patient record of using the specified feature. If a certain level is reached, the score can generate awards for the patient.

The work *Model for Ubiquitous Care of Noncommunicable Diseases* [19] is a second model for caring of NCCDs. This model focuses on supporting self-management, and communication among patients, community members, and health organizations. Support for self-management is carried out in the model through the generation of incentive reminders for healthy foods, physical activities, and for tasks that must be carried out continuously.

A modular framework for ambient health monitoring [1] is not specifically focused on chronic diseases, but presents an interesting proposal in this context: a framework integrating several medical and physical sensors to create a care and monitoring environment. The authors discuss the difficulty in integrating various sensor technologies however, the use of gamification is not discussed.

2.1   Gamification and Chronic Diseases

Several works have related gamification and non-communicable chronic diseases. *Neubeck et al.* [9] studies mobile applications to aid cardiovascular diseases treatment, and discusses the components of games capable of increasing motivation for medical treatment. The work concludes that the most important components should be the personalization, rewards and presentation of content in a simple and clear way.

Schoeppe et al. [16] present a systematic review on the efficacy of interventions that use applications to improve diet, physical activity, and sedentary behavior in children and adults. The results show that application-based interventions are indeed effective. Among the characteristics of more effective interventions were observed, among others, self-monitoring, performance feedback, rewards and rewards.

Goh et al. [5] and *Geelan et al.* [4] discuss the use of virtual characters as an incentive to adhere to health treatment activities. In general, the experiments showed that, in fact, the creation of virtual characters that react to activities carried out in the real world, promotes changes in users' behavior and encourages them to perform these activities.

From involving gamification and chronic diseases, it is possible to highlight some important game components in this context:

- Achievements: Physical, virtual, or status rewards for accomplishment.
- Collections: Set of items, skills, wealth, accumulated by the player.
- Points: Actions in the game environment reward player participation. With enough points, the player can, for example, improve their skills, equipment or get resources.
- Virtual Goods: Virtual resources, such as items, equipment, talents, skills, which may or may not be exchanged for real resources.
- Environment responses: Reaction in the game environment to the health care actions performed by the player, contributing to the patient's self-monitoring.
- Characters: Take roles as tutors, demonstrators or even user representations. Works have pointed out that the activities of care and interaction with the characters, generate a strong connection and involvement with the players.

2.2   Virtual Characters and Health

In general, gamification elements aim to increase the involvement between activities and the user. In this way, when it comes to involvement, one of the main elements is the virtual characters, due to communication and proximity relationship that is established.

In psychology, it is well known that the relationship between caring for something, real or virtual (plant, pet, child, for example) and motivation to live and perform activities. In general, studies conclude that the perception of health, success and development of a cared for entity is a great source of motivation and contributes positively to physical and psychological health and overall well-being of caregivers [17].

In recent years studies have linked these two concepts (virtual characters and care) as a way to motivate, proposing applications based on the care of virtual characters, aimed at the incentive and education in several areas. *Pollak et al.* [13] for example, conducted an experiment on seventh and eighth grade children, in which a virtual pet reacted positively to images of the child's healthy breakfast and negatively to an unbalanced or non-existent breakfast. The study found that children who participated in the game ate a healthy breakfast 52% of the time, while those who did not play approximately only 20% of the time. A strong sense of attachment to the virtual pet has also been identified. Other works have applied this strategy in several areas, such as encouraging exercise in children [6], incentive to carry out day-to-day activities in the elderly [8, 10], and also incentive to study in adolescents [23]. In general, the experiments conducted in studies on care and, also, creation of virtual characters are conclusive. They have shown that, in fact, the creation of a virtual character, which reacts to activities carried out in the real world, promotes changes in people's behavior and encourages them to carry out these activities.

3   Methodology

The use of virtual characters as an incentive to perform activities important in the treatment of chronic diseases is known in the literature and has been explored. In the development of the proposed framework, this relationship is used to generate an experience where the patient cares for a character through treatment activities as he takes care of himself. When health care activities are performed, the character changes his physical and emotional state.

This type of experience is not based on a scoring scheme and simple rewards. It allows for deep involvement with the character, feelings of fellowship and cooperation, acquisition of resources through activities that take place in the real world, and feedback to these actions. With these characteristics, we seek to create an exciting, motivating environment that promotes behavioral changes, and that allows a greater involvement with health treatment activities.

For the definition of a framework able to address the identified problems character-based modeling, a flexible architecture based on smartphone was designed (Fig. 1). The actors involved are highlighted in the first level, at an intermediate level the applications, the framework just below, and in the lower level the infrastructure consisting of services, such as the resources present in the smartphone, a data persistence service, and other integrated medical sensors.

The Gamification module is responsible for the game elements, and is intended to create involvement, concentration and motivation in the patient. It promotes mechanisms for creating, interacting and managing characters, their physical and emotional states, score and resource management, and also rewards.

The Services module is responsible for the acquisition and storage of the medical and control data obtained by the framework. It also contains features for generating history from stored medical data, and sending sound and visual notifications to the patient's smartphone. This module interacts with the storage service and sensors built into the smartphone.

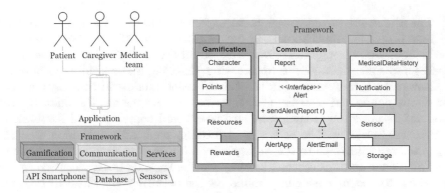

Fig. 1. Architecture of the proposed framework

The Communication module provides functionality for creating and sending reports to caregivers and the patient's medical staff. It is also responsible for generating alerts when medical data are identified outside normal ranges.

The characters are implemented in the Gamification module and are virtual entities that need patient care. They can be configured in the application in a variety of ways (eg. a baby, a city, a car), as caring activities are generic and extensible (eg. to feed, medicate or clean).

The performance of health treatment activities interferes with the physical and emotional state of the character, so that if the patient is performing the treatment activities (taking measurements through the sensors, taking medication at the correct times, performing physical exercises), the character becomes healthy. Likewise, if the patient is not performing the activities or is not at healthy levels, the character worsens his health.

The character's physical and emotional states are managed through an internal control of current states. The framework provides for the applications a state list that can be extended, containing, for example, the states Hungry, Sick and Dirty.

Food and hygiene are abstractions of the physical needs of the character. In the case of a city, for example, the food could be the set of raw materials needed for its maintenance. Hunger and the need for hygiene increase with time until a maximum level. When an intermediate level is reached, the patient is notified to perform the corresponding activity, and the related state (hunger or dirty) is added to the control of character states. This intermediate level, and also the speed with which the character needs to feed and clean are parameters passed to the framework by the application.

Medication flow is performed similarly to feeding. Drugs are registered in the framework, along with a schedule. Sound and visual notifications are issued as soon as a the patient medication time is reached initiating a flow that decreases the health value of the character until medication is taken. Upon reaching an intermediate value (configured by the application) the character receives the status *Sick*.

The framework provides some features for the character, food, medicines and hygiene products. Applications can create other types of resources, interesting to their context, through the resource classes of the Gamification module. These classes implement mechanisms for creating, deleting, and obtaining resources.

Rewards are implemented in the Gamification module in two ways: character status points and rewards. Character status rewards are the main rewards provided because they function as an instant and contextualized feedback for health care activities. The activities of feeding, medicating, cleaning, and also input sensors integrated to the smartphone influence the state of the character.

The second rewards scheme, scoring, can be used by the application in several ways, for example assigning points for the performance of health treatment activities. In other words, it is possible to inform the framework, for example by the detection sensor of hiking, so that with each step given in the real world, the score is increased. This example could be used to encourage the patient to practice regular exercises, very important in controlling health levels and fundamental in the treatment of NCCDs.

From a caregiver's point of view, points scored may not be such important rewards compared to gains in the welfare and health of the dependent and the perception of favorable outcomes. Even though this element of games is part of the framework as it contributes to the development of more complex game mechanics such as the marketing of features and improvements to the character.

The Framework supports the acquisition of medical data from various sensors, such as heart rate, weight, height, body mass index, respiratory rate, temperature, blood pressure, oxygen saturation, and glucose level. Support for these sensors is implemented in the Services module, through a very flexible architecture, involving interfaces and management classes.

4   Results and Discussion

In order to evaluate the proposed framework, the scenario validation strategy is used. This type of validation has been used by the scientific community to evaluate software tools for chronic diseases [12].

The first scenario describes the development of an application aimed at monitoring hypertension in adults. In this application the user must take care of a garden containing some plants, so that: *(i)* Plants need fertilizer and water, purchased with virtual money; *(ii)* new "medications" can be registered in the application, and their use at the correct times contributes to the overall health of the plants, generating money. When the system does not detect the correct medication, the health of the plant worsens; *(iii)* when a blood pressure sensor is used, the data obtained are stored allowing further inquiry, and the wealth increases. If the data obtained are considered normal, the health of the plants improves, otherwise the health of the plants worsens and an alert is sent to the caregiver.

The garden can be created in the framework through the Person class of the Gamification module. The need for fertilization and water can be modeled as feeding the character using the features of this class, informing the expected time until the character becomes hungry.

Fertilizer and water can be implemented with the help of the framework through the resource classes in the textit Gamification module (Food, Drugs and Hygiene Products). They can be considered Food and created informing this class the name of the resource, the initial amount available for the character, the nutritional value and the price points for the purchase in the store.

As with food creation, user's medications can be modeled through the Medications class. The main difference lies in the fact that the foods are defined by the developer of the application, while the medicines are created at a different time, through the data reported directly by the user. Figure 2 shows the character configuration and the creation of food and water in the medicament by the application.

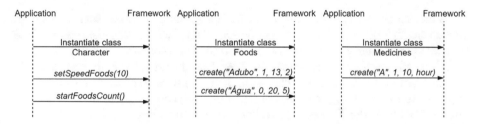

Fig. 2. Character configuration and resource creation.

Applications may require extra features for the character in relation to feeding, medication and hygiene. In this case, it is possible to extend the Character class and implement the new characteristics. Similarly, applications can extend the Resources class, defining a new resource type with additional attributes, and retaining the superclass's resource attributes: name, initial amount available to the character or price.

The money described in the scenario can be modeled on the score of the Gamifications class. In the scenario the punctuation does not observe a sensor or an event generator, being increased, only, by the medication and the use of sensors. Walk detection sensors, exercises, or even other observables could be passed to the Points class for the acquisition of coins in a contextualized way in relation to treatment activities.

The blood pressure sensor is controlled by the framework through the Blood-PressureManager class of the Services module. This class receives an implementation of the interface pressure interface, which treats the specifics of the sensor and functions as a driver (Fig. 3).

Figure 3 shows that the data is obtained by the application using the *getData* *()* method of the *Manager*. The other sensors supported by the *framework* follow the same structure of configuration and obtaining the data.

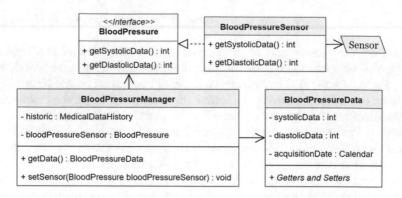

Fig. 3. Configuring blood pressure sensor in the framework.

When medical data is obtained through the blood pressure sensor, they are automatically added to listings in the MedicalData class of the *Services* module, facilitatin the generation of reports to monitor the evolution of the patient.

Alerts are configured through the AlertEmail and AlertApp classes of the *communication* module. These classes make it possible to send e-mails and messages to an external application, configuring two types of external services: the Simple Mail Transfer Protocol (SMTP) server for sending e-mails, and a MOM (Message Oriented Middleware) server to send messages, so that external applications can get these messages from the server asynchronously. The Alert interface of the same module can be implemented by the application to send messages in different ways.

In general, in this scenario, the application receives from the framework several aspects of the character, such as feeding, medication and health levels, a scoring scheme, and blood pressure sensor support. Several aspects could have been used in a more elaborate way, for example, scoring based on observation of chronic disease treatment activities, extended characters with new characteristics, new types of resources, and other medical sensors.

The characteristics used in the scenario, as well as those not used, can demonstrate the flexibility of the framework, in order to facilitate the development of applications aimed at chronic diseases. They also demonstrate that several scenarios of NCCDs fit into the proposed functionalities, such as:

- An application to combat childhood obesity: when the child walks, she receives points, which can be used to buy rations and props for a virtual pet. A walk makes the pet happy, and the sedentarism makes him sad.
- An application to help controlling diabetes in the elderly: the acquisition of data from the glycosimeter, triggers motivational and explanatory videos in relation to the disease are displayed.

5   Conclusion

This work proposed a framework for the development of applications aimed at the treatment of chronic diseases using gamification elements.

Scenario based tests showed that the framework is generic and supports different types of chronic diseases, as it presents a versatile structure focused on the care functions, and not on the specificities of a chronic disease. The description of the scenarios also showed that framework facilitates in many ways the development of gamified applications, focused on the treatment of NCCDs.

This paper presents contributions in relation to software applications and textures that already exist in the context of chronic noncommunicable diseases. The first contribution is in the use of the relationship between caring and motivation to live, focused on the health treatment of chronic diseases.

The use of this relation is not in itself innovative, since it was explored in other works that used virtual characters in the incentive to health activities. This important relationship, however, was not identified in related works that proposed *frameworks* for the development of chronic disease applications, in which the development of characters and components involved in their care activities were not explored. The framework proposed in this paper deals directly with these components, as well as several others highlighted in the literature, in order to simplify the process of development of gamified applications.

References

1. Burmeister, D., Schrader, A., Carlson, D.: A modular framework for ambient health monitoring. In: 2013 7th International Conference on Pervasive Computing Technologies for Healthcare and Workshops, pp. 401–404, May 2013
2. Deterding, S., Dixon, D., Khaled, R., Nacke, L.: From game design elements to gamefulness: defining gamification. In: Proceedings of the 15th International Academic MindTrek Conference: Envisioning Future Media Environments, MindTrek 2011, pp. 9–15. ACM, New York (2011). http://doi.acm.org/10.1145/2181037.2181040
3. Gama, A.C.C., Bicalho, V.S., Valentim, A.F., Bassi, I.B., Teixeira, L.C., Assunção, A.Á.: Adherence to voice therapy guidelines after discharge from vocal treatment in teachers: a prospective study. Scielo **14**(4), 714–720 (2013)
4. Geelan, B., Zulkifly, A., Smith, A., Cauchi-Saunders, A., de Salas, K., Lewis, I.: Augmented exergaming: increasing exercise duration in novices. In: Proceedings of the 28th Australian Conference on Computer-Human Interaction, OzCHI 2016, pp. 542–551. ACM, New York (2016). http://doi.acm.org/10.1145/3010915.3010940
5. Goh, D.C., Tan, A.C., Lee, J.S.: Gamification of heel raise plantarflexion physiotherapy. In: Proceedings of the 2nd International Workshop on Multimedia for Personal Health and Health Care, MMHealth 2017, pp. 35–43. ACM, New York (2017). http://doi.acm.org/10.1145/3132635.3132638
6. Johnsen, K., Ahn, S.J., Moore, J., Brown, S., Robertson, T.P., Marable, A., Basu, A.: Mixed reality virtual pets to reduce childhood obesity. IEEE Trans. Vis. Comput. Graph. **20**(4), 523–530 (2014)

7. Lister, C., West, J.H., Cannon, B., Sax, T., Brodegard, D.: Just a Fad? Gamification in health and fitness apps. JMIR Serious Games **2**(2), 9 (2014). http://games.jmir.org/2014/2/e9/
8. Nakashima, T., Fukutome, G., Ishii, N.: Healing effects of pet robots at an elderly-care facility. In: 2010 IEEE/ACIS 9th International Conference on Computer and Information Science, pp. 407–412, August 2010
9. Neubeck, L., Lowres, N., Benjamin, E.J., Freedman, S.B., Coorey, G., Redfern, J.: The mobile revolution – using smartphone apps to prevent cardiovascular disease (Report). Nature Rev. Cardiol. **12**(6), 350 (2015)
10. Obo, T., Kasuya, C., Sun, S., Kubota, N.: Human-robot interaction based on cognitive bias to increase motivation for daily exercise. In: 2017 IEEE International Conference on Systems, Man, and Cybernetics (SMC), pp. 2945–2950, October 2017
11. Oliveira, R., Moura, A., Barros, M., Cavalcante, A., Junior, F.: Gamificação e Crowdsourcing no Combate Sustentável ao Aedes aegypti. J. Health Inform., 390 (2016). http://www.sbis.org.br/biblioteca_virtual/cbis/Anais_CBIS_2016_Artigos_Completos.pdf
12. Paim, C.A., Victoria Barbosa, J.L.: Octopus: a gamification model to aid in ubiquitous care of chronic diseases. IEEE Lat. Am. Trans. **14**(4), 1948–1958 (2016). http://ieeexplore.ieee.org/document/7483539/
13. Pollak, J., Gay, G., Byrne, S., Wagner, E., Retelny, D., Humphreys, L.: It's time to eat! using mobile games to promote healthy eating. IEEE Pervasive Comput. **9**(3), 21–27 (2010)
14. Prakash, E.C., Rao, M.: Transforming Learning and IT Management through Gamification, chap. Introduction to Gamification, pp. 35–46. Springer, Cham (2015). http://dx.doi.org/10.1007/978-3-319-18699-3_3
15. Rocha, R., Reis, L.P., Rego, P.A., Moreira, P.M.: Serious games for cognitive rehabilitation: forms of interaction and social dimension. In: 2015 10th Iberian Conference on Information Systems and Technologies (CISTI), pp. 1–6, June 2015
16. Schoeppe, S., Alley, S., Bray, N., et al.: Efficacy of interventions that use apps to improve diet, physical activity and sedentary behaviour: a systematic review. Int. J. Behav. Nutr. Phys. Act. **13** (2016). http://search.proquest.com/docview/1855775131/
17. Smith, B.: The 'pet effect' health related aspects of companion. Aust. Fam. Physician **41**(6), 439–442 (2012). http://www.racgp.org.au/afp/2012/june/the-pet-effect/
18. Tavares, N.U.L., Bertoldi, A.D., Thume, E., Facchini, L.A., de Franca, G.V.A., Mengue, S.S.: Factors associated with low adherence to medication in older adults. Revista de Saúde Pública **47**(6), 1092–1101 (2013)
19. Vianna, H.D., Barbosa, J.L.V.: A model for ubiquitous care of noncommunicable diseases. IEEE J. Biomed. Health Inform. **18**(5), 1597–1606 (2014). http://ieeexplore.ieee.org/document/6676783/
20. Vianna, Y., Vianna, M., Medina, B., Tanaka, S.: Gamification, Inc: Como reinventar empresas a partir de jogos. MJV Press (2013)
21. Werbach, K., Hunter, D.: For the Win: How Game Thinking Can Revolutionize Your Business. Wharton Digital Press, Philadelphia (2012)
22. World Health Organization: Obesity and overweight. WHO Fact sheet (2016)
23. Wu, M., Liao, C.C.Y., Chen, Z.H., Chan, T.W.: Designing a competitive game for promoting students' effort-making behavior by virtual pets. In: 2010 Third IEEE International Conference on Digital Game and Intelligent Toy Enhanced Learning, pp. 234–236, April 2010

Impact of Parameter Tuning on Machine Learning Based Breast Cancer Classification

Ali Idri[1(✉)], Mohamed Hosni[1], Ibtissam Abnane[1], Juan M. Carrillo de Gea[2], and Jose L. Fernández Alemán[2]

[1] Software Project Management Research Team ENSIAS,
Mohammed V University, Rabat, Morocco
ali.idri@um5.ac.ma

[2] Software Engineering Research Group, University of Murcia, Murcia, Spain

Abstract. Breast cancer is one of the major causes of death among women. Different decision support systems were proposed to assist oncologists in order to accurately diagnose their patients. These decision support systems mainly used Machine Learning (ML) techniques to classify the diagnosis into malign or benign tumor. In this paper, we evaluate and analyze the accuracy of the parameters tuning on the accuracy of three well-known ML techniques: Support Vector Machines, Multi-Layer Perception and Decision trees. We investigate three parameters tuning techniques: Grid Search (GS), Particle Swarm Optimization (PSO) and the default parameters of the Weka Tool over four datasets obtained from the Machine Learning repository. The overall results suggest that using GS and PSO lead to build more accurate classifiers, and therefore can help oncologists to provide more accurate diagnosis.

Keywords: Breast Cancer · Machine Learning · Classification

1 Introduction

Breast cancer (BC) is a serious threat to women health all over the world and is growing annually in an unpredictable rate [1]. It is the most common type of cancer with around 25% of all type of cancer [2], is affecting about 1 of 8 women over the world, and is considered as the leading cause of cancer death among women [1]. However, even though this disease is very critical in women across the world, its reasons are not fully discovered yet [3].

Detecting and accurately diagnosing this disease at an early stage remain an important factor for women long-term survival [4]. Different techniques were used to detect BC such as mammography, ultrasound, thermography, biopsy and fine needle aspiration cytology [5]. Mammography is the most used and reliable by the physicians, and it may be followed by a further biopsy before the final decision is taken by the physician. We distinguish between two tumors benign or

© Springer Nature Switzerland AG 2019
Á. Rocha et al. (Eds.): WorldCIST'19 2019, AISC 932, pp. 115–125, 2019.
https://doi.org/10.1007/978-3-030-16187-3_12

malignant. However, the rate of detecting accurately the tumor is still limited to 60–70% which is considered as a low accuracy rate [4]; this results on consuming time and resources of both physicians and patients [6], since this rate does not allow the physician to pronounce the final diagnosis. Thus, a physician needs further biopsy which would results in a waste time and can cause mental discomfort for the patient especially if the biopsy reveals that the tumor is benign [6]. Within this context, researchers have proposed different techniques to classify the tumor using the information provided by the mammography aiming to assist the physician to quickly and accurately diagnosis the disease. Among these techniques, Machine Learning (ML) techniques were successfully used for BC classification and diagnosis [3]. These techniques classify the tumor as benign or malignant using the information extracted from the mammography. Even though ML techniques were applied massively in BC diagnosis, it can be noticed that their parameters have been often set with the same values over all the datasets used, or without clearly providing the rationale behind the choice made. Indeed, parameters settings have an important impact on the predictive performance of the ML techniques since the accuracy of ML techniques mainly depends on the context used. However, selecting the optimal values of the learning parameters is still challenging and several techniques have been proposed in the literature to address this issue, such as Grid Search (GS) [7], Particle Swarm Optimization (PSO) [8] and Genetic Algorithm (GA) [9]. These techniques have been widely investigated in many fields such as software engineering and pattern recognition to select the optimal parameter values of a given prediction technique in order to generate the best prediction performance [10–12].

Thus, this paper aims to apply and evaluate the accuracy of three well-known ML techniques: SVM, MLP and DT by optimizing their parameters using two techniques: GS and PSO. Thereafter, the impacts of GS and PSO on the accuracy of the three ML techniques were compared to the impact of the default parameters strategy of Weka Tool (Uniform Configuration: UC-Weka) over four datasets obtained from the Machine Learning repository. Hence, this paper aims at addressing two research questions (RQs):

- **(RQ1): Does the use of the GS and PSO optimization techniques instead of UC-Weka lead to build more accurate classifiers?**
- **(RQ2): Is there any ML techniques which distinctly outperform the other ML techniques?**

The main contributions of this paper are:

1. Tuning the parameters of three ML techniques according to the characteristics of each dataset.
2. Comparing the impacts of the use of two optimization techniques GS and PSO instead of UC-Weka on the accuracy of BC classification.

The rest of this paper is structed as follow: Sect. 2 presents an overview of the use of ML techniques in Breast Cancer diagnosis. Section 3 presents an overview of the three ML techniques used in this paper. Section 4 presents the experimental design of this paper. Section 5 presents and discusses the empirical results obtained. Section 6 presents the conclusions and future work.

2   Related Work

Table 1 presents an overview of related work investigating the ML techniques and their findings in BC classification. As it can be seen from Table 1, only one study [13] investigated the tuning of parameters of a given ML technique, while the other studies did not provide any information about the configuration of the ML techniques used. Similarly, by examining the pool of selected papers (403 papers) of Idri et al. in their systematic literature review of the use of datamining techniques in BC [3], we found that few papers investigated the configuration of the ML techniques, and that the majority of studies do not provide any information about the configuration used to build the proposed classifiers.

Table 1. Literature review of studies investigating ML techniques in BC field

Ref	ML techniques	Configurations	Findings
[13]	SVM	Grid search	Experimental results demonstrate the proposed SVM technique can achieve very high classification accuracy
[14]	DT, RBFN, Simple logistic	Default parameters of Weka	The experiments conducted claim that Simple Logistic generates more accurate results than the other techniques
[15]	CART	Not provided	The results show that a particular feature selection using CART has enhanced the classification accuracy of a particular dataset
[16]	SVM + Knn + J48 + NB	Not provided	The authors investigated different combination methods used (i.e. heterogeneous ensembles) on three BC datasets. The results show that the combination methods yield to better performance

RBFN: Radial Basis Function Network, CART: Classification And Regression Trees, Knn: K-Nearest Neighbors, NB: Naive Bayes

3   ML Techniques: An Overview

This section presents an overview of the three ML techniques used in this paper. These techniques were selected due to the high predictive performance shown in different fields [17].

3.1 Support Vector Machines

Support Vector Machines (SVMs) are a set of ML models characterized by the usage of kernels. The first formulation of SVM was proposed in 1992 by Vapnik, called maximal margin classifier [18,19]. The implementation of SVMs is based on statistical learning theory and structural risk minimization, which give these models a solid theoretical foundation. Moreover, SVM was proven to be effective in many fields such as handwritten digit recognition and text classification [20, 21]. When using SVM, there are three parameters to decide upon: the complexity parameter usually denoted with C, the round-off error denoted with Epsilon (ϵ) and the kernel with its parameters. This study used the RBFN kernel. Hence, another parameter to decide upon: gamma denoted by γ.

3.2 Multi-Layer Perceptron

The multilayer perceptron (MLP) neural networks are feed-forward neural networks used to solve both classification and regression problems [22,23]. MLPs have at least three layers [24]: input layer, one hidden layer and an output layer. The number of nodes in the input layer is specified by the number of features of the input patterns. As for the output layer, the number of nodes depends on the tackled problem [25]. Using the MLP requires setting 5 parameters: the number of hidden layers, the number of neurons of each hidden layer, the number of training epochs, and two parameters of the learning algorithm used (learning rate and momentum - i.e. term to the optimization). Generally, these parameters are often determined by means of a cross-validation method.

3.3 Decision Trees

Decision Trees (DTs) are the most frequent used supervised ML techniques, for both classification and regression problems [26]. A decision tree uses a tree-like model for solving the problem tackled. In fact, a decision tree is a tree where each node represents a feature from the data pattern, each branch is a decision rule, and each leaf represents an output which could be categorical (class) or continues (predicted) depending on the problem tackled. In this paper, the C4.5 technique was used [27]. Two main parameters to decide upon when using C4.5: The confidence factor used for pruning (Confidence) and the minimum number of instances per leaf (leaf).

4 Experimental Design

This section presents: (1) the two optimization techniques used in this paper, (2) the four datasets used in the evaluation of the proposed techniques, and (3) the performance measures used to assess the predictive performance of the proposed techniques.

4.1 Optimization Techniques

The predictive performance of ML techniques depends mainly on its parameters settings, and these parameters value differ from one dataset to another. In fact, this issue was addressed in regression problems [11,28]. The conclusion drawn suggest that optimizing the parameters settings of the ML models is necessary to achieve reliable results. Thus, the motivation of conducting this work. Choosing the optimal values of model parameters remains as an optimization problem since it results on a model with an optimal configuration which allows to build a classifier with high predictive performance. This paper uses, in addition to a uniform configuration; default parameters suggested by Weka tool (the rationale behind choosing Weka tool is due to the bar fact that it is the most used data mining tool among researchers), two optimization methods (GS and PSO) to optimize the parameters of the three ML techniques:

Grid Search: performs an exhaustive search by trying every parameter setting over a predefined range of parameters values (parameter sweep) and thereafter selecting the configuration that provides the **best** classifier. Generally, the evaluation process is performed through a cross-validation.

Particle Swarm Optimization: was developed in 1995 by Kennedy and Eberhart [8]. It purposes is to determine an optimal solution in a predefined search space for a given problem with respect to a fitness function. Every solution presents the parameters of a given classifier. Particle refers to a solution, while swarm denotes set of solution. The PSO assesses the fitness function at each particle and determines the new velocity for each particle. Then, PSO iteratively updates the particle locations, velocities and neighbors until a stopping criterion is reached [29,30]. Using PSO requires setting several parameters: number of iterations, swarm size, velocity (v_{min}, v_{max}), learning factors (c_1, c_2), and inertial weight (ω).

A preliminary round of executions was performed to determine the optimal parameters values of the three classifiers when using GS and PSO using all the parameters combinations of Table 2 for GS and the search space of parameters for PSO of Table 3. Afterward, the configuration of a given classifier that leads to generate optimal results (i.e. lower value of incorrectly classified through the ten-cross validation) with respect to GS and PSO is picked and used for the comparison in every dataset. As for the UC-Weka, this study used the default parameters values suggested by the Weka tool for each classifier over the selected datasets.

In order to shorten the names of the ML techniques, the following abbreviation rule was adopted:

Optimization technique used to set the parameters values-Technique Name
Where: **G** denotes the Grid Search method; **P** denotes the PSO method; **U** denotes the UC-Weka.

Example: PSVM refers to the SVM technique optimized by means of PSO.

Table 2. Search spaces of parameters values of each ML technique for Grid Search.

Techniques	Parameters with their search spaces
SVM	C = [5,200] with an increment of 5 Epsilon = [0.001, 0.1] with an increment of 0.001 Gamma = [0.001, 0.1] with an increment of 0.001
MLP	L = [0.01, 1] with an increment of 0.01 H = [1, 16] with an increment of 1 M = [0.1, 1] with an increment of 0.1 E = [100, 2000] with an increment of 100
DTs	Confidence = [0.01, 0.7] with an increment of 0.01 Leaf = [1, 20] with an increment of 1

Table 3. Search space for PSO algorithm.

Technique	Parameters	Min	Max
SVM	Complexity	5	200
	Epsilon	0.001	0.1
	Gamma	0.001	1
MLP	L	0.01	1
	M	0.1	1
	H	1	16
	E	100	2000
DTs	Confidence	0.01	0.5
	Leaf	1	20

4.2 Datasets Descriptions

To assess the predictive performance of the proposed ML techniques, four BC datasets were selected; these datasets were the most frequently adopted by researchers [3]. The four datasets were obtained from the online UCI repository. Table 4 summarizes these datasets. Note that the missing values presented in the BCD, Wisconsin and WPBC datasets were removed.

Table 4. Datasets description.

Datasets	#. Features	Missing data	Instances
Breast Cancer Data (BCD)	12	9	286
Wisconsin Diagnostic Breast Cancer (WDBC)	32	-	569
Wisconsin	11	16	699
Wisconsin Prognostic Breast Cancer (WPBC)	34	4	198

4.3   Performance Measures

Several performance criteria were proposed in the literature to evaluate the predictive capability of a given classifier [31]. In this paper three criteria were selected: accuracy, recall and precision.

The ten cross-validation method was adopted in all experiments raised in this paper.

5   Results and Discussions

This section reports and discusses the empirical results related to investigating the parameters tuning of the three ML techniques over the four selected datasets using ten cross-validation method.

To perform the empirical evaluations of this study, a software prototype based on Weka API was developed using Java programming language. The configuration adopted for the PSO algorithm was: (1) the swarm size was set to 30, the c_1 and c_2 were both set at 2; (2) the v_{min} and v_{max} were set at 1 and -1 respectively; (3) the inertia weight (ω) was fixed at 0.4; and (4) the number of iterations was set at 200. Note that these parameters values were set by performing many experiments based on the findings of several studies investigating the PSO algorithm [11, 29]. Note that three datasets: Breast cancer, Wisconsin and WPBC represent unbalanced datasets. To address this issue, the SMOTE [32] algorithm was used.

Table 5 shows the performance of the three ML techniques on the four datasets according to the three criteria (accuracy, recall and precision).

For SVM, we notice that:

1. In BCD, PSO gave the best results, followed by GS and then UC;
2. In WDBC, GS gave the best results, while PSO and UC behaved the same;
3. In Wisconsin, there was no difference between the three configurations and
4. In WPBC, PSO was by far the best, followed by GS and then UC.

For MLP, GS gave the best results, followed by PSO and then UC in all datasets.

For DT, PSO and GS gave the best results, followed by UC in BCD, while the three configurations gave similar results in the remaining datasets.

In order to explore the main reason behind the performance variation of the three techniques over the selected dataset, we provide in Table 6 the parameters settings for each technique according to each optimization technique over the four selected datasets.

The main finding is that the use of UC-Weka lead to different results over the four datasets; which underlines that the parameters setting of a given ML technique should be adjusted according to the characteristic of the dataset used. Moreover, we notice that given a dataset, GS and PSO choose different parameters for the techniques, which explain the difference in the accuracy results.

Besides, given a ML technique, GS and PSO choose different parameters for dif- ferent datasets. This confirms that the technique parameters must be adjusted differently for each dataset.

Table 5. Performance accuracy of the three ML techniques using different optimization techniques over the four datasets.

Tech.	BCD			WDBC			Wisconsin			WPBC		
	Acc	Recall	Prec	Acc	Recall	Prec	Acc	Recall	Prec	Acc	Recall	Prec
GSVM	80.74	80.7	81.5	97.51	97.5	97.5	97.71	97.7	97.7	88.85	88.9	89.8
USVM	78.76	78.8	78.9	96.92	96.9	96.9	97.89	97.9	97.9	74.66	74.7	74.8
PSVM	83.45	83.5	83.4	96.77	96.8	96.8	97.18	97.2	97.2	91.21	91.2	91.8
GMLP	83.2	83.2	83.3	97.21	97.2	97.3	98.06	98.1	98.1	91.89	91.9	91.9
UMLP	81.72	81.7	81.9	95.75	95.8	95.8	95.95	96	96	86.82	86.8	88.1
PMLP	82.46	82.5	82.4	96.92	96.9	97	96.83	96.8	96.9	90.87	90.9	91.3
GDT	82.96	83	83.4	96.33	96.3	96.4	94.02	94	94	85.47	85.5	85.5
UDT	80.98	81	80.98	96.04	96	96.1	92.97	93	93	84.12	84.1	84.1
PDT	82.22	82.2	83.1	96.33	96.3	96.4	94.02	94	94	85.47	85.5	85.5

In particular, we notice that the parameters chosen for the leaf in DT were the same for GS and PSO for three datasets (WDBC, Wisconsin, and WPBC) and led to the same accuracy values. However, in the Breast cancer data dataset, the leaf values were different and led to different results. Consequently, this confirms that the parameters variation is very impacting and leads to diverse results.

For MLP, we can explain the superiority of UMLP over PMLP by the number of iterations performed by PSO in order to obtain the final result: only 200 iterations compared to GS which tested more than 150.000 combinations. Hence, investigating the parameters setting of PSO is also required.

Table 6. UC-Weka, GS and PSO parameters for each ML technique in every dataset.

Tech.	Par.		BCD		WDBC		Wisconsin		WPBC	
		UC	GS	PSO	GS	PSO	GS	PSO	GS	PSO
DT	C	0.25	0.51	0.33	0.09	0.18	0.27	0.38	0.01	0.011
	Leaf	2	2	1	1	1	8	8	2	2
MPL	H	a	13	2	1	1	7	1	7	6
	L	0.3	0.58	0.6	0.01	0.54	0.01	0.82	0.08	0.99
	M	0.2	0.5	0.44	0.9	0.1	0.5	0.1	0.9	0.2
	E	500	1500	101	1100	100	1700	100	1500	100
SVM	C	1	20	5	80	6	15	63	175	157
	Epsilon	10 E-12	0.01	0.001	0.01	0.099	10E-03	0.045	0.1	0.09
	Gamma	0.01	0.1	0.46	0.001	4	0.1	0.2	0.1	0.71

In short, we found that:

- PSO and GS have a positive impact on the accuracy of the three ML techniques.
- Using optimization techniques rather than UC-Weka lead to build more accurate classification techniques.

Concerning the performance measures, we note that the three chosen criteria have the same trends of conclusion. Adopting different performance measures will help to draw a robust conclusion.

For the comparison of the ML techniques over the datasets, we notice that no ML technique was the **best/worst** over all datasets. PSVM was the best technique on the BCD while the worst technique was USVM. This is important since it shows the influence of the parameters tuning on the performance of ML techniques. Moreover, GSVM was the best technique on the WDBC dataset, while the worst technique was UMLP. In Wisconsin dataset, GMLP was the best technique and UDT was the worst. As for the WPBC dataset, GMLP was the best while USVM was the worst.

We notice that SVM is the most impacted by the parameters tuning, since using UC-Weka configuration significantly deteriorates its accuracy. More- over, we notice that the worst techniques in all datasets were those using UC-Weka configuration.

This is an important finding, since it asserts that, in order to obtain the best accuracy of a ML technique on a dataset, it is preferable to select its optimal configuration.

In short, the experiment performed in this paper states that there is *no best Classifier* in all context and building an accurate technique required to tune its parameters.

6   Conclusion and Future Work

This paper aimed to evaluate the impact of the parameters settings on the predictive performance of three classifiers: SVMs, MLP and DTs by using two parameters optimization techniques: GS and PSO. The optimized techniques were compared to the techniques build using the default parameters provided by Weka Tool. The three ML techniques were assessed over four datasets using three performance measures: accuracy, recall and precision. Toward this aim, two RQs have been discussed. Our findings are as follows:

(RQ1): Our findings assert that using GS and PSO led to build more accurate classifiers. We found that parameter tuning is suitable in order to obtain the best performance of an ML technique in a particular dataset.

(RQ2): We cannot claim that a particular ML technique is the best in all datasets. However, we found that the best results were obtained when GS or PSO were used. Moreover, the worst results were often obtained when parameter tuning was ignored (i.e. UC-Weka configuration used).

Ongoing work aims to investigate the impact of feature selection and parameters tuning on the accuracy of the classification techniques in BC.

Acknowledgments. This work was conducted within the research project MPHR-PPR1/09-2015-2018 . The authors would like to thank the Moroccan MESRSFC and CNRST for their support.

References

1. Solanki, K., Berwal, P., Dalal, S.: Analysis of application of data mining techniques in healthcare. Int. J. Comput. Appl. **148**(2) (2016)
2. Shajahaan, S.S., Shanthi, S., Manochitra, V.: Application of data mining techniques to model breast cancer data. Int. J. Emerg. Technol. Adv. Eng. **3**, 362–369 (2013)
3. Idri, A., Chlioui, I., El ouassif, B.: A Systematic map of data analytics in breast cancer. In: Australasian Computer Science Week (2018)
4. Luo, S.T., Cheng, B.W.: Diagnosing breast masses in digital mammography using feature selection and ensemble methods. J. Med. Syst. **36**, 569–577 (2012)
5. Chen, T.C., Hsu, T.C.: A GAs based approach for mining breast cancer pattern. Expert Syst. Appl. **30**, 674–681 (2006)
6. Kaushik, D., Kaur, K.: Application of Data Mining for high accuracy prediction of breast tissue biopsy results. In: 2016 3rd International Conference on Digital Information Processing Data Mining and Wireless Communication, DIPDMWC 2016 (2016)
7. Ma, X., Zhang, Y., Wang, Y.: Performance evaluation of kernel functions based on grid search for support vector regression. In: 2015 IEEE 7th International Conference on Cybernetics and Intelligent Systems (CIS) and IEEE Conference on Robotics, Automation and Mechatronics (RAM) (2015)
8. Kennedy, J., Eberhart, R.: Particle swarm optimization. In: Proceedings of the IEEE International Conference on Neural Networks (1995)
9. Fonseca, C.M., Fleming, P.J.: Genetic Algorithms for multiobjective optimization: formulation, discussion and generalization. In: Proceedings of the 5th International Conference on Genetic Algorithms (1993)
10. Das, H., Jena, A.K., Nayak, J., Naik, B., Behera, H.S.: A novel PSO based back propagation learning-MLP (PSO-BP-MLP) for classification. In: Proceedings of the International Conference on IEEE Symposium on Computational Intelligence and Data Mining (2014)
11. Hosni, M., Idri, A., Abran, A., Nassif, A.B.: On the value of parameter tuning in heterogeneous ensembles effort estimation. Soft Comput. **22**, 5977–6010 (2017)
12. Xiao, T., Ren, D., Lei, S., Zhang, J., Liu, X.: Based on grid-search and pso parameter optimization for support vector machine. In: 11th World Congress on Intelligent Control and Automation (WCICA) (2014)
13. Chen, H.L., Yang, B., Liu, J., Liu, D.Y.: A support vector machine classifier with rough set-based feature selection for breast cancer diagnosis. Expert Syst. Appl. **38**, 9014–9022 (2011)
14. Chaurasia, V., Pal, S.: Data mining techniques: to predict and resolve breast cancer survivability. Int. J. Comput. Sci. Mob. Comput. **3**, 10–22 (2014)
15. Lavanya, D., Rani, K.U.: Analysis of feature selection with classification: breast cancer datasets. Indian J. Comput. Sci. Eng. **2**, 756–763 (2011)
16. Makhtar, M., Yang, L., Neagu, D., Ridley, M.: Breast cancer diagnosis on three different datasets using multi-classifiers. Int. J. Comput. Inf. Technol. **32**, 2 (2012)

17. Wen, J., Li, S., Lin, Z., Hu, Y., Huang, C.: Systematic literature review of machine learning based software development effort estimation models. Inf. Softw. Technol. **54**, 41–49 (2012)

18. Vapnik, V.: Principles of risk minimization for learning theory. In: Advances in Neural Information Processing Systems (1992)

19. Vapnik, V., Bottou, L.: Local algorithms for pattern recognition and dependencies estimation. Neural Comput. **5**, 893–909 (1993)

20. Sadri, J., Suen, C.Y., Bui, T.D.: Application of support vector machines for recognition of handwritten Arabic/Persian digits. In: Second Conference on Machine Vision and Image Processing & Applications (MVIP 2003) (2003)

21. Tong, S., Koller, D.: Support vector machine active learning with applications to text classification. J. Mach. Learn. Res. **2**, 45–66 (2002)

22. Simon, H.: Neural Networks: A Comprehensive Foundation, 2nd edn. MacMillan Publishing Company, New York (1999)

23. Idri, A., Khoshgoftaar, T.M., Abran, A.: Can neural networks be easily interpreted in software cost estimation? World Congr. Comput. Intell. **2**, 1162–1167 (2002)

24. Nassif, A.B., Azzeh, M., Capretz, L.F., Ho, D.: Neural network models for software development effort estimation: a comparative study. Neural Comput. Appl. **27**, 2369–2381 (2015)

25. Braga, P., Oliveira, A., Ribeiro, G., Meira, S.: Bagging predictors for estimation of software project effort. In: Proceedings of International Joint Conference on Neural Networks (2007)

26. Wang, Y., Witten, I.H.: Inducing model trees for continuous classes. In: European Conference on Machine Learning (ECML) (1997)

27. Quinlan, J.R.: C4.5: Programs for Machine Learning. Morgan Kaufmann, San Mateo (1993)

28. Hosni, M., Idri, A., Abran, A.: Evaluating filter fuzzy analogy homogenous ensembles for software development effort estimation. J. Software: Evol. Process (2018)

29. Chen, K.H., Wang, K.J., Wang, K.M., Angelia, M.A.: Applying particle swarm optimization-based decision tree classifier for cancer classification on gene expression data. Appl. Soft Comput. J. **24**, 773–780 (2014)

30. Boeringer, D.W., Werner, D.H., Member, S.: Particle swarm optimization versus genetic algorithms for phased array synthesis. IEEE Trans. Antennas Propag. **52**, 771–779 (2004)

31. Sokolova, M., Lapalme, G.: A systematic analysis of performance measures for classification tasks. Inf. Process. Manag. **45**, 427–437 (2009)

32. Chawla, N.V., Bowyer, K.W., Hall, L.O., Kegelmeyer, W.P.: SMOTE: synthetic minority over-sampling technique. J. Artif. Intell. Res. **16**, 321–357 (2002)

Interactive Support System Using Humanoid Robot for Rehabilitation of Gross Motricity in Children

Piedad A. Semblantes[(⊠)] and Pilatasig Marco[(⊠)]

Universidad de las Fuerzas Armadas ESPE, Sangolquí, Ecuador
{pasemblantes,mapilatagsig}@espe.edu.ec

Abstract. The present work is based on the implementation of a human-robot HRI interaction system to support the sessions of gross motor rehabilitation in infants. The proposal is formed by a humanoid robot, which indicates the exercise of recovery that the child must imitate through movements of their extremities. At the same time, the mechanism visually and auditorily motivates the child to ensure that the proposed exercises are carried out in a correct manner and thus assess the child's progress in therapy. A Field-Programmable Gate Array (FPGA) is used for the architecture of the system, which is responsible for controlling the robot as well as the validation algorithm and an auxiliary processor for the acquisition and processing of the information taken from the sensors in real time. For the validation of the exercises proposed in the therapy, the system proposes an algorithm based on classifiers. The algorithm is based on the comparison of patterns of the exercises performed by the humanoid in contrast to the acquired movements of inertial measurement sensors subject to the user.

Keywords: Humanoid robot · FPGA · DTW · Rehabilitation systems ·
Gross motricity · Children

1 Introduction

According to the World Health Organization, (WHO) more than 1 billion people, or 15% of the world's population, have some kind of disability. In a more isolated case and depending on the type of limitation, children with disabilities usually have difficulties in carrying out daily activities, low educational performance, fewer economic opportunities, and a higher rate of poverty in adulthood compared to healthy individuals [1]. At the local level and according to records from the National Institute of Statistics and Census (INEC) of Ecuador, there is a growing demand in health centers in the area of physiotherapy, reporting in recent years a rate of four million treatments performed along the region [2, 3]. The need for therapy in people who have suffered an injury or illness causes a substantial increase in patients going to health centers, which requires a considerable investment of time and money. Another kind of problem is the rehabilitation of children, due to the emotional state of their age. In these cases, traditional rehabilitation sessions are often presented as unsuccessful and tedious therapies, because these processes tend to be routine and boring, causing children not accustom to them and therefore slow down their recovery [4]. For this reason, the use of new methods based on emerging, innovative, and motivating technologies in the traditional rehabilitation of children, promotes better results in their recovery. In this way, new methods to treat motor problems caused by cerebral palsy, brain injury or stroke have been effective. Virtual interfaces have been developed, which can be manipulated by means of innovative technological instruments such as gloves or robotic arms, thus forming systems capable of providing support in the recovery of motor disabilities and attracting the attention and curiosity of children [5–7].

On the other hand, robotics has been solving problems in different areas, both industrial and consumer. In this context, robotic prototypes aimed to physical assistance in an invasive way (prosthesis or exoskeletons) or social assistance in a non-invasive way (robots for guidance, supervision, and diagnostic of therapy) merge engineering with health areas [8]. In this context, the non-invasive robotic prototypes do not intend to apply forces on the extremities of children, but rather emphasize attracting the attention of the child, either by guiding or replicating the movements of the robot as he/she would be with other children. In this case, [9–11] propose humanoid robotic mechanisms to support the treatment of children with autism. In the same way, works dedicated to treating motor problems have been developed. For example, [12] presents a robot developed to rehabilitate the upper limbs by sequelae left by stroke (CVA). The work presents a technique of non-invasive therapy, in which the robot does not come into contact with the patient. The autonomy of the robot provides monitoring and recording of activities, besides encouraging the patient of the exercise routine of their recovery program. In conclusion, the programs of motor rehabilitation in children have proven, through exhaustive studies, to be very useful in the recovery of their functions and the improvement of their quality of life [13].

Therefore, this work proposes a novel alternative which provides support to the tedious rehabilitation sessions for minors, in order to present a game environment where the patient interacts physically with a humanoid robot, which motivates and guides the child in the execution of the therapy. Through a series of movements which the robot executes as guide exercises, it is intended to rehabilitate and strengthen the locomotor system coordinated by the cerebral cortex and secondary structures that modulate it, the same ones that have been affected by injuries left, either by a brain accident, illness or damage by physical activities. To achieve this, the proposed system carries out a validation algorithm for the exercises performed by the child, taking into account the techniques of pattern identification given by the DTW algorithm (dynamic temporal alignment). The purpose of this algorithm is to measure the homogeneity of the sequences obtained from a set of electronic armbands arranged in such a way in the injured extremities of the minor (3 SPACE SENSOR) versus the pre-recorded pattern sequence in the humanoid robot (BIOLOID ROBOT). Finally, several experimental results are shown to validate the proposed system.

2 System Structure

The developed system is based on a human-robot interaction (HRI) architecture, where the patient interacts directly with a humanoid robot. The rehabilitation proposal is based on the comparison between the therapeutic movements made by the robot versus the movements replicated by the patient. These activities are detected through the use of inertial measure sensors, which are located on the affected limbs which are to be rehabilitated in the child. On the other hand, the guiding movements which the robot executes are achieved through control criteria based on inverse kinematics. The Fig. 1 shows the scheme of the developed system. The proposal includes a rehabilitation strategy that can be used both in upper and lower extremities, where the progress of the therapy will be assessed by degrees of difficulty (moderate and mild) based on the initial medical diagnosis. Additionally, the system provides feedback through audio outputs which motivate and guide the child during the execution of the treatment.

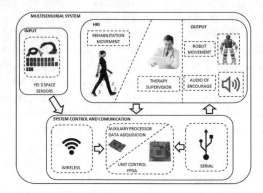

Fig. 1. Structure of the system for thick motricity rehabilitation

The interaction of the patient with the rehabilitation robotic system is based on the information acquired from the inertial measure sensors (IMU). On the other hand, the movements of the robot are in function to the previous diagnosis of the therapist, who evaluates the child and designates the exercises that must be carried out in the therapy. This operating scheme splits the system into five parts: system inputs, auxiliary processor architecture, control algorithm design, outputs, and user interface, as shown in Fig. 2.

The *(a) inputs of the system,* they are electronic devices in charge of capturing signals from a physical medium, to later be interpreted and execute actions. This research uses a set of 10 sensors 3-SPACE-SENSOR WIRELESS, formed by gyroscopes, accelerometers and triaxial compasses that give the system information about speed, acceleration and gravitational forces of the ex-extremes tested. The most relevant characteristics of this type of sensors are their autonomy and high performance since they use 2.4 GHz wireless communication, providing the system with low latency and high precision at the time of data capture in real time.

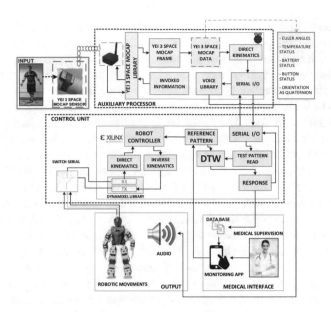

Fig. 2. Operative scheme

(b) architecture of the auxiliary processor, it has been considered to use a low cost card as a device responsible for carrying out the following processes: (i) acquisition of data, in this stage the information of an antenna which interconnects all the inertial sensors through wireless communication with spread-spectrum protocol (DSSS), obtaining temperature information, accelerometers, gyroscopes, and compasses (Fig. 3). (ii) Data processing, in this stage, through the programming of a library in Python the information provided by the antenna is read and unpacked for later processing and through direct kinematics obtain the position of the extremities with respect to a fixed point in real time. (iii) Voice feedback, this part uses free software based on Linux which can convert plane text into audio outputs, with the purpose of reproducing guide phrases and motivation during the execution of the therapy. (iv) Data output, finally in this stage the information of the sequences is framed as a function of position vectors in *x, y, z,* to be sent to the control unit commanded by an FPGA, by means of serial communication.

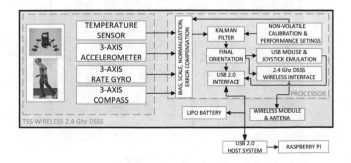

Fig. 3. Acquisition of data with 3-SPACE sensors

The (c) control unit, is presided over by an FPGA, where the control architecture is developed both for the positioning of the extremities of the humanoid robot and for the algorithm which validates the patient's movements. Based on the above and considering that there are vectors of reference movements pre-recorded in the FPGA, the following stages are detailed: (i) execution of the controllers for the extremities of the humanoid, where the motion reference vectors are used as desired inputs to the controllers. As detailed in Sect. 3, the controllers tend to reach the desired point through the use of basic techniques, where each of the extremities (each described with three degrees of freedom) must generate repeatable movements for the child. The FPGA includes internal libraries that command the chain of intelligent servomotors, as well as read positioning information to validate the execution and provide errors that will be corrected by the control strategy. On the other hand, the reference vectors are used by the DTW algorithm (an algorithm for each extremity) with the objective of (ii) compare them with the vectors generated by the patient, after a session is finished or when the application has reached a timeout. Finally, (iii) the FPGA can reply to serial communication commands, with the purpose of providing responses of the comparison of patterns, validating or not the successful execution of the session. Likewise, these responses are sent forcibly to the auxiliary preprocessor in order to generate sonorous commands that encourage the patient.

The elements used as *(d) outputs for this system* are described below: (i) the humanoid robot reproduces a series of movements by positioning its dynamixel motors, these movements are generated by the FPGA and are in function of the initial requirements of the therapist when evaluating the patient. (ii) Audio, it is used to guide and to encourage the patient while it develops the exercises of rehabilitation, these phrases are generated by the auxiliary processor and reproduced by a speaker located on the robot humanoid.

The *(e) user interface*, aims at the interaction of the therapist with the rehabilitation system through a menu of options which allow the user to select the routine to be performed, to save the results of the execution of the patient's exercises, as well as enable the storage of new rehabilitation movements, in order to provide flexibility to the proposed system (Fig. 4).

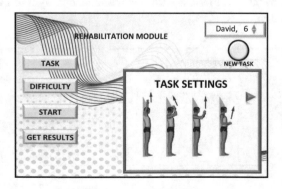

Fig. 4. User's interface

3 Proposal Development

3.1 Reading of Patient Movements

The location of the patient's points of interest (hands and feet) is established through the use of inertial sensors as shown in previous chapters. The placement of the sensors is shown in Fig. 5. By means of these sensors, it is possible to identify the angle in which the segment of the limb is located and by means of the geometric model the position of the points of interest is located. For this case, the sensors located in the upper extremities are placed on the biceps and on the forearm; while in the lower extremities they are located on the quadricep and the tibialis anterior. For each sensor, two Euler angles are obtained, the one which rotates on the X axis and the other one that rotates on the Z axis.

Fig. 5. Distribution of the 3 space sensors

Remark 1: The lower extremity (forearm or anterior tibial muscle) can rotate independently on the Z axis, therefore, it is considered as an additional parameter to obtain the final displacement of the point of interest.

The direct kinematic model allows to obtain the current position of the points of interest given the entry angles. Considering four degrees of freedom given both in the arms and legs, the displacement of the point of interest on all axes, for the right upper extremity is given by

$$p_x = dcsX - ad_2 \cos(\alpha_{uEr}) \cos(\beta_{uEr}) - ad_3 \cos(\alpha_{lEr}) \cos(\beta_{lEr}) \tag{1}$$

$$p_y = ad_2 \sin(\alpha_{uEr}) \cos(\beta_{uEr}) + ad_3 \sin(\alpha_{lEr}) \cos(\beta_{lEr}) \tag{2}$$

$$p_z = dcsZ + ad_2 \sin(\alpha_{uEr}) + ad_3 \sin(\alpha_{lEr}), \tag{3}$$

where $dcsX$ is the distance between the center of the thorax and the shoulder; $dcsZ$ is the distance between the abdomen and the center of the thorax; ad_2 is the distance between the shoulder and the elbow, ad_3 is the distance between the elbow and the point of interest (right hand); α_{uEr} it is a rotation angle of the shoulder, α_{lEr} is an angle of rotation of the elbow;α rotates on the Z axis y β rotates on the X axis. The negative sign for the X axis alternate the result of $\cos(\alpha_{uErmax})$, where $\alpha_{uErmax} = \pi$. For the left hand, the $\alpha_{uElmax} = 0$, therefore, the use of the negative sign is not necessary. The same criteria are used for the lower extremities.

In this way, the position of the point of interest can be determined in any of the axes of the space. The location of the operative end allows obtaining input data to the pattern comparator, which validates the correct execution of exercises.

3.2   Robot Extremities Controller

Considering the importance of the reference movements for the child to imitate them, a controller based on inverse kinematics is proposed to validate the movements of the extremities of the robotic mechanism. As shown in Fig. 6, the degrees of freedom of the uppers and lowers extremities of the Bioloid suggest the number of n equations and the number of m unknowns for the design of the controller. In this way, a system of three equations and three unknowns presented in (4), (5), and (6) solve the problem of positioning the points of interest of the upper extremities at any point in space.

Fig. 6. Degrees of freedom considered in the humanoid robot

The inverse kinematics is achieved with control purposes for the robot, in addition to ensuring that the end of the robot executes desired movements identifying some type of error in the intelligent motors that execute the movement.

$$p_x = dcsX + ad_2 cos(\beta_a) + ad_3 cos(\beta_a + \delta_a) \tag{4}$$

$$p_y = cos(\alpha_a).(ad_2 \sin(\beta_a) + ad_3 \sin(\beta_a + \delta_a)) \tag{5}$$

$$p_z = h_c + dcsZ + \sin(\alpha_a)(ad_2 \sin(\beta_a) + ad_3 sin(\beta_a + \delta_a)) \tag{6}$$

Remark 2: To obtain the required models, it is considered that all extremities are formed with the same number of degrees of freedom and similar behavior, with the only difference that the base of each limb has a displacement from a given reference center. Likewise, it is considered that the initial position of the robot has the upper extremities completely stretched on the X axis, while the lower extremities are stretched on the Y axis.

Being based on mathematical methods and trigonometric properties, the angles of α, β, and δ are shown in (7), (8), and (9).

$$\alpha_a = \tan^{-1}\left(\frac{p_z - h_c - dcsZ}{p_y}\right) \tag{7}$$

$$\delta_a = \cos^{-1}\left(\frac{(p_z - h_c - dcsZ)^2 + p_y^2 + (p_x - d_x)^2 - ad_2^2 - ad_3^2}{2ad_2 ad_3}\right) \tag{8}$$

$$\beta_a = \tan^{-1}\left(\frac{p_x - dscX}{\sqrt{(p_z - h_c - dscZ)^2 + p_y^2}}\right) - \tan^{-1}\left(\frac{d_3 \sin(\delta_a)}{d_2 + d_3 \cos(\delta_a)}\right) \tag{9}$$

The mathematical representations found are based on the position of the extreme of interest p_x, p_y y p_z, therefore, the displacement of a vector with positions will force the actuators to move the point of interest over the space, executing the desired therapy movements.

Bearing in mind that the vector of pattern movements is made up of a number of sequential points (movements in X, Y and Z desired), the objective of the controller is to modify the positions of the points of interest (hands and feet of the robot) depending on the new positions given by the reference vector. In this aspect, when a new desired point is proposed $\mathbf{p_d} = [p_{xd}, p_{yd}, p_{zd}]$, the controller reaches the objective through the calculation of errors $\mathbf{p}_{err} - \mathbf{p_d} - \mathbf{p_r}$, where $\mathbf{p_{err}} = [p_{errX}, p_{errY}, p_{errZ}]$, where, $\mathbf{p_{err}}$ represents the position error of the three components. The reduction of position errors is achieved with (10), where the use of a pair of constants is included $\mathbf{k_1}$ and $\mathbf{k_2}$.

$$\mathbf{p}_i = \mathbf{p}_{i-1} + \mathbf{p_{err}} * (\mathbf{k_1} + \mathbf{k_2}) \tag{10}$$

The response of the controller to achieve the desired new position can force the actuators exceedingly. To adjust the responses of the controller to rotation values that do not affect the structure, the use of the constants is proposed $\mathbf{k_1}$ and $\mathbf{k_2}$. For normal execution, $\mathbf{k_1}$ is included to provide the system with a basic speed, while through the use of $\mathbf{k_2}$, the user can modify the speed of execution to its need.

Remark 3: The implementation of the system includes four controllers, one for each extremity running simultaneously, which accelerates the error correction process given a vector of desired positions. For more information about the robotic platform used visit the following page http://www.robotis.us/bioloid-1/

3.3 Comparison of Patterns Through DTWv

The parallelism of the FPGA where the system is implemented facilitates the incorporation of four vector alignment and comparison algorithms. The cited vectors contain spatial positions X, Y, Z of the four points of interest (two hands, two feet), identifying as the pattern vector the number of desired spatial points, while the input vector contains the positions of the points of interest, obtained from the patient (by using inertial measurement sensors and processing by direct kinematics). For the development of the DTW in the FPGA, fundamental aspects are considered [7]: boundary condition, condition of monotony, adjustment window condition, and jump condition. *The boundary* condition states that the trajectory starts at the bottom right and ends at the top left; *the condition of monotony*, refers to the progress in the search for solutions, where in no case the trajectory can go backwards, only maintain or increase the values of index i and j; *the jump condition*, forces the trajectory to advance one step at a time; and finally, *the adjustment window condition*, establishes that the trajectory in shown Fig. 8. it should be not too far from the diagonal.

The programming of the algorithm yields a matrix of $2 \times n$ values, where the path that determines the similarity between both frames is defined. The results section includes the usability of the programmed algorithm. The determination of the path additionally allows to know a total of the weight matrix, whereby decision thresholds are defined to compare the input patterns.

4 Results

The results of the proposal are split into three parts, starting with the test of data acquisition on the patient's movements through direct kinematics and the use of inertial sensors. On the other hand, the second group of results presents the correct execution of reference movements by the robotic mechanism, validating the proposed controller. Finally, the alignment of the movement vectors is presented to determine the coincidences between the movements executed by the robot (reference movements) and the movements of the child.

The (i) *child's movement information acquisition tests* are shown in this subsection. Figure 7 presents the recreation of positions of each of the extremities of the child in Python, processing which takes place on a Raspberry processor. In this way and using the methods previously indicated, the movements of the points of interest (both hands and feet) are determined.

On the other hand, (ii) the correct execution of movements by the robotic mechanism is crucial to motivate the child to execute the exercises in the expected manner. The validation of the algorithm which controls the proposed limb is based on the results shown in Fig. 8, where the correction of positions through the convergence of errors to zero in each of the desired positions is observed.

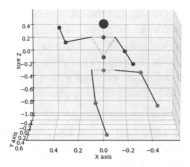

Fig. 7. Representation of the positioning of the extremities in the auxiliary processor

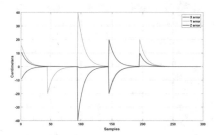

Fig. 8. Correction of errors in each of the axes

Likewise, the (iii) *alignment between the reference vectors and the input vectors* is presented in this section through the distance matrix, the total weight, and the aligned vectors. Considering that the comparison is between two experiments in which the satisfactory and wrong movements of the user's right hand are compared (Fig. 9), the alignment results are shown in this subsection.

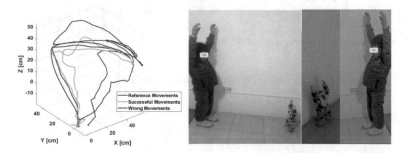

Fig. 9. Movements of the user's right hand

The representation of the distance matrix presented in Fig. 10 facilitate the interpretation of the results obtained, where the trend of dark colors in the diagonal of the first image indicates that the compared vectors have a certain similarity despite the fact that the amount of data is not similar. On the other hand, the second image shows a low tendency of negative colors on the diagonal, which indicates at first glance the non-similarity between the input vectors. This information is of great importance for the feedback of the system, since depending on the distance matrix, the total distance used to send commands to the auxiliary processor that interprets this information and generates audio outputs to motivate the child is obtained and comply with the exercises proposed in the therapy.

Remark 4: Based on experimentation of this work, the reference vectors of the three components form a single vector (in the order *x, y, z*), in the same way as the vectors read. In this way, a pair of vectors are formed and compared to obtain the result of the DTW as shown in Figs. 10, 11 and 12.

Remark 5: The topic is open for future research considering that the system has been tested with healthy children.

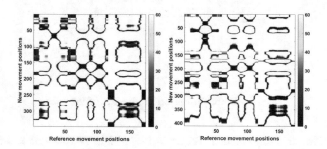

Fig. 10. Distance matrix

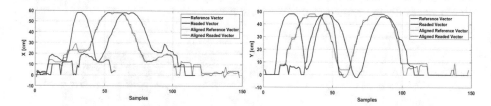

Fig. 11. DTW response to similar sequence

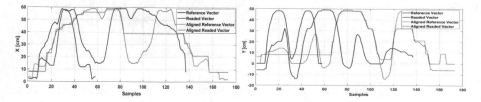

Fig. 12. DTW response to different sequence

5   Conclusions

This work presents a strategy of non-invasive rehabilitation for children, inte-grating a humanoid robotic mechanism, inertial sensors, and comparative and vector alignment techniques. The system is supervised by the therapist, being a support tool in reha-bilitation activities. The main purpose of this system is to provide the child with a more interactive environment through the use of the humanoid, where the user must replicate the movements of the robot to obtain a satisfactory result. To determine if the task is performed in the expected manner, the FPGA compares the pattern movements with movements generated by the patient through a temporal alignment algorithm (DTW), returning a sound response to motivate a new execution or validate the task done. Additionally, experimental results to validate the correct execution of movements of the limbs of the robot are presented, demonstrating the tendency of errors to zero in the positioning of the points of interest.

As future works, it is proposed to modify the structure of the humanoid in order to incorporate a camera in the head of the robotics mechanism and through artificial vision to implement the interactive rehabilitation system.

References

1. World Health Organization: World Disability Report (2011). http://www.who.int/disabilities/world_report/2011/summary_es.pdf?ua=1. Accessed 20 Sept 2018
2. INEC: Health Activities and Resources. Creative Commons (2015). http://www.ecuadorencifras.gob.ec/actividades-y-recursos-de-salud-2015/. Accessed 10 Nov 2018
3. CONADIS: National Council for the Equality of Disabilities (2018). https://www.consejodiscapacidades.gob.ec/estadisticas-de-discapacidad/. Accessed 4 Oct 2018
4. Domínguez, S.F.: Infant physiotherapy based on new technologies (2010). https://www.webpt.com/blog/post/advancements-pt-techemerging-trends. Último acceso 20 Oct 2018
5. Krista, C., Corinna, L., Kenton, K.: Development of an interactive upper extremity gestural robotic. In: Annual International Conference of the IEEE EMBS, pp. 5973–5976 (2009)
6. Pawel, P., David, W., Edith, C., Yves, H., Lisa, H., Ismael, F.: A paediatric interactive therapy system for arm and hand. In: 2008 Virtual Rehabilitation, pp. 127–132 (2008)
7. Semblantes, P.A., Andaluz, V.H., Lagla, J., Chicaiza, F.A., Acurio, A.: Visual feedback framework for rehabilitation of stroke patients. Inform. Med. Unlocked **13**, 41–50 (2018)
8. Acevedo, J., Caicedo, E., Castillo, J.: Application of robotics rehabilitation technologies. Revista Universidad Industrial de Santander Salud **49**(1), 103–114 (2017)
9. Billard, A., Robins, B., Nadel, J., Dautenhahn, K.: Building robota, a mini-humanoid robot for the rehabilitation of children with autism. Assistive Technol. **19**(1), 37–49 (2010)
10. Robins, B., Dautenhahn, K., Boekhorst, R., Billard, A.: Robotic assistants in therapy and education of children with autism: can a small humanoid robot help encourage social interaction skills? Univ. Access Inf. Soc. **4**, 105–120 (2005)
11. Syamimi, S., Hanafiah, Y., Luthffi, I., Fazah Akhtar, H., Salina, M., Hanizah, A.P.: Initial response of autistic children in human-robot interaction therapy with humanoid robot NAO. In: IEEE 8th International Colloquium on Signal Processing and its Applications, Malaysia (2012)

12. Eriksson, J., Matarić, M.J., Feil-Seifer, D.J., Winstein, C.J.: Socially assistive robotics for post-stroke rehabilitation. J. NeuroEng. Rehabil. **4**(5), 2–9 (2007)
13. Calderita Estévez, L., Bustos García, P., Suárez Mejías, C., Fernández Rebollo, F., Bandera Rubio, A.: THERAPIST: towards an autonomous socially interactive robot for motor and neurorehabilitation therapies for children. In: 2013 7th International Conference on Pervasive Computing Technologies for Healthcare and Workshops, pp. 374–377 (2013)

Online Peer Support Groups to Combat Digital Addiction: User Acceptance and Rejection Factors

Manal Aldhayan[1(✉)], Sainabou Cham[1], Theodoros Kostoulas[1], Mohamed Basel Almourad[2], and Raian Ali[1]

[1] Bournemouth University, Poole, UK
{maldhayan, scham, tkostoulas, rali}@bournemouth.ac.uk
[2] Zayed University, Dubai, UAE
basel.almourad@zu.ac.ae

Abstract. The obsessive usage of digital media may exhibit symptoms traditionally associated with behavioural addictions such as mood modification, salience, tolerance and conflict. The educational methods, interventions, and treatments available to prevent or control such a digital addiction are, currently, very limited. Digital Addiction (DA) is yet not formally recognised as a mental disorder by the Diagnostic and Statistical Manual of Mental Disorders. Recently, in 2018, the World Health Organization recognised gaming disorder. Fortunately, the nature of digital media can also help the hosting of methods and mechanics to combat DA, e.g. in the monitoring of online usage and enabling individuals to stay in control of it. One of the techniques proposed in the literature is Online Peer Groups platforms, towards allowing people to form a group and provide peer support to control and regulate their usage, collectively. Online peer support groups are meant to provide peer support, counselling, motivational and learning environment, and ambivalence reduction through sharing and hope installation. However, there is a lack of research about the factors influencing people with DA to accept or reject online peer support groups. In this work, we conduct user studies and explore the acceptance and rejection factors to join and participate in such DA regulation and relapse prevention method. This will help to design and introduce the method and increase its adoption.

Keywords: Online peer groups · Digital addiction · Behavioural change

1 Introduction

Digital media such as social networks, gaming and online shopping have become firmly established as part of our daily lives. Such media empowered social connectedness, information exchange and freedom of information exchange. However, despite the benefits, some usages of digital media can be considered compulsive and obsessive leading to negative consequences such as reduced involvement in real-life communities and a lack of sleep [1]. Recent studies have indicated that users who become addicted to digital technologies exhibit the same symptoms as other behavioural addiction such as salience, conflict and mood modification [2, 3].

© Springer Nature Switzerland AG 2019
Á. Rocha et al. (Eds.): WorldCIST'19 2019, AISC 932, pp. 139–150, 2019.
https://doi.org/10.1007/978-3-030-16187-3_14

The preventative, control and recovery mechanisms available for Digital Addiction (DA) are currently very limited. One of the reasons for such absence is that, with the exception of online games, DA is still not classified as a mental disorder in the latest 5th edition of the Diagnostic and Statistical Manual of Mental Disorders (DSM). Most of the existing research on DA focuses on the users' psychology, i.e. their reasons for the overreliance on social media and the relationship with personality traits [4]. Only few works placed software design at the centre of the DA problems, both in facilitating it and also in combatting it, e.g. the digital addiction labels and the requirements engineering for digital well-being requirements in [6, 7].

Despite the proliferation of software to assist with behavioural change, there appear to be several issues with the acceptance and efficacy of such application and whether they should be used as a primary or auxiliary mechanism. The perception of their role has changed following some failures and the recognition of associated risks [8]. This calls for further research to establish a deeper understanding of what role such software can play and how their design process would be conducted. A core element of the process for addressing those issues is to capture users' digital well-being requirements and to explore acceptance and rejection factors.

Behaviour change theories seek to link the intention to change the behaviour with the act of doing so and are used to predict and promote behavioural change [5]. Such methods have been used in the field of addiction and may provide useful insight into supporting changes in addictive behaviours. Peer groups are one of the approaches utilised to combat addictive behaviours by providing support and helping in the relapse prevention [9]. Peer groups can be constructed when a group of people share similar interests and in view of supporting and influencing each other's behaviour [10]. Alrobai et al. [11] were among the first to examine the peer group approach by utilising social computing techniques themselves to combat DA. Alrobai et al. [13] focused on the processes involved when running the group, e.g. the roles involved in doing so and the steps to be taken to prevent relapse. In that work, the authors only considered the operational phase of a peer support group and did not address the earlier stage of the lifecycle of peer groups, namely, peer's acceptance and agreeable protocols. Overall, there is a need for establishing a deeper understanding of acceptance and rejection factors of such software-assisted method by people with DA. This understanding will inform the strategies used to introduce such software, as well as its configuration and governance processes.

In this paper, we explore the acceptance and rejection factors of online peer support groups as a mechanism for combatting DA. Acceptance is vital as members of the group report their online use, emotions and intentions on a voluntary basis. Although technology can be designed to monitor digital usage, people can always find ways around it if they so desire, e.g. using different devices and accounts or claiming that the use was necessary for work reasons. As a method, we adopted qualitative research employing a secondary analysis of focus groups, originally conducted to explore the best design features of peer groups, and further conducting an interview study with 16 participants who self-declared to have DA, as a primary data collection technique. Our study is also intended to inform the introduction phase of technology-assisted behaviour change solutions to maximise acceptance and adoption.

2   Research Method

We adopted a qualitative method to explore acceptance and rejection factors of people with DA to join online peer support groups for combating their DA. We collected and utilised data from two studies to increase the credibility and coverage of the findings.

In the first study, we performed a secondary analysis of a focus group study of two sessions. The first session aimed at getting insight regarding what the participants think about an online peer group and what they needed to have in it. The second focus group served the purpose of identifying the design features of an online peer group. For this reason, mock design interfaces made based on the result of the first focus group were built and presented. The participants were asked about opinions regarding the mock design. The two focus group sessions were conducted with the same six participants three male and three females, aged between 20 and 26. Participants were selected because they identified themselves as persons who are having a problematic use of social networks. Some of the participants can be considered friends. Though this has some effects in the study, it is, also, beneficial, due to the fact that concerns regarding trust and privacy during the discussion process are suppressed.

The secondary analysis of the data collected from these focus groups was performed using thematic analysis [12]. This analysis revealed the main factors concerning the acceptance of this approach. The findings were used to construct the interview protocol for the primary study and provided a starting template for its analysis. The secondary analysis of the focus group and design session results explored five themes and notable aspects regarding the acceptance and rejection of online peer groups. These five themes related to (a) group moderation, (b) content, (c) governance and operation, (d) group coherence and trust and (e) goal setting and commitment.

In the second study, an interview study was undertaken in order to elaborate further on these aspects. The interview was conducted with 16 students self-declaring to experience problematic use of online services, e.g., obsessive or compulsive use, 8 males and 8 females, aged between 18 and 35. Each interview lasted between 30 and 40 min. The interview questions revolved around the acceptance and rejection factors discovered in the first phase as well as the design features which would support a desired operation of the online peer support group. The interviews were transcribed and analysed *via thematic analysis* following the theoretical position of Braun and Clarke [12]. The analysis explored different aspects that affect users towards accepting and rejecting an online peer group.

3   Results

In this section, we will present the factors which would affect both the acceptance and rejection of people with DA of the online peer support groups. We note here that some of these preferences are contradicted with each other. This would be expected from people with problematic behaviour who often have conflicting requirements about their health, on the one hand, and their desire to continue the problematic behaviour, on the other. Common attitudes and maladaptive behaviours which facilitate that conflict include denial, trivialization and cognitive dissonance [8].

3.1   Online Peer Support Groups to Combat DA: Acceptance Factors

Different aspects and perceptions explored during the interview contributed to the identification of the factors which affect the user's acceptance to join an online peer group focused on combating digital addiction. These factors should be considered when software and systems engineers design and introduce an online peer group for users who have a problem using digital media. In Fig. 1, we illustrate the acceptance factors that affect users regarding their decision in joining an online peer group. In the next subsections, we present these acceptance factors in details.

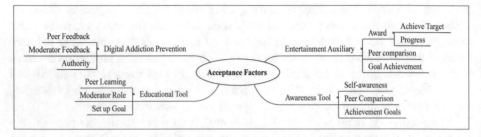

Fig. 1. Online peer support groups to combat digital addiction: acceptance factors

Accepting Online Peer Groups as an Entertainment Auxiliary. An important factor which motivates participants to accept an online peer group is its introduction as one of the "entertainment tools" which will ease the DA prevention and recovery processes. Participants suggested that these tools should include gaming elements which are implemented and adapted as a reinforcement function. This function corresponds to motivating group members to regulate and control digital media usage. That reinforcement function should be designed to be *"fun and look like a game"* by including *"rewards"* and *"comparison"*. The participants have three viewpoints in how to establish that: peer comparison, awards and achievement goals.

Peer comparison corresponds to the users comparing their performance and progress with other group members. For example, participants mentioned that comparing usage and progress with group members is *"fun"*. Also, comparing usage would motivate group members to regulate their own digital media usage. It is important to mention that the participants were concerned about comparing their progress with peers that have different levels of addiction since this kind of competition could impact self-esteem and self-efficacy. For example, one participant commented that *"when a group has members who have different levels of addiction and skills, and when the system compares progress with them... that might demotivate lower performing people, and then at the end, they might end up relapsing"*. Overall, the participants agreed that comparing performance and progress with other group members would motivate members to set up goals and targets for the usage and benchmark themselves to others. Additionally, some of the participants stated that they would enjoy sharing and comparing usage with other members and would find it essential and inherent to the sense of being in one group.

The second viewpoint is that participants prefer the online peer group to provide an award. Specifically, the participants thought that the platform should have a rewards mechanism to "*motivate them to achieve group goals*" and "*to regulate digital media usage*". They suggested that the platform could be like a game, with a target to achieve; if a user achieved the target, a reward would be given such as "*points or moving from one level to another level*". Most of the participants agreed that a reward using a points scheme would be useful and could motivate users to work hard to collect more points. As an example on how these points could be materialized, a participant suggested that a user who collected "*10 points could replace it and get gift or voucher*" or could "*upgrade their level in the group*". Other participants preferred to receive social recognition rewards. For example, members who achieve good progress could have their names or pictures displayed on the main page of the platform or could receive a "*congratulations*" message from the system, visible to others.

The last viewpoint is that of achieving the group goals. Besides individual and self-set goals, participants would prefer the group to have goals which are established with other group members. In this line, the participants argued that unified goals, applied equally on everyone, or a set of group-agreed goals allocated to members separately would create competition between members that would help them to achieve the goals. They felt that such group goals are "*fun by increasing competitiveness between group members*", although they argued that unified goals should be "*between peers who have similar levels of addiction*" and would be more effective with "*peers who share the same interests, such as working in the same organization or are post-graduate students*".

Accepting Online Peer Groups as a DA Awareness Tool. Participants appreciated the role of online peer group as an awareness raising and knowledge sharing. Such awareness revolved around self-awareness, peer comparison and ways for the achievement of goals. The first viewpoint corresponds to the expectation of some participants from the online peer group to help them becoming more conscious of their usage and the amount of time spent in "*each digital media app*". Moreover, they required the online peer group to have a monitoring system, which can track and monitor "*the frequency and the time spent using digital media applications*" by members. Moreover, the majority of participants commented that they need to be aware of "*their level of addiction to digital media usage*" because they may be thinking that their usage is "*normal, but maybe they are addicted*". This conflicts with other state-ment made by participant around the personal and context-dependent nature of the usage of digital devices and that judgmental approaches towards the claims of having DA are to avoid.

The second important viewpoint about the utilisation of online peer support groups as an awareness tool, related to peer comparisons. In this context, peer comparisons can help group members become aware of their digital media usage and their level of addiction through benchmarking. This can be done through various metrics including time and frequency and also the context in which digital devices are used, e.g. during work and meetings and meals. The comparison can be also non-usage related, e.g. emotions felt while detaching from social media and coping strategies used. Partici-pants agreed that the simple comparison amongst their usages would help increase awareness of their usage and if they "*used more than the rest, that would motivate them to reduce usage*". Also, participants preferred comparing their usage with the group members who have a similar level of addiction, share similar interest, or students enrolled in the same educational programme. A participant, who is a PhD student, mentioned that "*it is useful if the group members are PhD students so when the platform compared my usage with group members and the platform showed my usage is more than others then I have to be more aware of my usage*". They prefer that the platform sends a weekly comparison report and that the report compares "*their usage with past-usage*" and compares "*their usage with group members*". We note here that metrics for comparison around DA are to be investigated further in future research. Participants emphasised that their use of social media might be for work and hence shall be given a different weighting for its contribution to the problematic usage. They also explained that the calculation of usage shall be more sensitive to the context, e.g. festive season vs. work or sleeping hours.

The third viewpoint is the one of "achieving group goals". The participants prefer to set up group goals which help their commitment to achieving a more in control usage and become aware of their progress towards achieving their targets. Reminders and notification messages seemed to be highly needed as participants stated they might become unconscious of the usage and its amount and context. The group moderator or system could send a notification message to notify the user of the amount of "*time spent using digital media*" that would help members notice their usage. Furthermore, the system or group moderator could send a warning notification message which "*makes the user aware of usage*" and when the user "*exceeds the time limit of usage*". Moreover, the group moderator "*would block the digital media apps for a day because the user exceeded the limits*". However, it is important to note that the participants were concerned about the "*notification time*" and "*how many messages to send a day*", in the sense that they are not in favour of the system exceeding the notification messages more than once a day. We note here again the conflict between being looked after by the system and the requirements of privacy and non-obtrusiveness.

Accepting Online Peer Support Groups as an Educational Tool. The participants considered online peer groups as an educational platform. They generally preferred that such platforms provide functionalities that would help them learn how to control their digital media usage and find life alternatives. They have three viewpoints regarding where to obtain this knowledge from; from peers, group moderators, and by setting up group goals and learn how to achieve them. Regarding the first viewpoint, the participants mentioned that they can learn from peers who had successfully achieved the group goals before by "*asking them questions & receive advice regarding how they reduce usage*". Also, peers can learn from each other and "*share strategies they follow*" to help them control digital media usage while they are all trying in the same time as this can have both educational and motivational value. Moreover, participants mentioned that they can learn from peers' personal and real-world stories thus they prefer interacting with any member who is an ex-addict or one that has successfully achieved the group goals. Such share adds to the relatedness and sense of belonging in the group and acts as a hope installation mechanism. Gaming addiction would be one clear example here as participants who used to play games heavily found it difficult to find alternatives to games especially after building their online community around it.

The second viewpoint that affects acceptance of an online peer support group as an educational tool is the moderator's role. The group's moderator has been seen as an educational one and it is expected that the moderator has knowledge and experience in DA. In this sense, the moderator would deliver this *"knowledge to the group's members by providing advice"*. To empower this educational role, the group moderator should be enabled *"monitor the group member's usage"* and, based on the monitoring result, would then be able to *"know their level of addiction and provide support and guidelines suit to them on how to reduce usage"*. The participants mentioned that they could learn from the moderator's *"advice and guidance"* which would help them control and combat addiction. Moreover, the participants believed that the online peer group could use some kinds of *"role-playing"* which is similar to *"game learning"* [14] as a way for changing behaviour. They suggested that the moderator's role could *"rotate"*, meaning that after a period that *"any member who has accumulated high points"* could be a group moderator for a period of time. The moderator *"could provide advice and rewards to the members"* and *"set up the group goals"*.

The last viewpoint around education corresponded to setting up usage goals and learn how to achieve them. Goals seem to have the added value of being an additional motivation to learn. Participants agreed that setting up achievable and realistic goals is also an important factor to sustain the motivation to learn how to achieve them. Despite the fact that some of the participants preferred to set up their own goals, they also mentioned that the group moderator should be able to check if the goals are reasonable and achievable and, in case they are not, the moderator should *"explain how they can set up achievable and reasonable goals"*. In other words, the education can be also around goal setting skills.

Accepting Online Peer Support Groups as a Digital Addiction Prevention Tool.
Using online peer support groups as a mechanism for the prevention of digital addiction seems to be one of the acceptance factors. The participants agreed that such platforms should have monitoring and feedback features administered by the group moderator, peers or the system automatically. In this line, the participants mentioned that *"feedback is an important tool for prevention digital media"*. This feedback can be based on monitoring performance and adherence to the set goals. The participants accepted that a moderator should have the authority and ability to access group-members' digital usage and enact precautionary measures. Possibly, members would accept that this access is only from the moment of appointing the moderator, i.e. in the case of rotation based assignment policy. The moderator is expected to observe group members' performance and progress, and, based on that, make a judgment and send feedback and advice to the corresponding member. Also, the moderator should have the authority to take corrective measures if any member does not adhere to the group goals, e.g. *"lock some digital media application"*. Moreover, it is acceptable for a group moderator to observe whether the group members achieve their goals and, if any member struggles to achieve the goals, the moderator is expected to provide supportive information or *"amend the goals"*. Such amendments can be done through dialogue with users or by analysing their performance and profile.

With regards to how the moderator handles and makes use of the access to the digital usage of the group members, the moderator was expected to (a) make a judgment and (b) send feedback and advice on how to deal with addiction or guidance regarding the member's performance. Some participants preferred the feedback to be "*strict, formal and in order*" and were in favour of a "*warning message*" if they exceeded the usage limit or use in an inadequate context, e.g. during work hours. Moreover, participants mentioned that they preferred "*moderator feedback to be positive*", such as "*Congratulations, but you'll need to improve on this and that*". Same participants had the two different preferences while others were clear in their specification. This suggests the importance of personality and contextual information around the feedback tone and timing.

3.2   Online Peer Support Groups to Combat DA: Rejection Factors

In this section, we will present factors that would lead the users with DA towards rejecting an online peer support group. Figure 2 presents a summary of the rejection factors. These factors are detailed in the following paragraphs.

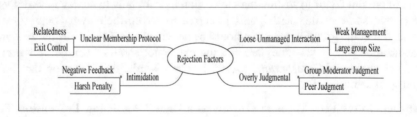

Fig. 2. Online peer support groups to combat digital addiction: rejection factors

Rejecting Online Peer Support Groups When Seen as an Intimidation Tool. Intimidation is one of the essential reasons for rejecting an online peer group platform. Participants have two viewpoints regarding the rejection due to this reason; the harsh penalty and the negative feedback. Penalties as anticipated by the participants included "*blocking from the group*" and "*writing member name in the main page of the platform as a loser*" even in a playful and gameful format. The harsh penalty seems to be affecting members "*self-esteem or make member leave the group*". Some participants mentioned that harsh penalties mean a "*threat*" to them and that this would affect their motivation to participate, truthfully, with the group and could lead to leaving the group or adopting workarounds such as using a secret device to access social media in a way that is not monitored by the online peer group software. They commented that "*members join the group because they would like to control their digital usage, so they do not prefer to have a penalty*" as this seems to be against the free spirit of membership and joining process. A participant stated that she "*would not accept a harsh penalty even when I do not achieve the group goals*". This again shows the delicacy of implementing rewards and penalty for problematic behaviours and the conflict between users' preferences which necessitates a process of consensus and commitment building when configuring the online per groups platforms and specifying their interaction and governance protocols.

Participants were much concerned regarding negative feedback messages received from group moderator or peers even if the feedback is factual such as "*you compare less favourably to other peers*". They prefer motivational feedback and reject the reception of a negative one. A participant commented that "*some group members could not achieve the group goals for different reasons such as setting difficult goals or simply could not control usage*", and that the feedback from a moderator or peers "*should not be harm*" additional to what they have already felt and should not "*underestimate the user*". Overall, participants mentioned that harsh feedback could affect their feeling of membership and relatedness and affect their self-esteem and this could lead them to reject the group. They would prefer to receive a message that motivates and reinforces them to reduce their problem. Having critical feedback is different from negative feedback, and it seems it is a matter of framing and language problem here. Participants did not like the highly serious feedback as although they view DA as a real problem; they reject a framing it as a formal addiction; "*I do not like to receive a message says that using digital media for a long-time lead to a mental health problem such as depression*". Besides building census and commitment, denial and trivialisation of issues are common attitudes of people with problematic behaviour and need to be dealt with when configuration and starting online peer groups.

Rejecting Online Peer Support Groups When Seen as Overly Judgmental. Participants tend to refuse to be judged by others in the group, especially when the judgment is coming from a peer and automated software. The situation can lead even to a harder reaction when peers know each other in person as the judgment will expand to the real personality. A participant mentioned that "*if the group members are my friends maybe if I meet them, they will judge me through my digital media usage*". Also, participants did not prefer receiving feedback in the form of questions such as "*why you are always on Facebook*". The main reason for the participants to reject online peer group is that they all have different usage styles and intentions of use when it comes to social media. This observation is an important parameter, which appears that software and peers can not simply judge it. Although less of a concern, participants also had issues by being overly judged by the group moderator. While they tended to accept feedback from the moderator in general, they preferred that the feedback shall include advice or guidance regarding their usage, rather than pure judgment messages. For example, a participant commented that "*sometimes the group goals are too hard and I could not achieve the goal*", and in this sense, moderators should send feedback to support them and show them strategies rather than sending scores and judgements.

Rejecting Online Peer Supports Group When Hosting Unmanaged Interactions. Participants prefer the group interaction to be run and overseen by the group moderator. They had two viewpoints to reject a group in that regard; (a) if it has weak management and (b) if it is a large size so management is hard. Participants would reject a group with a weak moderator who cannot make a decision, such as banning members who are not adhering to the group norms, e.g. in the conversations, and sending warning messages. Participants prefer a group moderator who is "*able to control the group connections and oversee messages sent by members*". There were concerned about weak management that is unable to stop members annoying others by sending feedback against the "*group aims*" such as "*friendship requests*", or "*jokes*". The participants preferred the moderator shall be able to delete any message that does not follow the group aim and send a "*warning message*". A participant mentioned: "*I joined the group because I would like to control my digital media usage and I do not want to receive any friend request from the group members*" and the group moderator "*has to warn any member who sends a friend request*".

The other reason to reject an online peer group is its large size. The participants expressed concerns regarding group size and feared that its management is "*difficult*" if it is a large one. Specifically, some participants preferred the group size to be small, i.e. from five to ten members. They argued that a big group would be "*massive and do not help to achieve the group goals*". Furthermore, a large group would be "*annoying and would receive many messages*", meaning that members could not focus on the group goals. Also, they argued that, in a large group, it "*is not easy to track all members and the competition in a small group would make more sense*".

Rejecting Online Peer Groups Due to Unclear Membership Protocol. Participants expressed concerns regarding (a) relatedness of group members and (b) conditions to exit from the group. Some of the participants rejected to join a group whose group members are friends or relatives. Some of the participants prefer to join a group with people who do not know each other's as that would "*make members feel more comfortable and confident*" and they would then accept to "*share usage, comments, and feedback with peer*" because they know "*no one would judge [them] in real life*". Other participants preferred to join a group of people known to each other's, but they prefer to be "*semi-anonymous*" and reject to provide their "*real name and picture*" when it comes to monitoring usages. This can be solved through messages like: one of your friends is having difficulty with games today, what do you like to tell them? Participants agreed that a group of friends "*make them trust the group and they will not worry about privac*y" as they already know that their usage is problematic. It can be noted here that participants had a paradox between trust and privacy here.

Participants rejected that the group can have conditions or regulation regarding the exit from the group because that would affect "*members feeling*" and that "*they will feel stress and that the group control their freedom*". Participants generally agreed that they should be free to leave the group whenever they like, but they would not approve a "*member leaving the group without giving notice and explanation why they decide to leave the group*". They suggested that when somebody leaves a group, they should invite somebody else to join, especially when the "*group size is small*" and when leaving may adversely "*affect the morale of members to achieve the group aims*". Again, we can note the conflicting preferences requiring a resolution process.

4 Conclusion and Future Work

In this paper, we explored the factors which affect the acceptance and rejections of people with DA regarding online peer support groups for combating their DA. A range of factors seems to be conflicting. For example, while people like the group to provide a friendly environment where game elements are used, e.g. regarding challenging each other regarding reduction time, they like not to be monitored and judged. Similarly, while they appreciate the freedom to join and to leave, they were concerns that this may affect their members and lead to trivialising the process. These observations call for methods to reconcile amongst these requirements. Although participants emphasised a wide range of factors around DA and its judgement, e.g. context and purpose of usage and non-usage related factors such as emotion and preoccupation, their examples tended always to be around time and frequency of using digital devices. This would be expected due to the recent nature of the DA concept and the debate around it. Our future work will look at building a consensus process so that group members can configure agreeable settings of their online platform utilised for peer support groups.

Acknowledgement. This work has been partially supported by the EROGamb project funded jointly by GambleAware and Bournemouth University, SSCoDA project funded by Zayed University and H2020-MSCA-RISE-2017 project, under grant agreement No. 778228 (IDEAL-CITIES) and Shaqra University in Saudi Arabia.

References

1. Hampton, K., Goulet, L.S., Rainie, L., Purcell, K.: Social networking sites and our lives. Pew Internet Am. Life Proj. **16**, 1–85 (2011)
2. Griffiths, M.: A 'components' model of addiction within a biopsychosocial framework. J. Subst. Use **10**(4), 191–197 (2005)
3. Widyanto, L., Griffiths, M.: Internet addiction: a critical review. Int. J. Ment. Health Addict. **4**(1), 31–51 (2006)
4. Winkler, A., Dörsing, B., Rief, W., Shen, Y., Glombiewski, J.A.: Treatment of internet addiction: a meta-analysis. Clin. Psychol. Rev. **33**(2), 317–329 (2013)
5. Webb, T.L., Sniehotta, F.F., Michie, S.: Using theories of behaviour change to inform interventions for addictive behaviours. Addiction **105**(11), 1879–1892 (2010)
6. Ali, R., Jiang, N., Phalp, K., Muir, S., McAlaney, J.: The emerging requirement for digital addiction labels. In: Fricker, S., Schneider, K. (eds.) Requirements Engineering: Foundation for Software Quality. REFSQ 2015. LNCS, vol. 9013, pp. 198–213. Springer, Cham (2015). https://doi.org/10.1007/978-3-319-16101-3_13
7. Alrobai, A., Phalp, K., Ali, R.: Digital addiction: a requirements engineering perspective. In: Salinesi, C., van de Weerd, I. (eds.) Requirements Engineering: Foundation for Software Quality. REFSQ 2014. LNCS, vol. 8396, pp. 112–118. Springer, Cham (2014). https://doi.org/10.1007/978-3-319-05843-6_9
8. Alrobai, A., McAlaney, J., Phalp, K., Ali, R.: Exploring the risk factors of interactive e-health interventions for digital addiction. Int. J. Sociotechnology Knowl. Dev. **8**(2), 1–15 (2016)

9. Davidson, L., Chinman, M., Kloos, B., Weingarten, R., Stayner, D., Tebes, J.K.: Peer support among individuals with severe mental illness: a review of the evidence. Clin. Psychol. Sci. Pract. **6**(2), 165–187 (2006)
10. Alrobai, A., McAlaney, J., Phalp, K., Ali, R.: Online peer groups as a persuasive tool to combat digital addiction. In: Meschtscherjakov, A., De Ruyter, B., Fuchsberger, V., Murer, M., Tscheligi, M. (eds.) Persuasive Technology. PERSUASIVE 2016. LNCS, vol. 9638, pp. 288–300. Springer, Cham (2016). https://doi.org/10.1007/978-3-319-31510-2_25
11. Alrobai, A., Dogan, H., Phalp, K., Ali, R.: Building online platforms for peer support groups as a persuasive behavior change technique. In: Ham, J., Karapanos, E., Morita, P., Burns, C. (eds.) Persuasive Technology. PERSUASIVE 2018. LNCS, vol. 10809, pp. 70–83. Springer, Cham (2018). https://doi.org/10.1007/978-3-319-78978-1_6
12. Braun, V., Clarke, V.: Using thematic analysis in psychology. Qual. Res. Psychol. **3**(2), 77–101 (2006)
13. Alrobai, A.: Engineering social networking to combat digital addiction: the case of online peer groups. Doctoral dissertation, Bournemouth University (2018)
14. Sousa, M.J., Rocha, Á.: Game based learning contexts for soft skills development. In: Rocha, Á., Correia, A., Adeli, H., Reis, L., Costanzo, S. (eds.) Recent Advances in Information Systems and Technologies. WorldCIST 2017. AISC, vol. 570, pp. 931–940. Springer, Cham (2017). https://doi.org/10.1007/978-3-319-56538-5_92

Automated Computer-aided Design of Cranial Implants Using a Deep Volumetric Convolutional Denoising Autoencoder

Ana Morais[1,4] (iD), Jan Egger[2,4] (iD), and Victor Alves[3(✉)] (iD)

[1] Department of Informatics, School of Engineering, University of Minho,
Braga, Portugal
a70484@alunos.uminho.pt
[2] Institute of Computer Graphics and Vision, Graz University of Technology,
Graz, Austria
egger@icg.tugraz.at
[3] Algoritmi Centre, University of Minho, Braga, Portugal
valves@di.uminho.pt
[4] Computer Algorithms for Medicine Laboratory, Graz, Austria

Abstract. Computer-aided Design (CAD) software enables the design of patient-specific cranial implants, but it often requires of a lot of manual user-interactions. This paper proposes a Deep Learning (DL) approach towards the automated CAD of cranial implants, allowing the design process to be less user-dependent and even less time-consuming. The problem of reconstructing a cranial defect, which is essentially filling in a region in a skull, was posed as a 3D shape completion task and, to solve it, a Volumetric Convolutional Denoising Autoencoder was implemented using the open-source DL framework *PyTorch*. The autoencoder was trained on 3D skull models obtained by processing an open-access dataset of Magnetic Resonance Imaging brain scans. The 3D skull models were represented as binary voxel occupancy grids and experiments were carried out for different voxel resolutions. For each experiment, the autoencoder was evaluated in terms of quantitative and qualitative 3D shape completion performance. The obtained results showed that the implemented Deep Neural Network is able to perform shape completion on 3D models of defected skulls, allowing for an efficient and automatic reconstruction of cranial defects.

Keywords: Cranial implant · Medical imaging ·
Computer-aided Design (CAD) · Deep Learning · 3D shape completion ·
Denoising Autoencoder

1 Introduction

Cranial defects can be congenital or acquired due to trauma, tumours, infections or even compressions due to brain oedema [1, 2]. The reconstruction of such defects is mostly performed to assure protection of intracranial structures and also for aesthetic reasons. Cranioplasty refers to the surgical procedure where a bone defect in the skull is repaired

© Springer Nature Switzerland AG 2019
Á. Rocha et al. (Eds.): WorldCIST'19 2019, AISC 932, pp. 151–160, 2019.
https://doi.org/10.1007/978-3-030-16187-3_15

through the use of cranial implants [3]. Over the past decade, there has been an increase in the number of performed cranioplasties. However, this is still a complex procedure in neurosurgery since each implant must be patient-specific and a precise fit of the implant to the cranial defect is undoubtedly difficult but crucial to achieve for a successful reconstruction. Otherwise, gaps in the fracture area can lead to endochondral ossification and consequently delay healing [4]. Obtaining precise and satisfactory clinical results requires plenty of surgical experience and can be very time-consuming [3].

Recent advances in medical imaging and the development of systems combining Computer-aided Design (CAD) with Computer-assisted Manufacturing (CAM) have enabled surgeons to design and manufacture patient-specific implants at an acceptable cost and in a reasonable time [5], while also increasing the reliability of the reconstruction and the quality of life for patients [2].

2 Related Work

Designing a virtual 3D model of a cranial implant is a challenging task. Existing software typically requires manually placed markers to design the implant [3]. Therefore, a fully automated CAD solution is needed so that the design process can be less-user dependent and less time consuming. The most commonly used method towards this automatization is the mirroring technique, that consists on replacing the defected part with a mirrored surface from the opposite and healthy counterpart of the skull. However, this method is only suitable for small and unilateral defects [6]. Another approach consists on using surface interpolation based on Bézier surfaces [7], or on radial basis functions [8]. The main limitation is that the actual anatomical structure is not considered. A third approach consists of creating a reference model and applying a deformation algorithm. *Fuessinger et al.* proposed a data-driven approach for reconstruction of cranial defects based on a Statistical Shape Model that captures the shape variability of the cranial vault [4]. However, this method requires manually placed anatomical landmarks.

In this work, creating a cranial implant, is addressed as a problem of 3D shape completion which consists of reconstructing damaged or missing parts of a volume [9]. Leveraging the recent success of DL and the availability of 3D shape databases, such as *ShapeNet* [10] and *ModelNet* [11], several machine learning-based approaches to shape completion have been proposed [12]. *Wu et al.* proposed an architecture, called *3D ShapeNets*, in which the input shapes, are given as input to a Convolutional Deep Belief Network that learns a probabilistic distribution from 3D volumes for 3D reconstruction. Despite the results on 3D shape completion, this type of network is very difficult to train [11]. *Dai et al.* proposed a 3D-Encoder-Predictor Network to predict missing data in partial 3D models, producing a complete but low-resolution model as the output. This output is then correlated with shape priors, being processed by a 3D shape synthesis method to generate a high-resolution output. This is a supervised learning approach to shape completion and even though it enables high-resolution outputs, it is not an end-to-end procedure [9]. *Sharma et al.* proposed a Volumetric Convolutional Denoising Autoencoder (*VConv-DAE*) [13], suitable for inputs represented as voxel occupancy grids. It consists on the extension of the regular Denoising

Autoencoder (DAE) to volumetric data, learning a volumetric representation from noisy data by estimating voxel occupancy grids. Moreover, *VConv-DAE* is trained from scratch and end-to-end, and has shown to outperform the supervised approach of *3D ShapeNets* [11].

3 Materials and Methods

The proposed Deep Neural Network was implemented by performing transfer learning and domain adaptation on the *VConv-DAE* proposed by *Sharma et al.* [13]. This unsupervised approach requires more training time than a fully supervised one. However, this results in a more deterministic learning and allows the network to be robust to any type of noise pattern. All source code of the methods described in this paper are publicly available[1].

Dataset - Computed Tomography (CT) is a more appropriate imaging modality for processing bone structures when compared to Magnetic Resonance Imaging (MRI). Also, extracting a 3D model of the skull from a CT head scan is possible almost through simple grey-value thresholding [14]. However, despite being more expensive to acquire, there are more publicly available MRI datasets. Thus, the implemented Denoising Autoencoder (DAE) was trained and tested on 3D models of skulls extracted from MRI images from the *1200 Subjects Release* (S1200), an open-access dataset provided by the *Human Connectome Project* (HCP). This dataset contains high-resolution structural MRI scans from 1113 healthy young adults [15].

Skull Mesh Generation - With *BrainSuite*, it was possible to produce surface mesh models of the inner and outer skull from the T1w MRI scans, in an automated way. Using *Matlab* functions from the mesh generation and processing toolbox *iso2mesh*, these files were processed in order to merge both surfaces into one single mesh. The inner and outer skull surface files of all the 1113 subjects in the dataset were processed, resulting in 1113 mesh files of 3D models of healthy skulls.

Voxel Grid Representation - Following the approaches of *3D ShapeNets* [11] and *VConv-DAE* [13], the chosen input representation for the autoencoder is the voxel grid representation. Each 3D skull mesh is represented as a binary tensor, where 1 indicates the voxel belongs to the mesh's surface and 0 indicates it does not belong to the mesh and represents empty space. Moreover, zero-padding is applied to the borders of each voxel grid to reduce the convolution border artifacts and avoid the information at the borders from vanishing too quickly. Experiments with the Volumetric DAE were carried out for inputs with three different voxel resolutions: 30^3, 60^3 and 120^3. These resolution values already take padding into account. The chosen padding size was kept at 3, therefore for each experiment the skull models were represented as voxel cubes of resolution 24^3, 54^3 and 114^3, respectively. Following a practice commonly found in literature [16], from the 1113 skulls in the dataset, now represented as binary voxel grids, 891 (80%) were used for training and 222 for testing (20%).

---

[1] https://github.com/ritaamorais/skull-complete.

Noise Injection to Simulate Cranial Defects - Due to the low availability of 3D models of real cases of defected skulls, masking noise was injected to simulate cranial defects on the skull models of the testing sets. For each instance, the location restricted to the neurocranium (upper part of the skull) of where a voxel cube will be removed from the skull model is randomly selected. Then, a voxel region with random dimensions is generated, but these dimensions are restricted to be at least 15% of the skull model's resolution value in order to avoid too small holes from being generated and to avoid selecting empty regions to be cut away. The generated voxel region is then placed in the skull model and the voxels where the region is placed have their occupancy value changed to 0, while the rest is left untouched, resulting in an artificially introduced defect in the skull.

Architecture

The proposed model was implemented using the DL framework *PyTorch* and is able to learn an embedding of shapes in an unsupervised manner for a subsequent shape completion. Moreover, this model can be trained end-to-end from scratch without the need of layer-wise pre-training.

The architecture of the implemented Volumetric Convolutional DAE that can process inputs with a voxel resolution of 30^3 is presented in Fig. 1. The input of the network is firstly fed into a Dropout layer with a probability value set at 0.5, which means that the network is trained for reconstruction by only observing a random 50% of the input voxels. Thus, the autoencoder is trained on a virtually infinite amount of data, which serves the purpose of input data augmentation and enables the network to generalise better by learning a skull shape representation that is more robust to different noise patterns, consequently avoiding overfitting. The dropout layer is followed by two 3D Convolutional Layers, which constitute the Encoder and gradually compress the input, enabling the extraction of more complex features at each level. The first convolutional layer has 64 filters (or channels) of size 9^3 and stride 3, while the second has 256 filters of size 4^3 and stride 2. In addition, each convolutional layer is followed by a ReLU activation function. The output of the last convolutional layer consists of 256 feature maps of size 3^3, which are then reshaped into a flattened one-dimensional vector of length 6912 (= $256 \times 3 \times 3 \times 3$). This vector is given as input to a Fully-Connected (FC) Layer, which embeds this condensed representation into latent space. This fixed size encoded vector is then reconstructed back into the proper dimensionality of the output feature space with two 3D Deconvolutional Layers, which constitute the Decoder and perform transpose convolutions until the original size of the input (30^3) is recovered. The first deconvolutional layer contains 64 filters of size 5^3 and stride 2 and the second deconvolutional layer, with 1 filter of size 6^3 and stride 3, finally merges all 64 cubic feature maps back into a voxel grid of size 30^3. This last deconvolutional layer is followed by a sigmoid activation function that maps the network's output values to a range of [0, 1] that represent the probability for the presence of each voxel.

In order to process inputs with a voxel resolution of 60^3, the network's architecture was essentially kept the same but some adaptations had to be made regarding the size and stride of the filters in the convolutional and deconvolutional layers and the size of the FC layer. With the increase in the voxel resolution, the number of the network's learnable parameters also increases. For the model that processes inputs with a voxel resolution of 30^3 the number of parameters is 50 940 097, while the model that takes in inputs with a voxel resolution of 60^3 has 1 031 203 585 trainable parameters.

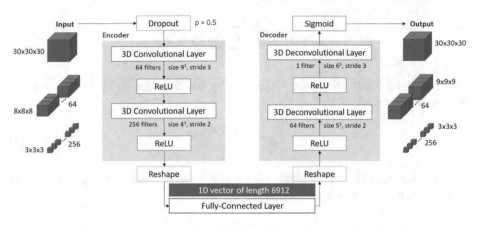

Fig. 1. Architecture of the Volumetric Convolutional DAE that processes input data with a voxel resolution of 30^3.

Training

Due to the satisfactory results obtained by the work of *Sharma et al.* [13] decisions regarding the loss function and the optimization were kept the same. The Binary Cross Entropy (BCE) loss function was the chosen one to measure the error of the reconstruction when training the autoencoder. In this case, the BCE function is calculated between the autoencoder's output and the uncorrupted input. Also, for all experiments, the chosen optimization algorithm was the Stochastic Gradient Descent (SGD) with a momentum term set to 0.9 and the learning rate set to 0.1. Mainly due to GPU memory limitations, the batch size was set to the lowest possible value: 1.

Evaluation

To evaluate the model, the defected instances are passed as input to the trained model and the output is then qualitative and quantitatively evaluated on shape completion. Besides the BCE loss, a more concrete evaluation metric is proposed - the reconstruction error - which represents how different the reconstructed skull is from its correspondent intact version (before being corrupted by masking noise).

Due to the sigmoid function, the output of the network is essentially a probability model for the presence of each voxel. In order to convert this output to a binary voxel representation, for each probability value p_i in the output, a threshold of 0.5 is set such that the following applies for each corresponding point o_i in the voxelized output model:

$$o_i = \begin{cases} 0, \text{ if } p_i \leq 0.5 \\ 1, \text{ otherwise} \end{cases} \tag{1}$$

Considering that O is the voxelized binary model consisting of all the o_i values, the total number of voxels in O that differ from those in H (the correspondent healthy skull model) is computed. Since O and H are volumetric grids of size N^3, the total number of erroneous voxels in a single reconstruction can be computed as follows:

$$d(O, H) = \sum_{i=1}^{N} \sum_{j=1}^{N} \sum_{k=1}^{N} |O - H|_{ijk} \tag{2}$$

Then, the reconstruction error is obtained by dividing $d(O, H)$ by the total number of voxels in the grid (N^3):

$$err = \frac{d(O, H)}{N^3} \tag{3}$$

The average reconstruction error for the attempted reconstructions on all defected instances can be calculated as follows:

$$err_{avg} = \frac{\sum_{i=1}^{T} err_i}{T} \tag{4}$$

where T denotes the number of instances in the testing set.

4   Results and Discussion

For the model that is able to process input data with a voxel resolution of 30^3, training was performed for 500 epochs with iterations through 891 instances of skull models with a size of 30^3 voxels, which took 5 h in total and required 1.51 GB of GPU memory. As for the model that processes input data with a voxel resolution of 60^3, training was performed for 500 epochs with iterations through 891 instances of skull models with a size of 60^3 voxels. With 5.88 iterations per second, training for this number of epochs took 28 h in total and required 21.08 GB of GPU memory.

Due to memory limitations, it was not possible to adapt the model's architecture to process input data with a voxel resolution of 120^3. With the available computational resources, the proposed model is only suitable for input volumes with a resolution not much greater than 60^3 voxels. Therefore, the skull models with a voxel resolution of 120^3 were divided into smaller but equally sized voxel grids of size 60^3 so that they could be fed into the model that processes input data with a voxel resolution of 60^3. Instead of training from scratch with this new data, the saved state from the model trained with skull models with a voxel resolution of 60^3 was loaded and training was resumed with the skull models with a voxel resolution of 120^3 split into 8 voxel grids of size 60^3 each, which resulted in 7128 instances to be used for training. The model was trained for 591 epochs, which took a total of 178.17 h and required 21.08 GB of GPU memory.

The graphs presented in Fig. 2 show the evolution of the BCE loss function during training for the different experiments that were carried out with skull models with varying voxel resolutions of 30^3, 60^3 and 120^3. The first graph (a) corresponds to training the model that processes input data with a voxel resolution of 30^3 with skull models with 30^3 voxels, the second graph (b) is regarding training the model that is able to process input data with a voxel resolution of 60^3 with the skull models with 60^3 voxels and lastly the third and fourth graphs (c and d) refer to training the model that processes input data with a voxel resolution 60^3 with the chunks of size 60^3 from the skull model models with a resolution of 120^3, with the difference between these two graphs resulting from an adjustment in the learning rate.

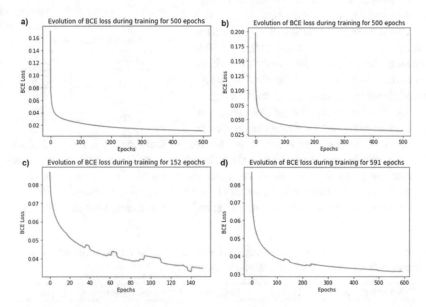

Fig. 2. Evolution of BCE Loss during training for the different experiments.

For the experiment with the skull models with a voxel resolution of 30^3, by the 500th epoch, it can be noticed that the model started to converge, with BCE loss values around 0.011. It was expected that, due to the dropout layer, a high number of epochs would be needed for the model to converge.

A similar behavior was observed for the experiment with the skull models with a voxel resolution of 60^3, in which by the 500th epoch the model had converged to BCE loss values around 0.031. This loss value is slightly higher when compared to the one verified in the experiment with skull models with a voxel resolution of 30^3. This is a consequence of the number of learnable parameters to optimize being much higher in the model that processes input data with a voxel resolution of 60^3, therefore learning is a much more challenging task.

As for the experiment with the skull models with a voxel resolution of 120^3, some oscillations in the BCE loss function started to be noticed when training with a learning rate of 0.1. Training was stopped at the end of the 152nd epoch, with a BCE loss value of 0.035. These oscillations were due to the learning rate being too high. Therefore, training was restarted with a new learning rate set at 0.08, which led to notably less oscillations in the loss function. Training was continued until the 591st epoch, when the model converged to BCE loss values around 0.032.

After having trained the models, it was possible to evaluate them by feeding all the instances of skulls with simulated cranial defects as input to each respective model. Then, shape completion is done by performing a forward pass on the network with each instance. The reconstruction error is computed for each instance of the testing set and by averaging these values, the average reconstruction error for the experiment is obtained. Table 1 presents the average reconstruction errors for each experiment.

The lowest average reconstruction error was observed in the experiment with skull models with a voxel resolution of 30^3, since the learning task is not as challenging in this experiment when compared with the others. Nevertheless, the obtained values are relatively low and therefore considered as satisfactory.

A qualitative evaluation of the proposed autoencoder on the task of shape completion of defected skull models can be done by taking an instance of a defected skull and performing one forward pass through the network with it. The output is a reconstructed version of the given defected skull. An example of the obtained qualitative results for shape completion on defected skull models for each voxel resolution of 30^3, 60^3 and 120^3 can be visualized in Table 2. As observed in the presented qualitative results, the

Table 1. Average reconstruction error for each experiment with skull models of different voxel resolutions

Resolution	Error
30^3	2.627%
60^3	3.237%
120^3	3.581%

Table 2. Examples of qualitative results for shape completion on skull models with different voxel resolutions, and the respective reconstruction error

Resolution	Input	Output	Ground-truth	Error
30^3				2.394%
60^3				2.471%
120^3				3.515%

implemented Volumetric DAE can successfully perform shape completion on skull models with cranial defects.

The quality of the reconstruction appears to be visually slightly worse on the skull models with a voxel resolution of 120^3. Also, due to the division that was performed on the 120^3 skull models, gaps can be observed in the regions where the smaller 60^3 chunks interconnect. This visual effect could have possibly been avoided if zero-padding was applied to the borders of each 60^3 volume, preventing the information on the borders to be washed away by the convolutional layers.

Nevertheless, the qualitative results are considered satisfactory and the completed skull surfaces can provide a good approximation for the reconstruction of the cranial defects in an automatic way, showing that reconstructing a cranial defect is possible by taking a 3D model of the defected skull and performing a single forward pass through the trained Volumetric DAE.

5 Conclusion and Future Work

With the work presented in this paper, it is possible to conclude that implementing a DL model to perform 3D shape completion on defected skull models can indeed be a promising approach towards the automated reconstruction of cranial defects.

As expected, the model adapted to process inputs with a voxel resolution of 30^3 was by far the one that required less time and GPU memory to train, since it is the one with the lowest amount of learnable parameters. This was also the model that obtained the lowest average reconstruction error. Nevertheless, the average reconstruction errors for the other models were relatively low, with values under 4% and therefore considered satisfactory. The obtained qualitative results are also considered satisfactory.

Since the main limitation of the voxel occupancy grid representation is its cubic growth in memory requirements with respect to the voxel resolution, due to limitations in available GPU memory, adapting the proposed Volumetric DAE to enable it to process inputs with voxel resolutions higher than 60^3 turned out to be unfeasible. However, an increase in the voxel resolution was still attempted, namely for the resolution of 120^3. Other ways to enable processing inputs with even higher voxel resolutions (as 256^3 or 512^3, for example) can be interesting possibilities for future work. A possible approach lies in adopting a different 3D data representation, such as the octree representation, which exploits the sparse nature of 3D data.

Acknowledgements. This work was supported by COMPETE: POCI-01-0145-FEDER-007043 and FCT – Fundação para a Ciência e Tecnologia within the Project Scope: UID/CEC/00319/2013, CAMed (COMET K-Project 871132) which is funded by the Austrian Federal Ministry of Transport, Innovation and Technology (BMVIT) and the Austrian Federal Ministry for Digital and Economic Affairs (BMDW) and the Styrian Business Promotion Agency (SFG), the Austrian Science Fund (FWF) KLI 678-B31 and the Erasmus+ Programme. We gratefully acknowledge the support of the NVIDIA Corporation with their donation of a Quadro P6000 board that was used in this research.

References

1. Chen, X., Xu, L., Li, X., Egger, J.: Computer-aided implant design for the restoration of cranial defects. Sci. Rep. **7**, 3–12 (2017)
2. Jardini, A.L., Larosa, M.A., Filho, R.M., Zavaglia, C.A.D.C., Bernardes, L.F., Lambert, C. S., Calderoni, D.R.: Cranial reconstruction: 3D biomodel and custom-built implant created using additive manufacturing. J. Cranio-Maxillofacial Surg. **42**, 1877–1884 (2014)
3. Egger, J., Gall, M., Tax, A., Ücal, M., Zefferer, U., Li, X., Von Campe, G., Schäfer, U., Schmalstieg, D., Chen, X.: Interactive reconstructions of cranial 3D implants under MeVisLab as an alternative to commercial planning software. PLoS ONE **12**, e0172694 (2017)
4. Fuessinger, M.A., Schwarz, S., Cornelius, C.-P., Metzger, M.C., Ellis, E., Probst, F.: Planning of skull reconstruction based on a statistical shape model combined with geometric morphometrics. Int. J. Comput. Assist. Radiol. Surg. **13**(4), 519–529 (2017)
5. Parthasarathy, J.: 3D modeling, custom implants and its future perspectives in craniofacial surgery. Ann. Maxillofac. Surg. **4**, 9 (2014)
6. Xia, J., Ip, H.H.S., Samman, N., Wang, D.: Computer-assisted three-dimensional surgical planning and simulation: 3D virtual osteotomy. IJOMS **29**, 11–17 (2000)
7. Mohamed, N., Majid, A.A., Piah, A.R.M., Rajion, Z.A.: Designing of skull defect implants using C1 rational cubic Bezier and offset curves. 050003 (2015)
8. Marreiros, F.M.M., Heuzé, Y., Verius, M., Unterhofer, C., Freysinger, W.: Custom implant design for large cranial defects. Int. J. Comput. Assist. Radiol. Surg. **11**, 2217–2230 (2016)
9. Dai, A., Qi, C.R., Nießner, M.: Shape completion using 3D-encoder-predictor CNNs and shape synthesis. In: Proceedings of the 30th IEEE Conference on Computer Vision and Pattern Recognition, CVPR 2017, pp. 6545–6554, January 2017
10. Chang, A.X., Funkhouser, T., Guibas, L., Hanrahan, P., Huang, Q., Li, Z.: ShapeNet: an information-rich 3D model repository. arXiv Prepr. arXiv:1512.03012 (2015)
11. Wu, Z., Song, S., Khosla, A., Yu, F., Zhang, L., Tang, X., Xiao, J.: 3D ShapeNets: a deep representation for volumetric shapes (2014)
12. Stutz, D., Geiger, A.: Learning 3D shape completion from laser scan data with weak supervision. In: Proceedings of the CVPR, vol. 1, pp. 1–10 (2018)
13. Sharma, A., Grau, O., Fritz, M.: VConv-DAE: deep volumetric shape learning without object labels. Lect. Notes Comput. Sci. (including Subser. Lect. Notes Artif. Intell. Lect. Notes Bioinformatics). LNCS, vol. 9915, pp. 236–250 (2016)
14. Lee, G.H., Chang, Y., Kim, T.-J., Lee, G.H.: Introduction to biomedical imaging (2014)
15. WU-Minn Consortium Human Connectome Project: HCP 1200 Subjects Dataset. https://db. humanconnectome.org/data/projects/HCP_1200. Accessed 22 Feb 2018
16. Buduma, N., Locascio, N.: Fundamentals of deep learning (2015)

Computational System for Heart Rate Variability Analysis

Larissa Fernandes Marães[1](✉), Milton Ernesto Romero Romero[1],
Bruna da Silva Sousa[2], and Vera Regina Fernandes da Silva Marães[2]

[1] Federal University of Mato Grosso do Sul, Campo Grande, Brazil
larissa.maraes@gmail.com, romero@facom.ufms.br
[2] University of Brasília, Brasília, Brazil
sousabrunadasilva@gmail.com, veraregina@unb.br

Abstract. The study of heart rate variation (HRV), also called cardiac frequency variability, is used to identify the range of samples where the respiratory sinus arrhythmia occurs and presents information of low frequency (LF) and high frequency (HF) densities. Cardiofrequency meters (Polar WearLink) are widely used heart rate monitors in sports and clinical practice. The sensor registers the values of the R waves (RR) intervals of the electrocardiogram (ECG) to be analyzed by the specialist. This process can introduce errors and deletion of important data due to human error. This work aims at designing a computational system to analyze the non-stationary signals to aid the specialist in cardiovascular diagnosis. The methodology is performed in three steps: (1) pre-processing with a compact support median filter to get rid of the outliers, then passing through a mapping of RR intervals to beats per minute (bpm) to verify maximum frequency, tachycardia and bradycardia in the time domain. To determine the outliers, a threshold is chosen to be applied to the bpm values according to the minimum and maximum values of the normal RR intervals, which depend on genre and age; (2) process with Spectrograms and continuous and discrete Wavelet transforms to improve the diagnosis; (3) visual interpretation of these results by the specialist by verifying that these results are consistent. The results, with the available data from the volunteers, show that the system proved to be consistent to help the specialist improve the accuracy in the cardiovascular diagnosis.

Keywords: R waves intervals · Heart rate variability ·
Respiratory sinus arrhythmia · Wavelet transforms

1 Introduction

The cardiac muscle has the capacity to maintain its own rhythm properly. The heart rate variation (HRV) may vary depending on the physiological conditions, i.e., body position, physical and pathological conditions. The study of HRV, also called cardiac frequency variability, is important in cardiovascular health [1–3].

To collect the HRV, cardiac frequency sensors, as cardiofrequency meters (Polar WearLink), are utilized in sports and clinical practice. The sensor is coupled to an elastic strap that is adjustable to the shoulder of the patient to send the signal to the

© Springer Nature Switzerland AG 2019
Á. Rocha et al. (Eds.): WorldCIST'19 2019, AISC 932, pp. 161–170, 2019.
https://doi.org/10.1007/978-3-030-16187-3_16

computer. Since the analysis of these signals is often performed visually and manually, this process depends on the experience of the specialist and can introduce errors and deletion of important data due to human error, therefore, computational systems are currently proposed to improve diagnosis, as the frequency-time analysis based on different (basis) Wavelets [4–6]. Analysis of HRV results obtained by using different wavelets is performed in [7]; comparison between Fourier Transform and Wavelet packets are shown in [8]; HRV processing with Fourier and Wavelet transforms is presented in [9]. Artificial intelligent methods for arrhythmic classification based on the RR-interval are addressed in [10].

The results obtained by utilizing different basis for the Wavelet analysis cannot define which is the most suitable Wavelet and Wavelet basis to perform the HRV processing. Taking into account these results, the main idea proposed here is to perform a time domain and frequency-time domain processing in the input signal and the specialist verifies consistency among these results in order to improve his/her diagnosis. It is expected that an abnormal heart behavior must be detected by the majority of these computational processes.

This work aims at designing a computational system to perform more automatic techniques for filtering and analyzing the non-stationary signals collected by sensor to develop a methodology to aid in cardiovascular diagnoses. The important information in this system is found in the amplitude, frequency, and time in which the frequencies occur. the solution method must be performed in three steps: (1) pre-processed with median window filter to get rid of the outliers, then passing through a mapping of RR intervals to bpm, (2) process with spectrograms and continuous and discrete Wavelet transform, (3) deliver to the visual interpretation phase. In order to preserve these invariants, since the noises do not occur in bursts, a median filter suffices to suppress the outliers. To determine the outliers a threshold is chosen according to the minimum and maximum value of the normal RR intervals, which depend on genre and age. Following into the next processing step, the RR intervals are converted to frequency values in beats per minute (bpm). To verify maximum frequency, tachycardia and bradycardia in the time domain, the threshold is applied to the bpm values. Frequency analysis in time domain is difficult to perform and, then, to determine the frequency domain invariant and time is which the particular frequency occurs, Frequency-time analysis with three transforms is performed. Firstly, short-time Fourier transform (STFT) spectrograms is used for the events of respiratory sinus arrhythmia that identify the range of samples where the arrhythmia occurs and present information of low frequency (LF) and high frequency (HF) densities. The STFT presents the difficulty of defining the window support for non-stationary signals, therefore, Wavelet (continuous and discrete) transforms is a suitable alternative to identify the arrhythmia event intervals providing the frequency interval and the time interval where these events occur. These three transforms results permit to improve the diagnosis due to the fact that they measured the amplitude, frequency, and time in which the frequencies occur and the specialist will verify if they are consistent to define the diagnosis.

The methodology to solve the problem proved to be consistent and this processing helps the specialist to improve the accuracy in the cardiovascular diagnosis.

The rest of this paper is organized as follows: Sect. 2 addresses the basic concepts; Sect. 3 describes the cardiovascular diagnosis computational system, results and discussion; and Sect. 4 summarizes the concluding remarks and future work.

2 Basic Concepts

The ECG is a noninvasive examination of the electrical activity record of the heart. It is formed physiologically by the waves P, Q, R, S, and T, representing the events of polarization and depolarization of the cardiac muscle [2]. The RR interval is the distance between two R waves of the ECG, and its analysis reflects the heart rate variability (HRV). In the normal diagnosis process, visual inspection of the RR intervals is performed by the specialist and the abnormal intervals are excluded. If this manual filtering approach excludes sample RR and the rest of the signal is shifted to the left, as a result, the signal undergoes phase changes, as well as moving the remaining samples. In addition, for very noisy signals, a large part of the signal is lost in this procedure. In [3] are considered only those with more than 95% of the original signal, thus, this filtering approach produces many losses of information of the HRV signal. A computational system addresses this issue. The processing is performed in time and frequency-time domains.

In time domain we sought to identify events in which the patient reaches maximum frequency. The maximum frequency is an indication that the heart is in its maximum effort, and therefore, varies according to the patient's age and sex [2] as given by Eq. 1 for males, and Eq. 2 for females, where "age" is the age of the patient.

$$f_{max(males)} = 220 - age \tag{1}$$

$$f_{max(females)} = 226 - age \tag{2}$$

In the analysis, we sought to identify events in which the patient reaches maximum frequency, sinus tachycardia, sinus bradycardia and respiratory sinus arrhythmia.

For better interpretation of the signal, the RR interval values (in milliseconds: ms) are converted to frequency values in beats per minute (bpm), as given by Eq. 3. For this, the value of the RR interval in ms is taking as 1/RR in order to obtain its frequency value.

$$y(t)_{bpm} = (1000 . 60) / y(t)_{ms} \tag{3}$$

The term tachycardia is defined as values above 100 bpm in adults and the term bradycardia is defined as values below 60 bpm [2], and this is important because very long events of bradycardia or a certain frequency of these events may lead to an infarct and even a cardiac arrest. The term sinus arrhythmia is applied when the signal may reveal a compass that seems, in most respects, to be a normal rhythm, except for being discreetly irregular. Generally, term sinus arrhythmia is defined as values near to normal.

The signal filtering is necessary to get rid of peak values mainly, not in burst, caused by noise (outliers), and then, non-parametric statistical methods, such as the median filter is sufficient for this process. The median filter puts the samplings in order and takes the sampling in the middle to define the value under consideration. A median filter with a support of size n = 10 samples traverses the signal verifying if any sample inside this window exceeds a value of the threshold, if yes, the RR interval is replaced by the median value, otherwise, the value is maintained. A threshold (200) is selected taking normal heart rate variations from 60 to 100 bpm, that are converted into RR intervals by Eq. (2) yielding 1000 ms and 600 ms, respectively, and the difference between these extremes (range) is 400 ms, i.e., ±200 ms.

The support of the filter is chosen by taking into account the number of consecutive outliers, that are assuming to be less than three consecutive outliers in the captured signal.

The frequency-time domain analysis is carried out by means of spectrograms and continuous and discrete Wavelets transforms since the arrhythmia produces frequency variations in a particular time. The main idea is similar to ask for a second medical opinion regarding the diagnosis. Note that, the spectrogram has the problem of choosing correctly the window which defines the time in which the frequency occurs [5]. The Wavelet has two dimensions: the scale (inverse of the frequency) and the translation which defines the time in which the particular frequency occurs. The continuous wavelet has a redundant basis and the discrete has an orthogonal basis. The Wavelet basis is the Mexican Hat and the detail coefficients of the discrete wavelet transform (DWT) [4] (from high-pass filters) contain important information for the diagnosis.

For the interpretation step, the specialist takes into account: amplitude, frequency, and time in which this frequency takes place and must verify if the time and frequency-time domain processing are consistent, if yes, the diagnosis is defined, if no, further analysis must to be performed.

3   The Computational System, Results and Discussion

The cardiofrequency meters strap is positioned in the xiphoid of the volunteer, so that, the collection could be terminated at any time if the volunteer has any discomfort. The HRV data collection has been performed with the subject in the supine position, for approximately 10 min. Ten volunteers have been instructed, at the first stage of collection, to maintain their breathing in a normal and calm manner, to remain silent, not crossing their legs or arms, not changing their position or sleeping through the entire collection procedure and, in the next stage of collection, to perform the respiratory sinus arrhythmia, that is, to reduce the respiratory cycle to a maximum of 5 respiratory cycles per minute.

The system has three stages: preprocessing, processing and interpretation. The RR signal is the input of the pre-processing step which is converted to bpm and filtered (median filter). The filtered signal is the input of the processing step, where it is processing in time domain and in frequency-time domain. The results of the processing are the input of the interpretation stage, where the diagnosis is defined.

In Fig. 1, the median filter result is shown and the filtered signal coincides with the original except for the outliers, demonstrating good filtering properties. Note that these are discrete values that are plotted continuously for clarity purposes only. As presented in Basic Concepts section, manual filtering [3] produces a loss of information and distortion of the signal, as well as moving the remaining samples, then, experiments are carried out in order to compare the manual filtering method and the filtering approach proposed here.

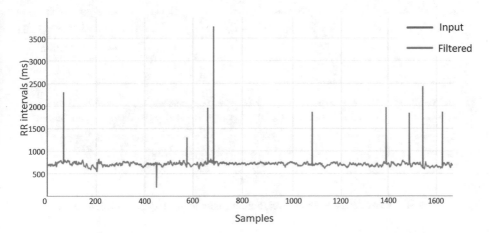

Fig. 1. Filtering using the median window approach (volunteer A).

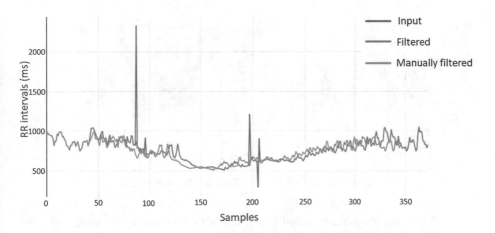

Fig. 2. Comparison between median filter and manual filtering approach (volunteer B).

In Fig. 2 the original signal in blue coincides with the filtered signal with the median filter except for the outliers. Note that, the green signal is manually filtered and presents phase and frequency differences with respect to the original, while the approach proposed here does not produce loss of samples in the signal, and keeps the signal filtered with the same amount of samples of the original signal.

The time domain analysis is performed to detect events of maximum frequencies, tachycardia and bradycardia. In Fig. 3 shows the detection of tachycardia events indicated in orange dots and the volunteer reached the maximum frequency values (FMax) and the normal values, as shown by red, blue, respectively.

In Fig. 4, the detection of bradycardia events, indicated in yellow, are shown.

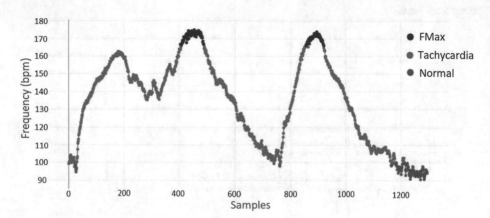

Fig. 3. Events detection of tachycardia and maximum frequency (volunteer A).

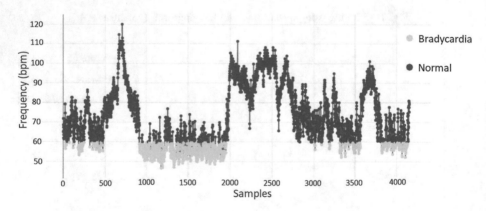

Fig. 4. Detection of bradycardia events (volunteer B).

In Fig. 5, the respiratory arrhythmia event, which occurs in the sampling range from 250 to 500, is shown. In Fig. 6, the spectrogram shows (in red) a peak at the 0.05 Hz frequency in the sample range coincident with the arrhythmia event. However, this peak is distorted because of the high values presented at very low frequencies (near 0 Hz). This is due to the impulse generated by the first FFT windows, see the bottom of the spectrogram near 0 Hz frequency for all samples which corresponds to the rectangular region in red. In this way, the discrepancy values (high values in red presented near 0 Hz) of the spectrogram are removed, as shown in Fig. 7, and displayed in 3D for better visualization and, therefore, it is possible to note that the peak of this spectrogram identifies the interval in which the arrhythmia occurs. In addition, with the side view of this graph, in the right, one can analyze the density of the LF and HF frequencies (and the interval in frequency at which these frequencies occur). The analysis of the density of these frequencies brings information about the operation of the sympathetic and parasympathetic nervous system in the heart [2].

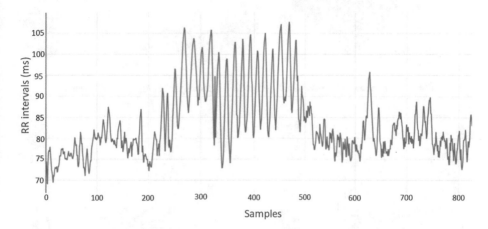

Fig. 5. Cardiac frequency signal with respiratory sinus arrhythmia event between samples 250 and 500 (volunteer C).

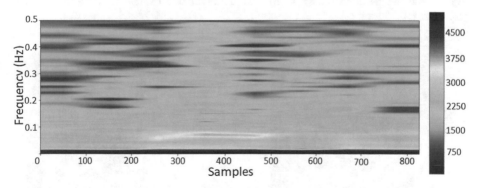

Fig. 6. Spectrogram, discrepant samples in the bottom near 0 Hz in red (volunteer C).

In Fig. 8a, the continuous wavelet transform is shown, in which the scaled graphs presents the Wavelet transform results for a signal without respiratory sinus arrhythmia and in Fig. 8b, a signal with respiratory sinus arrhythmia. Visual inspection permits to see clearly the differences.

In Fig. 9a, for a signal without respiratory arrhythmia, the Wavelet analysis is constant in most of its trajectory. Some peaks are identified, but with lower magnitudes compared to Wavelet analysis peaks for a signal with respiratory arrhythmia.

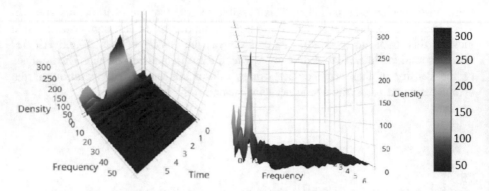

Fig. 7. 3D spectrogram with the removal of discrepant samples (volunteer C).

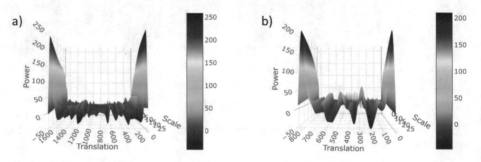

Fig. 8. Wavelet transform of a signal: **(a)** without arrhythmia (volunteer D); **(b)** with arrhythmia (volunteer C).

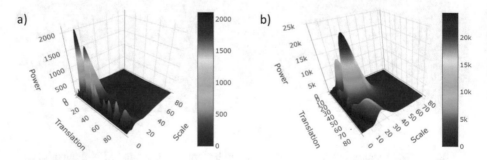

Fig. 9. Wavelet analysis graph of a signal: **(a)** without arrhythmia (volunteer D); **(b)** with arrhythmia (volunteer C).

In Fig. 9b, the wavelet analysis graph for a signal with respiratory arrhythmia shows that the power scale of this graph is greater than the scale of the previous graph. In addition, the highest power peak should be considered in the sample interval coincident with the interval at which the arrhythmia event occurs. Therefore, visual inspection shows that in this range of samples, the signal has high amplitudes.

For signals with respiratory arrhythmia, the Wavelet transform presents greater deformations in relation to the results obtained in normal signals, note the difference in the color scale and the oscillations with respect to the translation. The scales generated by the continuous Wavelet transform presented differences in their formats between arrhythmia and non-arrhythmia signals. However, to extract information of frequencies and amplitudes, it was necessary to analyze the graphs of the detail coefficients of the discrete wavelet transform. In the signal with the presence of respiratory arrhythmia, the peak was positioned in the range of samples where the arrhythmia occurred; indicating that in this interval the signal has higher amplitudes [5, 6].

Visual inspection performed by the specialist checks the consistency among the time domain and frequency-time results in order to improve the diagnosis, in case of discrepancy, further analysis must be conducted.

4   Conclusions

The computational system to improve cardiac diagnosis based on both time and frequency-time domains have been proved consistent.

Filtering with the median window approach has been efficient for the problem, as it performed cuts on the identified outliers without major distortions in frequency or in phase in comparison with the manual method which excludes the outliers and performs left shift of the samples. In addition, compared with manual filtering, this approach does not harm and does not cause loss of the original signal. Analyses in the time domain identified events of maximum frequency, tachycardia and bradycardia with efficiency. For arrhythmia events, it has been necessary frequency-time domain analysis, as it is noted that its invariants are contained in the frequency and amplitude of the signal.

For the detection of respiratory sinus arrhythmia events, the spectrograms show the range of samples in which the arrhythmia occurred. Moreover, with the spectrogram it is possible to analyze the density of the LF and HF frequencies. On the other hand, the spectrogram has the drawback that the window must be chosen carefully in order to obtain a suitable resolution in time and frequency and then, continuous and discrete wavelets transforms have been proved to be efficient to address the analysis of the RR signal to help in the health diagnosis by the specialists, based on the consistency of all the processing performed in time and in frequency-time domain. Given the above, it is clear that this methodology contributes to an improvement in cardiovascular analysis and diagnosis. Future work is related with the automatic diagnosis defining by a majority vote among the results of several convolutional neural networks.

Acknowledgment. Authors acknowledge the Technology Center of Electronics and Informatics, Federal University of Mato Grosso do Sul (UFMS), Laboratory of Clinical Physiology of University of Brasília (UnB) and National Council for Scientific and Technological Development (CNPq).

References

1. Marães, V.: Frequência cardíaca e sua variabilidade: análises e aplicações. Revista Andaluza de Medicina del Deporte **3**(1), 33–42 (2010)
2. Guyton, A., Hall, J.: Textbook of Medical Physiology, 13th edn. Elsevier, Philadelphia (2016)
3. Vanderlei, L.C.M., Pastre, C.M., Hoshi, R.A., de Carvalho, T.D., de Godoy, M.F.: Noções básicas de variabilidade da frequência cardíaca e sua aplicabilidade clínica. Revista Brasileira de Cirurgia Cardiovascular (Braz. J. Cardiovas. Surg.) **24**(2), 205–217 (2009)
4. Pale, U., Thürk, F., Kaniusas, E.: Heart rate variability analysis using different wavelet transformations. In: 39th International Convention on Information and Communication Technology, Electronics and Microelectronics (MIPRO), pp. 1649–1654 (2016)
5. Acharya, U.R., Joseph, K.P., Kannathal, N., Lim, C.M., Suri, J.S.: Heart rate variability: a review. Med. Biol. Eng. Compu. **44**(12), 1031–1051 (2006)
6. Vetterli, M., Herley, C.: Wavelets and filter banks: theory and design. IEEE Trans. Signal Process. **40**(9), 2207–2232 (1992)
7. Tzabazisab, A., Eisenriedac, A., Yeomansa, D.C., Moore, H.: Wavelet analysis of heart rate variability: impact of wavelet selection. Biomed. Signal Process. Control **40**, 220–225 (2018)
8. German-Sallo, Z.: Wavelet transform based HRV analysis. Procedia Technol. **12**, 105–111 (2014)
9. Ranganathan, G., Bindhu, V., Rangarajan, R.: Using wavelet transform for mental stress measurement. J. Theor. Appl. Inf. Technol. **11** (2010)
10. Tsipouras, M.G., Fotiadis, D.I., Sideris, D.: An arrhythmia classification system based on the RR-interval signal. Artif. Intell. Med. **33**(3), 237–250 (2005)

Information Technologies in Education

Principles for Design of Simulated Cases in Teaching Enterprise Modelling

Martin Henkel[(⊠)] and Ilia Bider

Department of Computer and Systems Sciences, Stockholm University,
Stockholm, Sweden
{martinh, ilia}@dsv.su.se

Abstract. Apprenticeship Simulation (AS) is a form of case-based learning where the students follow a virtual expert who selects and hands to them information sources with which they should work. A case as presented to the students can be quite complex, consisting of numerous types of diverse sources, including recorded interviews and links to real-world dynamic sources such as web pages. A teacher who designs an AS case needs to decide on the structure of the case presentation, including which sources should be used and how they and associated tasks should be presented to the students. In this paper, we present a set of principles for the design of AS case presentations. The principles are based on our experiences in applying AS in four courses in the area of enterprise modelling.

Keywords: Case-based learning · Modeling skills · Learning environment

1 Introduction

Apprenticeship simulation (AS) is an augmented version of Case-Based Learning (CBL), the essential part of augmentation being the substitution of text descriptions of a business case with its multi-media presentation [1, 2]. Moreover, AS is built upon that students can act as apprentices following a combination of virtual and real-life expert/master. Learning in AS is arranged around group (or individual) projects where the students follow the modelling master and help him/her to do some part of the work on building models. More specifically, the master chooses the information sources to be used for building a model, and hands the work of building the model to the students. Sources may include: (a) recorded interviews with stakeholders, e.g. CEO or CIO, which simulates actual participation of the apprentices as listeners in the interview sessions, (b) samples of relevant documents, e.g., meetings protocols, forms for managing orders, (c) web-based sources, e.g. a company website, a link to real product site, or simulated results of Twitter search on company name. Compared to a traditional case description using text, AS replaces the text with a multitude of diverse sources.

AS is aimed at solving the problem of teaching the skills needed for enterprise modelling. Modelling skills consist of two parts: (a) formal skills - knowledge on syntax and semantics of modelling languages, and (b) informal skills - knowledge on how to analyze the business reality in order to build a model. A problem in teaching and learning (enterprise/business) modelling in the university environment of today is

Á. Rocha et al. (Eds.): WorldCIST'19 2019, AISC 932, pp. 173–183, 2019.
https://doi.org/10.1007/978-3-030-16187-3_17

that it mainly focuses on acquiring the formal part of modelling skills; the students learn the syntax and, partly, the semantics of the formal languages used for modelling. The more important informal part of modelling – the knowledge on how to capture the reality to build a model through making field observations, interviewing people, and analyzing diverse documents - often remains outside the scope of the modelling courses.

While the formal part of the skills is quite suitable for acquiring in the classroom, the informal part is not, as it belongs to the area of tacit knowledge [3] (term coined by Polanyi), or Ways of Thinking and Practicing [4] (WTP - in the terminology of modern pedagogical studies). The best-known solution for acquiring this type of knowledge is apprenticeship where the students follow and help a modelling master in a real business case. However, in the university classroom setting this is difficult, if possible, to arrange, especially for the large undergraduate classes. AS could be considered as a good enough approximation to real apprenticeship.

The main AS concepts and our experience of using AS have been described in our previous works [2, 5]. In summary, we have found that AS provides several benefits compared to a traditional course design:

- *More realistic*. The students perceive that they get a more realistic understanding of the work that systems analysts do.
- *Preferred by the students*. Apprenticeship simulation is preferred by the students, compared to traditional case descriptions [2].
- *Are reusable*. With small changes a case has been reused on 3 courses [5].
- *Improves grades*. Gives slightly less low grades (F/FX/E) and more mid-level (D/C) grades [2].
- *Gives a holistic understanding*. Reusing the same case in 3 courses gave a perceived better understanding on how the subjects taught are related [1].

In this paper, we focus on the principles for designing a coherent case presentation. These include, for example, the creation of websites for simulating companies' and products web sites and deciding on how the modeling tasks should be presented to the students and linked to the sources. Each case is presented to the student in the form of a website that uses a combination of simulated and real sources, e.g. links to real company's website. Several cases designed for different courses can share some parts, which creates an interconnected environment. We call it Case-Based Immersive Learning Environment or CBILE for short, where term immersive in the name points to the use of a combination of real and simulated sources of information.

2 Overview of Related Research

As we have already mentioned in the introduction, the usual way of presenting a business case, real or imaginary, is by using only one type of media, a text in most cases. Using *simulation* for presenting a case is less common, but is reported in the literature. In the works that could be found in the literature, a case simulation in the computer is based on simulation of the object of investigation, e.g. an enterprise, not the situation of apprenticeship. For example, simulation of a patient in the medical profession has shown good results [6, 7]. However, we found only one example of using simulation of an enterprise for teaching systems analysis, HyperCase, which appeared as early as 1990. According to its designers, HyperCase showed to be more appreciated by the students than traditional methods [8]. Though HyperCase was introduced in 1990, it is still in use [9] as an accompanier for an IS course book [10].

IT-based *learning environments* are often used in courses to provide the students with course materials, including description of cases and associated tasks. By using these generic environments, it is possible to create, manage, and update learning materials in a structured way [11]. The main benefits of these environments are that they provide a generic way of presenting and managing learning objects. However, development of learning objects and associated tools has focused on the technical side in terms of platforms and exchange formats [12]. As a part of that work, it has been proposed to include all material required for a pedagogical goal, including student tasks, into a single learning object [13]. In this paper, we discuss principles tailored to the presentation of simulated cases, and are thus not targeting generic situations or the technical interchange of course material.

The literature, also, reports on tools for teaching specific types of enterprise modelling. One such example is MERODE [14], an environment built to teach students conceptual modeling. The main idea is that the students get prompt feedback on their models, which speeds up the learning cycle. This is similar to tools used for teaching SQL, such as [15], where feedback is given directly.

The tools discussed above are rather complementary to CBILE, they are not competing with it. What we aim at with the CBILE environment is realistic case presentations, generic learning environments and tools for teaching enterprise modeling can be integrated with CBILE technically or/and conceptually as needs arise.

3   Principles for Design of Business Cases for AS

In this section, we present principles for design of business case presentations for AS. They are derived from our own practice related to preparing four cases for four different courses. In our work, we used WordPress as a technical platform, but any other generic content management system, e.g. Drupal, can be used as well. For each case, an integrated project site is built, using a web-based content management system makes it easily available to many students working both from the university campus and from home.

Design principles can be split into two major categories:

1. Simulation – principles for creating a multi-media presentation of a business case for AS based teaching/learning
2. Packaging – principles for packaging the case presentation to make it available for the students.

The principles are described in the sub-sections that follow. Note that the principles concern the creation and distribution of case descriptions, and our experience with applying the principles. Other aspects of apprenticeship simulation are described in our previous work, such as the reuse of cases between courses [5] and the cost of applying apprenticeship simulation [2].

3.1 Simulation Principles

Design principles for simulation concern how the case is to be presented in terms of a combination of sources. We distinguish between three main types of sources:

1. *Video recordings.* Video recordings are used to capture interactions between the modeling master and the case participants – such as performing interviews targeting a certain business process. These sources represent a backbone of the case presentation.
2. *Web-based sources.* Video recordings are complemented with web-based sources for static information – such as information about products, organizational roles and other facts about the case. Web-based sources are constructed to look as real web pages.
3. *Document-based sources.* These are essentially documentation of internal events in the organization, such a board meeting protocols, and mails.

In our experience, the combination of the source types is what makes a case realistic, while using recorded interviews is the type most appreciated by the students. In a survey among the students of a course where all three-source types were used, 59% considered video interviews to be the most interesting to work with (Fig. 1), while 33% considered all sources equally interesting. At the same time, 30% of the students consider video recording as the most difficult type to work with, which is lower than we expected, while 36% consider all types equally difficult. The survey was performed on a large bachelor level course, where 49 out of the 210 students completed the survey. More results from the survey are reported on in [17].

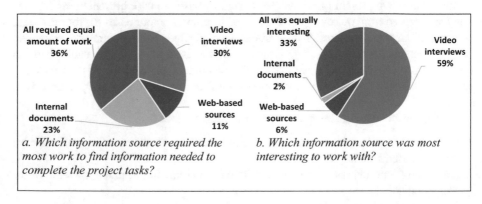

a. Which information source required the most work to find information needed to complete the project tasks?

b. Which information source was most interesting to work with?

Fig. 1. Student effort (a) and engagement (b) for the three types of sources

What helped the construction of sources in our practice was the use of a combination of real and fictitious sources. In this respect, we can introduce three categories of sources:

1. *Real world sources.* A real-world source is something that exists in a real world and is used "as is" without introducing any changes. A typical example here is information about what kind of software the case organization is using, which can be provided by referring to the real product, or vendor company web site.
2. *Prototype-based sources.* A prototype-based source is a real-world source where some elements have been replaced with fictitious ones. This included taking a real-world company's balance sheet and replacing the name of the company to fit the case. Another example is recording interviews with real people from a real company and changing the names of the company and people to the fictitious ones.
3. *Fictitious sources.* A fictitious source is an artificially constructed source. Fictitious sources are used to cover aspects that are not found in a real-world case. This included, for example, staging interviews with stakeholders, were teachers play both interviewees and interviewers. Another example, which we have used in a course based on recorded prototype-based interviews, is adding fictitious emails that contain crucial information that has been left out of the interviews.

The decision on which category of sources to employ depends on the level of the course and availability of the material. For example, in an introductory course for bachelor students, it might be wiser to use staged interviews, as it would be easier to provide information needed for modeling. For advanced courses, especially for master students, interviews with real people are preferred. As to the choice between real sources and prototype-based ones, the latter has an advantage to the former in that the material can be used a number of years without any change, while the "real-world" continues to evolve and may become not suitable for the case in question.

Summarizing the above, the design principles for simulation, in essence, are a loosely coupled set of guidelines on how to construct and combine sources of different types and categories (listed above) in order to represent a case. The combination of principles will be demonstrated on an example from our teaching practice in Sect. 4.

3.2 Packaging Principles

Design principles for packaging concern how the created sources are combined with tasks that the students should perform. There are three decision areas/dimensions for packaging:

1. Case presentation - how to present a project business case to the students
2. Task presentation - how to present the project tasks to the students
3. Task-sources relationships - how to connect tasks to the sources that need to be taken into consideration when completing the tasks

As far as the first dimension of case presentation is concerned, so far we used only one alternative. Namely, each project has a separate website where sources are arranged in groups according to their types: recordings, web-based sources, and document sources. Each type, if needed, could be subtyped, for example, internal and external documents. Each source has its own web-page, which in turn may have links to subpages, e.g. if a source is a company's website. A menu of any type can be used to provide links to the sources, e.g., a horizontal menu on the top of the project site as in Fig. 2, or a slide bar menu. In addition to sources provided in the project site, the students can be asked to find alternative sources themselves, e.g. possible competitors of the case company.

Regarding the *task presentation*, there are two alternatives, either they are represented in the case website as in Fig. 2, or outside it, for example in a university Learning Management System (LMS) that supports communication between the teachers and students. The advantages of the second alternatives are that a task definition can be easily connected to the time frame and to a submission box for grading. In our practice, we used both principles, dependent on the course and the preferences of the teachers involved. If the first principle is chosen, each task has its own page, and a menu is added to the project site which helps to navigate to a particular task, see, for example, the secondary upper menu in Fig. 2.

Regarding *task-source relationship*, there are two alternatives, either the students are required to find themselves which sources are relevant to a particular task, or the list of relevant sources is added to the task descriptions, as in Fig. 2 where it is done through including direct links to the relevant sources in the "Sources" section in the bottom part of the page. The first alternative is more appropriate for advanced courses, e.g. MS level courses; the second alternative is more appropriate for the introductory courses, e.g. BS level courses.

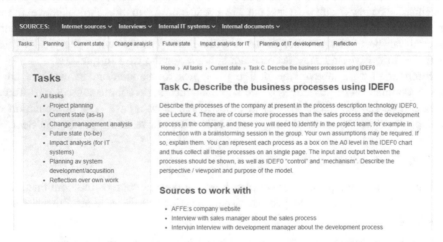

Fig. 2. Task directly related to simulated sources in a unified page

4   An Example of Using the Principles

The SERDES course will be used as an example of how the CBILE and the described principles have been used. SERDES aims at teaching the students software service design in the public sector. It contains elements of theories on service-oriented organizations, such as service-dominant logic (SDL). In addition, the course covers the use and design of e-government service platforms, and the design of service API:s. In the SERDES course, a case called "Harmony Inside" is used for creating enterprise models, including service value networks and conceptual models. The Harmony Inside case is based on a real-world business that provides dietary recommendations to their clients. The CBILE learning environment for the case consists of the following parts:

A Company Website. WordPress was used to create the "Harmony Inside" company site. The prototype-based category was chosen for this end – the content of the site was based on the real-world organization website, but the style had been changed (Fig. 3).

A Project Web Site. Again, WordPress was used to create a website for a project for which the students were supposed to work as apprentices (Fig. 4). On the project site, the expert "master" has gathered sources that are of interest. In the SERDES course, the *prototype-based* category was used for interviews – they were recorded with the real stakeholders, but the name of the company had been changed (Fig. 4).

Furthermore, the project site also used *fictitious* sources that augment the prototype-based ones. For example, a screen mockup was provided in order to create the feeling of that the students participated in an on-going design project (Fig. 5).

On the project site, links were also provided to the company website, and external web pages describing dietary illnesses and standards regarding information encoding of healthcare data.

For providing the students with tasks, the course utilized clear *task-source* relationships. In the final design, three tasks were given to the students – where each task had clearly identified sources. Thus, in the SERDES course, there were clear task-source relationships. The reason for doing this was to speed up the tasks by making them more focused. There was also the desire to mimic a typical project where information is added as the project progresses. It was furthermore decided that the *task presentation* should be separated from the project site. This was based on convenience – the university provided a learning environment with the possibility to have hand-in boxes close to the task descriptions.

Fig. 3. The company web site, created using a prototype approach

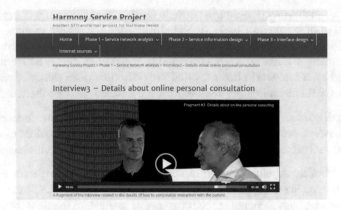

Fig. 4. Project web site, interviews with real stakeholders

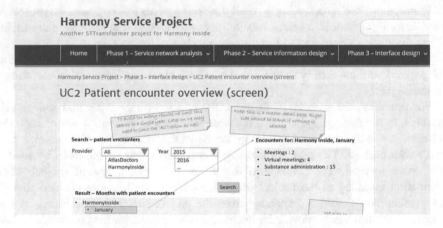

Fig. 5. Project web site, fictitious screen mockup

5 An Overview of the Use in Courses

The CBILE environment has been used in four courses, ranging from large (250+ students) to small (20 students). Table 1 contains a summary of the courses were CBILE has been used.

In all courses, WordPress was used for creating projects and case companies' web sites. The courses SERDES, SYSTOIT and BPCM were based on the same real-world case, the "Harmony Inside" case. While SYSTOIT and BPCM used the case as-is, SERDES added fictitious elements. ITO used a fictitious case, but employed the prototype-based category for creating some sources based on real data from real cases.

Having the same case for several courses was intentional, as we wanted to achieve a tacit-based connection between the subject matters taught in these courses based on the common case. The usage of AS and CBILE was evaluated through surveys and the results were positive, see [1, 2, 17].

Table 1. CBILE deployment in courses

#	Course name	Level (ECT)	# of rounds & students	Year	Types of models & sub-subjects
1	ITO - IT in Organizations	BS, 1st term (7.5)	5 * 250	2013–2018	Vide range of modelling notations and sub-subjects BPMN, IDEF0, BMM, etc. Requirements Engineering
2	BPCM - Business Process and Case Management	MS, 3rd term (7.5)	3 * 30	2016–2018	Business Process Modelling (mainstream & non mainstream)
3	SERDES - Citizen Centric Service Design	MS, 3rd term (7.5)	2 * 30	2016–2017	Service Value Network modelling, Conceptual modelling
4	SYSTIOT - System Theory, Organizations and IT	MS, 2nd term (7.5)	2 * 20	2017–2018	Viable System Model (VSM), Variety Engineering, Organizational goals, organizational learning

A combination of different source types and categories was used in most of the courses were CBILE was introduced. Table 2 contains a few examples of such usage.

Out of the four courses where CBILE was used, two opted for having a strong *task-source relationship* (Table 3), guiding the students to using specific sources for a certain task. For SERDES, the choice was based on the desire to introduce the case in a gradual fashion, while for ITO, the rationale was it being a bachelor level course where students needed more guidance. The SYSTOIT and ITO courses used a project site for both case presentation and task presentation. All courses also used links to external sites.

Table 2. Examples of used simulation principles – in form of source types and categories

Source type	Source category		
	Real	Prototype	Fictitious
Web	Link to external companies sites	Company web site	Project web site
Document		Balance sheet	Mail
Video		Stakeholders interviews	Staged interviews

Table 3. Use of packaging principles in the courses

Dimensions	SERDES	SYSTOIT	BPCM	ITO
Case presentation	Project site	Project site	Project site	Project site
Task presentation	Separate site	On project site	Separate site	On project site
Task-source relationships	Links	No links	No links	Links

6   Conclusion

In this paper, we have presented principles for construction of an environment that support a case-based learning approach called Apprenticeship Simulation (AS). AS aims at teaching enterprise modeling by using a set of multimedia presentations, rather than a traditional text-based case presentation. The core of AS is based on the student being viewed as an apprentice that follows a modeling master. The presented principles concern the design of a case presentation in terms of different source types (video, web, document), and the construction of these by the use of source categories (real, proto-type, fictitious). In addition, principles for packing source and task presentations has been introduced, which include several options for presenting source-task relationships.

The principles have been derived from our experience of using AS in four courses, for which several case presentations have been created using a set of interlinked WordPress sites.

References

1. Bider, I., Henkel, M.: Using the structure of tacit knowing for acquiring a holistic view on IS field. In: Proceedings of 11th IADIS International Conference on Information Systems, Lissabon, Portugal, pp. 19–26 (2015)
2. Bider, I., Henkel, M., Kowalski, S., Perjons, E.: Simulating apprenticeship using multimedia in Higher Education: a case from the information systems field. Interact. Technol. Smart Educ. 12(2), 137–154 (2015)
3. Polanyi, M.S.: Knowing and Being. University of Chicago, Chicago (1969)
4. McCune, V., Hounsell, D.: The development of students' ways of thinking and practicing in three final-year biology courses. High. Educ. 49(3), 255–289 (2005)
5. Henkel, M., Bider, I., Perjons, E., Mårtensson, F., Zainali, M.: Reusing cases for teaching enterprise modelling - feasibility study and reality check. In: Symposium on Conceptual Modeling Education (SCME 2017), International Conference on Conceptual Modelling (ER 2017), CEUR Proceedings, vol. 1954, pp. 4–14 (2017)
6. Bergin, R., Youngblood, Y., Ayers, M., Boberg, J., Bolander, K., Courteille, O., Dev, P., Hindbeck, H., Stringer, J., Thalme, A., Fors, U.: Interactive simulated patient: experiences with collaborative e-learning in medicine. J. Educ. Comput. Res. 29(3), 387–400 (2003)
7. Bergin, R.A., Fors, U.: Interactive simulated patient—an advanced tool for student activated learning in medicine and healthcare. Comput. Educ. 40, 61–376 (2003)
8. Kendall, J., Kendal, K.E., Baskerville, R., Barnes, R.J.: An empirical comparison of a hypertext-based systems analysis case with conventional cases and role playing. Data Base Adv. 27(1), 58–77 (1996)
9. Kendall, J., Kendall, K.E., Schmidt, A., Baskerville, R., Barnes, R.J.: HyperCase: a hypertext-based case for training systems analysts. http://www.pearsonhighered.com/hypercase/hypercase2.9. Accessed 2018
10. Kendall, J., Kendall, E.: Systems Analysis and Design, 9th edn. Pearson, London (2013)
11. Tono, L.: Learning objects: implications for instructional designers. Int. J. Instr. Media 38(3), 253–260 (2011)
12. Jesukiewicz, P.: Sharable Content Object Reference Model (SCORM), 4th edn. Content Aggregation Model. Advanced Distributed Learning (2009)

13. Rodríguez-Artacho, M., Verdejo Maíllo, M.F.: Modeling educational content: the cognitive approach of the PALO language. Educ. Technol. Soc. **7**(3), 124–137 (2004)
14. Sedrakyan, G., Snoeck, M., Poelmans, S.: Assessing the effectiveness of feedback enabled simulation in teaching conceptual modeling. Comput. Educ. **78**, 367–382 (2014)
15. Bider, I., Rogers, D.: YASQLT–Yet Another SQL Tutor. In: International Conference on Conceptual Modeling, pp. 197–206. Springer (2016)
16. Henkel, M., Bider, I., Kowalski, S., Perjons, E.: Reuse of simulated cases in teaching enterprise modelling. In: 5th Symposium on Conceptual Modelling Education (SCME 2015). CEUR (2015)
17. Bider, I., Henkel, M., Kowalski, S., Perjons, E.: Teaching enterprise modeling based on multi-media simulation: a pragmatic approach. In: International Conference on E-Technologies, pp. 239–254. Springer (2015)

Effectiveness and Fun Metrics in a Pervasive Game Experience: A Systematic Literature Review

Jhonny Paul Taborda[1(✉)], Jeferson Arango-López[1,2] [iD],
Cesar A. Collazos[1] [iD], Francisco Luis Gutiérrez Vela[2] [iD],
and Fernando Moreira[3,4] [iD]

[1] Faculty of Electronics Engineering and Telecommunications – FIET,
University of Cauca, Cl. 5 # 4–70, Popayán, Cauca, Colombia
{jptaborda, jal, ccollazo}@unicauca.edu.co
[2] E.T.S. of Computer and Telecommunication Engineering,
University of Granada, c/Periodista Daniel Saucedo Aranda, s/n,
18071 Granada, Spain
jeferson@correo.ugr.es, fgutierr@ugr.es
[3] REMIT, IJP, Universidade Portucalense,
Rua Dr. António Bernardino Almeida, 541-619, 4200-072 Porto, Portugal
fmoreira@upt.pt
[4] IEETA, Universidade de Aveiro, Aveiro, Portugal

Abstract. The progress of video games, hand in hand with the technological development, has been exponential in recent years. Its diffusion in society through multiple existing formats (computer, console, mobile devices, etc.) is growing. The number of video game players increases every year. As a result, the technical and user experience requirements also increase, and the behavior of the players with the game experience is unknown. Therefore, it is necessary to have the elements that allow evaluating and measuring this behavior from several perspectives. Therefore, it is necessary to have elements that allow evaluating and measuring the behavior of players with the game experience from different perspectives. It is important to have these different points of view to continuing the contribution in this aspect. Having that in mind, it has been made an analysis of measurement of effectiveness' measure and entertainment that will serve as a basis to approach studies in search of improvements in the user experience, a topic of great interest among the current generators or manufacturers of these video games.

Keywords: Effectiveness · Fun · User experience · Game experience · Metrics

1 Introduction

The emergence of new platforms and genres of video games shows us a problem from two perspectives: the analysis of the user experience is approached? And the second, how can we measure the properties that identify the level of "effectiveness" and "fun" of a gaming experience? [1, 2]. In addition, it is necessary to know which elements

© Springer Nature Switzerland AG 2019
Á. Rocha et al. (Eds.): WorldCIST'19 2019, AISC 932, pp. 184–194, 2019.
https://doi.org/10.1007/978-3-030-16187-3_18

within a video game are more related to the development and that can improve the game experience [3].

The general objective of developing a videogame is to be pleasant and rewarding for all possible players, but its design and development is a long and demanding process as well as complex considering the diversity of possible players, but over time it has been oriented to different efforts to design and evaluate usability aspects of videogames in the design and evaluation of usability aspects of these.

On the other hand, the user experience is considered a fundamental part in the evaluation of the video game performance with the user. This should start from the design of the game experiences, using evaluation techniques and repetitive adjustments in such a way that the development of complex processes and costs have greater acceptability in the user [4, 5]. For example, applying gamification patterns that help strengthen the relationship among the game experience and satisfaction [6].

In the videogame industry, effectiveness and entertainment is a subject treated by different authors [7, 8]. Although it lacks of analytical procedures and metrics to obtain a better estimate in the degree of acceptability and appropriation of the game by the player [2]. In this sense, pervasive games want the user to use and see the world in a different way, using the elements that are already integrated in our lives and that are part of the game.

The gaming experiences that have a higher level of gamification generate greater motivation and have greater acceptance by users as identified by Garcia in 2009 [4]. For this reason, there is a need in the video game industry to develop research that generates techniques and tools that allow the objective evaluation of the player's experience in games [9]. This article uses a thematic analysis of research for the classification of articles that have been published since 2009 to the present. These articles were searched in four databases: Scopus, IEEExplorer, ACM and Springer.

This article is structured as follows: Sect. 2 provides definitions and related works. In Sect. 3 we show the used methodology to carry out the systematic review of the literature. Section 4 shows the obtained results during the extraction and analysis of data. Finally, Sect. 5 describes the conclusions and future work.

2   Background

Metrics have been an important part since the beginning of software development [10], but as expressed in the literature, the measurement of user experience in video games is different due to the special characteristics of this type of software [11]. For a better understanding of this document, where was considered the works carried out and published that relate to the purpose of this review and the most important terms to be used are defined as effectiveness, metrics, and entertainment; as well, a review and detailed description of the theories that support the investigation.

2.1   Pervasive Games

According to [12], The Pervasive games (PG) are the kind of games that have the greatest complexity due to the diversity of characteristics and evolution of their rules and dynamics. From the perspective of user experience, a PG offers the player an enriched game experience through an evolution of the game dynamics, expanding the game space according to the context where it is played. This allows breaking the boundaries of the game world, making reality part of it and that the elements in that reality have an influence during the game [11].

The growing presence of games that combine the real world and the virtual world (game world) has the objective of improving the user experience. These are based on new technologies, context and multiple media with the existence of a narrative focused on the use of the interaction with the game for the appropriation of knowledge. Thus, there is evidence of the need to have a broad study about the possible metrics to be used in the evaluation of a game (whether or not it is pervasive). Everything, with the aim of improving the user experience that the player has with the gaming experience.

2.2   Game Metrics

Game metrics are interpretable measures of something related to games. Specifically, they are quantitative measurements of the attributes of the objects. A common source of game metrics is the telemetry data of the player's behavior [12, 13]. According to [14], there are three types of indicators used by game user experience researchers: (1) user metrics (related to the player), (2) performance metrics (related to software or hardware) and (3) process metrics (related to developing the game). When the game is measured, aspects of the game must be included, but there is something more important, the player's experience. In this line, there are metrics known as "playability metrics" [15] that allow measuring the fun of the player.

2.3   Effectiveness

A definition of effectiveness given by the famous video game designer Sid Meier, would be "the degree in which an interface facilitates the user to fulfill the task for which it was designed". This usually refers to the degree in which errors are avoided and tasks are successful, measured by "success rate" or "completion rate of tasks". By contrast, a measure of "error rate" is the amount of errors committed, and when are used to guide the design, those errors are often classified by cause [16].

2.4   Fun

Fun is recreation, rest, entertainment, hobby, joy, etc. The words "to have fun" have the meanings of entertaining, recreating and also those of diverting, distancing, and separating from routine. In military art, a secondary enterprise is called diversion which is carried out far from the main area of operations in order to get the attention of the enemy and separate it from its main objective, or force it to distract forces from the bulk of its army, weakening it. Historically, each era has had different ways of having fun [17].

3 Systematic Review of Literature

A systematic review of the literature is a method to analyze, evaluate and interpret each study relevant to a particular research question, specific area or phenomenon of interest [1]. This process was originated in the medical science due to the increasing amount of research in each area [18]. Consequently, it was necessary to identify and guide the research towards an uninvestigated subject [19]. The scientific community has proposed some steps for the application of these protocols, more specifically in the area of software engineering.

Kitchenham and Charters [18] propose a series of steps that are used in this document, which are adapted to our needs. The process of this methodology is presented in the following sections. The main objective of this systematic review of the literature was to obtain important data on scientific production considering the approach by academics and researchers in journal documents and/or conference proceedings, in order to identify the current state of effectiveness and fun metrics in video games. For this purpose, it was planned to search the different databases for relevant articles and we consider that the following questions are important for the investigation:

- **RQ1**: How to measure effectiveness and fun for the evaluation of a gaming experience?
- **RQ2**: Are there metrics that consider the characteristics of pervasive games in terms of effectiveness and fun?

3.1 Search Terms

It was necessary to evaluate different topics to select the terms to use in the search chains and their synonyms. In the case of pervasive games, the word pervasiveness was removed because many metrics applied to traditional games can apply in pervasive games. The terms to take into account in the systematic review are:

Mandatory words:	Optional words:
1. Metrics	1. Gamification
2. Effectiveness	2. Video games
3. Fun	
4. Gameplay	

3.2 Search Strategy

Considering the terms of Sect. 3.1, we have built a query string and this is complemented by logical operators to improve the execution results. We limited the search process to documents that had been published in journals, conference proceedings or book chapters since 2009. The sequence was executed on August 12th, 2018. For each database, it was necessary to build a specific query because each one has a different syntax; an example of a resulting query is shown below.

acmdlTitle:(+gameplay metrics measure) OR (+gameplay effectiv fun) AND
recordAbstract:(+gameplay metrics measure) OR (+gameplay effectiv fun) AND
keywords.author.keyword:(+game* effectiv* fun) AND (+game* metrics* measure)

3.3   Exclusion Criteria

1. Articles discarded by name.
2. Articles discarded by the summary.
3. Articles discarded by full text.

3.4   Inclusion Criteria

1. Articles included in the databases of Table 1.
2. Articles Published in Spanish and English.
3. Articles as a result of conferences, congresses, journals, book chapters.
4. Articles published since 2009.

3.5   Collected Information

We consider different databases to execute the search strings. Access to databases is private; the databases are shown in Table 1.

Table 1. Databases used in the search

Name	Acronym	URL
ACM digital library	ACM	https://dl.acm.org/advsearch.cfm
IEEE Xplore digital library	IEEEXplore	http://ieeexplore.ieee.org/
Springer link	Springer	https://link.springer.com
Scopus preview	Scopus	https://www.scopus.com/

When executing query strings in the databases of Table 1, 960 documents were found, which are in Appendix A; the results are shown in Table 2.

Table 2. Results of the run query for each database

Data base	Results	%
ACM	441	46
IEEEXplore	14	2
Springer	386	40
Scopus	119	12
	Total 960	

4 Data Analysis and Results

Table 2 shows the general results of the search. It was analyzed and evaluated the title and the abstract for each paper. Table 3 shows the percentages for the results of analysis.

Table 3. Title, abstract and full-text analysis results in the databases

Database	Total	D[a]	%D	A[b]	%A	TR[c]	AR[d]	RFT[e]	R[f]	%R
ACM	441	1	0,2	4	0,91	426	8	3	437	99
IEEExplore	14	0	0	0	0	9	5	0	14	100
Springer	386	0	0	2	0,5	366	12	6	384	99,5
Scopus	119	0	0,0	0	0	101	15	3	119	100
Total	**960**	**1**	**0,1**	**6**	**0,6**	**902**	**40**	**12**	**954**	**99,4**

[a]Duplicate: When a document was included in the results list more than once
[b]Accepted: Document that meets the requirements of the exclusion/inclusion criteria
[c]Rejected by title: Documents excluded because the title indicates another area of study
[d]Rejected for Summary: Documents excluded because the summary indicated another area of study
[e]Rejected by full text: documents excluded because the content indicated another area of study
[f]Rejected: Documents that do not meet the exclusion/inclusion criteria requirements (sum of TR, AR, and FTR)

4.1 Process Description

Once the first exclusion criterion was applied (based on the title), the number was reduced from 960 to 58 documents. When applying the second exclusion criterion (based on the abstract), the number of articles was reduced to 18. These 18 documents were read to achieve the third exclusion criterion (based on the full text). Finally, 6 documents were selected to answer the research questions.

Then, the referenced papers in Appendix A were classified by the categories. These categories as shown in Table 4, which were supported by data from: [18–22].

Figure 1 shows the results of the classification of the articles. The subcategory of serious games has 98 related documents, followed by the subcategory of Ux-Learning.

4.2 Results

After performing all the analysis of the data and the description of the process carried out, the effectiveness and fun metrics found in the accepted articles were shown, these metrics were classified in the categories of Social Game Metric, Measuring Player Population, Online Advertising [23, 26] as well as all the papers that were found in the different databases, a number was assigned to each one (Table 5).

Table 4. Results according to paper category

Category	Subcategory	#	Studies
Academic	User Adapted Interaction	20	576, 891, 894, 664, 941, 811, 719, 819, 914, 642, 917, 847, 924, 583, 613, 730, 840, 748, 865, 892
	Collaborative Work	42	632, 783, 835, 836, 852, 928, 648, 620, 929, 930, 937, 734, 787, 922, 886, 944, 896, 863, 712, 919, 678, 791, 867, 720, 765, 885, 568, 831, 722, 689, 687, 729…
	Ux, learning	85	561, 562, 563, 567, 569, 570, 581, 584, 599, 604, 618, 636, 679, 707, 746, 843, 753, 574, 585, 606, 644, 681, 762, 845, 846…
Technological	Virtual Augmented	28	890, 564, 598, 652, 704, 805, 866, 923, 789, 717, 731, 793, 936, 875, 830, 875, 895, 724, 572, 603, 695, 901, 688, 898, 698, 738, 860, 940
	Intelligent Technologies	98	927, 737, 743, 760, 617, 673, 690, 699, 700, 775,818, 862, 908, 932, 596, 609, 623, 624, 643, 658,676, 795, 799, 832, 909, 591, 747, 751, 755, 759,767, 792, 826, 858, 861, 921, 942, 586, 701, 814,694, 639, 739, 822, 841, 655, 876, 571, 638, 668, 692…
	Ubiquitous Computing	26	580, 595, 645, 761, 802, 887, 889, 905, 915, 801, 602, 742, 807, 873, 630, 674, 682, 800, 812, 960, 565, 589, 650, 653, 659, 735
Entertainment	Computer, entertainment	37	573, 577, 579, 587, 601, 605, 625, 660, 66, 777, 696, 709, 711, 721, 768, 773, 804, 810, 815, 849, 900, 903, 904, 906, 938, 813, 611, 641, 662, 786…

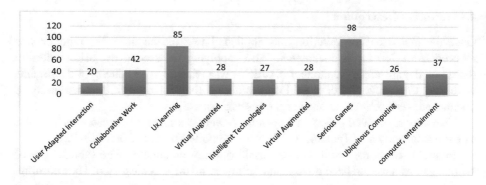

Fig. 1. Amount of papers classified by category

The information in Table 5 will be better interpreted in Fig. 2, which shows the number of metrics per category.

In Fig. 2 we can see that in the category Strong Narrative Structure, Meaningful Story Pieces, Interactivity, Skill Level is where there are more metrics found, then, according to the data thrown in the systematic review, the category of measuring

Table 5. Classification of metrics.

N° paper accepted	Category	#	Metrics
166	Social Game Metric	3	M001, M002, M003
285	Measuring Player Population	4	M004, M005, M006, M007
	Online Advertising	3	M008, M009, M010
	Measuring Monitization	5	M011, M012, M013, M014, M015
00	Strong Narrative Structure, Meaningful Story Piees, Interactivity, Skill Level	11	M016, M017, M018, M019, M020, M021, M022, M023, M024, M025, M026
561	Player Metrics	3	M027, M028, M029
604	Mechanics	1	M030
	Physical behavior	2	M031, M032
	Personally	2	M033, M034
	Game World Integration	4	M035, M036, M037, M038

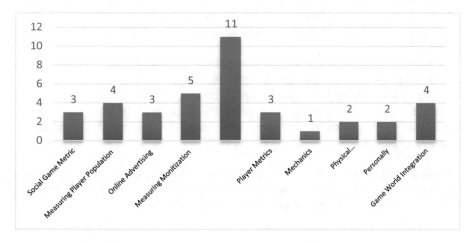

Fig. 2. Metrics contributed by the paper

monetization with a total of 5. Based on our systematic review of the literature and its results, provided answers to the research questions:

- **RQ1:** Through the effectiveness and entertainment metrics found in the review of the literature and taking advantage of their classifications, we can have the precision of knowing how effective and funny videogame can be. Furthermore, based on the findings, it can be concluded that the heuristics defined in the user experience can also be useful elements in the process of evaluating the effectiveness and entertainment transmitted by a gaming experience to the player.

- **RQ2**: According to this review, there are few studies on effectiveness and entertainment metrics in pervasive gaming experiences. As shown in Fig. 3, the metrics do not receive enough importance from researchers in this field. However, these metrics are fundamental in generalized computing environments due to limited hardware resources on mobile devices and the user experience that a person may have.

5   Conclusions and Future Work

This article presented a systematic review of the literature, which was aimed at answering research questions related to game effectiveness and fun metrics that could also be applied in a pervasive context. As a main conclusion, we can say that, through the systematic review, few metrics were identified that haunt the effectiveness and fun in pervasive games, with this we realize that there are still many possibilities of study in this area in order to improve the user experience.

On the other hand, we found a lot of study carried out in gaming experiences, but these are not measured through criteria that can be defined as metrics. Which leads us to conclude that these studies have not had the possibility of having tools that support the measurements that in many cases need to obtain better results in terms of the user experience.

It must be proposed efficiency and fun metrics based on the gameplay. And these metrics cover thematic of both general games and pervasive games. It will also be necessary to implement a prototype that serves as an experiment to apply the catalog of the metric set. This prototype will be a pervasive game that supports new students of higher education in the spatial adaptation of buildings, classrooms and main offices of a faculty or university.

Acknowledgments. This work has been funded by the Ministry of Economy and Competitiveness of Spain as part of the DISPERSA project (TIN2015-67149-C3-3-R) and the Excellence Project of the Regional Government of Andalucia (P11-TIC-7486).

Appendix A. Paper Identified

This information is available at http://goo.gl/QWDFwF.

References

1. López, J.A., Collazos, C.A., Velas, F.L.G., Moreira, F.: Using pervasive games as learning tools in educational contexts: a systematic review. Int. J. Learn. Technol. **13**(2), 93 (2018)
2. González, J.L., Vela, F.L.G.: Jugabilidad, Caracterización de la Experiencia del Jugador en Videojuegos (2010)

3. Boluda, I.K., Lozoya, V.C.: Efectos De Los Videojuegos En Las Marcas Emplazadas: La Transmisión De Emociones. Rev. Española Investig. Mark. ESIC **16**(1), 29–58 (2012)
4. Martínez García, A.: Patrones de Diseño aplicados a la organización de repositorios de objetos de aprendizaje, Des. Patterns Appl. to Organ. Learn. object Repos. (2009)
5. Gonçalves, R., Martins, J., Branco, F., Castro, M., Cota, M., Barroso, J.: A new concept of 3D DCS interface application for industrial production console operators. Univers. Access Inf. Soc. **14**, 399–413 (2015)
6. Ruiz, S., Taborda, J., Arango, J., Collazos, C.A., Guitierrez, F.: Catálogo de patrones de gamificación como contribución a la motivación del aprendizaje. Univ. del Cauca (2017)
7. Kiili, K., Moeller, K., Ninaus, M.: Evaluating the effectiveness of a game-based rational number training - In-game metrics as learning indicators. Comput. Educ. **120**, 13–28 (2018)
8. Sterkenburg, P.S., Vacaru, V.S.: The effectiveness of a serious game to enhance empathy for care workers for people with disabilities: a parallel randomized controlled trial. Disabil. Health J. **11**, 576–582 (2018)
9. Julianto, A., Anwar, N., Leslie, H., Spits, H.: Perfecting a video game with game metrics. Telkomnika **16**(3), 1324–1331 (2018)
10. Rodríguez, E.: Usabilidad, pp. 1–34 (2018)
11. Arango-López, J., Gutiérrez, F., Collazos, C., Valera, R., Cerezo, E.: Pervasive games : giving a meaning based on the player experience. In: Interacción 2017, pp. 1–4 (2017)
12. Arango-López, J., Collazos, C., Vela, F., Castillo, L.: A systematic review of geolocated pervasive games: a perspective from game development methodologies, software metrics and linked open data. In: Design, User Experience, and Usability: Theory, Methodology, and Management, pp. 335–346 (2017)
13. Drachen, A., El-nasr, M.S., Canossa, A.: Game Analytics – The Basics Take Away Points, pp. 13–40 (2013)
14. El-nasr, M., Durga, S., Shiyko, M., Sceppa, C.: Data-driven retrospective interviewing (DDRI): a proposed methodology for formative evaluation of pervasive games. Entertain. Comput. **11**, 1–19 (2015)
15. Sánchez, J., Vela, F., Simarro, F., Padilla-Zea, N.: Playability: analysing user experience in video games. Behav. Inf. Technol. **31**(10), 1033–1054 (2012)
16. Usability First - Usability Glossary - effectiveness | Usability First. http://www.usabilityfirst. com/glossary/effectiveness/
17. Roger Garzón, F.: El ocio, la fiesta, la diversión, pp. 1–13 (2018)
18. Kitchenham, B., Charters, S.: Guidelines for performing systematic literature reviews in software engineering. Engineering **2**, 1051 (2007)
19. Baptista, A., Martins, J., Gonçalves, R., Branco, F., Rocha, T.: Web accessibility challenges and perspectives: a systematic literature review, In: 2016 11th Iberian Conference on Information Systems and Technologies (CISTI), pp. 1–6. IEEE (2016)
20. Liu, K., Chen, S., Huang, H.: Development of a game-based cognitive measures system for elderly on the basis of mini-mental state examination, pp. 1853–1856 (2017)
21. Coleman, S., Menaker, E., Mcnamara, J., Johnson, T.: Communication for Stronger Learning Game Design, pp. 31–54 (2014)
22. Göbel, S., et al.: Serious games for health – personalized exergames, no. October, pp. 1663–1666 (2010)
23. Endrass, B., Hall, L., Hume, C., Tazzyman, S., André, E.: A pictorial interaction language for children to communicate with cultural virtual characters. In: Human-Computer Interaction. Advanced Interaction Modalities and Techniques, pp. 532–543 (2014)
24. Korhonen, H.: Playability heuristics for mobile games, pp. 9–16 (2006)

25. Emmerich, K., Masuch, M.: Game metrics for evaluating social in-game behavior and interaction in multiplayer games. In: Proceedings of the 13th International Conference on Advances in Computer Entertainment Technology, ACE 2016, vol. 2016, pp. 1–8 (2016)
26. Tychsen, A., Canossa, A.: Defining personas in games using metrics, pp. 73–80 (2008)

The Gamification in the Design of Computational Applications to Support the Autism Treatments: An Advance in the State of the Art

Gustavo Eduardo Constain M.[1] , César Collazos O.[2] ,
and Fernando Moreira[3,4(✉)]

[1] University of Cauca, National and Distance Open University,
Popayán, Colombia
gconsta@unicauca.edu.co
[2] University of Cauca, Popayán, Colombia
ccollazo@unicauca.edu.co
[3] REMIT, IJP, Universidade Portucalense,
Rua Dr. António Bernardino Almeida, 541-619, 4200-072 Porto, Portugal
fmoreira@upt.pt
[4] IEETA, Universidade de Aveiro, Aveiro, Portugal

Abstract. This paper presents the results found when consulting literature, with the purpose of elaborating the state of the art related to the use of gamification techniques in the development of computational applications that support the treatment of the Autism Spectrum Disorder (ASD). The research focuses on the results of using some models for the design of inclusive software, the steps to elaborate them and the possibility of adapting game techniques; so that its use can be modeled and adapted in a future framework for the design of specific applications for the autism management. Due to its characteristic, the game involves an intrinsic motivation, and a number of sensations that facilitate the development of emotions in the people who interact, that is why the study carried out seeks to find the importance of linking these techniques to the interior of the treatments of ASD, especially when it is intended to develop skills, such as self-recognition and social performance in children suffering from autism. With this article, we also want to propose an initial design framework for inclusive computational applications, based on gamification, related to the achievement of emotional skills in children with ASD.

Keywords: EmoTEA · Gamification · MDA Framework · 6D model ·
Autism Spectrum Disorder · MPIu+a model

1 Introduction

This paper is part of the development of a research project that seeks the creation of a framework, with some recommendations, for the design of inclusive computational applications applied to the treatment of Autism Spectrum Disorder (ASD). For the achievement of this purpose, it is important to identify the elements that are basic for

© Springer Nature Switzerland AG 2019
Á. Rocha et al. (Eds.): WorldCIST'19 2019, AISC 932, pp. 195–205, 2019.
https://doi.org/10.1007/978-3-030-16187-3_19

the inclusive computer design, but also, that what is implemented is part of the factors used in the treatment of autism, for example, the improvement of visual communication and social performance [1]. This is the case of 'play' as a basic element for the development of emotional skills in autistic children, as long as it is part of a professional treatment, applied among therapists, educators and the child's family.

The use of play strategies within the activities of the treatment of autism is quite usual; however, current therapeutic trends seek to improve the impact on the development of emotional and social skills with elements of serious games, implemented to through computer applications [1]. Therefore, by including some elements for treatment in computerized scenarios, called "gamification" or "serious games", has given positive results. However, it generates new concerns for therapists and the computer community. This is due to specific needs that vary from case to case of ASD, although they may appear similar.

In this paper, an approach to the state of the art presented that contributes to the construction of computational tools used as support in the treatment of autism, from a gamification based approach. It starts from the study of two models of computer application design for gaming and that related to the development of applications to support treatments of some type of disability. Among these are the MDA Framework and the 6D Method, which contain different frameworks for game design. While MDA provides the Mechanical, Dynamic and Aesthetic design of the application, 6D defines the phases by which we can build an inclusive application. Likewise, the MPIu+a model studied as a generic mechanism for the design of inclusive computing applications. Finally, we seek to propose an integrating model of design strategies for inclusive gamified applications and their use in inclusive environments that support the treatment of ASD, which corresponds to the EmoTEA Framework made by the authors of the document.

2 Related Work

The research work that originates this article seeks, on the one hand, the detailed knowledge of the characteristics of the ASD as well as the treatment options used; and on the other hand, the alternatives for the design of inclusive computational applications, investigating those that directly focus on cases of autism. Recently, a case study has shown us the direct relationship between the use of computer applications and the development of emotions in five children between 5 and 12 years of age (Fig. 1).

In this case, the study consisted of the evaluation of the impact of the use of computer tools in the generation of some emotional reaction in children diagnosed with autism. This study carried out with children from two educational institutions, with whom an exploratory study applied to find the relationship between autism and the use of learning applications for these cases.

Now, a next stage of the project requires an in-depth study of the types of inclusive computer applications most used in the treatment of autism and the determination of its functionality according to the most commonly used therapeutic requirements in the case study. This would allow identifying the characteristics of interaction, emotional

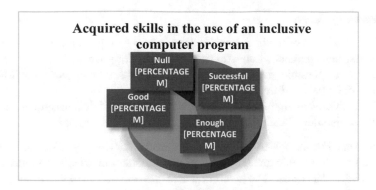

Fig. 1. Use of computer applications and emotional development

and social development, as well as communication between Therapist-Autistic Child-Family, this as part of an integrated model that improves the current treatment results.

3 Methodology

We opted for a systematic review (Fig. 2) to collect information of interest on technical issues considered important. The importance of the mapping of the systematic review is found in the structure and in the steps, it proposes to carry out the searches in an organized and methodological way, which helps to generate reliable results in the research [2].

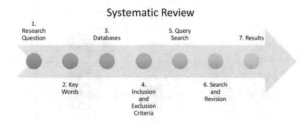

Fig. 2. Process for systematic review

In the process of systematic review [5], we follow the following steps: 1-Statement of research questions. 2-Definition of keywords for searches in English and Spanish. 3-Definition of databases to consult. 4-Definition of criteria for inclusion and exclusion of articles, and the search time interval. 5-Definition of advanced search chains in each of the databases. 6-Realization of search process and review of articles. 7-General review and Writing of found results. In this case, the research was specifically oriented to the design models of computationally inclusive applications and those that use gamification strategies and that could use as support for the treatment of the autistic spectrum syndrome.

Research Questions

RQ1: What types of treatment, with computational support, recently applied to the emotional management of autistic spectrum disorder - ASD?

RQ2: What possible gamification strategies used in computer applications to support the emotional treatment of autism?

RQ3: Is there any integration model between the clinical treatment of autistic spectrum disorder - ASD and the use of gamified applications?

Keywords: For the systematic review, the search for the key words in English and Spanish defined to include greater results of the searches and to allow a more complete revision in the databases. The words queries are the following: Abilities, Autism, Treatment, Emotions, Model and Gamification.

Databases: To make the research we use the following databases: Google Scholar, SCOPUS, Web of Science, IEEE Xplore and Science Direct.

Search Strings: We defined a general search chain based on the general concepts of the search time and that allowed us to answer the research questions posed. - *(("gamification*" OR "serious game") OR ("computation*")) AND ("treatment") AND ("autism*" OR "ASD" OR "syndrome").*

Search Process: Were found 496 articles in the databases and with them, the review process was started. In the first place, a general review carried out to find the articles, book chapters, etc., 53 articles repeated, originated by the different searches in the databases and by different search chains. Then, a revision of the titles and the summary of the works carried out. The general summary of the accepted works detailed in Fig. 3.

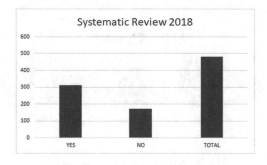

Fig. 3. Papers accepted in the systematic review

4   State of the Art

The review of literature allowed the identification of the most relevant studies for the purposes of the general project, which deepens in the knowledge of current treatments for autism, the existence of treatment activities based on the use of computer applications and the relationship between the use of gamification techniques and the development of emotional skills. In the same way, we study two models of design of gamified applications, viable of use for the particular purposes, and a model of design of inclusive applications.

4.1   Applied Studies

Related to the Treatment of ASD

It is clear that technology is a great ally to help in the development of people with disorders such as autism. In this context, there is a proliferation of initiatives aimed at the therapeutic support of this population, but adjusted to particular cases that have generated the formulation of projects in many cases, and collections of mobile applications aimed at communities that work in the treatment of this disorder.

The treatment alternatives experienced within the therapeutic circle has varied to look for complementary alternatives valid to the medicated management of children suffering from ASD [4]. Thus, for example, it been proven with the use of computer applications to facilitate the child's development of skills for concentration and learning activities in their daily lives. This therapy has used for its results in the medium or long term and it has varied in its application from the use of printed pictograms, digitized tablets, or computers to innovative therapies using trained dogs for children to copy their behavior in front of certain situations for which they want to appropriate their behavior, or simply because they calm their emotional sensations [5].

According to the above, we have explored the applications that most mentioned in the literature for work with autism, and that are known as Alternative and Augmentative Communication Systems (SAACs) [6].

In the initial exploration of the websites of the authors, it was possible to verify that, although there is a direct relationship in the conceptual management and presentation of the pictograms within the possibilities of use in cases of ASD, it is not directly visible its use as a complement to a therapeutic treatment accompanied by clinical experts. Consequently, we would have to resort to direct contact with the authors to request timely information about the technical details of implementation of these solutions and look for some type of case related to aspects of treatment design making use of each particular application.

Relationship Between Gamification and Development of Emotional Skills

The literature exploration, throws interesting deductions regarding the formulation of new therapeutic initiatives and that has originated programs such as the use of mobile devices with NFC technology [7]. Notwithstanding the above, no research have been found related to the combined application of the above elements, that is, the use of software tools that use NFC technology for the presentation of animated pictograms, in addition to the inclusion of elements of serious games that support the current treatment of children with ASD, or even, with other functional diversity, so it would be a contribution to new practices of inclusive education in this area of knowledge.

With respect to the model for the design of computational applications, that support the treatments of ASD, although in the selected literature the use of applications within some treatments is found with important results for the improvement of the behavior of children with ASD [4, 8], it is evident that it is not possible to generalize a treatment for various cases of autism that have been diagnosed, therefore, a specific treatment should be designed for each case and in this sense it would also have to make specific adjustments to the computational application that is designed from the focus of the game, interaction and accessibility principles [9].

Consequently, we seek to explore the advantages that the development of emotional skills involves the use of game techniques within therapeutic alternatives, this is the use of gamification within the design strategy of computational applications, where the use of these techniques for the innovation of therapeutic proposals facilitates the intrinsic motivation of the child with ASD (who learns) and increases the expected results [10, 12]. For the gamified design, the ARCS basic design model (Fig. 4) would be used to determine the levels of Satisfaction, Safety, Attention and Relevance [13] that are determined in a collaborative manner between the actors involved in the design: Therapist, Family members, Engineers and of course the factors identified in the child with ASD.

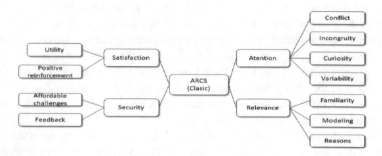

Fig. 4. Classic gamified design model

Related to the Design of Applications with the MDA Model

Because of the relationship between gamification and game, it is possible to use the same tools used for the development and creation of this type of applications. One such tool is the MDA Framework, one of the most common frameworks for game design [11]. This framework is a formal approach to understanding games, and its function is to be a design tool and a study tool. MDA allows approaching the design of the games in a methodical and iterative way. In addition, this approach allows conceptualizing the dynamic behavior of the games, thus facilitating the creation of iterative development techniques, allowing the early identification of undesirable results and rethinking development strategies according to what was expected (Fig. 5).

Fig. 5. MDA Framework

Related to the Design of Applications with the Model 6D

It is another working framework consisting of six (6) phases designed to design game type applications: (1) Definition of system objectives; (2) Definition of desired behaviors; (3) Description of the players; (4) Design of activity cycles; (5) Fun elements added; (6) Implementation of appropriate tools.

In addition to this type of architectural aspects, when gamified development is applied in contexts of support for therapies of syndromes such as ASD, it should be based on the collaborative gamification approach that should allow the mediation or accompaniment of an expert (therapist or specialized doctor) and guarantee the interactivity between all the participants of the experience: relatives and people supporting the computational development. All of them have the advantage of supported under HTML5 languages, CSS3, PHP and SQL support, which makes them compatible with many current internet devices and browsers.

Related to the Design of Applications with the MPIu+a Model

The Usability and Accessibility Engineering Process Model-MPIu+a, seeks to cover the aspects related to the Design of Interactive Systems Centered on the User (UCD) contemplating all its phases of realization (Fig. 6): Analysis, Design, Implementation, Launching, Prototyping and Evaluation [14, 15].

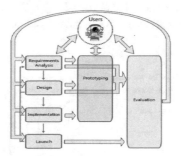

Fig. 6. MPIu+a model

Initially, the development of the model of inclusive computer applications from the concepts related to the disability linked to the ASD is framed and specifically in this case, it should focus on the development of intrapersonal skills (such as motivation) and interpersonal skills (for example, social skills) related to emotional intelligence that is what is sought in the research. This has to do with the conceptual organization of the project within the aspects of Software Engineering with the basic principles of Usability Engineering and Accessibility, providing a methodology that is able to guide development teams during the process of implementing a certain gamified, interactive and inclusive system.

The requirements of an interactive system usually refer to the functional component of the software leaving other aspects beyond the scope of the system, such as who will be the users and the use they make of the system. Due to the above, this phase of the process model is based on the Engineering Requirements and the quality model defined in the ISO/IEC 25010 standard, which describes the quality of the system requirements in the initial stages of the cycle life, referring mainly to the external view and view of the user (functionality, performance, security, maintainability, compatibility, portability and especially usability) rather than in reference to the internal or functional quality of the implementation, which is to which only the developers refer.

4.2   New Design Framework for Inclusive Applications for TEA (EmoTEA)

The engineering models of the requirements for the case of inclusive applications for the treatment of autism, add new factors to be taken into account, which should guarantee the development of applications with a much better degree from the functional point of view, that is, from its usability and its accessibility for children with ASD.

During the development of a computational application, once the functionalities that the system should cover along with the rest of its characteristics derived from the context of the interaction are resolved, the design of the activity and the design of the information are continued. These make up the general process of designing this interaction. In addition to this, the quality attributes mentioned by ISO 25010 and the playability characteristics required for gamified applications must be applied; therefore, within its architectural definition, a cycle of evaluation (or improvement) of the application to be developed must be included (Fig. 7). This is where the MPIu+a models and the already mentioned MDA and/or 6D design frameworks would be integrated.

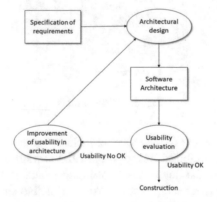

Fig. 7. Design of applications for usability

The above guarantees that both functional and non-functional requirements are met from the beginning of the design of the applications and not until the end that the tests are performed with children who immersed in an ASD treatment.

Once defined the functional requirements, a model of integration of the necessary elements is proposed to allow the design of computational applications with characteristics of development of emotional skills (Learning and treatment management system), tools for the co-creation of computer-assisted therapies, the product of gamified software and alternatives for interaction and communication between the therapist and the family of the child with ASD (Fig. 8).

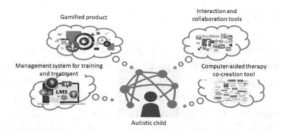

Fig. 8. Conceptual diagram of the EmoTEA Framework

5   Results

From a clinical approach, we learned that there are no two equal cases of ASD, therefore the definition of a treatment with computational elements must be specific for each child in particular. In this same sense, the development of emotional skills is a key factor in the treatment of autism, especially self-recognition and empathy, to facilitate the social interaction of the child. In this sense, there are diverse experiences of using specialized software tools for autism, which yield important results of usability and design, that provide key elements for a framework of design of specific computational applications in autism treatments, in the which requires the permanent monitoring of the therapist and the construction at home of the activities to be carried out for different periods of time.

For the above, there are computational applications design models, such as MDA, 6D and MPIu+a, that provide help for autism, but these have generally been used autonomously and without rigorous follow-up by the clinical specialist or without linkage of the child's family in the follow-up of the treatment.

These same models facilitate the development of computer applications for contexts of application of gamified elements, but taking into account the final user for whom they must designed, i.e. children with autistic disorder. Regarding the design of these specialized software applications, there are models, patterns and generalized recommendations for the design of inclusive software, but these lack some particular elements for the treatment of syndromes such as ASD. The state of art and technology, finds the existence of design models of inclusive applications (such as MPIu+a) from which you can start the proposal to create a specific framework, based on human-computer interaction, for the adaptation of inclusive computational applications, suitable to support the treatment of autism. This model of integration of inclusive computer applications and elements of autism treatment must be developed systematically and with a consequent validation of results from the most relevant usability and accessibility heuristics.

6 Discussion and Conclusions

Currently in Colombia is more evident the number of people with some type of disability, especially ASD, which requires a psycho-educational treatment tailored to their condition and especially configurable according to the progress of their therapy of cognitive stimulation. In this same context, we have identified treatments that make use of computer applications as complementary support, but not exclusive, within the therapeutic alternatives. These are mainly based on the training of emotional and social skills through the automated presentation of images (pictograms) or the mental exercise search of the child with ASD. With the above scenario, we solve the first question proposed for this review of the technique related to the treatment of autism.

However, it is found that the design of computer applications to support the disabled in Colombia is still incipient and it would be to verify if this is perhaps because the software community requires more conceptual tools that allow the design of inclusive applications orderly and quality for this population. The systematic review of computational applications for the ASD treatment shows an initial number of qualitatively evaluable software programs, however, it is not easy to identify validity as a support to established clinical treatments. Thus, a framework for the design of inclusive computational applications for the treatment of ASD can start from the use of already established architecture patterns and the use of models of application of game techniques, such as MDA or 6D; however, these must be adjusted to the extent that the therapeutic and technical conditions require it.

According to the above, it is possible to formulate a framework that seeks the parameterization of the design of computer applications that support the treatment of children with ASD, especially for the development of some specific predefined skills, which contributes to the achievement of our second question for this activity. Finally, in order to answer our last question, we can say that the MPIu+a model can be taken as a guide model for the usability and accessibility engineering process par excellence, however in particular cases of disability such as autism spectrum disorder could require some methodological adjustment to achieve the desired objectives, especially if specific requirements such as the linking of game techniques (gamification) are needed or if the computational application is an active part of the clinical treatment suggested by medical experts. The results of this study should verified and evaluated through their development and application in some case studies selected for experimentation within the next phase of the project that gives rise to this article.

References

1. Constain, G., Collazos, C., Fardoun, H.: Use of HCI for the development of emotional skills in the treatment of Autism Spectrum Disorder: a systematic review. Springer (2018)
2. Mattsson, M., Petersen, K., Feldt, R., Mujtaba, S.: Systematic mapping studies in software engineering, December 2018. http://www.robertfeldt.net/publications/petersen_ease08_sysmap_studies_in_se.pdf
3. Social Promotion Office: Situational Room for People With Disabilities. Ministry of Health and Social Protection, Republic of Colombia (2015)

4. Villalta, R., Sánchez Cabaco, A., Villa Estevez, J.: Design of digital applications for people with ASD. Int. J. Dev. Educ. Psychol. **4**(1), 291–297 (2013). National Association of Evolutionary and Educational Psychology of Children, Adolescents and Seniors, Badajoz, Spain
5. Sutton, H.: Therapy dogs can be successful motivating students with autism spectrum disorder. Disabil. Compliance High. Educ. **22**(7) (2017). Wiley Periodicals, Inc., A Wiley Company
6. Echenguia Cudolá, J.: Augmentative and alternative communication systems for the treatment of children with Autism Spectrum Disorder. **14**(28), (2016). Universidad Católica de Córdoba, Argentina
7. Constain, G., Moreira, F., Collazos, C.: Use of HCI for the development of emotional skills in the treatment of Autism Spectrum Disorder: a systematic review. In: 13ª Iberian Conference on Information Systems and Technologies (2018)
8. Lozano, J., Ballesta, J., Alcaraz, S., Cerezo, M.: Information and communication technologies (ICT) in the teaching and learning process of students with autistic spectrum disorder (ASD). Fuentes Mag. **14**, 193–208 (2016). University of Murcia, Spain
9. Vélez, M.: TEACCH program: proposal of psychoeducational intervention in people with ASD. Cadiz University (2017)
10. Constain, M.G., Mora, P.A., Santiago, M.J.: Use of gamification for the development of emotional intelligence skills: a theoretical and experiential approach. In: IV Ibero-American Conference on Human-Computer Interaction (2018)
11. Zicherman, G., Cunningham, C.: Gamification by Design. O'reilly Media Inc., Toronto
12. Pelegrina del Río, M., Tejeiro, R.: The psychology of video games: a research model. Aljibe. Accessed 10 Jan 2019
13. Hunicke, R., LeBlanc, M., Zubek, R.: MDA: a formal approach to game design and game research. In: Proceedings of the Challenges in Game AI Workshop, 19th National Conference on Artificial Intelligence. AAAI Press, San José. http://www.cs.northwestern.edu/~hunicke/MDA.pdf. Accessed 10 Jan 2019
14. Liu, W.: Natural user interface-next mainstream product user interface. In: 2010 IEEE 11th International Conference on Computer-Aided Industrial Design & Conceptual Design (CAIDCD). Accessed 10 Jan 2019
15. Granollers, T.: MPIu+a a methodology that integrates software engineering, human-computer interaction and accessibility in the context of multidisciplinary development teams. Accessed 10 Jan 2019

Using the SPAR Ontology Network to Represent the Scientific Production of a University: A Case Study

Mariela Tapia-Leon[1]([⊠]) [iD], Janneth Chicaiza Espinosa[2] [iD],
Paola Espinoza Arias[3] [iD], Idafen Santana-Perez[3] [iD],
and Oscar Corcho[3] [iD]

[1] Universidad de Guayaquil, Guayaquil, Ecuador
mariela.tapial@ug.edu.ec
[2] Universidad Técnica Particular de Loja, Loja, Ecuador
[3] Universidad Politécnica de Madrid, Madrid, Spain

Abstract. Research is commonly used to measure the prestige of universities. Currently, many universities register some of their scientific production (such as thesis, articles) in open access repositories using technologies like DSpace or ePrints. Likewise, scientific production is available in different and overlapping databases such as Scopus, Web of Science, Google Scholar, Crossref. Connecting these datasets, with the application of ontologies, will potentially increase their value and help discover interesting relationships amongst them. The present study aims to use the SPAR Ontology Network, which allows representing scholarly publishing, in order to check whether it is possible to represent the scientific production of universities. For that, we propose competency questions with the purpose of measuring scientific production. Likewise, we obtained data from Scopus regarding an Ecuadorian university (University of Guayaquil) and transformed it to RDF using the SPAR Ontology Network and other well-known ontologies to build it semantically. We used SPARQL queries to answer the competency questions. We concluded that the SPAR Ontology Network is a wide-ranging solution for scholarly publishing. Nevertheless, it is necessary to build an extension to one of their ontologies to provide a complete representation of a university's scientific production.

Keywords: RDF · Ontology · Scientific production · Scholarly publishing · SPAR Ontology Network · SPARQL

1 Introduction

According to the Oxford dictionary,[1] a university is a high-level educational institution in which students study for degrees, and academic research is done. Research is an essential activity inside universities, hence the scientific production of a university is a

---

[1] https://en.oxforddictionaries.com/definition/university.

© Springer Nature Switzerland AG 2019
Á. Rocha et al. (Eds.): WorldCIST'19 2019, AISC 932, pp. 206–215, 2019.
https://doi.org/10.1007/978-3-030-16187-3_20

crucial piece in the scoring of universities. Some internationally recognized organizations that measure and score universities, such as Times Higher Education-QS World University Rankings (www.timeshighereducation.com), Academic Ranking of World Universities (www.shanghairanking.com), Webometrics (www.webometrics.info) include research as one of the evaluation criteria.

Universities are now generally using open access repositories technologies like DSpace (duraspace.org) or ePrints (www.eprints.org) to register the scientific production (thesis or papers), which has helped to visualize and centralize the information in institutional repositories. Besides, they provide support for publishing stored contents in the form of Linked (Open) Data [1] and establish arbitrary relations between objects or provide additional metadata [2]. Likewise, information about the scientific production is available in different and overlapping databases (such as Scopus, Web of Science, Google Scholar, Crossref). Connecting these datasets, with the application of ontologies, will potentially increase their value and help discover interesting relationships amongst them [3].

The Semantic Web presents a new perspective on the association of the information contained in different databases [4] through the use of ontologies. An ontology is an engineering artifact constituted by a specific vocabulary to describe a particular reality [5]. Among their benefits, ontologies provide machine-readable metadata for data sources, using agreed standards that permit computers to assist in the tasks of information discovery and integration [6]. The Semantic Publishing and Referencing Ontologies, also known as the SPAR Ontology Network, is a suite of orthogonal and complementary ontology modules for every aspect of semantic publishing and referencing [7]. Their ontologies represent several aspects of scholarly publishing.

The present study aims to check whether the SPAR Ontology Network allows representing the scientific production of a university (particularly the University of Guayaquil in the 2017 as a case of study). For that, we used competency questions with the purpose of measure its scientific production. We transformed a series of datasets in CSV format (obtained from Scopus) to RDF and used SPARQL (Protocol and RDF Query Language) to answer the competency questions.

The rest of the paper is organized as follows. In Sect. 2, we introduce the notion of scientific production, and we describe the SPAR Ontologies briefly. In Sect. 3, we explain the methodology to evaluate the SPAR Ontology Network for scientific production. In Sect. 4 we expose de results of SPARQL queries. Lastly, in Sect. 5 we present our conclusion and future work.

2 Background

2.1 Scientific Production

Scientific production is the materialized part of the knowledge generated by research [8]. It is the result of research carried out by scientists generally belonging to a research group, department, center or university. It is usually collected through written supports. In this production, they disseminate the research to contribute to the growth and evolution of science. Scientific production is often measured by the number of publications produced by researches, institutions or countries. Other measures of productivity include the number of researchers by discipline and the number of citations received by their publications [9].

Some international organization such as Times Higher Education-QS World University Rankings, Academic Ranking of World Universities, Webometrics use scientific production as a criterion for evaluation and ranking. Likewise, there are government entities in charge of ensuring the quality of their universities based on the same criterion. In Ecuador, for instance, this organization is the CEAACES (http://www.caces.gob.ec/). They ensure that Ecuadorian universities contribute to the development of universal thinking, the deployment of scientific production and the promotion of transfer and technological innovations [10]. Hence, the importance to evaluate the scientific production at universities considering, for example, the publication of articles in databases (such as Scopus, Web of Science, Emerald, Pro-Quest, Scielo, Redalyc), books, books chapters, or publications at conference proceedings, workshops, seminars, forums or similar academic or scientific events.

2.2 Scholarly Publishing

There are some authors like Chauí [11] that support the idea that scientific production has a much broader scope, separating the production from the publication. For instance, a thesis in preparation, a thesis defended that has not yet been published, lab works completed and unpublished, fieldwork. All that is scientific production as well.

The guide to scholarly publishing and scholarly communication activities at Himmelfarb Library [12] explains that scholarly publishing is the result of research from which scholarly writings are created. Scholarly publishing exists to:

- Describe the research.
- Evaluate its reliability and reproducibility.
- Disseminate it through multiple channels.
- Preserve what has been done for future use.

Outwardly, the publication is the slight difference between scientific production and scholarly publishing. So, in the case of this article, we understand "scholarly publishing" as the publication of research results and "scientific production" as the measurement of scholarly publishing.

2.3 Ontologies for Representing the Scholarly Publishing Domain

The Semantic Publishing and Referencing (SPAR) Ontology Network (www.sparontologies.net) is a complete project to describe the scholarly publishing domain. It is a suite of orthogonal and complementary OWL 2 DL ontology modules for creating comprehensive machine-readable RDF metadata for all aspects of semantic publishing and referencing [4, 7].

Before SPAR there have been other attempts such as the well-known and widely used Dublin Core Metadata Terms (DCTerms - dublincore.org/documents/dcmi-terms), Functional Requirements for Bibliographic Records (FRBR - vocab.org/frbr/),

Publishing Requirements for Industry Standard Metadata (PRISM - www.idealliance. org/prism-metadata). All those models are actively used so far, but they lack the concepts of journal, article, book chapter, conference paper, reference list, citation, editor, and similar entities that are useful for describing the scholarly publication world in detail [7]. SPAR differs from other bibliographic ontologies like the Bibliographic Ontology Specification (BIBO - bibliontology.com), one of the first serious attempt towards providing an OWL-native scholarly-oriented publication ontology [7], in the fact that FaBiO (as part of the SPAR Ontology Network) is structured according to FRBR, providing the greater expressivity required for unambiguously describing the various essences of a bibliographic object. Besides, SPAR has a more comprehensive collection of classes and properties than BIBO, permitting more precise descriptions of bibliographic entities [14].

The current SPAR Ontology suite is composed of some ontologies. Each group of ontologies is addressed to a specific function like descriptions of textual publications (e.g., books, conference proceedings, journals); description of document components both structural (e.g., paragraph, section, chapter) and rhetorical (e.g., introduction, discussion, reference list, figure, appendix); characterization of the roles of agents (authors, editors, reviewers, publishers) in the publication process, bibliometric data and workflow processes, among others. Figure 1 depicts the ontologies involved in each functionality.

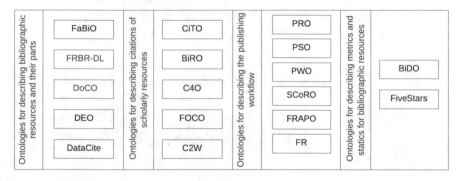

Fig. 1. SPAR Ontology Network

3 Methodology

For this study, we prepared ten competency questions with the purpose of measuring scientific production at the University of Guayaquil in 2017. These questions are divided into two groups. The first one (CQ1–CQ5) aims to obtain descriptions of textual publications like title, authors, affiliations, ISSN, DOI, volume, issue, publication date, among others (information useful for scholarly publishing). The second one (CQ6–CQ10) aims to obtain quantitative measures about the number of publications by research group them by books, journals, quartiles and conferences proceedings. Furthermore, to obtain some bibliometric data from authors or papers like h-index, the number of cites and number of publications (information useful for scientific production).

Some of these competency questions were based on the criteria "Scientific Production" of the "Model Generic of Evaluation of the Environment of Learning of Careers in Ecuador" by CEAACES – Ecuador [15].

- CQ1: What kind of publication is it?
- CQ2: With which other organizations the publication was made?
- CQ3: What is the article's bibliographic metadata?
- CQ4: What is the conference paper's bibliographic metadata?
- CQ5: What is the book's bibliographic metadata?
- CQ6: How many articles, books and conference papers have the researchers published?
- CQ7: How many citations have researchers' publications received?
- CQ8: How long has the researcher been publishing?
- CQ9: How many researchers have published in Qx (Qx = Quartile 1 or Quartile 2 or Quartile 3 or Quartile 4) journals and in which area?
- CQ10: What is the h-index, number of citations and number of publications for each researcher?

The following paragraphs explain the process carried out, from obtaining the dataset to make the queries in SPARQL. Figure 2 illustrates this process.

Our GitHub repository[2] contains all relevant material (e.g., files, queries, images, instructions) that can be used to reproduce our case study.

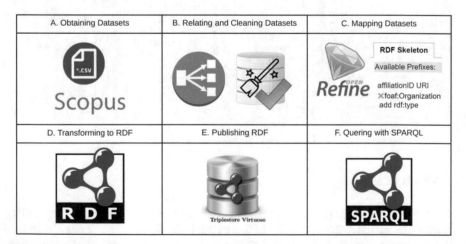

Fig. 2. The process from obtaining the dataset to making the queries in SPARQL

[2] https://doi.org/10.5281/zenodo.2536258.

3.1 Obtaining Datasets (A)

A public API (https://dev.elsevier.com/sc_apis.html) from Scopus was used to get the datasets. Scopus's API enables the consumption of data from all scholarly journals indexed by this database. We created some Python scripts for calling Scopus services with the objective to obtain data from the University of Guayaquil in the 2017 four our case study. The data was obtained in October 2018. Specifically, we obtained papers, authors, the source (journals, conferences, books) where they appear, and in order to obtain the necessary inputs to satisfy the question answering, we also obtained the ranking of Scimago Journal and Country Rank.

3.2 Relating and Cleaning Datasets (B)

We use MySQL and Open Refine to store and consolidate data obtained in the previous step. Table 1 shows the entity's name and their primary and foreign keys with which it was possible to relate the information. The column "Main Fields" displays the name of fields that provided necessary information about the scientific production.

Table 1. Main fields from the datasets provided by Elsevier

Entity	Primary and Foreign Keys	Main Fields
Authors	authorID, affiliationID	citations, docNumber, h-index
Papers	paperID, sourceID	cited by count
Authors-Papers	authorID, paperID	
Affiliations	authorID, affiliationID	Country
Sources	sourceID	SJR, h-index
Ranking2017	sourceID, areaID, disciplineID	SJR quartile, rank
AreasAndDisciplines	areaID, disciplineID	

Using MySQL, we merged the tables: Papers, Papers-Authors, AuthorsUG, AuthorsNotUG, Source and Affiliations; and directly using Open Refine, we joined the tables: Area&Disciplines with Ranking2017. Two Open Refine projects were created (union and ranking2017). We decided not to join them into one because the file ranking 2017 has sources that do not appear in the union file, and it could be useful the whole representation in RDF in the future. Finally, in Open Refine we cleaned the affiliation name of the University of Guayaquil, and we standardized the name to *Universidad de Guayaquil*.

3.3 Mapping Datasets (C)

For modeling the datasets into triples, we used the RDF extension from Open Refine. The RDF extension admits the reference of each component of the ontology to each column of data to be able to generate triplets. Some ontologies from the SPAR Ontology Network and other terms from well-known vocabularies were used to make the schema. Table 2 summarizes the ontologies re-used with their prefix and URI.

Table 2. The prefix of ontologies reused

Prefix	URI
bido	http://purl.org/spar/bido-core/
dcterms	http://purl.org/dc/terms/
fabio	http://purl.org/spar/fabio/
foaf	http://xmlns.com/foaf/0.1/
prism	http://prismstandard.org/namespaces/1.2/basic/
schema	http://schema.org/
time	http://www.w3.org/2006/time#
tvc	http://www.essepuntato.it/2012/04/tvc/

3.4   Transforming into RDF

Once the mapping of the data finished, that is when the columns of the table were related to its corresponding ontology term creating a schema, we proceeded to export an RDF file.

3.5   RDF Publishing

For publishing the RDF, we employed the triple database OpenLink Virtuoso (virtuoso.openlinksw.com).

The URI to access our endpoint to do any query to our data is http://spar.linkeddata.es/sparql, and the graph URI is http://spar.linkeddata.es/graph/ug.

3.6   Querying with SPARQL

For answering the competency questions, we used SPARQL. For each competency question, we wrote a query sentence. All these queries can be resolved in our endpoint. Also, it is possible to execute them from our repository.

4   Results and Discussion

Nine of the ten competency questions were successfully answering using three ontologies from the SPAR Ontology Network. The answer to competency questions is resolved like is shown in Table 3.

Table 3. Competecy questions resolved

CQ	Answer	SPAR Ontology	Observation
CQ1	Article, conference paper, letter, review, book chapter	FaBio	Total answers
CQ2	Universidad Agraria del Ecuador, Universidad Andres Bello, Universidad Autonoma de Baja California, Universidad Central del Ecuador, Universidad Nacional Autonoma de Mexico, Universidad Militar Nueva Granada	FaBio	Random extract from universities
CQ3	Issued: 7/11/2009 Publisher: Royal Society URL: https://goo.gl/f2cgtQ Volume: 276 Issue: 1674 https://doi.org/10.1098/rspb.2009.0998 ISSN: 09628452, 14712970	FaBio FRBR	The answer for the random publication 70449723371
CQ4	Issued: 23/5/2018 Publisher: IEEE Computer Society URL: https://goo.gl/fdXhvD Volume: 2018-April https://doi.org/10.1109/educon.2018.8363385 ISSN: 21659559, 21659567 ISBN: 9781538629574	FaBio FRBR	The answer for the random publication 85048143725
CQ5	Issued: 1/1/1989 Publisher: Balkema URL: https://goo.gl/xAebDy	FaBio FRBR	The answer for the random publication 24939296
CQ6	Articles: 296 Conference Papers: 98 Books: 4 Additional we obtained 24 reviews and eight letters	FaBio	Total answers
CQ7	Not was possible to represent the following information: paper citation count		
CQ8	Ten years		The answer for the random author Briones Claudett K
CQ9	Not was possible to represent the following information: area of study, quartile		Not resolved
CQ10	H-Index: 4 Number of citations: 41 Number of articles: 7	FaBio BiDO	The answer for the random author Flores L

5   Conclusions

The SPAR Ontology Network is a complete set of ontologies that allows representing information about scholarly publishing, as well as, some measures about scientific production. FaBiO, FRBR and BiDO ontologies from SPAR Ontology Network were used for mapping our Scopus data to RDF. Union Open Refine project was completely mapping. However, for Ranking2017 Open Refine project not was possible to find some concepts and properties to represent the information.

Six of ten competency questions were mostly resolved using the SPAR Ontology Network. For CQ2, we used FOAF and Schema vocabularies. So, the SPAR Ontology Network did not take part in the query sentences.

Concerning the scholarly publishing, it was possible to obtain:

- (CQ1) The kind of expression (an expression is the specific intellectual or artistic form that a work takes each time it is realized [16]): article, conference paper, letter, review, book chapter.
- (CQ3, CQ4, CQ5) The bibliographic metadata (DOI, ISSN, ISBN, URL, publisher, volume, issue, issued date).

Concerning the scientific production measures, it was possible to obtain:

- (CQ6) The number of articles, books, and conference papers that the University of Guayaquil had produced.
- (CQ8) The time in years that a researcher has published.
- (CQ10) The number of citations and h-index per author.

However, it was not possible to find any class and property that allow representing the following:

- (CQ7) The number of citations per researcher's publication.
- (CQ9) The Quartile and the area of study of journals.
- (CQ10) The Number of publications per researcher.

For that reason, our future work will consist in proposing an extension for BiDO ontology with the missing terms to provide a complete representation of a university's scientific production.

Acknowledgment. The first author of this article is very grateful to the members of the Ontology Engineering Group (OEG) of the *Universidad Politécnica de Madrid* who gave her the opportunity to be part of their group. Likewise, a special thanks to *the Universidad de Guayaquil* who financed her stay within the OEG group.

Funding. This work was partially supported by DATOS 4.0: RETOS Y SOLUCIONES – UPM Spanish national project (TIN2016-78011-C4-4-R).

References

1. Pascal-Nicolas, B.: Linked (Open) Data - DSpace 5.x Documentation. DuraSpace Wiki (2015). https://wiki.duraspace.org/display/DSDOC5x/Linked+%28Open%29+Data
2. ePrints Repository Softoware: New Features in EPrints 3.2. EPrints Documentation (2018). https://wiki.eprints.org/w/New_Features_in_EPrints_3.2#Linked_Data_Support
3. Omitola, T., et al.: Integrating public datasets using linked data: challenges and design principles. In: Future Internet Assembly Future Internet Assembly, pp. 1–10 (2010)
4. Santarém Segundo, J.E., Coneglian, C.S., Lucas, E.R.O.: Concepts and technologies of the semantic Web for academic-scientific cooperation: a study within the Vivo platform. Transinformaco **29**(3), 1–13 (2017)
5. Guarino, N., Oberle, D., Staab, S.: What is an Ontology?. Springer, Berlin (2009)
6. Shotton, D.: Semantic publishing: the coming revolution. Learn. Publ. **22**(2), 85–94 (2009)
7. Peroni, S., Shotton, D.: The SPAR ontologies. In: 17th International Semantic Web Conference, pp. 119–136 (2018)
8. Piedra Salomón, Y., Martínez Rodríguez, A.: Scientific production. Ciencias la Inf. **38**(3), 33–38 (2007)
9. Spinak, E.: Diccionario enciclopédico de Bibliometría. Cienciometría e Informetría. UNESCO, Caracas (1996)
10. Asamblea Nacional: Ley Orgánica de Educación Superior (LOES) (2010)
11. Guimar´es Pompío de Camargo, M.V.: Pesquisador científico: Avaliação de produçao (1997)
12. The Himmelfarb Health Sciences Library: Scholarly Publishing (2018). http://libguides.gwumc.edu/scholarlypub
13. Constantin, A., Peroni, S., Pettifer, S., Shotton, D., Vitali, F.: The Document Components Ontology (DoCO). Semant. Web **7**, 167–181 (2016)
14. Peroni, S., Shotton, D.: FaBiO and CiTO: ontologies for describing bibliographic resources and citations. J. Web Semant. Sci. Serv. Agents World Wide Web **17**, 1–15 (2012)
15. CEAACES: Modelo genérico de evaluación del entorno de aprendizaje de carreras en Ecuador (2017). https://bit.ly/2Dvp5iK
16. Ciccarese, P., Peroni, S.: Essential FRBR in OWL2 DL (2018)

Factors Affecting Adoption and Use of E-Learning by Business Employees in Cameroon

Marie Florence Abanda Maga[1(✉)], Jean Robert Kala Kamdjoug[1], Samuel Fosso Wamba[2], and Paul Cedric Nitcheu Tcheuffa[1]

[1] FSSG, GRIAGES, Université Catholique d'Afrique Centrale, Yaoundé, Cameroun
marieflorenceabanda22@gmail.com, jrkala@gmail.com, paulnitcheu5@gmail.com
[2] Toulouse Business School, Université Fédérale de Toulouse Midi-Pyrénées, 20 Boulevard Lascrosses, 31068 Toulouse, France
s.fosso-wamba@tbs-education.fr

Abstract. The aim of this study is to investigate the adoption and use of e-learning by business employees in Cameroon. To this effect, it seeks to: (1) identify the variables that influence the intention to use of e-learning; and (2) determine the impacts of the usage behavior on employees. To achieve these objectives, a mixed method approach combining quantitative and qualitative analyses was used. The quantitative approach is an analysis of research model obtained from a Technology Acceptance Model 2 (TAM2) that was modified by adding two other constructs: facilitating conditions from UTAUT, and well-being to measure the impact of eLearning. Then, with a sample of data collected from 159 business employees in Cameroon by means of a questionnaire, structural equation was measured for this analysis. The qualitative approach was applied concomitantly through the analysis of data collected during an interview with 10 employees who had always experimented eLearning before. The main result obtained from this study is that, in Cameroon-based enterprises, well-being at workplace is determined by usage behavior irrespective of gender and educational attainment when there is any intention to use e-learning.

Keywords: E-learning · Technology acceptance model · Well-being · Enterprise · Cameroon

1 Introduction

The development of skills has become a major issue for almost all companies in the world. Additionally, executives must provide their employees with two other important levers, namely (1) the means to perform more and more varied tasks, and (2) the right tools and cutting-edge knowledge at just the right time [1]. A laudable solution to such needs is the development of the eLearning, which can be defined as the delivery of course content via electronic media, such as the Internet, Intranets, Extranets, satellite broadcast, audio/video tapes, interactive TV, and CD-ROMs [2]. Organizations may well deliver consistent training to their employees in the workplace, update training content where necessary, reduce travel costs to outside training facilities, and provide training to employees on demand, anytime, and anywhere [3].

As a good number of African states and universities have been lagging behind in terms of socio-economic and educational performance, it is in their high interest to promote these third generation modes of training to meet their needs in economic environment skills [4]. According to Ambient Insight Regional Report, Africa is the most dynamic e-Learning market on the planet. The e-learning revenue in the continent reached $250.9 million in 2011 and went further than doubling by 2016 ($512.7 million). Cameroon has also fully realized the opportunity offered by this technology [5]. For example, through the E-National Higher Education Network project, which aims to cope with the exponential surge in student demand, the government intends to set up infrastructures and develop the skills need to facilitate e-learning [6]. However, in Cameroon, eLearning implementation in enterprise contexts is still in the early phases of adoption. ELearning implementation is prevalent only among established and large organizations. This can be justified by the fact that executives are still casting doubt on the ability of eLearning to effectively develop employee's skills according to enterprise needs.

Many studies have discussed the adoption and usage of eLearning. Most of them are based on the Technology Acceptance Model (TAM), the Diffusion of Innovation Theory (DOI) and the Unified theory of acceptance and the use of technology (UTAUT) models to explain the adoption and use of the e-learning. The findings of the major part of these studies reveal several factors that can positively influence the adoption and use of eLearning: perceived ease of use, perceived usefulness, compatibility, complexity, relative advantage, trialability, performance expectancy, effort expectancy, social influence, and facilitation conditions [7]. Beyond the continued adoption of various eLearning solutions in Africa, their actual usage is still mitigated. In fact, about 49% of 413 respondents from 42 countries who attended the 2013 eLearning Africa conference indicated that this training solution had been implemented but with a low level of usage.

It is only recently that eLearning has emerged in sub-Saharan African countries (including Cameroon), and as a result, the body of literature on the topic is very limited for this part of the world. It has therefore been very difficult for us find out relevant research on eLearning in the Africa's economic context. Under such circumstances, the present study is an attempt to fill this gap in the literature on eLearning in sub-Saharan Africa, while measuring the impact on the well-being of employees in their professional setting. Additionally, we seek to measure the moderating effect of the gender and scholar variables on the relation between the intention to use and the usage behavior, as well as on the relation between usage behavior and the well-being at work.

2   Theoretical Background

TAM has been widely used as the theoretical basis for many empirical studies on acceptance and use of eLearning. Park has stated that TAM was a good theoretical tool to understand the acceptance of eLearning by users [8]. His study has concluded that the subjective norms and the perceived ease of use are very good determinants to explain the acceptance of e-learning. In the same strain, Selim has conducted an empirical study on student acceptance of course websites, which validates the positive influence of perceived ease of use and perceived usefulness on the intention to use eLearning [9]. However, the majority of studies on the adoption of eLearning focus the acceptance of eLearning by students and very few are concerned with acceptance by employees in the enterprise context. It should be noted that for several authors, basic TAM is not suitable to cope with all the aspects of eLearning system acceptance by end-users within enterprise [10, 11].

Many empirical studies recommend the integration of TAM with other theories to cope with rapid changes in IS/IT which can improve their specificity and explanatory power. So, Lee, Hsieh and Hsu have presented a research model combining the innovation diffusion theory (IDT) with the technology acceptance model (TAM). The results of this study show that five perceptions of innovation characteristics significantly influence employees' eLearning system behavioral intention [12]. The research model of our study uses TAM 2 as the main theory to which we add the facilitating conditions of the UTAUT theory to try to explain the intention to use e-learning. Then, we use "well-being" to measure the impact of using eLearning on employees in enterprise context.

3   Research Model

Drawing on the TAM and UTAUT theories, we proposed the research model in Fig. 1 below, which enabled us to measure the validity of the links between the following constructs: subjective norms, output quality, perceived ease of use, image, perceived usefulness, and the intention to use and usage behavior.

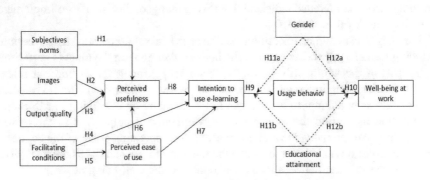

Fig. 1. The proposed model of study

Subjective Norms refer to "a person's perceptions that most people who are important to him think he should or shouldn't perform the behavior in question" [13]. Regarding the eLearning use, the individuals who can influence the behavior of an employee are notable colleagues and supervisors, amongst others. As soon as supervision is concerned, the perceived usefulness of eLearning by an employee seems to be boosted [14]. Then, we suggest the following hypothesis:

H1: Subjective norm will have a positive direct effect on image.
H2: Subjective norm will have a positive direct effect on the perceived usefulness of e-learning.

Image refers to "the degree to which use of innovation is perceived to enhance one's image or status in one's social system" [15]. The most important motivation for an individual to adopt a technological innovation is their desire to gain social status [16]. In this respect, employees who decide to engage in eLearning to develop their skills expect to earn notoriety among their colleagues. Then, we set forth the following hypothesis:

H3: Image will have a positive direct effect on the perceived usefulness of e-learning.

Output Quality is defined as "the degree to which an individual judges the effect of a new system" [20]. The perceived quality of a technology refers to its ability to facilitate the performance of work tasks. Thus, the use of a new technology must guarantee improvement of the results after use [20]. By analogy, the use of eLearning must guarantee the acquisition of new knowledge and the development of learners' skills, while having a direct relationship with the improvement of work performance. We, therefore, suggest the following hypothesis:

H4: Output quality will have a positive direct effect on the perceived usefulness of e-learning.

Facilitating conditions refer to the "organizational support that facilitates the use of an IT" [17]. Studies focusing on technologies define enabling conditions such as support or assistance to users. Support, information and training received by the user are therefore facilitating factors that can influence the intention to use the Internet [24]. Then, we suggest the following hypothesis:

H5: Facilitating conditions will have a positive direct effect on the perceived ease of use of e-learning.
H6: Facilitating conditions will have a positive direct effect on the intention to use to e-learning.

Perceived Ease of Use refers to "the degree to which the prospective user expects the target system to be free of effort". The efforts required are essentially determined not only by the perceptions of the usability of e-learning tools (including ICTs), but also by the ease of access to these technologies, provided they are available. Thus, the more the tool is perceived as easy to use, the more the perceptions of its utility increase for the employees. Therefore, the following hypothesis is formulated:

H7: Perceived Ease of Use will have a positive direct effect on the perceived usefulness of e-learning.
H8: Perceived Ease of Use will have a positive direct effect on the intention to use e-learning.

Perceived Usefulness is defined as "the prospective user's subjective probability that using a specific application system will increase his or her job performance within an organizational context". The perceived usefulness of eLearning can be defined as the employees' perceptions of the performance gained as a result of the use of this tool. Then, we suggest the following hypothesis:

H9: Perceived Usefulness will have a positive direct effect on the intention to use e-learning.

Intention to use refers to "a decision to make total use of an innovation as the best course of action available" [13]. For the purpose of this study, the intention of employees to use e-learning will be measured, and the close relationship between the intention to use eLearning and the actual usage behavior will be verified. To this effect, the following hypothesis was set forth:

H10: Intention to use will have a positive direct effect on usage behavior of e-learning.

Usage Behavior is defined as a consumption behavior, since the individual uses a tool (eLearning), thereby consuming it and the experience thereof. Well-being at work is defined as a state of mind characterized by a satisfactory harmony between the skills, needs and aspirations of the worker and the constraints and opportunities of the workplace. So, we suggest the following hypothesis:

H11: Usage behavior will have a positive direct effect on well-being at work.

In Cameroon, a good number of studies have been conducted on the moderating effects of the "gender" and "educational attainment" variables on the intention to adopt and use a technology [18, 19]. Some of them indicated that the educational status is a significant moderator of the intention to use a technology. Higher educational attainment was found to have a strong effect on these relationships [20]. The moderating effect of Gender on the relation between performance expectancy and behavioral intention showed that this relationship is stronger among men than women. Finally, the implications of gender as a moderator for the UTAUT model are discussed. Several studies suggest that gender is an important demographic variable with direct and moderating effects on the behavioral intention, adoption and acceptance of technology. This led us to suggest the following hypotheses:

H11 (a-b): Gender and educational attainment have a moderating effect on the relationship between the intention to use and usage behavior.
H12 (a-b): Gender and educational attainment have a moderating effect on the relationship between usage behavior and well-being at work.

4 Methodology

In this study, we propose a mixed approach consisting of a quantitative study and a qualitative study. For the quantitative study, we used the survey method, with a questionnaire as an instrument of data collection. A questionnaire was formulated based on a structural model integrating TAM 2 and UTAUT constructs, and our variables were measured on 7-point Likert scale. The survey was disseminated online with an URL address, and a questionnaire on social media platforms such as LinkedIn, WhatsApp and Facebook was sent to business employees in Cameroon. Data collection took about a 3 weeks (from 19 February 2018 to 12 March 2018). At the end of this survey, we obtained a total of 159 responses from participants that were all exploitable. Finally, we used SMART PLS as data processing software at the quantitative analysis stage.

Concerning the qualitative study, it came out to complement the quantitative study in this sense that it allowed us to better understand, with more details, the subject developed in this study. Thus, for the qualitative study, we designed an interview guide with main questions focusing on other driving factors of the use of eLearning, the difficulties encountered in the use of this technology and the benefits of eLearning. In this vein, we interviewed 10 employees from different companies. The results allowed us to better argue our discussions.

5 Data Analysis and Results

This section presents the results of our study following a proper processing of data collected in the field.

5.1 Respondents Profile

According to the demographic profile of our 159 respondents, only 13% are women. This is consistent with some studies that concluded that the rate of technology adoption by women in Cameroon was low [21]. Concerning age, the largest proportion (75%) of respondents belonged to the age group ranging from 25 to 34 years old, followed by those aged between 18 and 25 years (13%), while 12% of the respondents were above the age of 35. The surveyed respondents were generally well educated, with over 83% of them holding a Master's degree, 9% with a Bachelor's degree, and 8% with more than a Master's degree.

5.2 Reliability and Validity Test

Reliability and validity are used to assess the measurement model. For each construct, the reliability is measured (Composite Reliability and the Cronbach's Alpha). The acceptable value of these measures must be greater than 0.70 [22]. As for the convergent validity measured by the Average Variance Extracted (AVE), the preferred value should be greater than 0.50. The results of the CR, Cronbach's alpha and AVE are shown in Table 1.

As observed from the table above, all the constructs meet the conditions for internal consistency and convergent validity. Table 3 below gives the results for the discriminant validity of the constructs, with correlation among constructs and the square root of average variance extracted (AVE) on the diagonal. All indicators load more highly on their own constructs than on other constructs. Moreover, the correlation between the various construct is low, indicating that multicollinearity is not a problem among the variables. All these results point to the fact that the convergent and discriminate validity of our instruments items are valid.

Table 1. Reliability results

Constructs	Number of items	Mean	Standard deviation	Cronbach alpha	DG rho	AVE
SN	3	4.487	1.635	0.755	0.862	0.670
IMG	3	3.998	1.524	0.856	0.913	0.776
OQ	3	5.174	1.269	0.904	0.940	0.839
FC	4	5.029	1.688	0.837	0.836	0.541
PEOU	4	5.571	1.132	0.860	0.905	0.703
PU	4	5.515	1.116	0.856	0.903	0.700
IU	3	6.126	0.950	0.907	0.836	0.842
UB	3	2.044	0.571	–	–	–
WBaW	2	4.531	1.533	0.808	0.812	0.560

Table 2. The formative construct ≪ usage behavior ≫ of e-learning

	Correlations	Weight	Weight with Bootstrap	T	FIV
Frequency	0.090	−0.257	−0.142	0.567	1.121
Framework	0.307	0.291	0.313	1.967	1.015
Level of experimentation	0.934	1.00	0.972	2.873	1.107

Table 3. Discriminant validity results

Constructs	1	2	3	4	5	6	7	8
1. SN	**0.670**							
2. IMG	0.506	**0.776**						
3. OQ	0.319	0.697	**0.839**					
4. FC	0.319	0.462	0.301	**0.541**				
5. PEoU	0.463	0.359	0.455	0.456	**0.703**			
6. PU	0.316	0.523	0.634	0.268	0.636	**0.700**		
7. IU	0.403	0.373	0.529	0.376	0.721	0.647	**0.842**	
8. WBaW	0.232	0.513	0.502	0.550	0.283	0.236	0.184	**0.560**

Table 2 shows that the "usage behavior" of e-learning is constituted by the following items: frequency of use (per month), framework of use (personal, in relation with work, only at work), level of experimentation (beginner, intermediate, expert). The use behavior was introduced into our analysis as a formative variable. One of the main differences between formative and reflective constructs is that indicators are not necessarily positively correlated [23]. They are not supposed to co-vary even if it can happen, it is an exception and not a condition [24]. The need to ensure the reliability of a formative index by using Cronbach Alpha type homogeneity analyzes is therefore of no use. To analyze the use of e-learning, we have formalized an evaluation process following the recommendations of [25] and the procedure followed by [26]. The results show that the frequency does not contribute to the formation of the usage behavior (W < 0).

5.3 Results of Hypotheses Testing

This section contains the results of the testing of the research model through the use of structural equation modelling (SEM). SEM is a statistical technique for simultaneously testing and estimating causal relationships among multiple independent and dependent constructs. In this study, the Partial Least Square (PLS) method was used to assess the specifications of the model, using SmartPLS. The results of the path coefficients in the model are shown in Table 4.

Table 4. Partial least square results

Dependant variable	Hypothesis	Path coefficient (β)	R^2	Result
PU	IMG → PU	0.206*	0.573	Accepted
	SN → PU	−0.077*		Rejected
	OQ → PU	0.302*		Accepted
	PEOU → PU	0.457***		Accepted
IU	PEOU → IU	0.513***	0.619	Accepted
	FC → IU	0.082		Rejected
	PU → IU	0.308***		Accepted
PEOU	FC → PEOU	0.457***	0.209	Accepted
UB	IU → UB	0.131*	0.003	Accepted
WBaW	U → WBaW	0.389***	0.135	Accepted

Note: *$p < .05$ **$p < .01$ ***$p < .001$

Table 4 showed that PU and PEOU had the greatest effect on IU. Thus, hypotheses H10 and H11 were also supported. Furthermore, PU was significantly influenced by tree variables: IMG, OQ, and PEOU, of all which support hypotheses H2, H3 and H6. However, the effect of H1 was in contrast to what was hypothesized. On the other hand, hypothesis H8 (FC → IU) is the only one that is not supported. U influenced significantly WBAW, thereby supporting H10. Finally, FC and IU significantly influenced PEOU and UB. Therefore, hypotheses H5 and H9 were supported.

The PLS-MGA method is an important non-parametric test for the comparison of the group-specific bootstrapping PLS-SEM results.

Table 5. Multigroup analysis for the group "gender"

	Path(Men)	Path(Women)	Diff(M-W)	T-Value	P-Value
H11a: IU → UB	0,035	0,008	0,027	2,334	0,023
H12a: UB → WBaW	0,173	0,178	−0,005	0,660	0,783

Table 5 shows that the two gender groups are significant only for the relationship IU → UB (p-value = 0.023 < 0.05). Gender has no influence on the relationship UB → WBaW.

Table 6. Multigroup analysis for the group "Scholar level"

	Path(Master)	Path(Bach)	Diff(M-B)	T-Value	P-Value
H11b: IU → UB	0,026	0,014	0,015	0,918	0,353
H12b: UB → WBaW	0,194	0,186	0,008	0,503	0,821

Table 6 shows that the two scholar level groups are not significant for any of the relationships (p-value > 0.05). This suggests that scholar level has no influence on the relationships IU → UB, UB → WBaW.

6 Discussions, Implications and Future Research

In this study, we need to identify factors affecting his adoption by business employees. It should certainly contribute to enriching the literature on IT adoption research, especially by disseminating relevant experience from a developing country context like Cameroon. At the main two objectives of our study were to analyse some factors that may influence the adoption and use of e-learning and to measure the potential impact of e-learning usage on the well-being of employees. To this effect, we designed a research model inspired by the extended model of TAM to which we added the variable "facilitating conditions" of the UTAUT model. According to the results of this study, PU and PEOU had significant positive effects on IU. This observation is quite consistent with many previous researches based on TAM. However, the variable FC did not show any significant influence on the IU. Gender has a positive influence on the relationship IU → UB, and men are more involved in the use of e-learning than women, this hypothesis was confirmed by the qualitative analysis. Discussions with managers also reveal that the educational attainment has an influence on the intention to use e-learning. Quantitative results shows that subjective norms have no effect on the perceived usefulness, but the qualitative research explain that employees are more alike to use e-learning when the top management notice it to them. The demonstrability of the results is highly dependent on the ease of use of its tools. According to these studies, perceived usefulness and perceived ease of use are critical determinants of acceptance of e-learning for users. Qualitative analysis has shown that the poor quality of the Internet do not facilitate access to e-learning by employees.

Several implications can be drawn from the findings of this study. First, we could suggested to e-learning platform designers to improve the ease of manipulation and navigation of e-learning systems so that it will promote their adoption, as employees' efforts in this regard will be highly alleviated. It is also necessary to provide support, employee training and technological resources to facilitate the use of the technology that will be implemented In addition, our study reveals that employees tend to evaluate whether e-learning systems can meet their business needs or fit into their work. So, the needs of users must be evaluated before the implementation of e-learning system. The research explains that the absence of a relationship does not mean that subjective norms have no effect on the perceived usefulness, but that this effect can be complex and may only work in certain situations. The demonstrability of the results is highly dependent on the ease of use of its tools. According to these studies, perceived usefulness and perceived ease of use are critical determinants of acceptance of e-learning by business employees. As for the Internet, its quality should be improved, by providing appropriated telecommunications equipment. The poor quality of the Internet do not facilitate access to e-learning by employees. The public authorities are expected to engage firmly with internet providers (telecoms entities) in order to put in place equipment enabling an improved access to high-speed internet services. It would be wise for managers to consider occupational health as a key aspect of the concept of well-being at work.

An important contribution is the use of a preeminent intention-based model in the business organizations, which differs considerably from an educational context ordinarily studied in previous research. From a managerial standpoint, the findings of this study reveal that, in order to foster individual intention to use a technology, positive perception of the technology's usefulness is crucial. Training and information sessions on e-learning need to focus primarily on how the technology can help improve the efficiency and effectiveness of employees' learning process rather than on the procedures of actual use of the technology. It would be interesting for human resources managers to promote the method and show its advantages compared to other training methods for employees, particularly through awareness and communication actions. When implementing the e-learning project, it is necessary to accompany it with a change management project. It is therefore important to communicate on the expected benefits of e-learning. Thus e-learning must meet the specific needs of employees.

References

1. Noyé, D., Piveteau, J.: Guide pratique du formateur. INSEP Ed, Paris (1997)
2. Urdan, T.A., Weggen, C.C.: Corporate Elearning: Exploring a New Frontier. W.R. Hambrecht, San Francisco (2000)
3. Burgess, J.R., Russell, J.E.: The effectiveness of distance learning initiatives in organizations. J. Vocat. Behav. **63**(2), 289–303 (2003)
4. Emmanuel, B. État des lieux de la recherche sur les formations ouvertes et à distance en Afrique subsaharienne francophone. Distances Médiations Savoirs. Distance Mediat. Knowl. 14 (2016)
5. Adkins, S.S.: The Africa market for self-paced eLearning products and services: 2011–2016 forecast and analysis. Monroe WA Ambient Insight (2013)
6. Minesup Cameroun. Fiche Technique Sur Le Projet E-National Higher Education Network (2015)
7. Mtebe, J.: Acceptance and use of Elearning solutions in higher education in East Africa. Acta Electronica Universitatis Tamperensis, Finiland (2014)
8. Park, S.Y.: An analysis of the technology acceptance model in understanding university students' behavioral intention to use e-learning. Educ. Technol. Soc. **12**(3), 150–162 (2009)
9. Selim, H.M.: An empirical investigation of student acceptance of course websites. Comput. Educ. **40**(4), 343–360 (2003)
10. Lau, S., Woods, P.C.: An investigation of user perceptions and attitudes towards learning objects. Br. J. Educ. Technol. **39**(4), 685–699 (2008)
11. Ong, C.-S., Lai, J.-Y., Wang, Y.-S.: Factors affecting engineers' acceptance of asynchronous e-learning systems in high-tech companies. Inf. Manage. **41**(6), 795–804 (2004)

12. Lee, Y.-H., Hsieh, Y.-C., Hsu, C.-N.: Adding innovation diffusion theory to the technology acceptance model: supporting employees' intentions to use e-learning systems. J. Educ. Technol. Soc. **14**(4), 124–137 (2011)
13. Fishbein, M., Ajzen, I.: Belief, Attitude, Intention, and Behavior: An Introduction to Theory and Research. Addison-Wesley Pub. Co, Reading (1975)
14. Bagozzi, R. P.: The legacy of the technology acceptance model and a proposal for a paradigm shift. J. Assoc. Inf. Syst. **8**(4) (2007)
15. Moore, G.C., Benbasat, I.: Development of an Instrument to measure the perceptions of adopting an information technology innovation. Inf. Syst. Res. **2**(3), 192–222 (1991)
16. Rogers, E.M.: Diffusion of Innovations. Free Press/Collier Macmillan, New York/London (1983)
17. Venkatesh, V., Bala, H.: Technology acceptance model 3 and a research agenda on interventions. Decis. Sci. **39**, 273–315 (2008)
18. Bakehe, N.P., Fambeu, A.H., Piaptie, G.B.T.: Internet Adoption and Use in Cameroon. African Economic Research Consortium, RP_336, April 2017
19. Tamokwe Piaptie, G.B.: Les déterminants de l'accès et des usages d'internet en Afrique Subsaharienne. Réseaux **180**, 95–121 (2013)
20. Abu-Shanab, E.A.: Education level as a technology adoption moderator. In: 2011 3rd International Conference on Computer Research and Development, vol. 1, pp. 324–328 (2011)
21. Onguéné Essono, L.-M., Béché, E.: Genre et TIC dans l'école secondaire au Cameroun: Au-delà des progrès, des disparités. Educ. Afr. (2013)
22. Hair Jr, J.F., Hult, G.T.M., Ringle, C., Sarstedt, M.: A Primer on Partial Least Squares Structural Equation Modeling (PLS-SEM). SAGE Publications, Beverley Hills (2016)
23. Lacroux, A.: Analyse des modèles de relations structurelles par la méthode PLS : Une approche émergente dans la recherche quantitative en GRH (2009)
24. Bollen, K., Lennox, R.: Conventional wisdom on measurement: a structural equation perspective. Psychol. Bull. **110**(2), 305–314 (1991)
25. Diamantopoulos, A., Siguaw, J.A.: Formative versus reflective indicators in organizational measure development: a comparison and empirical illustration. Br. J. Manag. **17**(4), 263–282 (2006)
26. Ringle, C.M., Sinkovics, R.R., Henseler, J.: The use of partial least squares path modeling in international marketing. In: New Challenges to International Marketing, vol. 20, 0 vols. Emerald Group Publishing Limited, pp. 277–319 (2009)

Result of the Methodology for Learning English Distance with the Use of TICs. Case Study: Central University of Ecuador

Nelson Salgado[1(✉)], Javier Guaña[1], Charles Escobar[1], and Alvansazyazdi Mohammadfarid[2,3]

[1] Pontifical Catholic University of Ecuador, Quito 170135, Ecuador
{nesalgado,EGUANA953,CESCOBAR637}@puce.edu.ec
[2] Universidad Central del Ecuador, Av. Universitaria, Quito 170129, Ecuador
alvansaz@gmail.com
[3] Universidad Laica Eloy Alfaro de Manabí, Manta 130803, Ecuador

Abstract. The study presents the results of the implementation and evaluating of the methodological and technological proposal for the learning of English as a foreign language at the Central University of Ecuador (UCE) Distance, making use of information technologies and based on a previous analysis from the pedagogical and technological. For the result was based on the analysis of the marks for English skills of the students that make up the sample, which are 784 students and 302, for level 6, an A2 + that represents the Common European Framework [1], Assumed for the investigation and attending to criteria of viability and resources with which it counts for the development of the intervention.

Keywords: English learning · Methodology · Sample · Common European Framework

1 Introduction

In the case of the Ecuadorian Universities, one of the current features of the English learning is the insufficient development of communicative competition for the understanding and oral expression, also the writing, by a big number of students of higher education. In this regard, we will plan the exist of several factors, that they have affected the teaching-learning process, some of these, the ignorance of the importance of dominating the English language as a foraging language, the lack of motivation in students, the application strategies and traditional methodologies. That they do not attract the student's attention, the idea of getting a certificate of the language only like a graduation requirement, all of this coupled with the non-application of updated motivational strategies that invite the student to enter the classroom as a positive experience.

The domain of that language is also linked to the Gross Domestic Product (GDP) per capita. In others words, the better the English of a country, the higher the average income per individual, it is not a mystery, so, some many Occidental European countries occupy high positions in the English aptitude, according to the IEF report.

Á. Rocha et al. (Eds.): WorldCIST'19 2019, AISC 932, pp. 227–233, 2019.
https://doi.org/10.1007/978-3-030-16187-3_22

The English language is the foreign language that is officially studied in Ecuador. However, so far the results have not been especially good in his teaching. This nation ranks 35th out of 63 as one of the low English countries according to the English Fitness Index [3] (Fig. 1).

NIVELES Y SUBNIVELES EDUCATIVOS

	NIVELES	DENOMINACION	EDAD
EDUCACIÓN INICIAL	Inicial 1	no es escolarizado	3 años
	Inicial 2	comprende a infantes	3 a 5 años
EGB	Preparatoria	corresponde a 1.º grado de EGB.	5 años
	Básica elemental	corresponde a 2.º, 3.º y 4.º grados de EGB.	6 y 8 años
	Básica Media	que corresponde a 5.º, 6º. y 7.º grados de EGB	9 a 11 años
	Básica Superior	corresponde a 8.º, 9.º y 10.º grados de EGB	12 a 14años
BACHILLERATO	Bachillerato • Ciencias • Técnico	tiene tres (3) cursos Primero de bachillerato Segundo de bachillerato Tercero de bachillerato	15 a 17 años

Fig. 1. Level and sub-level education

As a measure to reverse the situation, the Ministry of Education presented in March 2015 the agreement 0052-14 which provided that the teaching of the foreign language (English) from 2016–2017 (Sierra regime) and 2017–2018 (Costa regime) would be compulsory from second grade of Basic General Education until third year of Bachelor [4].

2 Methodology Used in the Investigation

For the implementation of the methodological and technological proposal for the English language learning in the UCE, we started with the analysis of the notes for English skills of the students who make up the samples. Based on the design of non-equivalent groups [5] developed in the methodology of the present work, the results were analyzed independently in the students enrolled in the non-face-to-face.

For the group, it was possible to quantitatively analyze the teaching/learning methodology in the UCE based on the skills qualifications of the students, taking as reference the data of four consecutive semesters of level 6 as a reference is a A2 + of the Common Frame Euro-MCE. Each of these semesters, it has duration around 124 h [7].

At the non-face-to-face level, the enrollment is per semester, as indicated in the following table, since first at sixth level (Table 1).

Table 1. Number of students enrolled in distance English

Level	Student enrolled in distance English			
	2016	2017		2018
	Second semester	First semester	Second semester	First semester
First	132	143	154	166
Second	128	138	149	161
Third	125	135	146	158
Quarter	98	106	114	123
Fifth	75	81	87	94
Sixth	67	72	79	84
Total	625	675	729	786

3 Results Before and After Methodological and Technological Proponents

The objective of the diagnostic table detailed, it is to establish the results achieved before applying the proposal and after using the new methodology and technology, according to the Evaluation criteria established by the Institute of Languages of the UCE, a grade above 7 is considered approved while an inferior grade of it does not reach the levels of skill required for its approval. In Ecuador each of the Universities has its own form of evaluation and score to approve each subject, in the UCE the score for a student is approved is 7 points, that is why in the nominal variables has The following quantitative notes as: 1 to 6 (Bad), 7 (approve), 8–9 (Good), 10 (excellent).

For the sample of the students, a sample of 139 students from the pre-intervention group was selected from stratified random sampling [6], which were evaluated in a diagnostic manner, from the second semester of the year 2016 and first semester of 2017.

In a second part of the project the methodological and technological proposal was implemented; Evaluating the results in a sample formed by 163 students of the non-face-to-face course and selected in the same way as described above; But taking as a population the 79 students of the second semester of 2017 and 84 students the first semester of 2018.

In order to guarantee, the veracity of the statistical tests was taken into account that the two groups formed by these samples, although different in size, were homogeneous, these were selected by random sampling and comparable since the teachers were the Same classrooms, the same technological platform, the same exams and the same beginning of each semester.

Table 2 shows the distribution of the sample of "non-presential" students evaluated before and after the proposal in the Speaking skill.

Before the intervention, 46.04% of the sample is failing, however, we can observe that when applying the proposal this percentage decreases to 32.47%. In the case of "approved" students and those in the "good" and "excellent" categories, these percentages increased from 34.53 to 45.73; 12.95 to 14.53 and 6.47 to 7.27 respectively; So the proposal introduced an improvement in the grades associated with this ability.

Table 2. Evaluations students non-presential in the years 2016–2018 in the skill "Speaking"

Escalas	Speaking			
	Before the intervention		After the intervention	
	No	%	No	%
Between 1–6 (Bad)	64	46,04	53	32,47
Between 7 (Approve)	48	34,53	74	45,73
Between 8–9 (Good)	18	12,95	24	14,53
10 (Excellent)	9	6,47	12	7,27
Total	**139**	**100**	**163**	**100**

Table 3. Evaluations students non-presential in the years 2016–2018 in the skill "Reading"

Escalas	Reading			
	Before the intervention		After the intervention	
	No	%	No	%
Between 1–6 (Bad)	65	20,86	53	32,65
Between 7 (Approve)	52	50,36	73	44,99
Between 8–9 (Good)	17	15,83	23	14,13
10 (Excellent)	6	12,95	14	8,60
Total	**139**	**100**	**163**	**100**

Regarding reading comprehension (Table 3), the percentage of students evaluated before the proposal with notes minor a six is 20.86% after application of the same. Likewise in amount is minor, the percentages of approve, good and excellent presented an increase of 44.99%, 14.13% and 8.60% respectively.

Table 4. Evaluations students non-presential in the years 2016–2018 in the skill "Listening"

Escalas	Listening			
	Before the intervention		After the intervention	
	No	%	No	%
Between 1–6 (Bad)	64	20,86	54	33,00
Between 7 (Approve)	51	50,36	65	39,88
Between 8–9 (Good)	16	15,83	28	17,19
10 (Excellent)	8	12,95	17	10,44
Total	**139**	**100**	**163**	**101**

In the listening comprehension (Table 4) before the proposal, 20.86% fail, 50.36% approve with the minimum grade, 15.83% is in the rank of Good and only 12.95% is in Excellent. However, when analyzing the notes in the sample studied after the intervention, we have an improvement because only 33.0% fail, 39.88% approve with the minimum grade, and 17.19% are categorize as Good and 10.44% excellent.

Table 5. Evaluations students non-presential in the years 2016–2018 in the skill "Writing"

Escalas	Writing			
	Before the intervention		After the intervention	
	No	%	No	%
Between 1–6 (Bad)	64	20,86	52	32,09
Between 7 (Approve)	56	50,36	72	44,10
Between 8–9 (Good)	13	15,83	24	15,04
10 (Excellent)	6	12,95	14	8,83
Total	**139**	**100**	**163**	**100**

Table 5 shows the results obtained in the evaluation applied before and after the proposal developed in the "writing", taking into account the results, we can say that the percentage of students evaluated as poor in written production, decreases in 32.09% when implementing the proposal.

In the same way as the previous cases, the students approved with the minimum grade and those evaluated in good and excellent registered changes that represent improvements after applying the intervention, as the percentage of these increased by 44.10% 15.04% and 8.83%, respectively.

3.1   Analysis of the Results Before and After the Intervention

The analysis of the results of the non-presence students (distance) by dexterity has led us to confirm that they had a real problem in oral production skills, allowing us to determine the urgency of proposing the Moodle technological tool incorporating the plugins. Students write the English language and at the same time that they speak it aloud, Clearning, audio recording, bigblebutton etc. to develop these skills in these students who do not receive classroom classes. Also, the fact that they have problems in written production reflects both the relevance of tools to improve this skill in the Moodle platform, which helps educators to create learning communities-distribution on-line and achieve with the use of TICS a construction of knowledge with a communicative approach. Same that will serve for the improvement in the other skills.

In addition, the improvement in each of the skills is satisfactory, a fact that could be the result of the interaction with exercises in the new platform Moodle that promote the oral reading of written texts. It is important to emphasize that in the elaboration of these tasks we tried to do them in such a way as to complement the learning of different skills, such as pronunciation, vocabulary, etc. The use of the Online Audio Recording tool allows the student to record reading texts to practice their pronunciation, as well as the BigBlueButton tool through which you can chat with comrades and the teacher. The Moodle platform helped us significantly improve each skill with the accomplishment of tasks, exercises, pilot tests, diverse information, etc., more in the non-presential mode.

4 Hypothesis Testing

In order to determine that the methodological proposal allowed for improvements in English learning in the students of the Equinoctial Technological University, we proceeded to make a verification of the hypothesis used in this study. Initially we tested the normality of each variable, Represented by the different skills. Then, to test the hypothesis, we used the Mann-Whitney U test, which was performed in each of the skills and for each modality; as shown in the results. They present a significant variation in the improvement of the academic performance of the students after applying our proposal.

Assumptions: Each group is an independent random sample from a normal population. The analysis of variance is robust to deviations from normality. Groups should come from populations with equal variances (homocedastici-bility).

4.1 Testing of Non-classroom Students Hypotheses

H1 = average of the score obtained by the students Non-presence in the X skill is significantly different between the students who use the methodological proposal designed in this thesis and those who do not follow it.
H0 = average score obtained by the non-classroom students in the X skill is not significantly different between the students who use the methodological proposal designed in this thesis and those who do not follow it (Table 6).

Decision
Decision's rule is: If $p \leq 0.05$ we reject H0.

Table 6. Application of the Kolmogorov-Smirnov normality test.

Destrezas	Z de Kolmogorov-Smirnov			
	Antes	p (sig. asintótica bilateral)	Después	p (sig. asintótica bilateral)
Speaking	0,492	0,000	0,443	0,000
Reading	0,260	0,000	0,252	0,000
Listening	0,298	0,000	0,278	0,000
Writing	0,485	0,000	0,467	0,000

In the same way as in the previous case, the normality test is significant for all described abilities (in all cases $p = 0.000 < 0.05$); Thus justifying the use of parametric tests [2].

4.2   Analysis of the Hypothesis Test

With the antecedents indicated in a general way, and with the new tools used in the Moodle platform, it is possible to observe that the students non-presential modality had better performance in the skills to solve complementary exercises for their learning with a good percentage of acceptance for each of them.

In this way, it can be said that the average of the qualification obtained by the presential and non-presential students in the X skill is significantly different between the students who use the methodological proposal.

5   Conclusions

The application of the methodological proposal presented was carried out in two consecutive courses of level 6: the course taught in the second semester of 2017 and first semester 2018. The students enrolled for each semester are different, 48 additional activities were proposed that were designed to develop all the skills in each student, these were incorporated as a complement to improve English language learning in the UCE. During the year, the Moodle platform for learning was placed at the service of the students, in which the additional activities for all the skills were placed, because the designed exercises only have the objective to complement the teaching, settled in the Platform, the complements of other providers that are easily incorporated as audio, video, recording, etc., for their best use in each of the skills, the evaluations were carried out through this platform, as indicated in the.

Results achieved when analyzing in each skill the before and after using the methodological and technological proposal even the platform offers a greater dedication of time on the part of the non-presential for the learning of the English language, as a complement for their learning used the platform UCE in line and Cambridge, in the institute of idiomas of the UCE.

References

1. Cambell, D., Stanley, J.: Experimentales y cuasiexperimentales en la investigación social. Rand McNally & Company, Buenos Aires (2000)
2. Cochran, W.: Técnicas de muestreo. Editorial Continental S.A., México (2000)
3. https://es.wikipedia.org/wiki/Marco_Com%C3%BAn_Europeo_de_Referencia_para_las_lenguas. consulta 21 Nov 2017
4. http://www.eluniverso.com/noticias/2015/02/11/nota/4547176/ecuador-tiene-nivel-bajo-ingles-segun-informe-education-first. consulta 11 Feb 2015
5. Education First, «Informe anual de Índice de Aptitud en Inglés» (2014). http://www.ef.com.ec/__/~/media/centralefcom/epi/v4/downloads/full-reports/ef-epi-2014-spanish.pdf, Último Acceso 22 Nov 2017
6. https://educacion.gob.ec/wp-content/uploads/downloads/2014/03/ACUERDO-052-14.pdf, consulta 21 Nov 2017
7. Salgado, N.: Result of proposed methodology for learning English with the use of TICs. Case study. In: 2017 12th Iberian Conference on Information Systems and Technologies (CISTI), pp. 1–6. Universidad Tecnológica Equinoccial (2017)

Is It All About Frequency?

Students' Digital Competence and Tablet Use

Margarida Lucas[1]([⊠]) [iD], Pedro Bem-haja[2] [iD], António Moreira[1] [iD],
and Nilza Costa[1] [iD]

[1] CIDTFF, University of Aveiro, 3810-193 Aveiro, Portugal
{mlucas,moreira,nilzacosta}@ua.pt
[2] CINTESIS, University of Aveiro, 3810-193 Aveiro, Portugal
pedro.bem-haja@ua.pt

Abstract. Being digitally competent requires appropriate knowledge, skills and attitudes towards the use of digital technologies for different aspects of today's society. The role Education plays in equipping students with such competence is increasingly important, but despite the growing number of studies on the topic, little is known about the impact the use of digital technologies has on students' digital competence. This paper presents and discusses findings regarding the impact of tablet use on the development of lower secondary students' digital competence. A retrospective evaluation was made resorting to a questionnaire applied to students from two schools. Findings suggest that the frequency of tablet use seems to explain the development of digital competence as well as students' perceptions of their digital competence proficiency level.

Keywords: DigComp · Impact evaluation · Mobile learning · Tablets

1 Introduction

In the past decade, there has been an increase in the number of mobile technology-driven projects and initiatives implemented in schools, but little is known about the impact of the use of mobile technologies, specifically tablets, on the development of students' digital competence.

Digital competence is the set of knowledge, skills and attitudes needed to use digital tools and media to perform tasks, solve problems, communicate, manage information, collaborate, create and share content, and build knowledge effectively, critically, creatively, autonomously, flexibly, reflectively for work, leisure, participation, learning, socializing, consuming, and empowerment [1]. Its study may involve indirect measures, such as the application of models or frameworks that give students the opportunity to evaluate and judge their own competence. One of such frameworks is DigComp [2], which enables the development of indicators and tests to measure digital competence. Since its first launch in 2013 [2] DigComp has gone through two updates [3, 4]. Its latest version is structured around five dimensions. Dimension 1 outlines five competence areas: "Information and data literacy", "Communication and collaboration", "Digital content creation", "Safety" and "Problem solving"; Dimension

Á. Rocha et al. (Eds.): WorldCIST'19 2019, AISC 932, pp. 234–243, 2019.
https://doi.org/10.1007/978-3-030-16187-3_23

2 specifies 21 competences[1]; Dimension 3 presents the proficiency levels foreseen for each competence[2]; Dimension 4 outlines examples of knowledge, skills and attitudes[3]; and Dimension 5 provides examples of the applicability of the competence for different purposes, including learning. Table 1 provides an overview of the competences specified for the competence areas "Information and data literacy" and "Communication and collaboration", as these are the ones that were examined during our study.

Table 1. Overview of the competences outlined for the competence areas "Information and data literacy" and "Communication and collaboration"

Competence areas Dimension 1	Competences Dimension 2
1. Information and data literacy	1.1 Browsing, searching and filtering data, information and digital content 1.2 Evaluating data, information and digital content 1.3 Managing data, information and digital content
2. Communication and collaboration	2.1 Interacting through digital technologies 2.2 Sharing through digital technologies 2.3 Engaging in citizenship through digital technologies 2.4 Collaborating through digital technologies 2.5 Netiquette 2.6 Managing digital identity

Research on the use of DigComp to measure lower secondary school students' digital competence is still limited. The study by Siiman et al. [5] describes the development of a self-report questionnaire to measure how often students, from grade 6 and 9, use a smart device to perform a digitally competent activity in four different contexts. Results show that the use of devices outside school for purposes other than school learning achieve the highest frequency responses and that such use is higher for "Information and data literacy" and for "Communication and collaboration" than, for instance, "Digital content creation". Pérez-Escoda, Zubizarreta and Fandos-Igado [6] studied the extent of the use of technological devices as well as the Internet by primary school students and the levels of their digital competence. They found that simple exposure to, use of and coexistence with media and technology do not imply the development of digital competence. Data indicate heterogeneity among students' level of digital competence, which is in general low, and the need to address digital competence in schools in a way that increases acquisition in a gradual and progressive manner.

Despite the limited number of studies employing DigComp as an instrument to measure lower secondary students' digital competence, different studies have been focusing on measuring it, in particular on the competence areas under study (cf. Table 1) [7]. In such cases, even if the comparison between results may be hindered

---

[1] The name of the competence areas and competences was updated in version [3].

[2] The number of proficiency levels was updated to 8 in version [4].

[3] Only available in version [2].

due to the application of different instruments, it is worth mentioning that all tests include competences that can be aligned with the DigComp framework [7]. Hatlevik, Ottestad and Throndsen [8] found diversity in digital competence among 9th grade students and advance that family background can help explain the lack of digital equity. Van Deursen and Van Diepen [9] conclude that students aged between 11 and 16 have problems in determining information needs and in selecting and specifying appropriate search terms. Moreover, students demonstrate difficulties in evaluating information found, as well as in judging its reliability and relevance. Other studies also observe that not all students are competent in organizing and managing information [10, 11]. As to communication related competences, Calvani, Fini, Ranieri and Picci [10] found that students are little aware that online behavior needs to be adequate for their safety and respectful of privacy, in line with Livingstone [12] who found students are not adequately aware of the risks online activities may pose, for instance when sensitive information is not protected.

In general, we may conclude that lower secondary students lack competences as far as "Information and data literacy" and "Communication and collaboration" are concerned. Nevertheless, as referred before, little is known regarding the impact the use of digital technology has on the development of such competences. With a view to fill this gap, the present study aims to answer the following research question: What is the impact of tablet use on lower secondary students' digital competence, regarding the competence areas "Information and data literacy" and "Communication and collaboration"?

It uses a mobile technology-driven initiative implemented in two Portuguese schools as a background context for the study that provided students and teachers with tablets and digital contents to use both at school and at home. The impact was evaluated retrospectively through a questionnaire that participating students filled in at the end of the project second year.

2 Method

The study was conducted in two lower secondary schools in central Portugal and involved four classes (two per school). Data for the retrospective evaluation were collected from students who completed their second year of participation in the tablet initiative with a view to understand the impact of tablet use on the development of their digital competence in the areas of "Information and data literacy" and "Communication and collaboration". The data were collected through an online questionnaire filled in by students during class hours. The project coordinator at the school level was responsible for administrating the questionnaire and ensuring that all students had the option to respond.

Eighty students (35 boys and 45 girls, $Mage = 14.19$; $SDage = 0.82$) answered the questionnaire. Apart from the socio-demographic information, the questionnaire main measures included (i) the frequency of tablet use at home and at school, (ii) the perceived level of digital competence proficiency and (iii) the retrospective impact evaluation on the development of "Information and data literacy" and "Communication and collaboration" related competences. The frequency of tablet use was measured using a five-point Likert scale (Every day, Three or more times per week, One or 2 times per week, Occasionally and Never). According to results obtained, percentiles 60 and 40 were calculated. Students who scored below the cut-off point of the 40th percentile were classified as non-frequent tablet users and students who scored above the cut-off point defined by the 60th percentile were classified as frequent users. This segmentation resulted into two groups: one with 33 participants (14 boys and 19 girls; *Mage* = 13.91; *SDage* = 0.67), who were classified as non-frequent users, and the other with 34 participants (15 boys and 19 girls; *Mage* = 14.50; *SDage* = 0.89), who were classified as frequent users. There were 13 participants who had intermediate results and were therefore discarded as they did not fit the extreme groups created.

To measure the proficiency level, a set of statements were presented, each one corresponding to a specific competence and proficiency level[4]. For example, to measure competence 1.1 (see Table 1) three statements were presented: 'I can use a search engine to find information on a specific topic' corresponded to level A (foundation); 'I can use proper key words and apply search filters to refine my results list' corresponded to level B (intermediate), and 'I can subscribe to feeds to be updated on information that interests me' corresponded to level C (advanced). The same strategy was employed for the remaining competences. Students' responses were scored algorithmically taking into account the importance of each level (A, B and C) for actual proficiency according to the following formula[5]:

$$\left(\left(\sum A \text{ level statements}\right) * 0.2\right) + \left(\left(\sum B \text{ level statements}\right) * 0.3\right) + \left(\left(\sum C \text{ level statements}\right) * 0.5\right)$$

It should be noted that students' frequency of tablet use at home and at school, and level of digital competence proficiency had already been measured at the beginning of the project, in order to evaluate the real state of proficiency and frequency of use in target groups. Although the experimental design of the present study is not a within subject one, due to the fact that, for reasons of ethical nature, we could not guarantee the measurement of such variables for each student at different times, it is a within group design. And this is because we are able to measure both variables in the same group of students in two different moments: pre and post project.

The retrospective impact evaluation, the main focus of the present article, was measured using 14 statements (Table 2), inspired by DigComp, against which the respondents had to position themselves using a five-point Likert scale (Totally agree - Totally disagree). It being a five-point ordinal scale allows us to assume statistical continuity and therefore use measures of central tendency. Before performing statistical inference tests using the mean, a visual analysis of the results was carried out using four classification bands. As such, the scores between points 1 and 2 were grouped in the "Totally disagree" band, the scores between points 2 and 2.9 in the "Disagree" band, the scores between 3.1 and 4 in the "Agree" band and the scores between 4 and 5 in the "Totally agree" band. The midpoint (3) was removed from the analysis of the classification bands.

---

[4] Proficiency levels form version [2] were used, as this was the version available when instruments for the data collection were developed.

[5] Due to the priority of the multiplication relatively to a sum, the outside parentheses are redundant, but they are represented for visual clarity of the formula.

Table 2. Statements presented to students to enable the retrospective impact evaluation, prompted by the initial statement 'After participating in the project, I started to …'

Label	Statements	Competence areas
A	Filter information more carefully	Information and data literacy
B	Search for information more effectively	
C	Evaluate the credibility and reliability of websites better	
D	Select information more critically	
E	Be more organized in storing and managing the information that interests me	
F	Backup all my files using the cloud	
G	Communicate with teachers and colleagues more often	Communication and collaboration
H	Be more confident communicating online	
I	Share the assignments I do with my class	
J	Check the property right of content	
K	Better understand the potential of technologies for civic participation	
L	Work at a distance with colleagues using online collaborative tools	
M	Be more aware of netiquette rules	
N	Be more aware of the risks and benefits related to my digital identity	

3   Findings

For a better understanding of the results and their visual analysis, we decided to use graphs to present the mean scores obtained for each area of competence. The average scores obtained by the sample regarding the statements that compose the area of digital competence "Information and data literacy" are presented in Fig. 1.

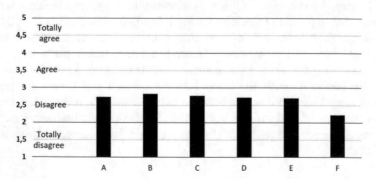

Fig. 1. Mean scores obtained by the participants regarding statements pertaining to the competence area "Information and data literacy"

When analyzing Fig. 1, we can verify that the participants' positioning regarding the statements presented fell into the "Disagree" classification band, i.e., on average, participants do not agree that they have improved their digital competences after starting using the tablet within the scope of the EduLab project. The means obtained by the sample regarding the statements that compose the area of digital competence "Communication and collaboration" are presented in Fig. 2.

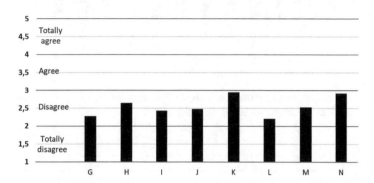

Fig. 2. Mean scores obtained by the participants regarding statements pertaining to the competence area "Communication and collaboration"

Just as for the competence area "Information and data literacy", the average of answers provided by the participants also fell into the "Disagree" classification band regarding the competence area "Communication and communication", i.e., on average, participants do not agree that they have developed digital competences after having participated in the project under study.

Given the (somehow) unexpected results reported above, we decided to compare the perception of change using two groups of students: those who used the tablet frequently and those who did not (see Method). Figure 3 illustrates results obtained for the statements that comprised the competence area "Information and data literacy", and Fig. 4 illustrates those for the competence area "Communication and collaboration".

Interestingly, we have found that the frequency of tablet use seems to be modulating the retrospective response of participants regarding what has changed in relation to their digital competence. In fact, the averages obtained by the group that reports a frequent tablet use reach the classification band "Agree" in almost all items of the competence area "Information and data literacy".

This means that students who use the tablet more often agree that they have improved their information related competences. To verify if the differences observed in Fig. 3 reach statistical significance and considering a left skewed distribution of the majority of the variables we applied Mann-Whitney U tests. Statistically significant differences were obtained for all items pertaining to this competence area: $(UA = 264.500,$ $Z = -3.974$, $p < .001$; $UB = 226.000$, $Z = -4.432$, $p < .001$; $UC = 279.500$, $Z = -3.766$, $p < .001$; $UD = 211.000$, $Z = -4.662$, $p < .001$; $UE = 251.000$, $Z = -4.113$, $p < .05$; $UF = 422.000$, $Z = -1.985$, $p < .05$).

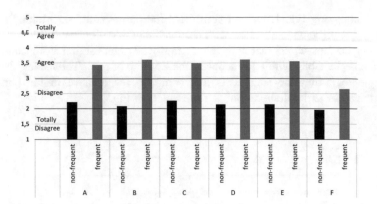

Fig. 3. Mean scores obtained by the two groups of participants regarding statements pertaining to the competence area "Information and data literacy"

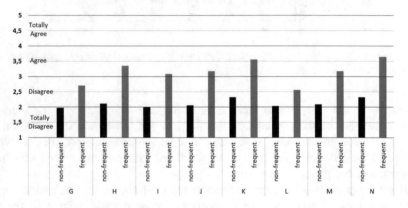

Fig. 4. Mean scores obtained by the two groups of participants regarding statements pertaining to the competence area "Communication and collaboration"

After verifying the impact of the frequency of tablet use, we tested if that same variable was related with the perceived level of digital competence proficiency. After calculating the Proficiency variable (see Methodology), we applied a Mann-Whitney U analysis to check for differences regarding proficiency using the two groups of students. The descriptive and inferential results are presented in Table 3.

Results of this analysis show significant differences between the two groups, i.e. frequent users rate themselves as more proficient than non-frequent users. Although we did not design a pre-post study, we asked students to self-evaluate their level of proficiency and frequency of tablet use when they began the EduLab project (see Methodology) and have, therefore, performed the above-mentioned analysis and procedures using that data. The descriptive and inferential results are presented in Table 4.

Table 3. Descriptive and inferential results regarding students' perceived level of digital competence at the end of the project

		Group	N	Mean	SD	SE
Proficiency	non-frequent		33	4.542	2.630	0.458
	frequent		34	5.691	2.196	0.377

				95% CI for Rank-Biserial Correlation	
W	p	VS-MPR*	Rank-Biserial Correlation	Lower	Upper
392.500	**0.035**	3.135	-0.300	-0.529	-0.031

Note: Mann-Whitney U test.

*Vovk-Sellke Maximum p -Ratio: Based on a two-sided p -value, the maximum possible odds in favor of H_1 over H_0 equals $1/(-e\ p \log(p))$ for $p \le .37$

Table 4. Descriptive and inferential results regarding students' perceived level of digital competence at the beginning of the project

		Group	N	Mean	SD	SE
Proficiency	non-frequent		36	4.411	2.040	0.340
	frequent		37	4.854	1.514	0.249

				95% CI for Rank-Biserial Correlation	
W	p	VS-MPR*	Rank-Biserial Correlation	Lower	Upper
536.500	0.153	1.281	-0.195	-0.434	-0.069

As far as students' perceptions regarding their level of digital competence proficiency is concerned, results show that there were no significant differences between the two groups, i.e., at the beginning of the project, the difference in proficiency between the frequent users' group and the non-frequent users' group was not significant.

4   Discussion and Conclusion

This paper addresses the impact of tablet use on the development of lower secondary students' digital competence, specifically in the areas of "Information and data literacy" and "Communication and collaboration". When we look at students as a whole, the findings reveal that, on average, they do not agree they have improved their digital competences after two years of using the tablet. However, when we look at students grouped according to levels of frequency of tablet use, findings reveal significant different perceptions. Students labelled as frequent users tend to agree that the use of the tablet had an impact on the development of their digital competence, both in the areas of "Information and data literacy" and "Communication and collaboration". On the contrary, students labelled as non-frequent users disagree as to that impact.

Although results may seem, somehow predictable, they partially contradict the ones found by Pérez-Escoda, Zubizarreta and Fandos-Igado [6], who concluded that students do not develop digital competence by simply using media and technology. However, they are in line with the assumption made by the same authors, and others [9, 10], regarding the need to address digital competence in schools in a way that digital technology is used with a view to increase acquisition in a gradual and progressive manner. In this respect, teachers also need to be digitally competent in order to infuse digital technologies in their practices and facilitate their learners' digital competence. Nevertheless, their ability to do so is often questioned and the need to get more insights on how they can achieve it is claimed [13]. Results are also in line with those found by previous studies [14], conducted within a similar context, that suggest students who use the tablet more often learn more and reach higher academic achievements.

At this point, some limitations of the study need to be considered as well as ways to overcome them in the future. First, the fact that it used an indirect measurement instrument. Several authors [7, 15] point at the weaknesses of such measurements, arguing that students' self-reports capture individual beliefs, self-confidence and self-efficacy, which are not always a good or faithful representation of their actual performance level. Considering the inclusion of direct measures, such as performance-based tests, should be taken into account in future studies. Second, the fact that other variables did not come into play during the analysis, which could have helped to deepen the findings achieved. These could include: (i) the type of activities students performed in school and at home and knowing whether teachers prioritized digital competence during lessons and school activities or not. Research points at the role teachers play in promoting students' digital competences [16]; (ii) the socio-economic background of the students. Different studies refer family background, cultural capital or academic achievements and aspirations as predictors of digital competence [8, 10, 11, 13]. These are directions that future research in the field should approach. Despite the limitations, this article contributes to the existing literature in several respects, such as the unveiling of students' perceptions regarding the impact of tablet use on the development of their digital competence, which is an under-researched topic, or the opportunity to further research on emerging aspects that may contribute for positive impacts.

Acknowledgements. This work is supported by the Portuguese Foundation for Science and Technology (FCT) under grant number SFRH/BPD/100367/2014 and project UID/CED/00194/2019.

References

1. Ferrari, A.: Digital Competence in Practice: An Analysis of Frameworks. Publications Office of the European Union, Luxembourg (2012)
2. Ferrari, A.: DIGCOMP: A Framework for Developing and Understanding Digital Competence in Europe. Publications Office of the European Union, Luxembourg (2013)

3. Vuorikari, R., Punie, Y., Carretero, S., Van den Brande, L.: DigComp 2.0: The Digital Competence Framework for Citizens: Update Phase 1: The Conceptual Reference Model. Publications Office of the European Union, Luxembourg (2016)
4. Carretero, S., Vuorikari, R., Punie, Y.: DigComp 2.1: The Digital Competence Framework for Citizens With Eight Proficiency Levels And Examples of Use. Publications Office of the European Union, Luxembourg (2017)
5. Siiman, L., Mäeots, M., Pedaste, M., Simons, R-J., Leijen, Ä., Rannikmäe, M., Võsu, K., Timm, M.: An instrument for measuring students' perceived digital competence according to the DIGCOMP framework. In: Zaphiris, P., Ioannou, A. (eds.) Proceedings of the International Conference on Learning and Collaboration Technologies LCT 2016, pp. 233–244. Springer, Cham (2016). https://doi.org/10.1007/978-3-319-39483-1_2
6. Pérez-Escoda, A., Castro-Zubizarreta, A., Fandos-Igado, M.: Digital skills in the Z generation: key questions for a curricular introduction in primary school. Comunicar 23(49), 71–79 (2016). https://doi.org/10.3916/C49-2016-07
7. Siddiq, F., Hatlevik, O.E., Olsen, E., Throndsen, R.V., Scherer, R.: Taking a future perspective by learning from the past—a systematic review of assessment instruments that aim to measure primary and secondary school students' ICT literacy. Educ. Res. Rev. 19, 58–84 (2016). https://doi.org/10.1016/j.edurev.2016.05.002
8. Hatlevik, O.E., Ottestad, G., Throndsen, I.: Predictors of digital competence in 7th grade: a multilevel analysis. J. Comput. Assist. Learn. 31(3), 220–231 (2015). https://doi.org/10.1111/jcal.12065
9. Van Deursen, A.J.A.M., van Diepen, S.: Information and strategic Internet skills of secondary students: a performance test. Comput. Educ. 63, 218–226 (2013). https://doi.org/10.1016/j.compedu.2012.12.007
10. Calvani, A., Fini, A., Ranieri, M., Picci, P.: Are young generations in secondary school digitally competent? A study on Italian teenagers. Comput. Educ. 58(2), 797–807 (2012). https://doi.org/10.1016/j.compedu.2011.10.004
11. Claro, M., Preiss, D., San Martín, E., Jara, I., Hinostroza, J.E., Valenzuela, S., Cortes, F., Nussbaum, M.: Assessment of 21st century ICT skills in Chile: test design and results from high school level students. Comput. Educ. 59(3), 1042–1053 (2012). https://doi.org/10.1016/j.compedu.2012.04.004
12. Livingstone, S.: EU kids online. In: Hobbs, R. (ed.) The International Encyclopedia of Media Literacy. Wiley-Blackwell, Oxford (2017)
13. Hatlevik, O.E., Guðmundsdóttir, G.B., Loi, M.: Digital diversity among upper secondary students: a multilevel analysis of the relationship between cultural capital, self-efficacy, strategic use of information and digital competence. Comput. Educ. 81, 345–353 (2015). https://doi.org/10.1016/j.compedu.2014.10.019
14. Ramos, J.L., Carvalho, J.M.: Tablets no ensino e na aprendizagem. A sala de aula Gulbenkian: entender o presente, preparar o futuro. Fundação Calouste Gulbenkian, Lisboa (2017)
15. Aesaert, K., van Nijlen, D., Vanderlinde, R., van Braak, J.: Direct measures of digital information processing and communication skills in primary education: using item response theory for the development and validation of an ICT competence scale. Comput. Educ. 76, 168–181 (2014). https://doi.org/10.1016/j.compedu.2014.03.013
16. Karaseva, A.: Pedagogy of connection: Teacher's experiences of promoting students' digital literacy. In: Erstad, O., et al. (eds.) Learning Across Contexts in the Knowledge Society, pp. 225–242. Sense Publishers, Rotterdam (2017)

Using Gamification in Software Engineering Teaching: Study Case for Software Design

Gloria P. Gasca-Hurtado, Maria C. Gómez-Álvarez[✉],
and Bell Manrique-Losada

Universidad de Medellín, Cra.87 No. 30-65, Medellín, Colombia
{gpgasca, mcgomez, bmanrique}@udem.edu.co

Abstract. Software engineering discipline needs promoting and responding to the demands of the software industry and their challenges, centered on the diversity and short delivery times in the projects. Looking to align with such demand, software engineering teaching has evolved incorporating newel strategies for increasing student motivation in the learning process. Gamification is one of them strategies centered on games principles, as the interactivity, ludic, and enhance engagement. This strategy, compared to other teaching techniques, improve the processes of conceptual understanding and learning. In this paper we propose a method, based on gamification, to design pedagogic instruments, comprising a strategy, techniques, and materials for teaching a specific subject. The goal of method is facilitate the gameful activities design in the classroom and then increase of motivation, cooperation and teamwork in participants, in the learning process of conceptual and practical subjects. The method was validated with the topic of software design in a course of a software engineering of the Universidad de Medellín (Colombia). In this pilot we establish improvement actions and recommendations incorporated in the final game version.

Keywords: Active learning strategies · Engineering teaching and learning · Pedagogic instrument · PSP

1 Introduction

Today software industry demands the development of high quality complex software. Such demand requires software engineers with excellent competencies, able to choose the suitable tools and processes to accomplish dynamic requirements. Actually, key challenges of software engineers are increasing diversity and the need to shortened delivery times while guaranteeing trustworthy quality [1]. Therefore, software engineering teaching should consider such diversity. Accordingly, research in this field incorporates active learning strategies, like gamification, for increasing student motivation in the learning process. Gamification is being used in the learning context because it's a strategy focused on the interactive and highly engaging character of games. This strategy motivates learners to "take responsibility for their own learning, which leads to intrinsic motivation" [2], and "enhance engagement and improve learning outcomes by means of integrated learning environments" [3].

© Springer Nature Switzerland AG 2019
Á. Rocha et al. (Eds.): WorldCIST'19 2019, AISC 932, pp. 244–255, 2019.
https://doi.org/10.1007/978-3-030-16187-3_24

Related approaches using gamification in software engineering teaching are focused on developing competencies in software engineering in several topics [4]. They indicate that the gamification compared to other teaching techniques, help to understand and learn concepts. Recent gamification research has suggested that game element-mapping to learning content may indeed facilitate processes good to software engineering [3, 4].

Both, academic and professional qualifications of software industry should promote and reflect the required competencies. For this reason, it is important to explore new ways of teaching and learning that support the development of management competencies, continuous improvement, and high performance, as are described in Personal Software Process (PSP) framework. These competencies should be aligned with the needs and requirements of the software industry about the professionals training [5].

In this paper, we propose the PID (pedagogic instrument design) method. The created instrument following PID comprises a teaching strategy, techniques and materials required for teaching a particular subject. The aim of the PID is facilitating the trainer work from a pedagogical point of view. PID is based on gamification as a strategy to design playful scenarios in the classroom, in order to encourage direct interaction of the participants (student or professional in training). The validation of PID is based on a pilot pedagogic instrument for teaching software design. The overall purpose is the concepts appropriation and promotion the creativity development by participants; the specific purposes are oriented to strengthening teamwork and cooperation [6, 7].

This paper is organized as follows. Section 2 describes background and related work. Section 3 introduces PID, the method to design pedagogic instruments. Section 4 describes the experimental study cased used to evaluate PID. Section 5 then combines the results and discussion about the pilot study. Finally, Sect. 6 presents our conclusions.

2 Background and Related Work

In this section, some approaches of application of gamification in software engineering teaching, and the reported benefits in terms of motivation and students performance are summarized.

Hazeyama [8] presents a learning environment for collaborative software development associating artifacts management with communication support. Such management is supported by a tool providing functions like file-based artefacts management, planning and progress report management, meeting minute management and announcement from teaching staff the student progress in the proposed activities. Meanwhile, Pieper [9] proposes the usage of simulation and digital learning games as a teaching strategy for a software engineering learning environment due to their potential to extend the learning experiences beyond lectures and class projects.

Dubois and Tamburrelli [10] promote the gamification usage for engaging, training, and monitoring students involved in the software products development from inception to maintenance phases. They propose a strategy based in three complementary activities: (1) analyze gamification approaches and identify the most appropriated to be applied in software development phases; (2) integrate such approach to the development process, and (3) evaluate the proposed activities. The preliminary results show an important increase in the quality of software artifacts produced by the students in the gamified approach regarding to the artifacts produced without it. In the same sense, Barata et al. [11] highlight the motivational power for incorporating game principles in non-games processes, like education. They compare a software engineering gamified course with his previous non-gamified version through different performance measures.

Berkling and Thomas [12] propose gamifying a software engineering course to promote independent learning in students. Such proposal is based on that games are good motivators and involve the participants in an environment where they want to reach their goals and obtain recognition among their classmates.

In the context of configuration management, de Melo et al. [13] apply gamification to recognize software developer collaboration and commitment. They use a tool for extracting information from a control version system where developers execute code commits. Such a tool shows a ranking of more active developers using metrics. The goal is determine if gamification usage encourages collaboration and commitment in team members. Moreover, Singer and Schneider [14] use gamification of version control for encouraging students to make frequently commits with a social software.

Concerning to software development process, La Toza et al. [15] propose crowd development as an option for organizing software development process into micro-tasks. Micro-tasks are short (few minutes), modular, and self-descriptive. Such proposal could motivate students to join and contribute to an open source real project developing new skills, sharing knowledge, and participating in collaborative work.

Finally, Prause et al. [16] describe an experiment of gamification of code quality in agile development. The experiment consists on form teams of ten experimented students in programming and one instructor doing agile development. The teams work for four weeks in a share space. The indicators used for measure software internal quality are: (1) understandability of source code, (2) reputation score of each developer because his uploaded files and, (3) number of bugs injected and removed for each developer.

In summary, in the previous approaches exist an interest for incorporating active learning strategies in the teaching and learning of software engineering process, looking for the development of technical and social competencies in the students. In this context, gamification is an option to promote motivation and engagement. However, its necessary formalize the gamification teaching experiences with elements adapted to new generations, spaces adopted for reflecting future improvement and guarantying participants wellbeing, among others. Additionally, all such approaches incorporate gamification for software engineering teaching but is also necessary include in these experiences the definition and assessment of competencies expected in students as a guide for teachers. Accordingly, the goal of this paper is present the PID method to use gamification as a teaching method in a software engineering course including the assessment of competencies to develop in students.

3   PID Method

In this section, we present the PID (Pedagogic Instrument Design) method to design pedagogic instruments for SE teaching. The goal of PID is guide the creation of pedagogic instruments, based on: (1) experience as key factor to learning, and (2) gamification in an educational environment, as a strategy to stimulate the classroom work and increase the participants motivation. The method is designed as a sequential path where a trainer can obtain a pedagogic instrument to guide the teaching of a specific subject. The elements of the method are presented in Fig. 1 and are as follows:

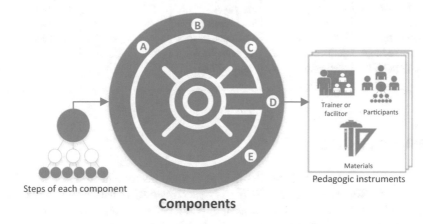

Fig. 1. Elements of the PID method

- **Components.** Comprises the following components: (A) Preparation, (B) Design, (C) Pilotage, (D) Scheduling, and (E) Assessment.
- **Steps.** Each component contains a step-by-step sequence to obtain a pedagogic instrument designed under the gamification strategy.
- **Pedagogic instrument.** Generated artifact because of the steps of each component. Each generated instrument comprises: *Participants* (groups of students or professionals in training; *Trainer or facilitator* (professors, trainers, or facilitators of an activity conducted with an instrument), *Materials* (set of necessary resources for the application of the gamified pedagogic instrument.

The method was designed considering the following pedagogical principles: (1) Planning [17]; (2) Environment gamification [18]; and (3) Experimentation [19]. The description of each component is shown in Table 1.

Table 1. Components of PID

Component				
A−PREPARATION	B−DESIGN	C−PILOTAGE	D−SCHEDULING	E−ASSESSMENT
Define goals to achieve with the instrument, based on the analysis of: (1) competencies to be developed, (2) learning goals, (3) profile of the population, (4) particular interests and age of the population	Outline elements of gamification to include in the instrument. *i.e.* be reward, status, achievement, and competition. Also the mechanics of instrument is defined, *i.e.* the rules and processes	Test instrument with a different audience to the target population (friends, family, and colleagues). Then adjusts of the game mechanics (rules, materials, or time for each activity) are executed	Provide spaces, resources, and materials required for the application of the final instrument	Identify participants perception about the instrument and facilitator performance, by using the assessment proposal presented in [20]

4   Pilot Study Case

According to the PID method, we design a pedagogic instrument for teaching *software design* under PSP, called *CAR DESIGN PSP*. Such instrument uses an analogy between 'car design' and 'software design', considering that before construct a car or a software product is necessary elaborate the structural models of a final product. Such instrument can be applied in a session of 1.5 to 2 h. Also, in CAR DESIGN PSP we use a checklist to evaluate the product quality, corresponding the PSP premise: "measure before improve". The results of the PID application are as follows:

A – Preparation. The development results of the steps included in this component are:

- **Learning Goals:** Identifying software bugs in early stages caused by software design models of poor quality; emphasizing the importance of software design in the software engineering process, and encouraging the implementation of good practices described in the PSP framework for software design.
- **Competencies to be developed by the student:** Understand the importance of using PSP in software design; improve how to design software implementing PSP good practices; and implement software design practices focused on PSP.

B – Gamified Instrument Design. The instrument comprises:

- **Rules:** Comprise the rules set required to achieve the game goal: (1) Teamwork, not interfere with the work of other teams; (2) Respect time assigned to create the car elements list; (3) Respect the turn of expert participants when they describe car elements; (4) Accept and follow the instructions given by the Car Inspector; (5) Car Inspector must not exhibit the checklist to experts; (6) Car Inspector must not indicate car design elements to participants; and (7) Team must use all puzzle pieces.

- **Game materials:** Materials for game are: (1) checklist for the Car inspector to register the mechanical, design, and technology elements for the car, (2) colored paper, (3) puzzle, (4) template of the 4 + 1 architectural view model [21], and (5) chronometer.
- **Game roles:** The roles for CAR DESIGN PSP are shown in Table 2.
- **Game steps**: Next, the step-by-step for developing this game are the following:

1. Participants conform teams of five persons where a participant assumes the role of car inspector and the other four will be experts from the automotive sector.
2. It has four quadrants corresponding to categories to generate checklists for car design: design, technical specifications, safety, technology, and comfort.
3. The inspector tells the team when start to fill the quadrant (distributed by experts) identifying items and assigning them a priority for generating checklists.
4. The inspector assesses the expert performance in team, and if he/she has a satisfactory performance and delivers puzzle pieces to assemble a car.
5. The winning team is the one that make the most detailed specifications of each category of car design and assemble the puzzle in the shortest time possible.

C – Pilotage. We developed a pilot, conforming a team of five students of different levels of a software engineering academic program of the Universidad de Medellín (Colombia). In this pilot, we establish improvement actions and recommendations incorporated in the final game version.

Table 2. Roles of CAR DESIGN PSP

Role	Responsibilities
Car inspector	1. Check if the team has all the resources necessary to execute the game activities 2. Check if the 4 + 1 architectural view model is consistent with the car elements 3. Prepare report of the game activities 4. Measure time for each game activity 5. Reward the team with the puzzle pieces, when is necessary 6. Check the template of the 4 + 1 architectural view model in each phase
Expert	1. Work in team 2. Participate in all game activities 3. Create a list of elements necessary for the car design 4. Assemble the puzzle
Facilitator	1. Help to teams in achieving the game objectives 2. Support every team to do their best in the different activities 3. Promote collaboration and try to achieve synergy 4. Indicate mission, challenge, instructions and rules of the game

D – Scheduling the Final Instrument. The pedagogic instrument was adjusted based on improvements and suggestions incorporated in the pilot. In this session was possible to evaluate the game using the assessment proposal mentioned in the component E.

E – Assessment. We developed an assessment of the proposed method, considering the three sections of the reference assessment proposal: (1) competencies of participants; (2) instruments and teaching techniques; and (3) levels of student satisfaction regarding the teaching process. Such assessment was performed using the survey as a support tool for evaluation. We design a detailed instrument for assessing the competences of participants, as a complement to the Sect. 1 from the reference assessment proposal.

Such a detailed assessment instrument is based on a rubric template. The template was designed to support the competence-based assessment, due to other assessment sections were more developed—instruments and teaching techniques, and satisfaction levels. The template is shown in Table 3, and the application level detail are as follows:

Table 3. Rubric template for supporting the competence-based assessment

Learning levels			Rubric features		Application level
#	Level	Activity category	Related abilities/skills	Weight	Excellent (E) Good (G) Fair (F)
1	**Knowledge**	**Become acquainted**	• Recognize the role of design software in the software engineering process **[10%]**	**30%**	(E)/(G)/(F)
		Prioritizing information	• Identify the needed information about: Errors in design quality; PSP for design software; and 4 + 1 View Architecture **[20%]**		(E)/(G)/(F)
2	**Comprehension**	**Understanding and inferring**	• Understand the meaning of using PSP for software design **[5%]** • Interpret the basic aspect for car design **[5%]**	**15%**	(E)/(G)/(F)
		Exploring alternatives	• Plan and analyze alternatives of a solution to given problem **[5%]**		(E)/(G)/(F)
3-4	**Application/Analysis**	**Structure analysis**	• Interpret elements, principles, and structure for car design **[10%]**	**15%**	(E)/(G)/(F)
		Matching	• Identify internal relationships and components from car **[5%]**		(E)/(G)/(F)
5	**Synthesis**	**Fulfillment of duties**	• Assign, compliance, and fulfill of responsibilities of design **[5%]**	**30%**	(E)/(G)/(F)
		Implementation strategy	• Apply a strategy for design and implement the solution **[10%]**		(E)/(G)/(F)
		Completion	• Complete the design given time and resource constraints **[15%]**		(E)/(G)/(F)
6	**Evaluation**	**Reflection**	• Recognize the missing features required in the design **[5%]**	**10%**	(E)/(G)/(F)
		Judgment compliance	• Accept judgments relating to external criteria to improve and complete the design **[5%]**		(E)/(G)/(F)

The application level is defined as a scale of achieved learning, in terms of: Excellent (10 points), Good (8 points), or Fair (6 points). The application level for each learning level from the template, is in Table 4:

Table 4. Application level for each learning level of assessment template

Level		Description
1: Knowledge	E	All-important major and minor elements for design software are identified and appropriately prioritized. All relevant information is obtained and exact information sources are consulted. Design recommendations are well supported by the information
	G	All major elements for design software are identified but one or two minor ones are missing or priorities are not recognized. Sufficient information is obtained and most sources are valid. Design recommendations are mostly supported by the information
	F	Many major elements for design software are not identified. Insufficient information is obtained and/or sources lack validity. Design recommendations are not supported by information collected
2: Comprehension	E	Among the alternatives analyzed they have been considered: prioritized criteria, improvement cycles, and identification of bugs. Three or more alternatives are considered. Each alternative is appropriately and correctly analyzed for technical feasibility
	G	Among the alternatives analyzed they have been considered: improvement cycles and identification of bugs. At least three alternatives are considered. Appropriate analyses are selected but analyses include some minor procedural errors
	F	Among the alternatives analyzed they have been considered only prioritized criteria Only one alternative is considered. Inappropriate analyses are selected and/or major procedural and conceptual errors are made
3–4: Application/Analysis	E	Car design has consistent aspects by categories. All conditions and rules are considered correctly
	G	Some mistakes are evident in the definition of criteria by categories. Some conditions and rules are considered correctly
	F	The criteria by all categories are not defined. The conditions and rules are not considered
5: Synthesis	E	Responsibilities have been delegated fairly, and each member contributes in a valuable way to the design. The work of all team members demonstrates the implementation of a design and implementation strategy of the solution The car design was correctly finished, at time and with resources constrains established
	G	Some minor inequities in the delegation of responsibilities. Some members contribute more heavily than others but all members meet their responsibilities The work of the half team members demonstrates the implementation of a design and implementation strategy of the solution. The car design was at least 80% finished, at time and with resources constrains established
	F	Major inequities in delegation of responsibilities. Team has obvious freeloaders who fail to meet their responsibilities or members who dominate and prevent others from contributing. The teamwork no demonstrate applying a design and implementation strategy of the solution. The car design was at least 50% finished

(continued)

Table 4. (*continued*)

Level		Description
6: Evaluation	E	Recognize all the missing features required in the design. All the judgments founded by the car inspector related to the external criteria are accepted to improve the design
	G	Recognize some of the missing features required in the design. At least the judgments founded by the car inspector related to the external criteria are accepted to improve and complete the design
	F	The missing features required in the design are not recognized for the team The judgments founded by the car inspector to improve the design, are not accepted

5 Study Case Discussion

The assessment was applied to 20 people (professionals from a software company at Medellín (Colombia) and systems engineering students at the Universidad of Medellín) during the 2nd semester of 2015. In this section we present the results obtained from the application of the instrument for the Sect. 2 'instruments and teaching techniques' and Sect. 3 'levels of student satisfaction', from component E. The survey contains 20 variables; the most representative ones according to each evaluated feature are presented in Table 5. The assessment scale is Very Poor (VP), Poor (P), Fair (F), Good (G) and Very Good (VG). According to the results the positive ratings are related to the features of the didactic technique used—gamified instrument; in most cases the highest percentage corresponds to good and very good insights on the evaluated feature.

Table 5. Frequency of teaching techniques

Teaching techniques (Features)	Frequency (%)				
	VP	P	F	G	VG
Instructions presentation	0	0	10	10	90
Time for development of the activity	0	0	5	30	65
Clarity of instructions	0	0	0	20	80
Teaching materials quality	0	0	0	0	100

Regarding the assessment of the student satisfaction level during the gamified instrument application we found that features as the enjoyment level and creative thinking stimulation are above 85% with ratings as very good or excellent; while the level of closeness to reality is below 60% with ratings as excellent and very good.

Finally, as a way for measuring the student competencies, the variables assessed are shown in Table 6. About the concepts learned, PSP gets a rate of 50%, this is one of the core concepts of the instrument. However, students identify a lesser extent (10%) the importance of software design and disciplined practices for software development. Another aspect to highlight is that 100% of participants manifest understand the software design concept and the intention to apply PSP in the area of SE in which they work. Such an indicator is important, since one of the problems with PSP teaching is the lack of real implementation in companies.

Table 6. Frequency of competences assessment

Participant competence		Frequency (%)
What are the main concepts learned during the activity?	• Concept and importance of PSP	50
	• Importance of software design	20
	• Detailed lists	10
	• Prioritization lists	10
	• Disciplined Practices	10
What is the most important activity that the systems engineer must do in the software design based on PSP?	• Detailed and prioritized list	65
	• Acquisition of skills	15
	• Importance of design	10
	• Checklist generation	5
	• Design activities planning	5

6   Conclusions

This paper presents a proposal of a pedagogic instrument design method based on experiential work as learning engine and gamification as a transversal motivating element. This method includes five components that cover the whole process of pedagogic instrument design, from preparation and design, going through the pilotage of application of the instrument to test his mechanics and goals, until reach the schedule and assessment of the instrument once it has been applied to the target audience.

The study case consists in designing a pedagogic instrument with the proposed method, for teaching software design under PSP. The results obtained from the survey support the conclusion that the didactic technique applied to the instrument is appropriate. In fact, the 100% of participants say they would apply the concept of PSP in the area of SE which they currently perform, showing that if other alternatives are sought to present the subjects, it is possible to obtain more receptivity of the public. In relation with the level of learning achieved by the participants, they recognize the importance of PSP (50%), to a lesser extent the importance of software design (10%) and the use and prioritization of checklists (10%). This means that at the level of competencies of the participants, we achieved to emphasize the importance of software design and the use of PSP framework in this area of knowledge.

Regarding to the competencies-based assessment, in this paper we present a rubric template as an instrument for the facilitator in the application of the game CAR DESIGN PSP. Such template includes the Bloom's taxonomy learning domains, activity categories corresponding with the main tasks expected in students and a set of rubrics for orientate the assessment of application level for each activity. Such template is an important contribution for gamified activities, since it is not only important the design process, but also the evaluation process of the competencies to be promoted.

As future work we identify: (1) apply and evaluate the proposed method for design gamified pedagogic instruments for others subjects of software engineering like software quality, design patterns, and effort estimation; (2) develop a digital tool to support the pedagogic instrument design and as a repository for the instruments, and; (3) develop a tool to guide the professor in the definition of rubrics for the gamified activities created.

Acknowledgement. This work has been partly founded by the Universidad de Medellín, the Universidad Politécnica de Madrid, and Procesix Inc., under the joint project "PSP/TSP Teaching and Learning Methodology as an initiative for quality and productivity levels improvement of software development teams".

References

1. Sommerville, I.: Software Engineering, 10th edn. Pearson, Boston (2015)
2. Akili, G.: Games and simulations: a new approach in education, 1st edn. IGI Global (2014)
3. Nah, F., Zeng, Q., Telaprolu, V., Ayyappa, A., Eschenbrenner, B: Gamification of education: a review of literature. In: International Conference of HCI in Business, pp. 401–409. Springer (2014)
4. Uyugari, F., Intriago, M., Jacome, E: Gamification proposal for a software engineering risk management course. Springer, Switzerland (2015)
5. Towhidnejad, M., Hilburn, T.: Integrating the Personal Software Process (PSP) across the undergraduate curriculum. In: 27th Conference Frontiers in Education, Pittsburgh, USA, pp. 162–168 (1997)
6. Chacón, P.: El Juego Didáctico como estrategia de enseñanza y aprendizaje ¿Cómo crearlo en el aula? Nueva aula abierta **16**, 32–40 (2008)
7. González, A.: Diseño de juegos y creatividad: un estudio en el aula universitaria. Opción **31** (4) (2015)
8. Hazeyama, A.: Collaborative software engineering learning environment associating artifacts management with communication support. In: 3rd International Conference on Advanced Applied Informatics, Kitakyushu, Japan, pp. 592–596. IEEE (2014)
9. Pieper, J.: Learning software engineering processes through playing games: suggestions for next generation of simulations and digital learning games. In: Proceedings of Workshop on Games and Software Engineering, Zurich, Switzerland, pp. 1–4. ACM (2012)
10. Dubois, D., Tamburrelli, G.: Understanding gamification mechanisms for software development. In: Proceedings of the 2013 9th Joint Meeting on Foundations of Software Engineering, Saint Petersburg, Russia. ACM (2013)
11. Barata, G., Gama, S., Jorge, J., Goncalves, D.: Engaging engineering students with gamification. In: Proceedings of the First International Conference on Gameful Design, Research, and Applications, Ontario, Canada, pp. 10–17. ACM (2013)
12. Berkling, K., Thomas, C.: Gamification of a Software Engineering course and a detailed analysis of the factors that lead to it's failure. In: 2013 International Conference on Interactive Collaborative Learning (ICL), Kazan, Russia, pp. 525–530. IEEE (2013)
13. De Melo, A., Hinz, M., Scheibel, G., Berkenbrock, C., Gasparini, I., Baldo, F.: Version control system gamification: a proposal to encourage the engagement of developers to collaborate in software projects. In: International Conference on Social Computing and Social Media, Crete, Greece, pp. 550–558. Springer (2014)

14. Singer, L., Schneider, K.: It was a bit of a race: gamification of version control. In: 2nd International Workshop on Games and Software Engineering, Zurich, Switzerland, pp. 5–8. IEEE (2012)
15. La Toza, T., Towne, W., Van Der Hoek, A., Herbsleb, J.: Crowd development. In: 2013 6th International Workshop on the Cooperative and Human Aspects of Software Engineering (CHASE), San Francisco, USA, pp. 85–88. IEEE (2013)
16. Prause, C., Nonnen, J., Vinkovits, M.: A field experiment on gamification of code quality in agile development. In: Psychology of Programming Interest Group Annual Conference (2012)
17. Colomo-Palacios, R., Tovar-Caro, E., García-Crespo, A., Gómez-Berbís, J.: Identifying technical competences of IT Professionals: the case of software engineers Professional Advancements and Management Trends in the IT Sector. Int. J. Hum. Cap. Inf. Technol. Prof. (IJHCITP) **1**(1), 31–43 (2010)
18. Oprescu, F., Jones, C., Katsikitis, M.: I PLAY AT WORK—ten principles for transforming work processes through gamification. Front. Psychol. **5**(14) (2014)
19. Shepard, L.: The role of assessment in a learning culture. Educ. Res. **29**(7), 4–14 (2000)
20. Manrique-Losada, B., Gasca, G., Gómez, M.: Assessment proposal of teaching and learning strategies in software process improvement. Revista Facultad de Ingeniería **77**, 105–114 (2015)
21. Krutchen, P.: The 4 + 1 view model of architecture. IEEE Softw. **12**(6), 42–50 (1995)

Geolympus - Cloud Platform for Supporting Location-Based Applications: A Pervasive Game Experience

Juan Luis Berenguel Forte[1(✉)], Daniel Pérez Gázquez[1],
Jeferson Arango-López[1,2] ⓘ, Francisco Luis Gutiérrez Vela[1] ⓘ,
and Fernando Moreira[3,4] ⓘ

[1] E.T.S. of Computer and Telecommunication Engineering,
University of Granada, c/Periodista Daniel Saucedo Aranda, s/n,
18071 Granada, Spain
{juanlubf, danielpg, fgutierr}@ugr.es,
jeferson@correo.ugr.es
[2] Faculty of Electronics Engineering and Telecommunications – FIET,
University of Cauca, Cl. 5 # 4 – 70, Popayán, Cauca, Colombia
[3] REMIT, IJP, Universidade Portucalense, Rua Dr. António Bernardino
Almeida, 541-619, 4200-072 Porto, Portugal
fmoreira@upt.pt
[4] IEETA, Univ Aveiro, Aveiro, Portugal

Abstract. The development of game experiences based on the player's location is becoming increasingly popular. This games are used to increase players' skills and knowledge about a particular topic in different contexts like education, health and tourism. When pervasive or narrative components are added, it becomes evident the need for a tool to manage the information in an appropriate and dynamic way. When we talk about pervasiveness, in addition to the elements of space and time, there is an additional component related to social interaction, which can be achieved through the exchange of information between various games to provide a better player experience. For this reason, in this article we present a platform enabled for the creation and edition of game experiences based on the player's location. Which has the ability to exchange information between projects, including narrative in the different games. A developed experience evidences the relevance of this platform in the education context.

Keywords: Geolocation · Pervasive game · Web platform · Cloud · Microservices

1 Introduction

Games based on players' location involve a component that is not found in console games: the locomotion of players through the outside world and, possibly, the realization of physical activities [1], taking the players into a highly pervasive environment. Geolocation games is one of the game genres in which pervasiveness is applied

Á. Rocha et al. (Eds.): WorldCIST'19 2019, AISC 932, pp. 256–266, 2019.
https://doi.org/10.1007/978-3-030-16187-3_25

the most, having become rather popular in recent years and being Pokémon Go the most prominent case [2, 3]. These games are played in the real world, supporting the player's actions within the virtual space through the game rules and real space [4].

The project described in this paper is a new framework for supporting geolocation applications, which facilitates their development and provides the necessary tools for their management and analysis. Different papers have been found describing tools and platforms with a similar objective [5–8]. However, they do not provide an effective solution to all the needs of this kind of applications. For this reason, the existence of a complete framework that facilitates the development of these systems and unifies their implementation seems quite necessary. One of the main components of these tools is to capture and analyze the information in real time. For this reason, it is necessary to consider different technologies that may support this processing.

Our proposed framework, Geolympus, is a platform that provides the necessary tools and means for the developer to implement, manage and analyze the geolocation data of their projects, being able to access them both directly from the client applications as well as from the backend of the projects. In the latter case, this allows for greater control over the data management carried out. In addition, it offers the possibility of sharing data between different applications or game experiences, which allows to contribute to the IoT (Internet of Things) and, possibly, to extend the concept of social pervasiveness.

This paper is structured as follows. Section 2 presents the work related to the development of PG. In Sect. 3, the design, implementation and functionalities of the Geolympus tool are shown. Section 4 presents the "Descubre-UGR" game experience. Finally, Sect. 5 shows the conclusions and the future work.

2 Works and Related Tools

As previously said, during the review of tools and related platforms some have been found that pursue a similar objective to the one our framework has. The most relevant ones are described below, together with a brief comparison with Geolympus.

2.1 CREANDO

It is a web environment tool that emerged to solve the problem of creating PG based on narrative in closed spaces. The main objective of CREANDO [9] is to provide a solution that allows users (with technical knowledge of programming) to design, build and execute pervasive game experiences based on the narrative and location of players in closed spaces within an educational context.

Geolympus is a more general tool, which only provides support for the design and management of the applications' location data (both in open and closed spaces). Thus, it does not only focus on an educational context, but it supports all kinds of geolocation applications.

2.2   LAGARTO

Web tool for building games based on the players' location, including Augmented Reality. Its main objective is to provide a way to easily build geolocation games in open spaces for non-programmer users.

Although it may make developing geolocation games easier than geolympus would, Lagarto is a more specialized tool, and as such it is more limited than Geolympus.

2.3   Play Visit

It is a web platform for the development of geolocation games [5]. This tool is located in external servers and the games are executed through the browser of the local devices. It allows the creation of 'City Hunter' games in open spaces.

Geolympus takes advantage of this platform like the previous ones, by allowing the player to be located in both open and closed spaces within the same application, by providing more configuration options with additional entities and attributes and by allowing the seamless sharing of data between different projects.

2.4   Backendless

It is a cloud platform for backend development [10]. Its goal is to allow the developers to create the functionality of their application backends with no need to make any deployment. Additionally, it has tools for supporting geolocation applications, such as a map to visualize the inserted elements, easy search for elements by radio and geofencing (delimitation of areas that will trigger events when the followed elements enter, remain or leave them).

This platform is the most similar to Geolympus. The management of geolocation data is only one of the Backendless' features, since its purpose is the development of backends in general. However, Geolympus, being more specialized, includes additional tools and mechanisms that extend its usefulness in geolocation: additional and/or automated elements such as routes, groups, users, etc., the possibility of sharing data between projects, support for indoor location, data analysis tools and pervasiveness support.

3   Geolympus

Geolympus is a cloud platform whose main functions are, as it has been described, to facilitate the development of geolocation applications and to provide a means of effectively managing their location data. Among other features, the easy sharing of data between different projects makes it propitious for the development of systems that interact with each other, such as pervasive game experiences. Geolympus is based on an entity system, each entity running as a microservice. The user of an application being geolocated in real time or points of interest existing within an application are examples of Geolympus entities.

The platform allows the definition of the different Entities that will exist in each project and to configure their behavior and the data to be stored for them and their Instances. "Entity", the base type of entity, contains the basic implementation all other entity types inherit: definition of the data schemas the entity and its instances will store, management of the data following these schemas and the public access for external agents. An "EventHandler" component triggers events in a subscription/publication system (Google Pub/Sub) so that all interested subscriber services get instant updates of the status of the entity and its instances.

From the base entity type there is a branching of more specific entity types, each one inheriting the base data schemas and functions of the parent entity types. Thus we have Geolocated Entities (GE, with various types according to the nature of the instances to be located), areas (that automatically detect if other instances are inside or outside them), Groups and Routes. The hierarchy can be consulted in Fig. 1.

After defining the Entities that will exist in an application, its main backend will be able to update and consult its data by means of calls to its API REST (other ways such as using websockets may be as well implemented in order to improve the communication efficiency). Also, as mentioned above, you can subscribe to a complete list of events, such as the creation, removal and modification of Instances among others. A series of libraries would facilitate the integration with Geolympus in different programming languages and some video game engines (Unity).

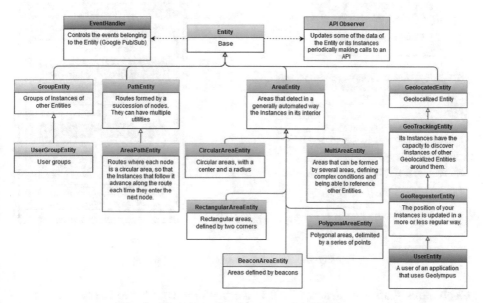

Fig. 1. Hierarchy of entities types in Geolympus.

Once up and running, Geolympus provides a map showing the existing located instances, areas, and routes in the project, and full search and filter functions, all seen in real time (it may be configured to ignore some sensitive entities, such as users, to comply with any privacy requirement). The platform allows the creation, deletion and modification of instances of the defined entities from the map as well as from a table of instances of each entity, and also triggering custom events for specific entities and instances. Furthermore, additional tools can be integrated according to the needs (e.g. data analysis tools such as heat maps).

As previously said, one of the main features of Geolympus is the ability to interconnect different projects so that, with the appropriate permissions, a project can access data from another project. This is useful, for example, in PGs where the entities' instances are known in different applications, producing an integration between these projects. For this, Geolympus allows the definition of several Interfaces for each project. These enumerate a list of permissions on the entities, their functions and the data that will be available through their access, as well as a list of users and projects that will have access permissions.

3.1   Design

Geolympus is divided into three main components: Geolympus Platform, GeolympusCore and the users. Figure 2 shows a diagram summarizing the structure made up of these components and the architecture will be described in depth in Sect. 3.2.

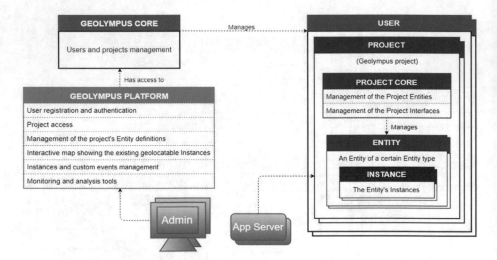

Fig. 2. Basic structure of Geolympus.

Geolympus Platform (Frontend). It is the platform available to Geolympus administrators. It allows the creation, management and deletion of said administrators and their associated projects. As mentioned before, it offers management and monitoring tools for the project Entities and Interfaces.

Geolympus Core (Backend). It constitutes the central core of the system, being accessed by Geolympus Platform to manage the Geolympus administrator users, to create and delete projects and to know the users linked to each project. A project can have three types of users linked to it:

- Owner: The user who created it and has complete control over it.
- Administrator: Users that can manage the Project through the platform.
- Observer: Users that may access some of the project interfaces, having limited access to their entities.

Users (Backend). This component's meaning is more abstract: it is the Geolympus administrator users. Each user is made up of a set of microservices, having for each owned project a ProjectCore microservice and as many other ones as Entities have been defined:

- ProjectCore: It is in charge of creating, managing and deleting the Entities and Interfaces of the project. Accessible by means of a REST API.
- Entity: Each one, running as a microservice, controls its data and Instances. Accessible through a REST API.

3.2 Architecture

Geolympus is built on the Google Cloud Platform, automatically creating projects and clusters and using the Google Kubernetes Engine to deploy the pods on which the different microservices run. A main Google Cloud project contains a cluster with two Kubernetes pods containing the publicly accessible components Geolympus Platform and Geolympus Core. Figure 3 shows a summary of the structure of Geolympus:

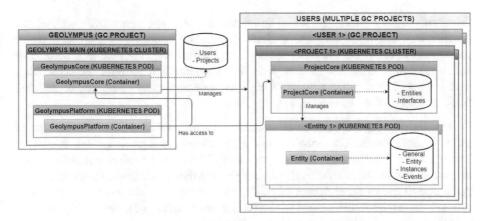

Fig. 3. Detailed structure and main commands.

Implementation of the Main Geolympus Elements. Table 1 shows the equivalence between the main geolympus elements and the real Google Cloud elements used for their implementation.

Table 1. Main elements of Geolympus.

Name	Description
Administrator user	Additional Google Cloud Project
User project	Project cluster. A MongoDB "Replica Set" is deployed on a pod using the Kubernetes "StatefulSets", which serves as the central BD, and the Project Core as another pod
Project entity	Additional pod, functions, data and public methods in your API depending on the type of Entity
Interface	Document in the central database with the basic data
Instance	Document in the collection "instances" of the BD of the entity

Data to Be Saved for the Entities. Each Entity has three associated collections in its database:

- **general**: Collection whose structure is the same for all the types of Entity, storing their basic data.
- **entity**: Stores the default data and the data defined by the project administrator relative to the Entity in general ('static' data with no relation to the Instances).
- **instances**: Contains a document for each existing Instance of the Entity. Saves the default data and the data defined by the project administrator relative to the Instances.

Communication Between Entities. Some entities need to communicate with others, whether they are from the same project or not. For example, area entities, which periodically check if the Instances of the target Entities are within the area defined by each one of their Instances, need to periodically access their position. To do so, they send requests to the public address of the corresponding entity using one of its interfaces, as long as they have access permission.

Events. Entities have a number of associated events, which depend on the type of entity. Google Pub/Sub, a Google Cloud Publishing/Subscription system, is used to implement this functionality. External servers will be able to subscribe to these events so that, each time one is published, they receive the data associated. Also, additionally to the default events for each type of entity, it is possible to configure "custom" events with any message schema, so that whenever the administrator or a service trigger them, the subscribers will receive the published data instantly.

Scalability. It is possible to increase the number of nodes associated to the cluster of a project, thus scaling will be automated in case more resources are needed.

3.3   Technologies

Table 2 shows the main technologies used for the development of Geolympus.

Table 2. Technologies used in the development of Geolympus.

Nombre	Descripción
Google Cloud Resource Manager	Automatic GC Project Management
Google Kubernetes Engine	Automatic deployment of components (containers in Google Compute Engine)
Google Pub/Sub	Entities events
Stackdriver Logging	Logging for the different components
Componentes	Node JS with Express (Javascript)
Bases de Datos	MongoDB, Mongoose, mongoose-dynamic-schemas (npm)

4   DescubreUGR: A Game Experience with Geolympus

A geolocated pervasive game experience is presented, which uses the components of Geolympus to generate its structure. With this experience, Descubre-UGR, the users can learn about the faculties and buildings of the University of Granada (UGR). This is achieved through a game that takes place through a story inspired by the well-known *Frankenstein*.

This experience allows the player to use the narrative to get historical information of each of the places visited, evaluating the knowledge acquired throughout the execution of the game. Once the player finishes the quest line associated to each facility, they are rewarded with a piece of the popular character featured by said quest line, which at the end will be used to create a unique 'Frankenstein monster'.

4.1   User Location

In Descubre-UGR the users must move physically to go to each of the buildings of the University of Granada, thus having pervasiveness through the geolocation of players. This geolocation is done through the internet connection of the mobile device and GPS.

However, the use of the GPS is only effective outdoors, so beacons and QR codes are used for indoor positioning. Additionally, the buildings are also geolocated to be found by the players. There is a difference between the two types of elements in the game. First, the players move physically, so their position must be upgradeable. Second, buildings will have static positions that do not change over time.

4.2   Implementation

Design in Geolympus: For outdoor location, each user is represented as an Instance of the type of Entity *UserEntity*. As this Entity type inherits from the *GeoRequesterEntity*, *GeoTrackingEntity* and *GeolocatedEntity* classes, it is able to detect other nearby geolocated instances (by *GeoTrackingEntity's definition*), such as the faculties, and to periodically update their position (by *GeoRequesterEntity's definition*). Buildings are positioned as instances of an entity of the *GeolocatedEntity* type, as they are elements with a specific static location, although they can eventually be moved if necessary.

For indoor location, positioning areas are defined with beacons, which will detect when any user gets in their range. For this, the entity type *BeaconAreaEntity* was used, which inherits the functionality of the class *AreaEntity*.

Obtaining Data from the Faculties. When starting the application, the first thing it does is to consult the existing facilities in the project. Once the list is obtained, the corresponding icons will be placed on the returned positions. The instances return all the needed data, stored within custom Instance attributes (defined by the administrator in Geolympus), such as their name, a link to a photo, etc. Also, they store their own story (JSON), so that it will be returned for the player as soon as they get to them. This story will serve as the script of the quest line, defining dialogs, questions and other items.

Visitoffing the Faculties. It is necessary to know when a user is close enough to a faculty to consider that they are visiting it (so that they can be allowed to request to start the faculty story). To do so, the user instances periodically check whether any of the faculty instances are near enough. If any instance is found, the system (both in the server and the client) will consider that the user is visiting the faculty, allowing the access to its story.

4.3   Information Management

Geolympus Platform is used to create all the elements described in Sect. 4.2. It makes possible to define and create in a simple and fast way the elements to be positioned. This greatly accelerates the development and facilitates the flexibility of the elements controlled through Geolympus.

In addition, Geolympus allows to monitor the elements created through the web platform. Therefore, it is possible to know in real time the number of connected players, their locations, if they have visited any of the buildings and when, etc. The data can be downloaded for further analysis and to improve the game experience or to carry out studies. Below, Fig. 4 shows screenshots of the Descubre-UGR gaming experience.

Fig. 4. Pervasive gaming experience Descubre-UGR developed with the use of Geolympus.

5   Conclusions and Future Work

This article presents a cloud platform for the management of location data for applications and a pervasive game experience that makes use of the Geolympus platform with the aim of providing the students with a tour through the faculties of the University has also been described.

Given the recent increase in popularity of geolocation video games, the need for tools to facilitate their development has become clear. To this purpose, the Geolympus platform has been proposed, which will facilitate the development of geolocation applications, their management and the sharing of data between different projects.

Future work focuses on optimizing the system to the maximum, including the access to databases, the flow of networking packets, use of Redis for the most frequently used queries, etc.; implementing security measures to detect whenever a user tries to illegally mock their location; and developing a number of libraries to easily integrate Geolympus with different programming languages and engines, while developing various projects that use the platform and making intensive testing.

Acknowledgements. This work has been funded by the Ministry of Economy and Competitiveness of Spain as part of the JUGUEMOS Project (TIN2015-67149-C3).

References

1. Avouris, N.: A review of mobile location-based games for learning across physical and virtual spaces. J. Univers. Comput. Sci. **18**(15), 2120–2142 (2012)
2. Rauschnabel, P.A., Rossmann, A., Claudia, M.: An adoption framework for mobile augmented reality games: the case of Pokémon Go. Comput. Hum. Behav. **76**, 276–286 (2017)
3. Paavilainen, J., Korhonen, H., Alha, K., Stenros, J., Koskinen, E., Mäyrä, F.: The Pokémon GO experience: a location-based augmented reality mobile game goes mainstream. In: Proceedings of the 2017 CHI Conference on Human Factors in Computing Systems - CHI 2017, May 2017, pp. 2493–2498 (2017). no. Figure 1
4. Schlieder, C., Kiefer, P., Matyas, S.: Geogames: designing location-based games from classic board games. IEEE Intell. Syst. **5**, 40–46 (2006)
5. Games, G.: PlayVisit (2017). https://playvisitstudio.geomotiongames.com/
6. Arango-López, J., Valdivieso, C.C.C., Collazos, C.A., Vela, F.L.G., Moreira, F.: CREANDO: tool for creating pervasive games to increase the learning motivation in higher education students. Telemat. Inform. (2018)
7. Santos, F., Almeida, A., Martins, C., Gonçalves, R., Martins, J.: Using POI functionality and accessibility levels for delivering personalized tourism recommendations. Comput. Environ. Urban Syst. (2017)
8. Gonçalves, R., Martins, J., Branco, F., Castro, M., Cota, M., Barroso, J.: A new concept of 3D DCS interface application for industrial production console operators. Univers. Access Inf. Soc. **14**, 399–413 (2015)
9. Maia, L.F., et al.: LAGARTO: A LocAtion based Games AuthoRing TOol enhanced with augmented reality features. Entertain. Comput. **22**, 3–13 (2017)
10. Backendless: BACKENDLESS PLATFORM (2013). https://backendless.com/

Information Technologies in Radiocommunications

Bio-Inspired Petal-Shape UWB Antenna
for Indoor Applications

Paulo Fernandes da Silva Júnior[1]([⊠]),
Ewaldo Elder Carvalho Santana[2],
Raimundo Carlos da Silvério Freire[2],
Paulo Henrique da Fonseca Silva[3], and Almir Souza e Silva Neto[4]

[1] Computer Engineering and Systems, State University of Maranhão,
São Luis, Brazil
paulo.junior@ee.ufcg.edu.br

[2] Electrical Engineering, Federal University of Campina Grande,
Campina Grande, Brazil

[3] Electrical Engineering, Federal Institute of Paraíba, João Pessoa, Brazil
phdafs@gmail.com

[4] Federal Institute of Education, Science and Technology of Maranhão, IFMA,
São Luís, Maranhão, Brazil
almir.neto@ifma.edu.br

Abstract. This paper presents a methodology for the development of the printed monopole antenna bio-inspired in orchid petal-shape, generated by Gielis formula for indoor wireless applications, operation in 4G, 4.5G, and wireless local area networks bands. The monopole measured and simulated have greater results, as indication for Long Time Evolution Advanced Proof 3rd Generation Partnership Project, with measured bandwidth of 4.42 GHz, omni-directional radiation pattern, maximum gain simulated of 5.43 dBi and half power beam width of 102°.

Keywords: Bio-inspired shape · Orchid petal · Ultra-wideband · Antenna · Indoor applications

1 Introduction

Several institutions have intensified yours researches with solutions biologically inspired. According [1, 2] nature can be used as solutions base in some problems found in the engineering. Plants used the light, electromagnetic waves, transforming it into chemical energy by the process of photosynthesis, analogue characteristics to the dish antennas [3]. Gielis formula is an attempt to generate the forms found in nature with a polar expression.

Some works have used the plant shapes in antenna development. In [4] has presented an antenna bio-inspired in a flower with four petals, with feeding by probe, with bandwidth of 4 GHz (4–8 GHz). In [5] an ultra-wideband (UWB) printed monopole antenna (PMA) with maple leaf shape, feeding of transmission line, and bandwidth of 11 GHz (3–14 GHz) was proposed.

© Springer Nature Switzerland AG 2019
Á. Rocha et al. (Eds.): WorldCIST'19 2019, AISC 932, pp. 269–277, 2019.
https://doi.org/10.1007/978-3-030-16187-3_26

A methodology biologically inspired in design of the leaf-shaped on PMA, with application in 4G band in 700 MHz and gain of 7.7 dBi, developed in [6]. In [3] a wearable textile printed monopole antenna built in denim covering the 2G, 3G and 4G systems (1.8–2.69 GHz), with bio-inspired shape of the ginkgo biloba plant, generated by Gielis formula, was presented. In [7] has presented a PMA bio-inspired in the jasmine flower, generated by Gielis formula for UWB band for meets of Federal Communications Commission (FCC), with bandwidth of 9.75 GHz and gain of 5.99 dBi, and density spectral power covering the required of FCC.

A standard referenced to UWB technology is the Federal Communications Com-mission (FCC's report 2002), which identifies the parameters required for the UWB technology. We can be highlight 7.5 GHz bandwidth with spectral mask of 3.1 to 10.6 GHz omnidirectional radiation diagram, large beam-width >60°, low density spectral power (less than −43 dBm), and compact antenna, with thickness of dielectric, preferably, much thinner than the wavelength (λ). The UWB antennas are generally uses in devices with small dimensions in medical and non-medical applications [4–6]. Researches in medical, and indoor wireless applications cover data telemetry, diag-nosis, treatment, and monitoring. Some works used several antennas types for used inside and outside human body, with specific characteristics for each case, and UWB band [6, 8, 9].

This paper proposes a methodology for the development of monopolo antenna, bio-inspired in orchid petal, generated by Gielis formula in UWB band for indoor wireless applications.

2   Materials and Methods

According [5] the PMA perimeter (p) in mm can be approximated by relation between the first resonance frequency in −10 dB (f_1) and relative permittivity (ε_r), defined by:

$$p = \frac{300}{f_1(GHz)\sqrt{\frac{\varepsilon_r+1}{2}}}, \tag{1}$$

from the perimeter is possible to designer antennas with different shapes, and used in the first dimensions values of ground plane. The other dimensions antenna dimensions can be estimated as variations of the effective wavelength (λ_{eff}) in m, indicated by:

$$\lambda_{eff} = \frac{c}{f_{r(Hz)}\sqrt{\frac{\varepsilon_r+1}{2}}}, \tag{2}$$

where c is speed light in vacuum, and fr is the central resonance frequency.

Figure 1 shows PMA examples with structure indications and different shapes of polygonal (rectangular, trapezoidal, etc.), circular, elliptical, and others, have been proposed for UWB applications [10].

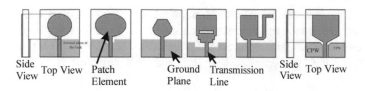

Side View · Top View · Patch Element · Ground Plane · Transmission Line · Side View · Top View

Fig. 1. Example of PMA for UWB applications.

The Gielis formula is a polar transformation that can be circle, square, rectangle, and ellipse geometries [11]. Members of superellipses group, they are limited to symmetric structures, defined by:

$$\left|\frac{x}{a}\right|^n + \left|\frac{y}{b}\right|^n = 1 \tag{3}$$

The Gielis formula given in (3) uses of polar coordinate $r = f(\theta)$. Replacing $x = r\ cos(\theta)$ and $y = r\ sin(\theta)$ in (3) and inserting the argument $(m/4)\theta$ create a specific rotational symmetry allowing the generation of similar geometries found in nature. The variables 'n_i', 'm', 'a', and 'b' are real numbers different from zero.

$$r = f(\theta)\frac{1}{\sqrt[n_1]{\left(\left|\frac{1}{a}\cdot cos\left(\frac{m}{4}\theta\right)\right|\right)^{n_2} + \left(\left|\frac{1}{b}\cdot sen\left(\frac{m}{4}\theta\right)\right|\right)^{n_3}}} \tag{4}$$

Patch of printed monopole antenna, with orchid petal was generated by Gielis formula, with the parameters: $m = 1$; $n_1 = 1/1.46$; $n_2 = 1/1.46$; $n_3 = 1/1.46$; a and $b = 1$; and structure ratio of 13.4 mm. Transmission line was calculated by [12].

The design methodology for the bio-inspired antennas, generated by the Gielis expression, was adapted from [13] and it is developed in 7 steps:

1 - Definition of the application and determination of operating frequencies;

2 - Choice of the antenna characteristics suitable for the application, for example, broadband or narrowband, etc.

3 - Selection of the conductor and dielectric materials, with characterization of the materials properties, using the technical data informed by the manufacturers or using some available characterization method;

4 - Perimeter design of the patch element;

5 - Selection of a bio-inspired shape with total perimeter close to the antenna with Euclidean geometry;

6 - Generation of the image using the Gielis expression with the use of computer aided (CAD) techniques in a format exportable to the full-wave simulation software;

7 - Simulation and optimization of the antenna characteristics, some adjustments are done to obtain the desired resonant frequency. Then, the bio-inspired antenna is produced; finally, validation tests are performed to compare the measured results with the simulated ones.

Figure 2 shows development of monopole antenna bio-inspired orchid petal. The patch of the monopole antenna generated by Gielis formula, Fig. 1(b), used similar shape of orchid petals, indicating for red traced, Fig. 1(a). From the bio-inspired shape was simulated a printed antenna, Fig. 1(c), and built prototype in fibber glass (FR4), Fig. 1(d).

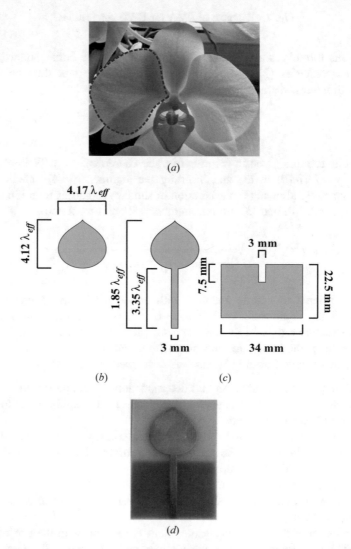

Fig. 2. Development of bio-inspired printed monopole antenna on orchid flower: (a) an orchid flower [14]; (b) generated image by Gielis formula; (c) simulated antenna; (d) prototype.

3   Printed Monopole Antenna Bio-Inspired on Orchid Petal

Figure 3 shows comparison of |S11| simulated and measured parameters of printed monopole antenna bio-inspired on orchid petal, and values of first () and second resonance frequencies, and bandwidth can be observed in Table 1. We noticed that the simulated and measured antennas obtained closed results, with similar curves of the resonance frequencies. The measured antenna presented difference of bandwidth of 16.37%, and 8.4% in the first resonance frequency.

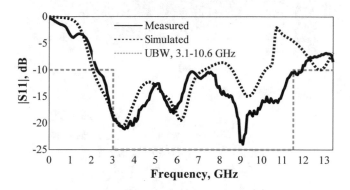

Fig. 3. Comparison of $|S_{11}|$ simulated and measured parameters of bio-inspired orchid petal antenna.

Table 1. Results of Bio-inspire orchid petal PMA

Bio-inspired PMA	f_1 (GHz)	f_2 (GHz)	Bandwidth (GHz)
Simulated	2.18	10.22	8.04
Measured	2.38	12.03	9.65

Figure 4 shows radiation pattern of the PMA orchid petal, with indication of current density (A/m^2), gain (dBi), and HPBW in resonance frequencies of 3.45 GHz and 9.21 GHz. Gain, current density and HPBW are important antenna parameters. By gain it is possible to determine the power radiated by the antenna at resonance frequency, when associated to the HPBW, it is possible to evaluate the scattering of this power in a given region, and the current density indicates the amount of energy circulating at the edges of the antenna. From these parameters, it is possible to identify the possible applications of the antenna used.

Current Density,

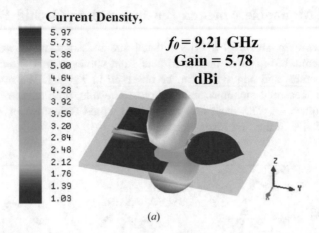

(a)

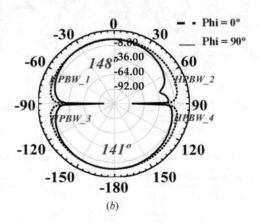

(b)

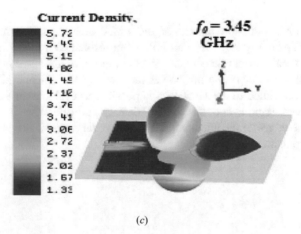

(c)

Fig. 4. Radiation pattern of the PMA orchid petal: (a) 3D with gain and current density in resonance frequency at 9.21 GHz; (b) 2D with HPBW indications in resonance frequency at 9.21 GHz; (c) 3D with gain and current density in resonance frequency at 3.45 GHz; (d) 2D with HPBW indications in resonance frequency at 3.45 GHz.

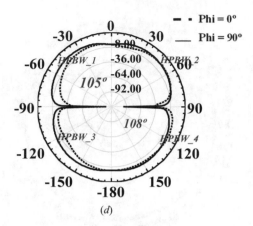

Fig. 4. (*continued*)

We noted than both results presents closed current density, HPBW more than 100°, Fig. 4(*b*) and (*d*), however, the maximum gain, in broadside direction, is observed in frequency resonance of 9.21 GHz (5.78 dBi), Fig. 4(*a*). In comparison to the results obtained in [7], it is possible to evaluate that DSP, in both cases is similar, indicating that the antenna can be used indoors, with low power intensity in a compact structure.

Figure 5 shows the simulated and measured impedance on Smith Chart from 2 to 3 GHz. In both results, a wide frequency band at the proximity of 50 Ω matching points and an inductive effect over the structure can be observed. The red line indicates the bandwidth with voltage standing wave ratio (VSWR) smaller than 2, i.e., the |S11| parameter smaller than −10 dB, Fig. 5(*a*).

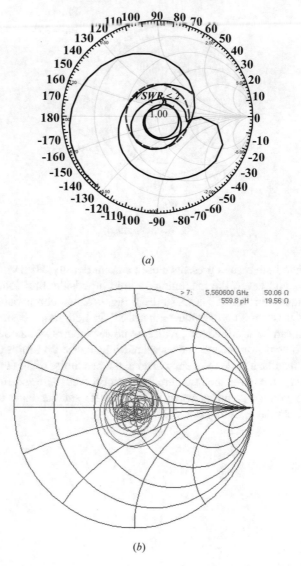

(a)

(b)

Fig. 5. Impedance on Smith chart of PMA orchid petal: (a) simulated with VSWR indication; (b) measured.

4 Conclusions

A printed monopole antenna bio-inspired of orchid petal-shape, generated by Gielis formula for ultra-wideband (UWB) band for indoor wireless applications is presented in this paper. The monopole greater measured and simulated results, covering the require of Federal Communications Commission for UWB technology, with measured bandwidth of 9.65 GHz, 36.82% more than FCC, maximum gain simulated of

5.78 dBi, half power beam width of 108°, current density of 5.72 A/m^2, demonstrated good results for indoor wireless applications.

Acknowledgements. We greatly appreciate the COPELE/UFCG, PECS/UEMA, PPGEE/IFPB and CAPES by support and funding these institutions, without which this work would not be possible.

References

1. Bar-Cohen, Y.: Biomimetics – Biologically Inspired Technologies. CRC Press, Boca Raton (2006)
2. Benyus, J.M.: Biomimicry, Innovation Inspired by Nature. William Morrow, New York (1997)
3. Silva Júnior, P.F., Freire, R.C.S., Serres, A.J.R., da Fonseca Silva, P.H., Silva, J.C.: Wearable textile bioinspired antenna for 2G, 3G and 4G systems. Microwave Opt. Technol. Lett. **58**(12), 2018–2023 (2016)
4. Lotfi Neyestanak, A.A.: Ultra wideband rose leaf microstrip patch antenna. Prog. Eletromagn. Res. **86**, 155–168 (2008)
5. Ahmed, O.M.H., Sebak, A.R.: A novel Maple-Leaf shaped UWB antenna with a 5.0–6.0 GHz band-notch characteristic. Prog. Electromagn. Res. **11**, 39–49 (2009)
6. Silva Júnior, P.F., Freire, R.C.S., Serres, A.J.R., da Fonseca Silva, P. H., e Silva, J.C.: Bio-inspired antenna for UWB systems. In: 1st International Symposium on Instrumentation Systems, Circuits and Transducers (INSCIT), Belo Horizonte, pp. 153–157 (2016)
7. Silva Júnior, P.F., da Fonseca Silva, P.H., Serres, A.J.R., Silva, J.C., Freire, R.C.S.: Bio-inspired design of directional leaf-shaped printed monopole antennas for 4G 700 MHz band. Microwave Opt. Technol. Lett. **58**(7), 1529–1533 (2016)
8. Rahmat-Samii, Y., Kim, J.: Implanted Antennas in Medical Wireless Communications. Morgan & Claypool Publishers, San Rafael (2006)
9. Taoufik, E., Nabila, S., Ridha, G.: The ultra wide band radar system parameters in medical application. J. Electromagn. Anal. Appl. **3**, 147–154 (2011)
10. Li, X.: Body Matched Antennas for Microwave Medical Applications. Scientific Publishing, Karlsruhe (2013)
11. Kishk, A.: Advanced in Microstrip Antennas with Recent Applications. Intech, Rijeka (2013)
12. Gielis, J.A.: A generic geometric transformation that unifies a wide range of natural and abstratic shapes. Am. J. Bot. **90**(3), 333–338 (2003)
13. Wentworth, S.M.: Fundamentals of Electromagnetics with Engineering Applications. Wiley, New York (2005)
14. Khaleel, H.: Innovation in Wearable and Flexible Antennas, p. 236. WIT Press, Boston (2015)
15. Patel, S.: With orchid (2011). http://www.sunilpatel.co.uk/2011/12/white-orchid/

Single-Step Approach to Phaseless Contrast-Source Inverse Scattering

Sandra Costanzo$^{(\boxtimes)}$ ⓘ and Giuseppe Lopez

Department of Computer Engineering, Modelling, Electronics,
and Systems Science (DIMES), University of Calabria, Rende, CS, Italy
costanzo@dimes.unical.it

Abstract. An enhanced phaseless approach to the inverse scattering algorithm, starting from the Contrast-Source formulation of the Electric Field Integral Equation, is presented in this work. It fully exploits the available measurements of the amplitude-only total field, combined with the known complex incident field, to implement a single-step procedure able to reconstruct the target profile by avoiding the usual phase-retrieval process.

Keywords: Electromagnetic (EM) inverse scattering problem ·
Phaseless Contrast-Source Inversion (P-CSI) ·
Phaseless back-propagation (P-BP)

1 Introduction

In recent years, microwave tomography has gained popularity and increasing scientific interest, due to the possibility to reconstruct the dielectric profile of unknown targets, starting from electromagnetic (EM) field measurements. A widespread field of applications can be included, such as target identification, biomedical imaging and material diagnostics. The inverse scattering problem is stated as a nonlinear inverse problem, which is usually ill-posed in sense of Hadamard [1]. Different approaches can be considered, such as the well-known Born approximation, by which constraints on the nature of the Object Under Test (OUT) are assumed (i.e. weak scatterer), leading to the linearization of the inverse problem. In the case of strongly inhomogeneous scatterers, the Born approximation results into an incorrect reconstruction of the dielectric distribution. Therefore, the proposed method tackles the problem in its full non-linearity, with no restriction to the nature of the scatterers, with particular focus on inhomogeneous targets. In several applications, full data availability accounting for amplitude and phase information cannot be guaranteed, due to the lack of accuracy in phase measurements, especially at higher frequencies, e.g. terahertz domain. In order to overcome the above drawback, some phase retrieval schemes, originally proposed in electron microscopy and optics, and starting from the Gerchberg-Saxton approach [2–4], were successfully reformulated and generalized to a variety of applications, i.e. crystallography, astronomy and microwave domain.

Traditionally, the inverse scattering problem is assessed as a two-step procedure. The first step aims at retrieving the full-data information of the scattered field, starting

© Springer Nature Switzerland AG 2019
Á. Rocha et al. (Eds.): WorldCIST'19 2019, AISC 932, pp. 278–283, 2019.
https://doi.org/10.1007/978-3-030-16187-3_27

from the knowledge of some *a priori information*, such as the amplitude of the total field, together with the knowledge of the incident field [5], that is the measured field in the absence of the OUT. Alternatively, novel approaches propose to consider multiple measurement of the total field on two different measurement planes or, as recently proposed in [6], by using a single measurement plane with a novel two-probes scheme [7]. The second step amounts to retrieve the dielectric profile starting from the reconstructed scattered field, by solving the inverse scattering problem in the classical sense.

Here, the final goal is to formulate a single-step procedure, able to reconstruct the dielectric profile of the OUT starting from the knowledge of the measured amplitude of the total field, together with the full (complex) incident field, without considering any phase retrieval step. Therefore, the inverse scattering problem can be recast as an optimization problem over variables, expressed in terms of the unknown electrical properties to be retrieved. Specifically, starting from the Contrast-Source (CS) formulation of the scattering problem, a data residual is redefined, in analogy with the procedure described in [8, 9], and a novel initial guess, based on the back-propagation (BP) method is considered, by exploiting the available information on the incident field and the amplitude-only total measured field. The complete method is denoted as Phaseless Contrast-Source Inversion method (P-CSI). The unknowns of the inverse problem, i.e. the contrast function and the contrast source, are iteratively updated, through the minimization of a properly defined cost functional, obtained as a combination of the discrepancy that affects both the Electric Field Integral Equations (EFIEs). The minimizers are determined by using the conjugate gradient method.

2   Phaseless Contrast-Source Inversion Method

Let us consider a two-dimensional inverse scattering problem, as shown in Fig. 1, where an inhomogeneous target $\mathbb{B}$ is bounded into an object domain $\mathbb{D}$ and immersed in a background medium of known electrical properties, here fully described by ε_b. The permeability of the background and that of the OUT are assumed to be equal to that of free-space, namely μ_0. The measurement points are located in a domain $\mathbb{S}$, generally a curve, outside $\mathbb{D}$, in which the source points are placed as well. The EM scattering problem is completely described by considering the following EFIEs, known as *data equation* and *state equation* respectively, which result as solution of the Helmholtz equation applied to the aforementioned setup:

$$u(r) = u^{inc}(r) + k_b^2 \int_{\mathbb{D}} G(r, r')\chi(r')u(r')dr' \quad r \in \mathbb{S} \tag{1}$$

$$u(r) = u^{inc}(r) + k_b^2 \int_{\mathbb{D}} G(r, r')\chi(r')u(r')dr' \quad r \in \mathbb{D} \tag{2}$$

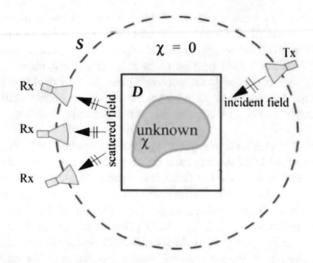

Fig. 1. Generic acquisition setup.

In the above equations, $G(r, r')$ stands for the Green's function of the background medium, with $r, r' \in \mathbb{R}^2$ denoting the observation and the source points respectively, $u(r)$ indicates the total field, while $\chi(r')$ denotes the contrast function normalized with respect to the wavenumber of the background k_b:

$$\chi(r') = \begin{cases} \dfrac{k^2(r')}{k_b^2} - 1 = \dfrac{\epsilon_r(r')}{\epsilon_b(r)} - 1, & r' \in \mathbb{B} \\ 0, & r' \notin \mathbb{B} \end{cases} \quad (3)$$

The inverse scattering procedure assumes the scatterer to be successively illuminated by a limited number of incident fields $u_v^{inc}(r)$, with $v \in \{1, V\}$, yielding to different distributions of the electric field for each transmitting setup. Therefore, considering the multi-view case and defining the equivalent contrast source $\omega_v(r) = \chi(r)u_v(r)$, the EFIEs into Eqs. (1)–(2) are re-arranged in a more compact form, according to the CS formalism, namely:

$$u_v(r) = u_v^{inc}(r) + G_S \omega_v(r) \quad r \in \mathbb{S} \quad (4)$$

$$\omega_v(r) = \chi(r)u_v^{inc}(r) + \chi(r)G_D \omega_v(r) \quad r \in \mathbb{D} \quad (5)$$

where $G_{S\text{-}D}$ denotes the same operator with limitation in the range of r.

2.1 P-CSI Formulation

The inverse scattering problem can be recast as an optimization problem [8–13], where a proper cost functional $F(\omega_v, \chi)$ is defined, according to the Van Den Berg approach [12], as a linear combination of the errors affected by the cost functional of the data equation and the state equation, denoted as $F_S(\omega_v)$, $F_D(\omega_v, \chi)$ and indicated as data and state cost functional respectively:

$$F_S(\omega_v) = \alpha_S \sum_v \|\rho_v\|_\mathbb{S}^2 \tag{6}$$

$$F_D(\omega_v, \chi) = \alpha_D \sum_v \|r_v\|_\mathbb{D}^2 \tag{7}$$

$$F(\omega_v, \chi) = F_S(\omega_v) + F_D(\omega_v, \chi) \tag{8}$$

Here, $\|\cdot\|_{\mathbb{S}-\mathbb{D}}$ indicates the norm defined on $\mathbb{S}$ and $\mathbb{D}$, $\alpha_{S\text{-}D}$ are the normalization factors, chosen so that $F(0, \chi) = 1$, while ρ_v and r_v denote the so-called data and state residual, respectively, relative to the v-th illumination setup:

$$\rho_v(\omega_v) = |f_v|^2 - \mathrm{Re}\{u_v^{inc} + G_S\omega_v\}^2 - \mathrm{Im}\{u_v^{inc} + G_S\omega_v\}^2 \tag{9}$$

$$r_v(\omega_v, \chi) = \chi u_v + \chi G_D\omega_v - \omega_v \tag{10}$$

where $|f_v|$ represents the measured amplitude of the total field in $\mathbb{S}$.

As shown in (9), both the total field (amplitude only) and the incident field (full-data) are used in the data residual expression, the latter resulting as function of the contrast source ω_v only, while the state residual is defined in terms of χ and ω_v.

Hence, an iterative minimization of the cost functional F is considered, by alternatively updating the contrast source ω_v and the contrast function χ, thus constructing iterative sequences in terms of n. Firstly, the contrast source ω_v is updated by considering the conjugate gradient scheme [8]:

$$\omega_{v,n} = \omega_{v,n-1} + \alpha_n^\omega d_{v,n} \tag{11}$$

where α_n^ω indicates the update step-length, whereas $d_{v,n}$ defines the direction update, expressed as:

$$d_{v,n} = g_{v,n} + \gamma_n d_{v,n-1} \tag{12}$$

Here, $g_{v,n}$ indicates the gradient of the cost functional F with respect to the contrast source, while the γ_n coefficients are computed according to the Polak-Ribiére update, as described in [12, 13]. When the direction update is completed, the update step-length $\alpha_{v,n}^\omega$ is determined by minimizing the cost functional $F(\omega_{v,n-1} + \alpha d_{v,n}, \chi_{n-1})$ [9].

Before updating the contrast function χ_n, the updated contrast source $\omega_{v,n}$ is exploited in the evaluation of the updated total field $u_{v,n} = u_v^{inc} + G_D\omega_{v,n}$. Then, the contrast function update is obtained by further minimizing the state cost functional F_D, since χ is involved in the state cost functional only. Therefore, the contrast function is obtained as minimizer of F_D, resulting, for the n-th step, in the following expression:

$$\chi_n(\omega_{v,n}) = \frac{\sum_v \omega_{v,n}\overline{u_{v,n}}}{\sum_v |u_{v,n}|^2} = \frac{\sum_v \omega_{v,n}\overline{(u_v^{inc} + G_D\omega_{v,n})}}{\sum_v |u_v^{inc} + G_D\omega_{v,n}|^2} \quad (13)$$

2.2   Improved P-CSI Implementation

The effective convergence of the minimization procedure results to be strongly dependent on the choice of the initial guess. However, only the CS requires the definition of a starting guess, since the contrast function expression in (13) does not involve any of its previous values. One of the most common method is the back-propagation (BP) solution, which is ordinarily implemented in the case of full-data measurements availability [12, 13]. The phaseless back-propagation (P-BP) scheme, considered in [8, 9], includes an arbitrary starting phase value for the total measured field. Due to the lack of reasoning for this choice, this approach is not further considered. Here, the P-BP starts from the evaluation of the gradient of the data cost functional F_S, computed for a null contrast source, $\omega_{v,-1} = 0$, and denoted as -1th step for notation convenience:

$$g_{v,ns}\big|_{\omega_{v,-1}} = g_{v,ns}\big|_0 = g_{v,0s} = -2\alpha_S G_S^*\left[u_v^{inc} \cdot \left(|f_v|^2 - |u_v^{inc}|^2\right)\right] \quad (14)$$

The P-BP solution can be obtained by applying the steepest descent method to the data cost functional F_S [14], limited to the first step, thus resulting in the determination of the initial guess $\omega_{v,0}$:

$$\omega_{v,0} := \omega_{v,-1} - \beta g_{v,0s} = -\beta g_{v,0s} \quad (15)$$

Specifically, the optimal step-length β can be obtained in a closed form by minimizing Eq. (6), valued in $(\omega_{v,0})$. Compared to [10], where only the incident field is taken into account in the starting guess evaluation, as explicitly denoted in (14), all the available a priori information are *fully exploited* in the initial guess estimation.

3   Conclusion

In this work, a phaseless inverse scattering method starting from the Contrast Source formulation, has been discussed. The proposed technique leads to retrieve the dielectric profile of the OUT starting from amplitude-only measurements of the external total field, combined with the full (complex) knowledge of the incident field. The analytical elaboration consists of an optimization procedure, starting from a phaseless data functional that is linearly combined with a state cost functional. The contrast source updates are found as a conjugate gradient step, while the contrast function is updated through the minimization of the state cost functional. Furthermore, an alternative starting guess that accomplishes for both the available a priori information, i.e. total field amplitude and the incident field, has been defined, promising to be a more accurate initial guess. Numerical validations of the proposed method will be discussed during the conference.

References

1. Noghanian, S., Sabouni, A., Desell, T., Ashtari, A.: Sequential forward solver. In: Microwave Tomography, pp. 21–37. Springer, New York (2014). https://doi.org/10.1007/978-1-4939-0752-6_2
2. Gonsalves, R.A.: Phase retrieval from modulus data. J. Opt. Soc. Am. **66**(9), 961 (1976)
3. Fienup, J.R.: Phase retrieval algorithms: a comparison. Appl. Opt. **21**(15), 2758 (1982)
4. Fienup, J.R.: Reconstruction of an object from the modulus of its Fourier transform. Opt. Lett. **3**(1), 27 (1978)
5. Crocco, L., D'Urso, M., Isernia, T.: Inverse scattering from phaseless measurements of the total field on a closed curve. J. Opt. Soc. Am. A **21**(4), 622 (2004)
6. Costanzo, S., Di Massa, G., Pastorino, M., Randazzo, A.: An inverse scattering approach for inspecting dielectric scatterers at microwave frequencies without phase information. In: 2013 IEEE International Conference on Imaging Systems and Techniques (IST), pp. 392–397 (2013)
7. Costanzo, S., Di Massa, G.: An integrated probe for phaseless near-field measurements. Measurement **31**(2), 123–129 (2002)
8. Li, L., Zheng, H., Li, F.: Two-dimensional contrast source inversion method with phaseless data: TM case. IEEE Trans. Geosci. Remote Sens. **47**(6), 1719–1736 (2009)
9. Hu, Z., Lianlin, L., Fang, L.: A multi-frequency MRCSI algorithm with phaseless data. Inverse Probl. **25**(6), 065006 (2009)
10. D'Urso, M., Belkebir, K., Crocco, L., Isernia, T., Litman, A.: Phaseless imaging with experimental data: facts and challenges. J. Opt. Soc. Am. A **25**(1), 271 (2008)
11. Crocco, L., D'Urso, M., Isernia, T.: Faithful non-linear imaging from only-amplitude measurements of incident and total fields. Opt. Express **15**(7), 3804–3815 (2007)
12. Van Den Berg, P.M., Kleinman, R.E.: A contrast source inversion method. Inverse Probl. **13**(6), 1607–1620 (1997)
13. Van Den Berg, P.M., Abubakar, A.: Contrast source inversion method: state of art. J. Electromagn. Waves Appl. **15**(11), 1503–1505 (2001)
14. Zakaria, A., Gilmore, C., LoVetri, J.: Finite-element contrast source inversion method for microwave imaging. Inverse Probl. **26**(11), 115010 (2010)

Technologies for Biomedical Applications

Wearable Approach for Contactless Blood-Glucose Monitoring by Textile Antenna Sensor

Vincenzo Cioffi and Sandra Costanzo[(✉)] [ID]

Department of Computer Engineering, Modelling, Electronics, and Systems
Science (DIMES), University of Calabria, 87036 Rende, CS, Italy
costanzo@dimes.unical.it

Abstract. In this work, a new approach for non-invasive monitoring of blood
glucose concentration is presented. This approach is based on the adoption of an
antenna sensor designed on a textile substrate, thus realizing a wearable sensing
device able to assure greater discretion and less discomfort to diabetes patients.
The antenna is numerically validated by assuming a stratified medium under test
well simulating the human body. As a further enhancement, an accurate
dielectric model is adopted for blood, which specifically considers its dispersive
behavior in terms of dielectric constant as well as loss tangent factor when
changing the glucose concentration. Some preliminary numerical results are
discussed to prove the effectiveness of the proposed wearable antenna sensor.

Keywords: Dielectric model · Blood-glucose concentration ·
Wearable antenna

1 Introduction

Devices for self-monitoring of blood glucose (SMBG) are essential tools for people
with diabetes. These devices should be easy to use and easily integrated in daily life.
Nowadays, one problem is due to the fact that usual monitoring tools, such as glu-
cometers, are mostly invasive, as they need a drop of blood taken from the fingertip to
work. To face this problem, the adoption of microstrip technology has been recently
proposed for the realization of low-cost microwave sensors able to exploit the correlation between the blood permittivity and the blood glucose levels (BGC) [1].

From this perspective, an idea is to design antennas on a textile substrate. In the last
years, textile antennas has gained great interest due to their application in wearable
systems [2, 3], and some examples of their feasibility assessments yet exist in the
framework of on-body communications and biomedical telemetry [4–6].

In this work, a novel approach to textile antenna for the monitoring of BGC is
presented. At first, the complex permittivity of the textile substrate (denim) is experi-
mentally characterized; then, the antenna is designed on the software CAD Ansys HFSS
by adopting an improved dielectric model to accurately characterize blood changes for
variable glucose concentration. In particular, a stratified medium well simulating the
human body is considered; furthermore, an enhanced dielectric model is assumed to

© Springer Nature Switzerland AG 2019
Á. Rocha et al. (Eds.): WorldCIST'19 2019, AISC 932, pp. 287–291, 2019.
https://doi.org/10.1007/978-3-030-16187-3_28

properly characterize the dispersive behavior of the biological radiation medium, as well as the complex permittivity changes versus the physiological parameter to be monitored (namely BGC), thus considerably increasing the validity of the tests. Preliminary numerical validations are reported in order to assess the application feasibility of the proposed wearable sensor.

2   Design Procedure

The antenna is designed with a slight curvature in order to take into account the real shape of human arm, so to be installed on the sleeve of a shirt or jacket. The sensor is designed to work in the ISM-band (around 2.4 GHz), and a basic configuration given by a standard coaxial probe fed patch antenna is assumed as test (Fig. 1).

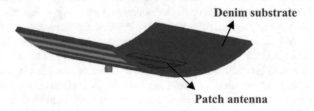

Fig. 1. Antenna configuration conformal to human arm.

As mentioned in the introduction, an accurate dielectric characterization of denim, namely the adopted substrate, is performed, by measuring its dielectric constant and tangent loss factor with a coaxial probe connected to a vector network analyzer, resulting in $\varepsilon_r = 1.6$ and $\tan\delta = 0.01$ at 2.4 GHz. Denim is chosen because it is durable, inelastic, low thickness, comfortable for user and low-cost.

The wearable antenna is optimized on CAD Ansys HFSS, by considering the structure layout reported in Fig. 2.

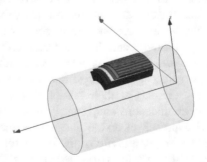

Fig. 2. Antenna layout on CAD Ansys HFSS.

Table 1. Characteristics of the stratified superstrate @ 2.4 GHz

Medium	Thickness [mm]	ε_r	tanδ
Dry skin	0.015	38.06	0.2835
Wet skin	0.985	42.92	0.2727
Fat	0.5	10.84	0.1808
Blood	2.5	Our model [9] for different BGC	Our model [9] for different BGC
Muscle	15.2	52.34	0.1893

For the accurate antenna design, a stratified medium, including also the blood layer to be sensed, is considered as superstrate [7]. The dielectric characteristics and thicknesses of the assumed layers are reported in Table 1.

In particular, the standard Cole-Cole model [8] is adopted to obtain the dielectric parameters for the tissues in the superstrate, while an enhanced dielectric model, recently proposed by the authors in [9], is used for blood. It specifically includes two correction factors for the real and the imaginary part of the complex permittivity [9], both depending on the BGC parameter (in mg/dl), as well as on the frequency value.

3 Preliminary Numerical Validations

In order to validate the antenna, a parametric analysis is performed by simulating the return loss for different BGC values, in the range $100 \div 300$ mg/dl, with the complex permittivity of blood derived from the enhanced dielectric model [9] and reported in Fig. 3.

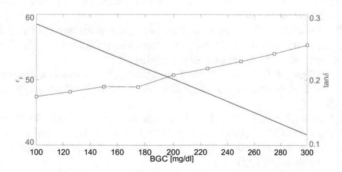

Fig. 3. Complex permittivity of blood vs. BGC (from model in [9]).

The computed return loss curves, reported in Fig. 4, reveal a frequency shift, and a more evident variation in amplitude when changing the blood glucose concentration. This is due to the sensitivity versus the loss tangent properly considered in the antenna design procedure [10].

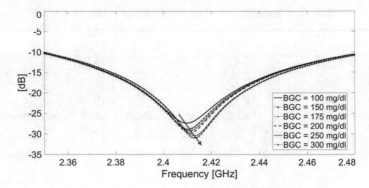

Fig. 4. Simulated return loss for the proposed textile antenna sensor at various BGC values.

4 Conclusions

A new approach of innovative non-invasive sensor for blood glucose monitoring, based on a textile antenna, has been presented in this work. The main novelty in the design procedure is due to the adoption of an accurate human phantom stratification, in conjunction with an enhanced dielectric model for blood, properly taking into account its dispersive behavior as well as the changes due to the variable glucose concentration. This results in a more accurate correlation between the antenna reflection response and the blood glucose level, exploiting also the amplitude level variation. Further studies will be addressed to a full experimental validation of the proposed approach.

References

1. Amaral, C.E.F., Wolf, B.: Current development in non-invasive glucose monitoring. Med. Eng. Phys. **30**, 541–549 (2008)
2. Grilo, M., Correra, F.S.: Parametric study of rectangular patch antenna using denim textile material. In: 2013 SBMO/IEEE MTT-S International Microwave & Optoelectronics Conference (IMOC), Rio de Janeiro, Brazil (2013)
3. Mantash, M., Tarot, A.C., Collardey, S.: Investigation of flexible textile antennas and AMC reflectors. Int. J. Antennas Propag. 10, Article no. 236505 (2012). https://doi.org/10.1155/2012/236505
4. Soh, P.J., Van den Bergh, B., Xu, H., Aliakbarian, H., Farsi, S., Samal, P., Vandenbosch, G. A.E., Schreurs, D.M.M., Nauwelaers, B.K.J.C.: A smart wearable textile array system for biomedical telemetry applications. IEEE Trans. Microw. Theory Tech. **1**(5) (2013). https://doi.org/10.1109/tmtt.2013.2247051
5. Samal, P.B., Soh, P.J., Vandenbosch, G.A.E.: UWB all-textile antenna with full ground plane for off-body WBAN communications. IEEE Trans. Antennas Propag. **62**(1), 102–108 (2014). https://doi.org/10.1109/TAP.2013.2287526
6. Dar, S.H., Ahmed, J.: Wearable textile antenna design in body centric wireless communications: a systematic literature review. Biomed. Res. **28**(8) (2017)
7. Kumar, S., Singh, J.: Measuring blood glucose levels with microwave sensor. Int. J. Comput. Appl. **72**, 4–9 (2013)

8. Gabriel, S., Lau, R.W., Gabriel, C.: The dielectric properties of biological tissues: III. Parametric models for the dielectric spectrum of tissues. Phys. Med. Biol. **41**, 2271–2293 (1996)
9. Costanzo, S., Cioffi, V., Raffo, A.: Complex permittivity effect on the performances of non-invasive microwave blood glucose sensing: enhanced model and preliminary results. In: WorldCIST 2018, Naples, Italy (2018)
10. Costanzo, S.: Non-invasive microwave sensors for biomedical applications: new design perspectives. Radioengineering **26**, 406–410 (2017)

Air Quality and Open Data: Challenges for Data Science, HCI and AI

Applying Safety Methods to Sensor Networks

Torge Hinrichs$^{(\boxtimes)}$ and Bettina Buth$^{(\boxtimes)}$

HAW Hamburg, Berliner Tor 7, 20099 Hamburg, Germany
{torge.hinrichs,bettina.buth}@haw-hamburg.de

Abstract. In this paper we propose the use of techniques typically employed for safety-critical systems for the identification of weaknesses in traffic control systems, specifically in systems with dynamically changing sets of sensors. The paper introduces the basic terminology of dependability engineering as well as typical architectures for sensor networks and common problems for their operation. Mapping both aspects to the application domain traffic control systems, the paper gives some ideas how these techniques could be used to evaluate the trustworthiness of decisions and to identify risks for these decisions.

Keywords: Dependability · Sensor networks · Safety methods

1 Introduction

According to the Council on Clean Transportation [5] the future city will use automated monitoring of air quality and active traffic control based on these data to ensure the overall quality of living in city. Similarly inbound and outbound traffic during rush hours will be rerouted based on congestion monitoring to reduce the overall pollution in the city. Such concepts heavily rely on the availability and consistency of data and frameworks for correct interpretation of these data and suitable decisions based on the available data. The technical foundation of these data are sensor networks consisting of fixed installations as well as flexible data sources.

In general sensor networks consist of large numbers of individual nodes. Nodes can be dedicated devices, often in the form of small battery-powered embedded systems as well as smartphones using specific build-in sensors or weather stations integrated in a smart home environment. These nodes provide heterogeneous data with regard to data types (temperature, humidity, noise) and with regard to transmission frequency and protocols. They can be permanent installations or may provide data from different locations over a period of time. In general, the number of nodes providing data may vary over time for specific areas. In addition, the validity of the data may strongly depend on the specific device.

For homogenous fixed installations typically adjustments and calibration are performed to improve the overall quality of the data and to identify erroneous data sources [4]. This will presumably not be possible to the same extent for

© Springer Nature Switzerland AG 2019
A. Rocha et al. (Eds.): WorldCIST'19 2019, AISC 932, pp. 295–303, 2019.
https://doi.org/10.1007/978-3-030-16187-3_29

temporary sources such as e.g. smartphones. On the other hand, the overall benofit of uoing a multitude of readily available sensors as a basis for decisions relies on the ability of the decision framework to deal with the uncertainties and potential deviations between data from these sensors.

With this paper we would like to start a discussion whether methods used in safety-critical systems design could help to identify potential risks for safety and security or in general dependability of such sensor networks. Techniques of the design and implementation of dependable systems are normally used in highly regulated and specialized areas such as aviation [7] or automotive industries [9]. In these application domains, the overall goal is the avoidance or at least mitigation of potential risks for people or the environment. Industry standards require the application of specific development, analysis and assurance methods for such systems, which influence development duration and cost significantly.

Sensor networks as such are not safety-critical systems, but their use in the context of traffic control at least requires to discuss their dependability. This includes the identification of potential risks as well as the analysis of specific aspects in the system architecture or implementation. This paper aims to provide initial ideas to transfer techniques from dependability engineering to the application area of highly dynamic distributed traffic control systems. To this end, the paper starts with an introduction to the essential terminology and an overview of techniques used in the domain of dependable systems. In Sect. 3 two basic architectural approaches for sensor nets are introduced. Section 4 provides an outline how the knowledge from the previous two sections can be transferred to the application domain traffic control systems. The paper concludes with a brief summary and open questions. Currently there is no concrete application of these ideas - the paper aims at providing a starting point for collaborations between the communities.

2   Dependability

This section provides a brief overview of terminology used in the area of dependability engineering as well as typical techniques applied for risk analysis. It also provides a suggestion how these dependability considerations can be related to traffic control systems.

2.1   Terminology

In a dependability context all terms are defined in standards. These documents specify the fundamentals of the corresponding topic. The appropriate standard for this context is ISO/IEC 25010 "Systems and software Quality Requirements and Evaluation (SQuaRE) – System and software quality models" [9], which replaced ISO/IEC 9126-1. For this paper we focus on the following criteria:

Functional Sustainability. This characteristic represents the degree to which a product or system provides functions that meet stated and implied needs when used under specified conditions.

Reliability. Degree to which a system, product or component performs specified functions under specified conditions for a specified period of time.

Availability. Degree to which a system, product or component is operational and accessible when required for use.

Maintainability. This characteristic represents the degree of effectiveness and efficiency with which a product or system can be modified to improve it, correct it or adapt it to changes in environment and in requirements.

Security. Degree to which a product or system protects information and data so that persons or other products or systems have the degree of data access appropriate to their types and levels of authorization.

These definitions are also used in risk assessment. A method to identify and rate potential risks and hazards that can occur during development and later on in operation. A risk is defined as:

$$Risk = Propability * Consequence \tag{1}$$

As a result, risks can be categorized in differed criticality levels based on the probability of the system to fail and the corresponding consequence. These categories are called "Safety integrity level" (SIL) defined in IEC 61508 [8]. The levels are ordered increasing order starting with "SIL1" as the lowest risk level and "SIL4" as the highest. To identify and mitigate the risks and the corresponding SILs an analytical approach will be used. The next section describes common techniques used in safety analysis.

2.2   Ensuring Dependability in Critical Systems

In general risks and hazards are identified through risk assessment. Standards such as ISO 9001:2015 [11] are used in the certification of products or development processes. ISO 9001 defines general guidance for development and services. For software development there is a more specific standard the ISO/IEC 90003:2014 [10]. Methods can be either analytical, techniques that gain insight in the system and its dependability characteristics or constructive, which mitigate risks in the design and ultimately for the operation.

Analytical Approach. A common analytical approach is a strict and rigorous review of specification, design and implementation. A team of domain experts examine all project documents starting with specification and design. As the project progresses the implementation will be reviewed as well. By having a closer look at these artifacts errors will most likely be found, due to the experience of the review team.

The review process can be enhanced with safety techniques such as Hazard and Operability Analysis (HAZOP), Fault Tree Analysis (FTA), Failure Modes, Effect and Criticality Analysis (FME(C)A). These methods provide a structured approach for identifying risks and potential problem causes.

HAZOP is typically applied to identify potential risks using a structured brainstorming approach. FTA and FME(C)A provide in-depth insight into the cause and consequences of failures. While FTA aims at identifying the chain of causality and potential root causes for known problems in a top-down approach, FME(C)A starts from known local problems (the so-called failure modes) of components and analysis the effect of fault-propagation through the system and thus is a bottom-up analysis. Both are strongly based on the knowledge of data flow through the system architecture. If performed adequately both techniques reach a suitable level of completeness, but this requires a good understanding of the system structure as well as the functional and non-functional requirements.

Constructive Approaches. These techniques and patterns can be used as a guideline to ensure safety during the design and implementation phase of a project, for example safety cases. These scenarios can be used to directly derive the design or even parts of the implementation of the system.

Another design principle is to design a fault tolerant system. Therefore redundancy concepts can be implemented to increase the reliability and availability of the system. In addition a fault containment strategy can be developed. If a fault occurs the consequences spread only to specific predefined boundaries, as a result, the system can stay intact [6].

From a more technical point of view dependability properties of a system can be improved by adding risk mitigation mechanisms like watchdog or brownout detection. A watchdog is a component that observes the behavior a system. The system needs report to the watchdog in a certain period of time. If the time elapses the watchdog will trigger an alarm. A brownout detection is used to sense intentional or unintentional drop of voltage in a power supply. If a voltage drop is detected an alarm is triggered.

2.3   Dependability Issues in Traffic Control

While Traffic Control Systems are not safety critical and up to now seldom exhibit critical situations resulting from security attacks, this may change with the progressing digitization in this area. Future railway switches could communicate via the internet with the railway control center - what if an intruder manipulates the switches or the train information? This may become even a risk to safety. Traffic Control Systems in Smart Cities are intended to react dynamically and timely to the overall traffic situation based on data collected and thus improve the overall throughput, prevent congestion and subsequently even reduce the overall pollution due to car exhausts. What if data is delayed and wrong decisions are made that lead to an even worse situation? Or if data is manipulated to prevent that traffic is routed through your neighborhood? Thinking further: if traffic control relies on very heterogeneous sets of sensors some of which are fixed installations and some of which are switching on and of or change their location - as smartphones would do - how do we determine how reliable the data basis at any given time is? How many data sets need to be consistent, which data should be dismissed as outliers, how do we determine that the overall information is strong enough to derive a rerouting decision?

Essentially such and similar questions need to be answered to decide on specific architectures and algorithms for traffic control. Questions similar to those related to safety critical systems. This encourages the evaluation of techniques from dependability engineering to this specific application areas.

3   Architecture of Sensor Networks

In this section typical architectures for sensor networks are introduced, which could be used for traffic control systems. These architectures have known reliability and security risks, which are summarized as well.

Sensor networks are typically used to gather environment data in a local area of deployment. These networks potentially contain a large number of nodes. Each node is a small low power embedded device with network connectivity and sensors attached to it. The nodes are typically self-sufficient by the use of batteries and solar energy. One essential characteristic of such nodes is that they may be deployed almost anywhere. In some cases they need additional hardware [2]. For different purposes such as temperature, humidity, light, even CO_2 or NO_2 readings [3] different sensors can be employed. They are usually connected via Wifi. Deployed in an urban area, the existing network infrastructure like local area (WLAN) or wide area networks (WAN) can be used to transmit the data to a storage and evaluation server. An example for such a network is the globally deployed crowd-sensing "Smart Citizen" -project [3]. In more isolated areas, e.g. agriculture, nodes can act as a relay to forward data to a sink, which then transfers accumulated data to a server [13]. In such scenarios no additional infrastructure is required aside from a slightly more powerful node. For the discussion in this paper, we consider two basic network types.

3.1   Star Networks

The "Smart Citizen Project" uses affordable, pre-assembled hardware to provide a sensor node to everyone, who wants to participate. The collected sensor data will then be contributed to the project. The system uses a star architecture, like shown in Fig. 1(a), to transmit the environment data to a server. The figure shows a schematic set up of a star architecture. The sensor nodes connect to a local gateway/router, which is connected to the Internet. It is possible to connect multiple sensor nodes to a local gateway. Nodes connect via a wireless connection with a local area network and therefore can reach a server collecting the information. An advantage of this architecture (especially in a crowd-sensing context) are the redundant gateways and therefore a fault tolerant setup. If one node fails, the rest of the system can stay operational. If one gateway fails, only the dependent nodes fail. On the other hand more network hardware is needed to set up the system. This architecture is useful in an urban environment where the surrounding infrastructure is already present. In more rural areas, like forests or agriculture a reliable network connection is less likely to be present, therefore another approach is necessary.

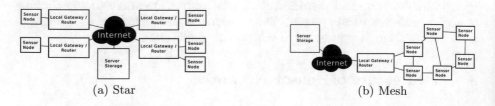

(a) Star   (b) Mesh

Fig. 1. Schematic: common IoT network infrastructures

3.2   Mesh Architecture

A mesh sensor network uses each node to relay the data between the nodes. Figure 1(b) shows a schematic setup of a mesh network. Each node is able to send messages to each other node in physical reach [12]. On the "edge" of the network some nodes are connected to a gateway, which is attached to the Internet and can access the data storage server. This architecture is favorable in rural areas due to its self-containedness. The network is also able to handle malfunction. If one node is not able to communicate a new route can be established and bypass the broken node. This is not the case if the malfunctioning node acts as a bottleneck to a part of the network. If this node fails the following nodes are isolated from the system. A temporal or permanent loss of single nodes is also possible during normal operation and can influence the interoperability of the components.

3.3   Potential Risks Using Sensor Networks

Besides the issues caused by the network infrastructure sensor networks have to deal with more general issues, e.g. wrong sensor data, dislocation or loss of connection/power. Typical issues are:

Inaccurate measurements: A sensor can provide incorrect data. Reasons may be a broken connection between the nodes or from sensor to the node. Another reason might be a wrong configuration or calibration of the sensor.
Dislocation or destruction: It is possible, that a node suffers from vandalism or extreme weather conditions. Therefore it can be destroyed or moved to a different place.
Loss of connection or power: Similar to "dislocation or destruction" e.g. extreme weather condition can lead to a communication loss of nodes. The Wifi antenna can fail or the solar panel could break. Charging the battery is not possible anymore.
Manipulation of Sensordata: It is a reasonable concern, that an attacker may adulterate the sensor data to corrupt the system or the decision it might make.

4   Applicability of Dependability Techniques to Pollution Detection Networks

The first German Cities are going to restrict the use of highly frequented roads to certain car types aiming to reduce the pollution in certain areas [1]. As mentioned above we have to consider, that the decisions are based on sensor networks run by the public incorporating highly heterogeneous nodes e.g. deployed on mobile installations like buses or taxis or incorporating data collected by smartphones Taking into consideration that these assumptions, it is important to ensure the dependability of such systems, especially its robustness with regard to partial failures and incorrect data and fluctuating sources. Starting from the typical issues summarized in Sect. 3, we should be able to analysis the overall dependability of the system. The following section tries to give some ideas which of the techniques described in Sect. 2.2 may be useful.

In the following, we discuss the potential risks operating a sensor network and their impact on the overall functionality of the traffic control system. Furthermore we relate these risks to the overall setting of dynamically changing sensor nodes.

"Loss of Connection or Power": The loss of power is in general considered to be a critical failure of the network or part of the network. If the battery fails the system is not longer operational. Using constructive approaches the likelihood of a total loss of the system can be reduced by adding redundant and independent power supplies. The loss of connection might not be as critical as the loss of power. The system can still sense data and save it on a storage device. With dynamically changing sets of sensor nodes the likelihood that a node, e.g. a smartphone, leaves the range of a subnet is increased. This situation is similar to a loss of connection or power. In both cases it needs to be established whether the data from a subnet is still complete enough to provide a reliable basis for decisions.

We can employ analysis techniques from safety engineering for two purposes in this context:

- in order to identify a failure of the system as a whole, a HAZOP analysis should be applied first followed by an FTA to identify the root causes (minimal cut sets in FTA terminology). This allows to determine critical components and suitable mitigation strategies such as redundancy or watchdog components
- on the other hand FME(C)A can be employed to derive the effect of local problems to the system as a whole - e.g. whether the loss of a number of nodes effects the overall reliability of the data.

"Inaccurate Measurements:" This is a typical issue in safety-critical system. For example an airplane has to be able to deal with inconsistent data and still make the correct decision. Assuming a situation like the smart citizen project, where

multiple sensor nodes are distributed in a single street, ideally they deliver relatively consistent data due to their approximately same location. The downside of this technique is that it can not deal with a malfunctioning node. If one of the nodes is sending false data, caused by inaccurate measurements or a broken sensor, the mean will be tampered by $1/(Number of Sensors)$. In dependable system this issue will be resolved by a voting mechanism. So if an outlier is detected it can be removed from further calculation.

In the context of dynamically changing sets of sensors of heterogeneous quality, the effect of deviating measurements is comparable to inaccurate measurements. While typically in fixed settings of sensors calibration is used to identify problematic nodes, calibration of nodes moving in and out of the range of a subnet will potentially not be a viable solution. As for loss of connection, FME(C)A could be used to identify the effect of such inconsistencies in the data and also to compare different algorithmic approaches to determine if Analogously to a malfunctioning sensor node it is possible, that one or more nodes are compromised by an attacker. This problem is more a security then a safety issue, but has to be taken into consideration. An attacker could be able to manipulate a multiple nodes and manipulate the sensor data. To deal with this concern monitoring could be implemented to observe the sensor nodes over time. If a node acts differently from the estimated behavior, it will be recognized and the corresponding nodes can be deactivated.

As these first ideas show, employing techniques commonly used in dependability engineering, there is a good chance to evaluate the effect of dynamically changing sensor settings on the reliability, availability and security of the overall decision process in traffic control. These techniques can also be used to compare the dependability of architecture variants, e.g. by changing the ratio between trustworthy nodes and less reliable nodes.

Unfortunately it will not solve all issues, e.g. dislocating or even destroying a node can only handled by increasing the physical security of the node. For example the node could be covered with a robust casing or installed in a placed in an unreachable place - which obviously is only a solution for fixed installations. On the other hand the analysis may be able to provide guidance up to which level a loss of nodes can be tolerated.

5   Conclusion

In this paper we presented some ideas how to employ well-known techniques form dependability engineering to analyse and compare sensor nets in the context of traffic control systems. This ideas as such have not yet been applied to realistic systems - the paper could be a starting point for such a project. In the previous sections, we introduced common terms and approaches to analyze safety-critical systems and help to increase the dependability of such systems. In addition we described some typical architectures of sensor networks. Finally we discussed how dependability methods can be used to identify potential risks within the design or operation of sensor networks and improve the robustness of the system. These techniques have the potential to reveal weaknesses in dynamically changing traffic control systems especially with nodes of varying quality and trustworthiness.

References

1. ADAC e.V.: Dieselfahrverbot: Alle Fragen und Antworten. https://www.adac.de/rund-ums-fahrzeug/abgas-diesel-fahrverbote/fahrverbote/dieselfahrverbot-faq/. Accessed Nov 2018
2. Akyildiz, I.F., Wang, X., Wang, W.: Wireless mesh networks: a survey. Comput. Netw. **47**(4), 445–487 (2005)
3. Barcelona, F.: Smart citizen. https://smartcitizen.me. Accessed Nov 2018
4. Boudriga, N., Marimuthu, P.N., Habib, S.J.: Trusting sensors measurements in a WSN: an approach based on true and group deviation estimation. In: Rocha, Á., Adeli, H., Reis, L.P., Costanzo, S. (eds.) Trends and Advances in Information Systems and Technologies, pp. 603–613. Springer International Publishing, Cham (2018)
5. Dallmann, T., Borken-Kleefeld, J.: Remote sensing of motor vehicle exhaust emissions. white paper, International Council on Clean Transportation (2018). https://www.theicct.org/sites/default/files/publications/Remote-sensing-emissions_ICCT-White-Paper_01022018_vF_updated.pdf
6. Saridakis, T.: Design patterns for fault containment. In: Proceedings of 2003 EuroPLoP Conference (2003)
7. Do-178b, software considerations in airborne systems and equipment certification. Standard, International Organization for Standardization, Geneva, CH (2010)
8. Functional safety of electrical/electronic/programmable electronic safety-related systems. Standard, International Organization for Standardization, Geneva, CH (2010)
9. Systems and software engineering – systems and software quality requirements and evaluation (square) – system and software quality models. In: Standard, International Organization for Standardization, Geneva, CH (2011)
10. ISO/IEC 90003:2014 software engineering – guidelines for the application of ISO 9001:2008 to computer software. Standard, International Organization for Standardization, Geneva, CH (2014)
11. Iso 9001:2015 quality management systems - requirements. Standard, International Organization for Standardization, Geneva, CH (2015)
12. Yick, J., Mukherjee, B., Ghosal, D.: Wireless sensor network survey. Comput. Netw. **52**(12), 2292–2330 (2008)
13. Yu, X., Wu, P., Han, W., Zhang, Z.: A survey on wireless sensor network infrastructure for agriculture. Comput. Stan. Interfaces **35**(1), 59–64 (2013)

Visualising Air Pollution Datasets with Real-Time Game Engines

Uli Meyer[1,2(✉)], Jonathan Becker[1,2], and Jessica Broscheit[1,2]

[1] CSTI, Hamburg, Germany
csti@haw-hamburg.de
[2] Hamburg University of Applied Sciences, Berliner Tor 7, 20099 Hamburg, Germany
https://csti.haw-hamburg.de

Abstract. Visualising Volunteered Geographic Information (VGI), including air pollution data, can be used as an explorative tool in the context of workshops and maker labs. This requires a technology that has a low entry-level, but provides a powerful interactive prototyping framework. We describe the potential of real-time computer game engines as visualisation tools for interdisciplinary cooperation between non-experts and experts. We discuss how properties of air pollution, including invisibility, pervasiveness and its ability to permeate organisms, can be visualised with particle systems, and outline two use cases for different output devices, including AR and VR.

Keywords: Visualisation · Air pollution · Real-time engine ·
Game engine · Particle system · Augmented reality · Virtual reality

1   Introduction

Visualising Volunteered Geographic Information (VGI), including air pollution data, can be used as an explorative tool in the context of workshops and maker labs. This requires a technology that has a low entry-level, but provides a powerful interactive prototyping framework. Ideally, this framework allows non-experts and experts to work together on interdisciplinary projects that support experiential learning [5]. In the context of the serious game movement and for interactive prototyping, real-time computer game engines have proven to be such a versatile tool [13,16,20]. While they cannot compute scientifically accurate simulations, during development processes they provide immediate and continuous real-time visualisations within the 3D view port.

Visualising air pollution data sets can have diverse functions, ranging from declarative to experimental [1]: Simplified visualisations can serve as a warning, for example in big cities, or for sensitive persons [15,17]. Scientific visualisations give insight into behaviours and compositions of air pollution over time. Visualisations can also raise awareness for air pollution in more pedagogic contexts [10], or they can be used as a tool for community activism and policy change [8].

© Springer Nature Switzerland AG 2019
Á. Rocha et al. (Eds.): WorldCIST'19 2019, AISC 932, pp. 304–312, 2019.
https://doi.org/10.1007/978-3-030-16187-3_30

Depending on the function, different types of visualisation, from highly abstract 2D diagrams to interactive 3D info graphics and immersive VR experiences, are available [4,6,11].

This paper describes the technological potential of game engines for experimental visualisations of complex air pollution datasets, and proposes possible use cases for such a visualisation.

Section 2 introduces visualisation terminology and discusses the specifics of air pollution visualisation. Section 3 gives an introduction to game engine technology for visualisation purposes and Sect. 4 outlines two potential use cases for experimental air pollution visualisation in a game engine.

2   Data Visualisation

Data visualisation implies a transfer of abstract information, usually via a numerical transformation, into a more concrete form such as geometric shapes (Fig. 1).

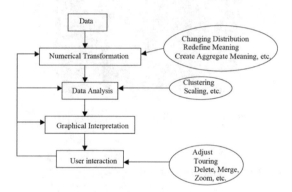

Fig. 1. Visualisation process [9].

2.1   Terminology

Depending on the function, Scott Berinato [1] classifies visualisations along two axes: from "conceptual" to "data-driven" and from "declarative" to "exploratory". According to his classification system, data-driven visualisations can be described as more "declarative" or as more "exploratory" (Fig. 2).

Traditionally, data visualisation meant static 2D diagrams [6]. With the advent of 3D graphics since the mid-20th century, elements such as movement, real-time rendering and interaction were introduced [4]. The development from 2D graphics to 3D graphics brought new possibilities, but also new problems: While 2D graphics are easier to read and more condensed, their "cleanness" and simplicity might be misleading, especially for organic data. 3D visualisations, on the other hand, can be harder to read, for example by creating visual overlap. They need very precise planning, or additional movement in the form of pre-rendered animation.

Since the late 1990s, real-time movement and interaction became available with the introduction of real-time engines. Real-time visualisations expose one or several parameters that can be influenced by the user in some way, for example camera position, object translation, rotation, scale, and so on. Data visualisation in virtual reality (VR) constitutes an extreme case of real-time interactivity, as the user can move through, or even climb on diagrams in virtual space.

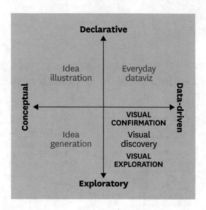

Fig. 2. Visualisation categories [1].

2.2   Visualising Air Pollution Data

If we follow Berinato [1], visualisations of air pollution datasets can be categorised along an axis from declarative to exploratory (Fig. 2). Declarative visualisations of air pollution would be what he calls "everyday dataviz", e.g. the 2D diagrams one finds in the media or in textbooks. These declarative visualisations are also used for warning systems in cases where the existence of air pollution is undisputed, and where concrete measures need to be taken, often within a very short time frame (Fig. 3). While declarative visualisations are usually rendered as static 2D graphics, exploratory visualisations often contain interactive elements, for example sliders or real-time animations [19] (Fig. 4).

Visualisations that exist on the boundary between declarative and explorative can be found where the degree or danger of air pollution is under debate. Their purpose is usually educational or activist. They often contain hybrid combinations of 2D abstractions and more concrete 3D elements, such as bar charts in 3D space. Or they visualise air pollution data as clusters of "balls" that can pile up (Fig. 5).

While these hybrid displays are often striking, and can have an immediate effect on the viewer, they somewhat obscure the properties of air pollution. By

Fig. 3. Air pollution monitoring in China [15], Air pollution warning app [17].

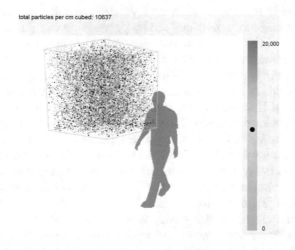

Fig. 4. Interactive visualisation of air pollution [3].

Fig. 5. Hybrid air pollution visualisations [3].

Fig. 6. WIFI visualisation [12], geospatial visualisation [7].

abstracting air pollution - an inhaled gaseous mixture that contains fluids and tiny particles - as solid objects, these images suggest a degree of visibility and potential control that does not in fact exist. But evoking simple solutions and a "lets-do-this" attitude might be a valid strategy for marketing purposes, which includes fund-raising or activism.

Air pollution is mostly invisible and volatile, and it permeates organisms. Actual or metaphoric visualisations of the pervasive nature of air pollution are relatively rare [2]. We can look at two augmented reality (AR) applications as models for visualising the invisible, in these cases WIFI density or gravity in space. They overlay images or real-time video of specific physical environments with areas or clouds of colour in space. Different colours indicate different types of quality (Fig. 6). But even these applications maintain a separation between the user and the coloured clouds. They are "over there", not all around the user. An example for the visualisation of real-time immersion in an air/water mixture would be the Rain art installation at the Barbican Centre in London [18]. There visitors can move freely through physical water drops in space, without getting wet. The effect is similar to "diving into" particle simulations in VR experiences such as The Blue [21] or Cosmic Sugar [14] (Fig. 7). While The Blue immerses users in a simulated underwater environment, filled with floating particles and sea life, Cosmic Sugar surrounds them with more abstract "gaseous" particles. Inside both VR experiences, the user can interact with the particles in real-time by using a hand controller.

As both AR and VR applications are built with game engines, it makes sense to create explorative visualisations of air pollution data with game engines.

3   Technology

3.1   Game Engines

Game engines are real-time engines that provide software-development environments, or frameworks, for building computer games and other real-time applications. Due to the popularity of computer games, and the fact that several game engines are free for non-commercial purposes, game engines are an ideal technology for entry-level development, prototyping and experimentation. By

Fig. 7. Rain installation at the Barbican, London (2013) [18], The Blue (2016) [21].

playing computer games, many people have already come in contact with games engines. They can provide an explorative framework for workshops, makerlabs or private houses, both as an agile and experimental prototyping environment, and for creating interactive applications and visualisations. 3D game engines that are currently free for non-commercial purposes include Unity 3D, Unreal Engine, and CryEngine.

Game engines consist of a 3D view port where the user can assemble 3D objects with simple drag-and-drop interactions. Other windows and tabs contain menus for influencing position, scale and rotation, or animation and lighting. All parameters can be made interactive, i.e. the user can influence them in real-time, either by typing in numbers, or by affecting them in the 3D view port with a mouse or controller.

For some engines, such as Unity, basic, entry-level coding in C# or Java Script is needed, while others contain node-based, visual scripting systems (e.g. Unreal Engine's Blueprint system) that provide easy access for beginners. Numerous ready-made scripts, add-ons, environment assets and tutorials are available online.

3.2 Particle Systems

As we have seen, complex air pollution data is usually visualised as either highly abstract shapes, such as bar charts, or as clusters of small spheres and cloud-like structures. Bars can be simply modelled as 3D blocks, and animated along the Y axis. Shape clusters are created with particle generators, i.e. small functions that "spout" a stream of forms or particles with specific properties over time. The user can influence parameters such as particle size, form, number, speed, lifetime, variation, colour, animation, and so on.

CPU rendering a high number of particles in real-time requires a lot of computing power, but both Unity and Unreal recently tackled that problem with the introduction of GPU rendered particle systems, within the VFX Graph in Unity [Fig. 8] or Niagara in Unreal, respectively. Both systems can render several million particles. Game engines also contain physics systems such as gravity that can be used to experiment with particle simulations. While they cannot generate scientifically exact data, they can give first impression of approximated behaviours of air pollution in space, such as weight, collision, penetration and so on.

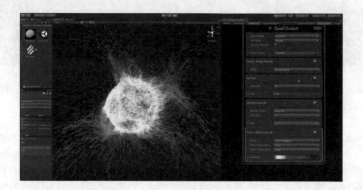

Fig. 8. Unity 3D: VFX Graph particle generator.

3.3   Hardware

Air pollution data can be streamed into the engine, for example via a simple and cheap sensor kit containing an ESP and/or a Raspberry Pi. In 2018 the cost for such a set-up is under 50 Euro.

4   Proposal

Visualising air pollution data sets in the context of experimental situations at workshops or makerlabs requires not only a game engine, but also an organisational structure that integrates non-experts. We describe so-called game jams as a model for such an open and experimental event.

4.1   Game Jams

Game engines are regularly and productively used in short-term experimental prototyping situations, so-called game jams: In university labs or game production studios, a group of people meets over a specific period of time (48 h, 2 weekends etc.) and builds an application within that short time frame. The engine usually contains the needed functionality to finish the task, especially in combination with the above-mentioned free assets.

While a game jam team should contain participants who have worked with the engine before, by "visualising" the prototyping process inside the 3D space and the different windows, game engines provide easy access for all participants. Apart from providing a context for experiments, a game jam can also kick off the production of a more polished application for visualising air pollution data at home. That way, the game engine becomes both a tool for experimental on-the-fly visualisation in 3D space, and for building a permanent visualisation application. The following section outlines two potential use cases for air pollution visualisation in the context of a game jam.

1. Particles. By reading air pollution data into a particle system, the actual properties and behaviours of air pollution in space can be visualised. Different visual properties, such as colours or forms, indicate different types of pollution.

Experiments with engine physics give an approximation of how pollution particles react to collision with solid or organic objects, to gravity, wind and so on.

The particle visualisation works on a static 2D screen. But as the particles fill a three-dimensional space and surround the user (=user camera), other display types may be more efficient. By building the application for an AR display, the user can move the display freely and explore the visualisation in space. The particle simulation can be overlaid and connected to a concrete environment such as a room or a street.

Building a simple VR application lets the user experience actual "body" immersion in air pollution. How does it feel when you can actually see the pollution that surrounds your body, and even permeates it?

2. Real-Time Interaction. Streaming the data into the engine in real time, the user can interact directly with the physical environment, for example by removing a source of air pollution, and observing the effect on the visualisation.

5   Conclusion

In this paper we discuss the potential of game engines as an explorative, low-entry tool for visualising complex air pollution data sets. We propose that game engines are suited for teams that contain non-IT and non-data experts in the context of experimental situations such as workshops or makerlabs.

We outline Beritano's model of data driven visualisation types, which he categorises along an axis from "declarative" to "experimental", depending on function. As the form of a visualisation follows its function, we then describe different graphic types of air pollution visualisation for different functions, ranging from more abstract to more concrete. We call attention to the specific properties of air pollution, including its invisibility, pervasiveness, and the fact that it permeates organisms, and propose to visualise them with particle simulations.

For experimental workshop situations that allow experts and non-experts to work together on air pollution visualisation, we describe the "game jam" model, a prototyping method from game design. We outline two use cases for air pollution visualisation in the context of a game jam.

In the next step, we will implement an air pollution visualisation in Unity to assess its robustness and flexibility for different functionalities and data sets.

References

1. Beritano, S.: Visualisations that really work (2016). https://hbr.org/2016/06/visualizations-that-really-work
2. Broscheit, J., Draheim, S., von Luck, K.: How will we breathe tomorrow? In: EVA Copenhagen 2018 - Politics of the Machines, Art and After (2018). http://dx.doi.org/10.14236/ewic/EVAC18.10
3. Carbonvisuals: Carbonvisuals website. http://www.carbonvisuals.com/all
4. Chen, C.: Information Visualisation and Virtual Environments. Springer Science & Business Media, London (2013)
5. Dougherty, J.P.: Information technology fluency at a liberal arts college: experience with implementation and assessment. J. Comput. Sci. Coll. **18**(3), 166–174 (2003)
6. Friendly, M.: A brief history of data visualization. In: Handbook of Data Visualization, pp. 15–56. Springer (2008)
7. GeoscienceAustralia: Data visualisation with the ' oculus rift' dk2. https://youtu.be/n0r9Q-zsPhs
8. Hsu, Y.C., Dille, P., Cross, J., Dias, B., Sargent, R., Nourbakhsh, I.: Community-empowered air quality monitoring system. In: Proceedings of the 2017 CHI Conference on Human Factors in Computing Systems, pp. 1607–1619. ACM (2017)
9. Kaidi, Z.: Data visualization. Retrieved **8**(22), 2010 (2000)
10. Kim, S., Paulos, E.: inAir: measuring and visualizing indoor air quality. In: Proceedings of the 11th International Conference on Ubiquitous Computing, pp. 81–84. ACM (2009)
11. Kirk, A.: Data visualisation: a handbook for data driven design. Sage (2016)
12. Lamm, N.: Wifi visualised. https://www.vice.com/en_us/article/9an9m7/heres-what-wi-fi-would-look-like-if-we-could-see-i
13. Lewis, M., Jacobson, J.: Game engines. Commun. ACM **45**(1), 27 (2002)
14. Lobser, D.: Cosmic sugar (vr experience). https://store.steampowered.com/app/559010/Cosmic_Sugar_VR/
15. Lu, W., Ai, T., Zhang, X., He, Y.: An interactive web mapping visualization of urban air quality monitoring data of China. Atmosphere **8**(8), 148 (2017)
16. Marks, S., Estevez, J.E., Connor, A.M.: Towards the holodeck: fully immersive virtual reality visualisation of scientific and engineering data. In: Proceedings of the 29th International Conference on Image and Vision Computing New Zealand, pp. 42–47. ACM (2014)
17. PlumeLabs: Plume air report (app). https://play.google.com/store/apps/details?id=com.plumelabs.air&hl=de
18. RandomInternational: Rain Installation: Barbican Centre London. https://www.dezeen.com/2012/10/04/rain-room-by-random-international-at-the-barbican/
19. San José, R., Pérez, J.L., González-Barras, R.M.: 3D visualisation of air quality data. In: Proceedings of the 11th International Conference "Reliability and Statistics in Transportation and Communication", Riga (2011)
20. Stone, R.: Serious games: virtual reality's second coming? Virtual Reality **13**(1), 1–2 (2009)
21. WeVR: The blue (vr experience). https://wevr.com/theblu

Digital Transformation

Process Innovation Supported by Technology – Making for Longer Injury-Free Careers in the Case of High Performance Musicians

Eliseu Silva[1], Manuel Au-Yong-Oliveira[2(✉)], Pedro Fonseca[3],
Rui Garganta[4], and Christopher Bochmann[1]

[1] University of Évora, Évora, Portugal
eliseu7silv@gmail.com, bochmann@uevora.pt
[2] GOVCOPP, Department of Economics, Management,
Industrial Engineering and Tourism, University of Aveiro, Aveiro, Portugal
mao@ua.pt
[3] LABIOMEP: Porto Biomechanics Laboratory, Porto, Portugal
pedro.labiomep@fade.up.pt
[4] Faculty of Sports of the University of Porto, Porto, Portugal
ruigarg@fade.up.pt

Abstract. Technology and its widespread usage has dictated a far-reaching revolution in many domains, including in the teaching of music, as described in this study. Using technology (costing €150,000 for the cameras – MoCap – Qualisys AB, Sweden; and €150,000 for the system for analysis – Visual 3D software – C-Motion, USA) it was possible to predict that the duration of the career of a violinist supported by such tools – correcting one's technique – may be prolonged significantly – as much as fifteen years or more. This approach may even mean the difference between being able to continue with one's chosen profession or having to end it prematurely. Such technology may be applied to even the youngest of performers – aged 6 or 7 years old – a number of whose careers we suggest be followed to provide further support for our study. The correct placing of the left hand will help avoid injury and support a quicker evolution. Less time will be wasted, for better results. The musician will be more efficient, more flexible, and will have a more natural position and less musculoskeletal problems. The technology used compared the data and it was verified that for each chord there was an ideal height – corresponding to a more relaxed and natural position of the left hand when playing. One may represent the study's findings as a model, as follows: OPIH (optimal performance and ideal height of the left hand) = function of {dimension of the instrument, length of the strings, dimension of the hand, dimension of the fingers}.

Keywords: Software · Biomechanics of the left hand · Injuries · Ergonomics · Violin

© Springer Nature Switzerland AG 2019
Á. Rocha et al. (Eds.): WorldCIST'19 2019, AISC 932, pp. 315–328, 2019.
https://doi.org/10.1007/978-3-030-16187-3_31

1 Introduction and Background of the Study

This study proposes an exploratory model on the placement of the left hand on the violin in the different technical-performative challenges, supported by technology. In this article we publish the results and discuss the model at which the scientific methodology has arrived. The study was in its sixth year of development, at the time of the writing of this article, and has been self-financed.

It should be noted that the theoretical model, verified with the experimental work performed in a controlled and monitored environment, and with the participation of a single internationally renowned violinist (the first author of this work, working in a multidisciplinary team including a specialist in technology – other violinists were also used, initially, however it was not deemed practical and so the experiment was finally concluded with only one expert violinist – the lead author – who was able to provide more stable results) – has already been put into practice with musicians of varied ages and levels of specialization. Our perspective is that we are already currently working to avoid injuries in musicians - including younger musicians - who will thus have an opportunity to have a longer, more fulfilling and more barrier-free career, avoiding damaging injuries that can shorten careers at the highest level [1].

We discuss how the study was done, with which guidelines, which exercises, as well as discussing the anatomical models under analysis. In addition, the study benefits from numerous original images, some of which are presented here. For example, we have images taken from Visual 3D software (C-Motion, USA) - which also measures distances. Everything is captured in 3D.

The best results were achieved with technique B (the second technique) proposed by the principal researcher. There is thus a higher finger speed, greater precision, a more natural position and greater flexibility and joint amplitude with this technique. The techniques are two "A" s, one "B" and one "C". The techniques were first played at a controlled speed and then as fast as possible, for the techniques "A", "B" and "C". It was found that the correct precise angulation and height of the hand related to the fingerboard, corresponds to an optimized left-hand performance. Detected by technology - Qualisys 3D LABIOMEP software. MoCap system. A total of 12 image capturing cameras were used (the investment was € 150,000 for MoCap cameras - Qualisys AB, Sweden and € 150,000 in the analysis system - Visual 3D software - C-Motion, USA - equipment made available for this study).

2 Literature Review - Injuries of Orchestra Musicians, Especially Violinists and Viola Players, and the Importance of Technology

In the research of [2] it is argued that, regardless of the instruments played, the body zone that is most affected by musculoskeletal problems is the left wrist. These results may relate to the fact that, of the various groups of musicians studied, most studies point to string players as those with a higher prevalence of musculoskeletal problems. They also show that by comparing the two arms of violinists and viola players, they have twice the prevalence of muscular disorders in the left upper limb compared to the right upper limb. This is because the right hand flexors do not have great mobility, just the task of holding and controlling the movements of the bow. [3] argue that the great majority of musculoskeletal problems in violinists and viola players stem from technical inaccuracies or muscular tensions that violinists develop in the performance of their instrument in a specific passage or a more generalized approach.

To address such an important issue in the life of string musicians, it is important to first talk about work-related musculoskeletal injuries or disorders – which are repetitive strain injuries [4]. There is a list of inflammatory and degenerative diseases, associated with repetitive movements, muscular overload and incorrect body posture that result in occupational risk factors [5]. These injuries are predominant in the upper limbs and in the spine associated to the various professions in which physical movements are very repetitive, as is the case of the performance of string instruments [6].

The types of injury can be characterized as muscular, tendon and/or as being connected to the nervous system. The muscular lesions come essentially from the laceration of the junction of the muscle with the tendon coming from it, or may even be caused by muscle fatigue, translated into degenerative inflammatory diseases [7]. Tendon lesions translate into tendinitis, which is a result of large loads. Tenosynovites are also tendon problems resulting from friction between them which creates synovial thickening [8].

In order to reinforce the importance of the quality of the technique when performing, so as to diminish the prevalence of injuries, we must mention that a great part of the injuries come from a lack of ergonomics in the performance itself and therefore from poorer technique, which involves great muscular effort to get the work done, with strained postures.

The importance of technology in solving this process is fundamental because the high-performance artist or musician will have a strong personality – that leads him/her in his/her activity, often solitary and without great help from third parties, in human terms and especially in commercial terms - and associating this aspect with the fact that positions, even if wrong, have in most cases hundreds or thousands of individual training hours behind them, the artist will resist the change, unless, as process, there is technology to report position and posture errors. In effect, technology may provide the necessary pre-arrangement for necessary change, to healthier techniques, for example with the violin. Resistance to change arises when there is the realization that change will be threatening to the individual. The success of change will depend on whether you can persuade and change mindsets [9]. It is necessary to persuade to think differently - perhaps with the long term and health in focus - in order to also work differently with a greater awareness of ergonomics of movement [9].

Technology refers to "new resource combinations" [10, p. 3], or to new combinations of resources. The innovation proposed in this study is process innovation based on technology [10], as defined in the [11, p. 49]: "A process innovation is the implementation of a new or significantly improved production or delivery method. This includes significant changes in techniques, equipment and/or software." We thus propose process improvements in order to increase the returns of the activity. We can even make a parallel to the manufacturing process in which there will now be more process quality control, faster response times, as well as the improvement of a number of dimensions that improve performance over time. A musician's career may even last for many years, until a very old age, with the motivation to do so, and better performance results and a career free from injury and subsequent stops associated with this can be decisive.

High performance work systems (HPWS), although they have not existed for a long time, and despite their definition not being a consensual one, they have been widely discussed in the literature [12]. As the objective is the improvement of individual performance, the system described in this study can be considered to be a HPWS. Our study focuses on training, human resource safety, career management and productivity enhancement, which can lead to important benefits in long-term development and learning, including greater satisfaction with the activity (Fig. 1).

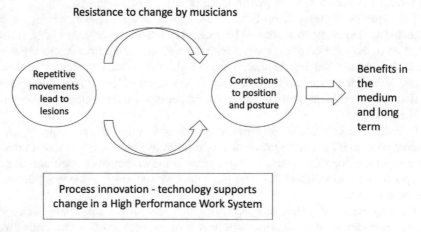

Fig. 1. Technology leveraging change in music and in a High Performance Work System (own elaboration)

3   Methodology

It should be noted that diverse technology is used in cycling and swimming, for example, so the approach is not entirely new. However, it should be noted that this technology in LABIOMEP to date has not been applied in this way to musicians (so-called "arts professionals", as opposed to sports professionals or high-level athletes). We see the performance of the violinist here as being analogous to that of the elite athlete. Thus, as mentioned above, in LABIOMEP they did this type of analysis for the first time with violinists (from the arts), although they have already done so in other activities, more specifically in sports. Everything is analysed "to the millimeter". They work, at LABIOMEP, with sports entities such as from the city of Barcelona, in neighboring Spain, and also with swimming federations from Qatar and Thailand. At the national level, they have worked with the three main clubs in Portugal – F.C. Porto, Sporting, and Benfica – having used advanced technology to analyze the performance in different sports, such as in football, cycling, swimming, waterpolo, and hockey, among others.

After an analysis of the strategies and scientific devices available and applicable in this scientific project, we came to the conclusion that the most promising method of analysis would be the motion capture in 3D Qualysis.

It is possible to measure with greater precision the most sensitive variation of the position, in terms of angulation and height of the hand in relation to the scale of the violin, and to analyze the biomechanical functioning of the left hand.

There is also the possibility to relate data and results of different parameters and in the three X, Y, Z axes that correspond to the three axes of the movement of the various joints that would be analyzed, as can be seen in image 1 (image taken from Visual 3D software).

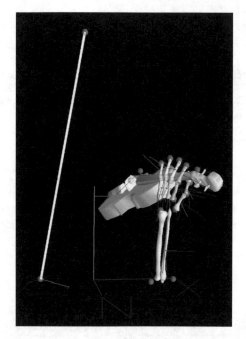

Image 1. Markers and vectorial system on the three axes X, Y, and Z

The exercises were elaborated in order to analyze the variation of several parameters:

– The variation of "hand height" (or distance from the metacarpophalangeal joint of the index finger to the edge of the violin scale) on the same string: deliberately high position, comfortable position, low hand position.

This height variation was performed in the first position and with a perfect fourth interval between the first and fourth finger.

– Variation of the angle of the hand in relation to the scale of the violin.

This exercise was performed at intervals of greater extension, such as the 5th perfect, generally used in exercises with fingered octaves, and intervals of the minor 6th generally used in intervals of minor 10th, when referring to double stops. The three types of exercises were chosen for the different types of opening of the left-hand fingers. The three forms of opening of the fingers correspond to the interval between the first (index finger) and fourth finger (little finger).

The "perfect fourth" interval - where we find an octave between two adjacent strings, will be translated with the letter A. The "perfect fifth" interval - which represents the technique of fingered octaves on adjacent strings - will be represented by the letter B. Finally, the intervals of greater extension, intervals of sixth, equivalent to the technique of greater extension that represents the position of tenths in double stops, that will be analyzed in C.

Different exercises: Exercises A1mi and A2mi (Fig. 2).

Fig. 2. Musical exercise used to perform A1mi and A2mi

The first exercise was performed with a deliberately high hand height, therefore with a smaller distance from the metacarpophalangeal joint in relation to the violin arm, called A1mi. The second exercise was performed with the same musical score but with the lower hand position than in A1mi, therefore with the most distant joint from fingerboard, called A2mi.

Exercises A1sol and A2sol (Fig. 3).

Fig. 3. Musical exercise used to perform A1sol and A2sol

The exercise performed with the low hand height in relation to the violin arm was called A1sol and with a slightly higher height and more comfortable position for the G string denominated A2sol.

Exercises B1 and B2 (Fig. 4).

The following exercises were performed at bigger intervals and the variation of the angulation and its relation to the biomechanical functioning of the hand were analysed, in particular. Therefore, the following exercise was performed with a small angulation in B1 and with a greater angulation between the metacarpophalangeal joint and the fingerboard in B2.

Fig. 4. Musical exercise used for the execution of B1 and B2

Exercises C1 and C2 (Fig. 5).

The last exercises were also performed on the G string, but with an even wider interval between the first and Little finger, called 6th minor.

The first exercise of this group was performed with a little angulation between the metacarpophalangeal joint alignment in relation to the violin fingerboard in C1 and with a more pronounced angulation in C2.

Fig. 5. Musical exercise used to perform C1 and C2

In each experiment, two sequences were performed. The first was controlled metronomically with the tempo at 1 beat per second and for 5 s. The second part was immediately performed for about the same time but at the highest possible speed.

4   Results

The technology applied has provided numerical analytical results that has allowed us to compare values of angulations and distances. Additionally, we were able to measure the height of the hand relative to the violin. There is an ideal height for each positioning - for the four strings of the violin - so there are four different ideal heights.

In the various exercises proposed, with the three different intervals between the index and little finger, of perfect 4th - exercise A - intervals of perfect 5th - exercise B - and interval of minor 6th - Exercise C - the results were very encouraging.

The proposed technique "2" (the second official exercise in each category), which imposed a specific position of the hand very precise in terms of angulation and height in relation to the violin fingerboard, showed very concrete and transversal results: higher finger speed; higher precision, lower error predisposition, demonstrated by a lower standard deviation; a more natural position of the wrist, hand and fingers, as the medial position is closer to the natural joint position; greater flexibility and joint amplitude. In general, we can say that better ergonomic performing results were achieved.

5   Analysis and Discussion of Objectives

In light of the results we sought to converge the objectives and results of the research into a more organized and thoughtful perspective of the functional mechanics of violin performance, so we may elaborate a functional anatomical model of left hand positioning. Therefore, we may respond scientifically to the three (research) questions, as portrayed below.

Question 1 - How to control the position of the hand in different strings, obtaining the same interval relation, with the minimum muscular tension and maximum efficiency?

The results obtained demonstrate that the biomechanical functioning of the hand in A2mi is quite similar in A2sol. The practical experiments analyzed the following aspects: the general movement of the little finger in the two planes of motion of the metacarpal joint, the functioning of the index finger on the sagittal axis of the adductions and abductions, and the movement of the wrist in its two axes of movement. We conclude that it is a clearly more ergonomic and efficient movement than in A1mi and A1sol.

These are parameters that indicate that there is actually an ideal specific height for each of the four strings, so that the digital operation and the wrist movement are as efficient and as free as possible. The technology used compared the data – data which was captured by the cameras utilized and translated into images and numbers – for the exercises A1mi, A2mi, A1sol, A2sol – and it was verified that for each chord there was an ideal height – corresponding to a more relaxed and natural position of the hand when playing – in accordance with Figs. 2 and 3 – please refer also to Table 1 – where A2 has values much closer to zero and is therefore better (see also Graphs 1 and 2). Supported by the functional peculiarity of the hand, which, due to the condyloid joints, in the metacarpophalangeal joint, at the moment of flexion of the fingers these touch the center of the palmar zone, the results demonstrate that, in the different strings, there is a specific position, in terms of distance of the metacarpophalangeal joint of the index finger and the superior edge of the violin fingerboard, which makes the biomechanical and performative functioning of the hand more efficient and natural. Thus, it can be deduced that there are four basal levels with four levels of distance from the metacarpophalangeal joint to the edge of the violin fingerboard. The basal level of

Table 1. Values for adduction and abduction of the little finger in A1mi and A2mi

A1 slow abduction	2.61°
A1 slow adduction	16.31°
A1 rapid abduction	17.41°
A1 rapid adduction	12.80°
A2 slow abduction	−4.48°
A2 slow adduction	9.30°
A2 rapid abduction	2.46°
A2 rapid adduction	8.75°

greater distance is that of the E string, followed by the level of the A string, the D string and finally the G string that has the smallest distance between the metacarpal-phalangic joint and the edge of fingerboard.

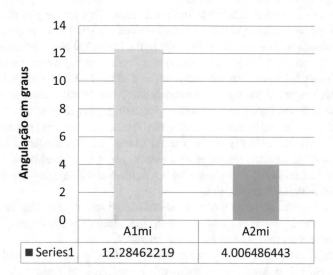

Graph 1. Little finger adduction and abduction on average in A1mi e A2mi

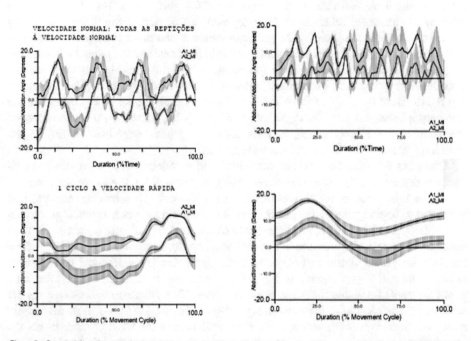

Graph 2. Adduction and abduction of the little finger. On the left at a normal speed and on the right at a rapid speed. The higher graphs showing play throughout the exercise and the lower graphs just for one cycle (one cycle is the average of all the cycles).

One can therefore consider four basal levels on which the hand moves, on the four different strings, E(mi), A(la), D(ré), and G(sol).

It should be remembered that in these exercises attention has been taken to maintain the same angle of the hand in relation to the violin's arm, and the height of the hand has been purposely changed in relation to the point edge. This leveling of the hand on the different strings must occur (as shown in the images captured and analysed), otherwise interpalm tensions will be created, especially in the lumbrical and interosseous muscles, responsible for the adduction and abduction of the fingers, which creates a greater tension in the performance, greater stiffness, and slower speed, especially in muscles that are small, are not as strong as the deep flexors, and because most of the time they are not the muscles referring to the motion plane of greater amplitude of the joints, as are the case of lumbricals and interosseous. These, as they are responsible for the movements of digital laterality in the frontal plane, are less powerful and extenuate themselves with great ease, potentiating a whole panoply of problems, being able to arrive, in the last instance, at performative musculoskeletal injuries like tendinitis and syndromes of nervous compression.

Question 2 - Will there be any mechanical-functional way of the hand capable of controlling, in the intervals of great extention, the opening of the fingers, which, being limited, generates a great palmar muscular tension?

Regarding the second question and referring to the objectives, we will consider the following exercises: B, with special focus on the intervals of perfect 5th, between the index and the little finger; and C reserved to minor 6th intervals. We will also compare the exercises in A that refer to the intervals of the perfect 4th. Based on the same mechanical-functional principle of the extension and flexion of the fingers, which by having a condyloid joint joining the proximal phalanx and the metacarpal, and the fact that the flexion of the fingers is directed towards the center of the palm by deduction, the greater the extension, the greater the opening of the fingers. In this way, we can consider different heights of the hand, relative to the basal level of a given string, so that the different openings of the 4th, 5th and 6th intervals are feasible. Distancing the metacarpophalangeal joint from the edge of the violin, allows us a greater general extension of the fingers which, in turn, allows a greater amplitude of the fingers especially between the index and the little finger.

Question 3 - Will there be any structural and positional way to control the progressive decrease of the interval relation, which happens along the violin fingerboard?

As for this question, which was initially used to mark this scientific research, it is important to begin by noting the fact that, on the violin fingerboard, as the fingers go up in the direction of the bridge, the distances between fingers gradually decrease. One can better perceive this phenomenon in a visual way, on the fingerboard of a guitar, where the frets are getting closer and closer to each other. The less optimized performative results in B1 and C1 are essentially due to the fact that the hand angle to the violin scale is not the most favorable for the desired exercises. Thus, since in these exercises the position of the hand with the most natural digital flexion is not the most favourably indicated, there is a string abduction or ulna deviation of the wrist in order to get the little finger closer to the string, and this in turn, at the moment of the attack on the string that represents the most flexible angle, a strong abduction is also performed in order to achieve the right pitch, creating great discomfort, and in turn, since the position is not the most free and natural, the index finger is strongly adducted in order to control the intended note. However, even if these exercises were not performed in other positions along the violin scale and by varying the angulation to perceive differences in performance (something that would make this research work even more exhaustive), we perceive that we are within the possibility of deducing that the angulation in both B1 (43.66) and C1 (38.44) could result in higher positions, such as the third or fourth position, where the intervals are closer.

The use of technology leads us to conclude that, based on an anatomical-functional analysis and the results in different interval amplitudes, varying the angulation in relation to the violin point, although only in the first position, we can also control in a detailed way the interval relation, that is to say, the gradual decrease in intervals along the scale.

6   General Considerations

To conclude this theoretical explanation, it would be important to mention a subject of the utmost importance. All this scientific study was a theoretical and practical attempt to be able to resort to an anatomical-functional mechanism to solve the various requirements in the violin performance of the left upper limb – especially of the hand – where a very high prevalence of musculoskeletal problems occurs, related to performance. At the end of this study (in the PhD thesis undertaken by the first author), some considerations are set forth based on exercises analyzed by three-dimensional motion analysis devices which led to the extraction of numbers and concrete data.

However, we have to consider that all hands are different. The dimensions of the hands, arms, fingers, and joint amplitude of the various levers are different in the many individuals who play, or may come to play, the violin. We would like to point out that this study may be perfectly adapted to viola players, since the mechanical phenomena of the instrument are the same as for violinists. With this study we intend to simply perceive, in a more systematized way, mechanically-functional systems of the hand capable of responding to different instrumental techniques and demands.

These conditions change and must, in each case, be adapted to the dimensions of the instrument, the length of the strings from the top nut and the bridge, the dimensions of the hands and the fingers.

One may represent the study's findings as a model, or formula, representing reality, as follows: OPIH (optimal performance and ideal height of the left hand) = function of {dimension of the instrument, length of the strings, dimension of the hand, dimension of the fingers}.

An example of the above is: the bigger the dimension of the hand, the smaller the angulation and distance between the metacarpophalange articulation and the fingerboard.

It is also important to realize that this approach based on the angulation and leveling of the height of the hand in relation to the violin fingerboard is executed and controlled by a phenomenon relative to the individual - it is the proprioceptive capacity, which has the power to regulate our positions and muscular force exerted on objects.

Image 2 shows reflectors placed on the metacarpophalangeal joints, and at the four extremities of the violin fingerboard. Image 3 shows the angulation of the metacarpophalangeal joint in relation to the edge of the violin point.

Image 2. Reflectors placed on the metacarpophalangeal joints, and at the four extremities of the violin fingerboard

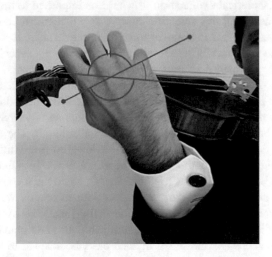

Image 3. Angulation of the metacarpophalangeal joint in relation to the edge of the violin fingerboard

7 Conclusion and Suggestions for Future Research

This study seeks to show the knowledge and depth to which musicians (and teachers in particular), in the laboratory, can descend in order to optimize their performance and aided by technology. It focuses on the left hand of the violinist, which executes one of the finest and most detailed jobs when performing at a high level, requiring a huge amount of strength, accuracy, speed, endurance, supporting the instrument, while flying over the finger board all together, and which is the most complicated point in terms of the health of the elite musician. It should be noted that this methodology applied to the reality of the violin is already bearing fruit in practice, and with students of the first author. It is with enthusiasm that we are close to the end of the project which, like other activities, such as high competition sports, promises to bring musicians to play for more time and with greater efficiency and naturalness. To point out that using this technology could boost a musician's career for a period of fifteen years (or more) may seem ambitious. However, one has only to look at the example of technology applied to other aspects of human life to realize the possible scope of improvement. In this way, we find that technology - the applied knowledge of the human being to solve certain problems - is fundamental and has surprising results - if well implemented. We suggest a longitudinal study over time in order to track the effects of the technology described herein (establishing a contrast between groups of violin players with correct and incorrect positions of the left hand).

In swimming the technology can be applied to the range of shoulder and arm movement in order to minimize, for example, the "swimmer's shoulder" – a recurring problem that "plagues" top swimmers. This technology could also be applied to the movement of the golf player who, at the end of his or her career (as has occurred with several renowned players) may require the total replacement of the hips by prostheses - due to the forcing of the amplitude of rotation, linked to a blocking of the front supporting foot, an unnatural movement with harmful effects. For example, we can consider the best golfer of all time, and still alive at the time of writing – the North American and former player Jack Nicklaus - who has two prosthetics on his hips; but there are more cases of this type of injury which have led to such an outcome. It will not be too much to stress the importance of technology to increase the quality of life and prolong careers at the highest level.

The techniques and methodology discussed herein deserve to be monitored so that, with a longitudinal study, over several decades, one can prove that the course indicated is indeed right - for this study that is exploratory, for the moment - but which may eventually become commonplace in the industry - alongside other technologies that have become mainstream - such as the use of smartphones, even by children. A use that, being prolonged, and from a very young age, may in turn lead to other types of injuries – still undiscovered. This will certainly lead to the need for other specific studies (see, for example, [13]).

Technology has been widely used in several activities [14–16], as the world is increasingly complex, which leads to changes in people themselves [17].

We hope this study has been able to elucidate how a violinist may be monitored in order to maximize his/her performance and well-being.

References

1. Silva, E., Au-Yong-Oliveira, M., Fonseca, P., Garganta, R., Bochmann, C.: The musician as entrepreneur – multidisciplinary innovation with performing violinists – achieving a sustainable competitive advantage via the biomechanical enhancement of the left hand. In: Costa, C., Au-Yong-Oliveira, M., Amorim, M. (eds.) Proceedings of the 13th European Conference on Innovation and Entrepreneurship, ECIE 2018, University of Aveiro, 20–21 September, pp. 984–994 (2018)

2. Frank, A., Muhlen, C.A.: Queixas músculo-esqueléticas em músicos: prevalência e fatores de risco. Revista Brasileira de Reumatologia **47**(3), 188–196 (2007)
3. Moraes, G.F., Papini, A.A.: Desordens musculoesqueléticas em violinistas e violistas profissionais: revisão sistemática. Universidade Federal de Itajubá Campus, Itabira (2012)
4. Serranheira, F., Sousa, A.U., Lopes, M.F.: Lesões músculo esqueléticas e trabalho: Alguns métodos de avalliação de risco. Escola Nacional de Saúde publica, Lisboa (2008)
5. Zaza, C.: Playing-related musculoskeletal disorders in musicians: systematic review of incidence and prevalence. Can. Med. Assoc. J. **158**, 1019–1025 (1998)
6. Sousa, L.F.: Lesões por esforço repetitivo em instrumentistas de cordas friccionadas. Universidade de Aveiro, Aveiro (2010)
7. Caldron, P., Calabrese, L.H.: A survey of muskuloskeletal problems encountered in high level musicians. Med. Probl. Perform. Art. **1**, 136–139 (1986)
8. Hoppmann, R., Patrone, N.A.: A review of musculoskeletal problems in instrumental musicians. Semin. Arthritis Rheum. **19**, 117–126 (1989)
9. Huczynski, A.A., Buchanan, D.A.: Organizational Behaviour, 8th edn. Harlow, Pearson (2013)
10. Burgelman, R.A., Christensen, C.M., Wheelwright, S.C.: Strategic Management of Technology and Innovation, 5th edn. McGraw-Hill International Edition, New York (2009)
11. Oslo Manual. Guidelines for Collecting and Interpreting Innovation Data, 3rd edn. Paris, France, OECD, Eurostat and the European Commission (2005)
12. Zhu, C., Liu, A., Chen, G.: High performance work systems and corporate performance: the influence of entrepreneurial orientation and organizational learning. Front. Bus. Res. China **12**(4), 1–22 (2018)
13. Au-Yong-Oliveira, M., Gonçalves, R., Martins, J., Branco, F.: The social impact of technology on millennials and consequences for higher education and leadership. Telematics Inform. **35**(4), 954–963 (2018). https://doi.org/10.1016/j.tele.2017.10.007
14. Gonçalves, R., Martins, J., Branco, F., Perez-Cota, M., Au-Yong Oliveira, M.: Increasing the reach of enterprises through electronic commerce: A focus group study aimed at the cases of Portugal and Spain. Computer Science and Information Systems **13**(3), 927–955 (2016). https://doi.org/10.2298/csis160804036g
15. Gonçalves, R., Martins, J., Pereira, J., Cota, M., Branco, F.: Promoting e-commerce software platforms adoption as a means to overcome domestic crises: the cases of Portugal and Spain approached from a focus-group perspective. Trends and Applications in Software Engineering, pp. 259–269. Springer (2016)
16. Gonçalves, R., Rocha, T., Martins, J., Branco, F., Au-Yong Oliveira, M.: Evaluation of e-commerce websites accessibility and usability: an e-commerce platform analysis with the inclusion of blind users. Univ. Access Inf. Soc. **17**(3), 567–583 (2018). https://doi.org/10.1007/s10209-017-0557-5
17. Au-Yong-Oliveira, M., Moutinho, R., Ferreira, J.J.P., Ramos, A.L.: Present and future languages – how innovation has changed us. J. Technol. Manag. Innov. **10**(2), 166–182 (2015). https://doi.org/10.4067/S0718-27242015000200012

New Lighting Representation Methodologies for Enhanced Learning in Architecture Degree

Isidro Navarro[1](✉) iD, Albert Sánchez[1](✉) iD,
Ernesto Redondo[1](✉) iD, Lluís Giménez[1](✉) iD, Héctor Zapata[1](✉) iD,
and David Fonseca[2](✉) iD

[1] Barcelona School of Architecture, Polytechnic University of Catalonia,
08028 Barcelona, Spain
{isidro.navarro,albert.sanchez.riera,ernesto.redondo,
lluis.gimenez,hector.zapata}@upc.edu,
[2] School of Architecture of La Salle, University Ramon Llull,
08022 Barcelona, Spain
david.fonseca@salle.url.edu

Abstract. Lighting conditions adjustment is one of the most difficult tasks in architecture visualization. This paper presents a new teaching methodology in the field of lighting representation for enhanced learning with students of architecture degree. It aims to assess how lighting balance competences are acquired in undergraduate and master's courses in architecture, urbanism and building technologies. The working hypothesis is that this new learning process will contribute to a better understanding and mastery of lighting control in architectural rendering scenes. It is based on comparing different widely used technologies such as building information modelling, virtual reality, augmented reality, photorealistic rendering software and game engines. The implementation will be evaluated including user experience analysis all along the process with quantitative and qualitative surveys. The research will try to demonstrate how this method can improve student's skills, especially in realistic lighting representation of projects of the subjects related to digital and graphic representation. The subjects will be from several courses: Architecture Representation II from second course, Architectural Representation III from third course, Multimedia and Techniques of Modelling and Digital Production Oriented to the Development of Constructive Solutions from fourth and postgraduate course respectively. This research will also address the issue of how students can implement their new acquired skills to other subjects of the Degree of Architecture at the School of Architecture of the Polytechnic University of Barcelona.

Keywords: Architecture · Enhanced learning · Lighting · Virtual reality

1 Introduction and State of the Art

One of the most difficult tasks in architectural rendering is the Lighting settings adjustment. Most of the times final images are presented with false exposure and camera parameters in order to make them more attractive. In this cases lighting balance is automatically generated and this compensation may not be used in real environments

© Springer Nature Switzerland AG 2019
Á. Rocha et al. (Eds.): WorldCIST'19 2019, AISC 932, pp. 329–335, 2019.
https://doi.org/10.1007/978-3-030-16187-3_32

especially indoor and in night scenes. In contrast, final renders in professional lighting software widely used for calculate number and type of lights, are usually represented using light intensity isolines and numerical parameters. These non-realistic images are inadequate resources to enable decision-making based thereon.

The study of artificial lighting conditions of buildings is a topic that has been applied for some time in the field of education of future engineers. [1–3] but it's a very rare event in the training of architects, except in those countries where the degree in Architecture is a long educational cycle. Students are then hardly trained in lighting variables as colour temperature, intensity, camera exposure parameters, etc. but once they try to replicate this parameters to their real projects, space perception completely changes.

In particular, that competence is usually addressed by other technicians involved in the field of construction [4]. It's often approached from a professional perspective, especially on studies of indoor environments [5] or daylight conditions [6–8]. This could be due to the more creative and visual profile of architects and interior designers. Now, in any case it is not addressed in the field of urban design, except lighting calculations of certain urbanization projects, in which case, the perceptual aspects are not taken into account. Our proposal aims to assess how these competences are acquired in several undergraduate and master's courses in architecture, urbanism and building technologies, either in buildings design, public spaces and urban areas, as they are understood in Spain.

2   Methodology

The research will take place in two periods of the degree of Architecture. The duration of the degree is six years, students finish their studies developing the final degree project at the last year. At the first period the assessed subjects are focused on lighting calculation and analysis to obtain visual and data results. In a second phase, the subjects improve the communication capabilities of the students for realistic images representation. At the same time, the students are attending other subjects where this practices will impact.

The proposal stars with an approximation to the representation of the daylight and shadows at the subject Architectural Representation II from second course of the degree. Traditionally, this process of representation was developed with Computer Aided Design software, but these methodologies are moving to Building Information System software. The results of the first approach to light representation is possible with the visualization tools provided by this software. The research will include virtual reality methodologies to expand the perception of lighting and shadows of the projects and will analyse the results of the VR implementation and the impact at the subject Projects II from second course.

The second field of study is the photorealistic representation of the projects and it is developed at the third course of the degree at the subject Architectural Representation III. Lighting and materials mapping are the main topics in order to get the most accurate render images. The tools used for this purpose are 3D modelling software and several render engines software. The research will include the data analyse of illumination based on calculation software of indoor/outdoor lighting and also the approach between both methods for the final project design.

After the second and third courses, the students are able to represent their projects with photorealistic images and graphic documentation where lighting provides visual and data results. Next steps will be focused on the communication phase of the projects, it means, the way how the students show the concepts, processes and final results of their projects.

At fourth course the subject Multimedia is centred into immersive representation with an approximation to videogame engines to create virtual reality and augmented reality presentations. The research is focused on the analysis of the learning process of these tools in the field of architecture visualization. However, this implementation requires mechanisms of optimization of lighting for a better performance of the experiences, and this becomes the focus of the methodologies to explore.

The last field of work will be the development of the final degree's project with technical documentation and presentations in photorealistic images and videos, and the subject is Techniques of Modelling and Digital Production Oriented to the Development of Constructive Solutions from postgraduate course. The practices include parametric and real-time rendering software. The impact of the results of this work will be explored during the development of the final degree's project.

2.1 Assessment

The experience is based on the hypothesis that the new tools of information and communication technologies (ICT) such as immersive Virtual reality, when they're used in an e-learning environment, can help to improve students learning processes. It increases their motivation and competences with a reduction of time and it's a cost-effective solution (we'll use for that free applications or educational licenses).

In a first phase we will study how the subject will be presented in the basic courses of the BIM Architecture degree and Architectural Representation applied to Urban Design. In a second level, it will be evaluated if the competences are undertaken in master courses and in Architecture and Building Engineering final works. Finally, highest competences acquirement in the design and calculation of artificial lighting systems will be evaluated in the master of Architecture, Urbanism and Building Technologies. Assessment will be carried out both for indoor environments and public and urban space. In adition interactive and realistic visualization strategies for their final adjustment will be conducted. The ultimate goal is the evaluation through analytical learning, if students assume such competencies combining scientific reasoning with visual education, intuition and creativity.

2.2 Phases

Research Design Process. This section is the first of the research process, understood as the one that will be carried out with a group of students with a similar profile. This group of students will define the control group of each subject and it will be the one in which the VR will be introduced for the study of lighting. The improvement of their skills will be compared with that of another group of students who use traditional methodologies. As previous studies in urban projects for education [9–11], the analysis of the data obtained in the pre-test, post-test surveys and bipolar laddering tests will allow us to evaluate the improvement through the learning analytics techniques [12–15]. This research evaluates subjective aspects of the perception of comfort and light safety of an exterior or interior architectural environment. For this reason, the tests will introduce a qualitative assessment as a complement to the quantitative one, which is understood as a method mixed.

The teaching experiment will be repeated in various courses of the degree and master, either in the design of interior environments such as public space and urban design [16–19]. That data will be available in various phases of the training of future architects, urban planners and building engineers, as it's demonstrated in previous projects for digital urban transformation [20, 21]. This process will evaluate then if the competencies improvements vary according to the previous preparation or are equivalent indistinctly, due to the use of the immersive VR.

Project Based Learning. The development of the contents will have a sequence of steps depending on the structure and academic contents of each of the courses of the implementation. For example, in the subject Architectural Representation III of the third course, the first step will be the generation of a first 3D digital model with a specialized software (Fig. 1), then the ambient lighting will be added, and subsequently it will lead to a 3d digital model that will be visualized with VR HMD. After this immersive experience, a readjustment is made using again photometric lights. The scene lighting is then rendered to textures and projected onto the surfaces, greatly improving both the performance of the simulation and the realism and faithfulness to the lighting conditions that would occur in reality (Fig. 2), the result will be a final design of the architectural or urban environment (Fig. 3).

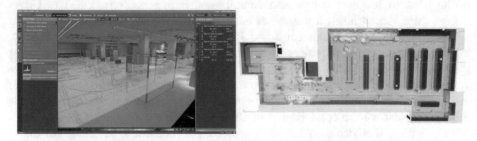

Fig. 1. Testing the lightning evaluation in indoor projects with free specialized software.

Fig. 2. Students testing parameters of the urban project lightning with a Virtual Reality HMD.

Fig. 3. Creating VR scenarios for testing lightning and visual analysis.

3   Conclusions

We intend to demonstrate the lack of background in the design of architectural lighting models taking into account the perceptive aspects as a complement to the calculations. It is necessary to incorporate perceptive and subjective factors in architecture studies. It is convenient to introduce elements of environment as the vegetation in the designs in the external environments. A real analysis of the materials in interior environments is required as key factors to generate comfortable designs that offer security. It is necessary to generate realistic virtual scenarios and immersive visualization with virtual reality HMDs (Head Mounted Displays) for an optimal understanding and perception of the projects. This experience will allow students to travel freely and consequently assume an integral design of the environments. It is a priority to introduce highly efficient and agile teaching processes, which allow to go from the initial designs with reliable calculation bases to fully controlled final designs, in a very short time rapidly increasing the students' skills, following validated previous works [22, 23].

Acknowledgments. This research was supported by the National Program of Research, Development and Innovation, Spain aimed to the Society Challenges with the references BIA2016-77464-C2-1-R & BIA2016-77464-C2-2-R, both of the National Plan for Scientific Research, Development and Technological Innovation 2013–2016, Government of Spain, titled "Gamificación para la enseñanza del diseño urbano y la integración en ella de la participación ciudadana (ArchGAME4CITY)", & "Diseño Gamificado de visualización 3D con sistemas de realidad virtual para el estudio de la mejora de competencias motivacionales, sociales y espaciales del usuario (EduGAME4CITY)". (AEI/FEDER, UE).

References

1. Taljaard, M.J., et al.: The importance of lightning education and a lightning protection risk assessment to reduce fatalities. In: 2017 Australasian Universities Power Engineering Conference, AUPEC 2017, 2017-November, 5 February 2018, Melbourne, Australia; 19 November 2017 through 22 November 2017, pp. 1–6 (2017)
2. Apse-Apsitis, P., et al.: Practically oriented e-learning workshop for knowledge improvement in engineering education computer control of electrical technology. In: 2012 EEE Global Engineering Education Conference, EDUCON2012, 17 April 2012 through 20 April 2012, Article number 620110, Marrakech, Morocco (2012)
3. Cobb, M., et. al.: Higher education Building efficient electrical design. In: Conference Proceedings - IEEE SOUTHEASTCON Volume 2016-July, 7 July 2016, Article number 7506719 SoutheastCon 2016; Norfolk; United States; 30 March 2016 through 3 April 2016 (2016)
4. Das, A., Paul, S.K.: Artificial illumination during daytime in residential buildings: factors, energy implications and future predictions. Appl. Energy **158**, 65–85 (2015)
5. Busch, J.F., Du Pont, P., Chirarattananon, S.: Energy-efficient lighting in Thai commercial buildings. Energy **18**(2), 197–210 (1993). https://doi.org/10.1016/0360-5442(93)90104-L. Cited 13 times
6. Chirarattananon, S., Chaiwiwatworakul, P., Pattanasethanon, S.: Daylight availability and models for global and diffuse horizontal illuminance and irradiance for Bangkok. Renewable Energy **26**(1), 69–89 (2002). https://doi.org/10.1016/S0960-1481(01)00099-4. Cited 47 times
7. Krarti, M., Erickson, P.M., Hillman, T.C.: A simplified method to estimate energy savings of artificial lighting use from daylighting. Build. Environ. **40**(6), 747–754 (2005). https://doi.org/10.1016/j.buildenv.2004.08.007. Cited 125 times
8. Li, D.H.W., Cheung, G.H.W., Cheung, K.L., Lam, J.C.: Simple method for determining daylight illuminance in a heavily obstructed environment. Build. Environ. **44**(5), 1074–1080 (2009). https://doi.org/10.1016/j.buildenv.2008.07.011. Cited 28 times
9. Navarro, I., Fonseca, D.: Nuevas tecnologías de visualización para mejorar la representación de arquitectura en la educación. Archit. City Environ. **12**(34), 219–238 (2017). https://doi.org/10.5821/ace.12.34.5290
10. Llorca, J., Zapata, H., Redondo, E., Alba, J., Fonseca, D.: Bipolar laddering assessments applied to urban acoustics education. In: World Conference on Information Systems and Technologies, pp. 287–297. Springer, Cham, March 2018
11. Fonseca, D. Villagrasa, S., Navarro, I., Redondo, E., Valls, F., Llorca, J., Calvo, X.: Student motivation assessment using and learning virtual and gamified urban environments. In: Proceedings of the 5th International Conference on Technological Ecosystems for Enhancing Multiculturality - TEEM 2017, pp. 1–7 (2017)
12. Amo, D., et al.: Using web analytics tools to improve the quality of educational resources and the learning process of students in a gamified situation. In: Proceedings of 12th Annual International Technology, Education and Development Conference, p. 5 (2018)
13. Fonseca, D., et al.: Informal interactions in 3D education: Citizenship participation and assessment of virtual urban proposals. Comput. Hum. Behav. **55**(2016), 504–518 (2016). https://doi.org/10.1016/j.chb.2015.05.032
14. Fonseca, D., et al.: Technological adaptation of the student to the educational density of the course. A case study: 3D architectural visualization. Comput. Hum. Behav. **72**, 599–611 (2017). https://doi.org/10.1016/j.chb.2016.05.048

15. Fonseca, D., et al.: Mixed-methods research: a new approach to evaluating the motivation and satisfaction of university students using advanced visual technologies. Univ. Access Inf. Soc. **14**(3), 311–332 (2015). https://doi.org/10.1007/s10209-014-0361-4
16. Redondo, E., et al.: Educating Urban Designers using Augmented Reality and Mobile Learning Technologies RIED. Revista Iberoamericana de Educación a Distancia. **20**, 141–165 (2017). https://doi.org/10.5944/ried.20.2.17675
17. Fonseca, D., et al.: Improving the information society skills: Is knowledge accessible for all? Univ. Access Inf. Soc. **17**(2), 229–245 (2018). https://doi.org/10.1007/s10209-017-0548-6
18. Calvo, X., et al.: Programming virtual interactions for gamified educational proposes of urban spaces. In: Lecture Notes in Computer Science (including subseries Lecture Notes in Artificial Intelligence and Lecture Notes in Bioinformatics), pp. 128–140 (2018)
19. Escudero, D.F., et al.: Motivation and academic improvement using augmented reality for 3D architectural visualization. Educ. Knowl. Soc. **17**(1), 45–64 (2016). https://doi.org/10.1021/ja003055
20. Sanchez-Sepulveda, M., Fonseca, D., Franquesa, J., Redondo, E.: Virtual interactive innovations applied for digital urban transformations. Mixed approach. Future Gener. Comput. Syst. **91**, 371–381 (2019)
21. Valls, F., Redondo, E., Fonseca, D., Torres-Kompen, R., Villagrasa, S., Martí, N.: Urban data and urban design: a data mining approach to architecture education. Telematics Inform. **35**(4), 1039–1052 (2018)
22. Pinto, M., Rodrigues, A., Varajão, J., Gonçalves, R.:. Model of functionalities for the development of B2B e-commerce solutions. In: Cruz-Cunha, M.M., Varajão, J. (eds.), Innovations in SMEs and conducting e-business: technologies, trends and solutions (IGI-Global.), p. 26. IGI-Global (2011)
23. Fonseca, B., Morgado, L., Paredes, H., Martins, P., Gonçalves, R., Neves, P., Soraci, A.: PLAYER-a European project and a game to foster entrepreneurship education for young people. J. Univ. Comput. Sci. **18**(1), 86–105 (2012). https://doi.org/10.3217/jucs-018-01-0086

The Future Employee: The Rise of AI in Portuguese Altice Labs

Bernardo Lopes[1], Pedro Martins[1], José Domingues[1],
and Manuel Au-Yong-Oliveira[2]([⊠])

[1] Department of Electronics, Telecommunication and Informatics,
University of Aveiro, Aveiro, Portugal
{bernardobentolopes,martinspedro,j.domingues}@ua.pt
[2] GOVCOPP, Department of Economics, Management, Industrial Engineering
and Tourism, University of Aveiro, 3810-193 Aveiro, Portugal
mao@ua.pt

Abstract. This case study was conducted in the scope of the rise of artificial intelligence (AI) in today's society and provides the answer to the following questions: How are companies following the AI technology revolution?, How are organizational culture and identity changing within companies? and Where are companies investing in order to prepare to the upcoming competitive future? To answer these questions a semi-structured interview was done with Jorge Sousa, the Head of M2M/IoT & API Management & Virtual Assistant - Network and Service Solutions Department in Altice Labs. Altice Labs is owned by the multinational Altice Group and is responsible for the innovation and development of new solutions, technologies and trends in the telecommunications area for all the Altice group. Having already won several awards such as the Dell EMC Award in 2018 and the international Technology Leadership Award in 2017. Altice Labs is in a privileged position that only a few companies are in, performing high tech R&D. From this case study we conclude that organizational Identity and Culture are changing due to the external stimulus that AI is having in the market. These changes are being implemented as a sense-giving function of the organizational leaders and spokesmen that recognize the need of staying competitive in the market. The impact on the labor force *status quo*, layoffs and company culture are presented, aligning the interviewee's opinion with the current theories on company culture, offering also an analysis on the tendencies of this revolution and summarizing the interviewee's perspectives.

Keywords: Technology · Revolution · Virtual assistants · Labour force

1 Introduction

Barely a day goes by without a reference to Artificial Intelligence (AI) or/and Machine Learning (ML) by the media. All it takes is accessing Facebook's web page and chances are that there will be a video in a news feed of the new intelligent robot performing some sort of "jaw-dropping" action or perhaps there will be news of a new interview with the humanoid robot Sophia. The business and employment-oriented

Á. Rocha et al. (Eds.): WorldCIST'19 2019, AISC 932, pp. 336–347, 2019.
https://doi.org/10.1007/978-3-030-16187-3_33

service *LinkedIn*, which at the time of writing counts with 562 million users [1], displays more than 12 million Artificial Intelligence related job opportunities. Media companies like *Forbes*, *The Guardian* and *The New York Times* discuss the ethics of AI [2–4] manifesting their interest in the new technological era. AI and ML are here to stay and everybody wants a piece of them. The impact that this kind of technology is having on our daily life is undeniable, whether we are students, engineers or house-keepers and, thus, there is no surprise as to how much we are exposed to this topic throughout our day. The market tendency is to make every product smart, intercon-nected and dependable so, by the time the general population will start to be aware of these new concepts most of the companies and firms that seek to thrive in tomorrow's world will have already started to reshape their own future. In this process some questions start to arise: *How are companies following these technological advances?*, *How is the organizational culture and identity changing within companies?* and *Where are companies investing in order to prepare for the upcoming competitive future?*. In this article, we will tackle these questions from the perspective of Altice Labs, a research and technology development company, located in Aveiro, Portugal, and owned by the Dutch multinational Altice Group. This article is structured as follows. In the current section, a brief introduction on the work done is presented. In the second section, the pertinent literature on the topic is reviewed emphasizing the themes of Organizational Identity and Culture. The third section introduces the background story and development of Altice & Altice Labs as companies. In the fourth section, the specific procedures used to obtain and analyze information about the usage of AI in Altice Labs are specified. The fifth section showcases the data retrieved from the interview conducted. The sixth section presents the discussion and the analysis done concerning the data obtained. Finally, section seven concludes the main questions of the research study and proposes future research topics framed with this work. In summary, this article aims to infer the impact that the usage of AI will have on the organizational identity and culture of technological companies.

2 Literature Review

Social identity theory draws from the psychology arena, namely on how "an individ-ual's identity is derived from being part of a group" [26, p. 405]. Groups need to be helped to form a collective identity, a task which should befall the leader [26]. The definition of organizational identity is a controversial theme among scholars and researchers which gives rise to different theories and statements. Ravasi and Schultz [7] suggest that these definitions can be divided in two major perspectives on Organiza-tional Identity: Social Actor versus Social Constructionist.

2.1 Social Actor Perspective (E.g. Firms Which Adapt with Difficulty)

According to this view, organizational identity resides in a set of institutional claims that explicitly states the views of what the organization is and what it does represent. One may see "the unique nature of organizations as distinct actors with emergent person-alities and enduring qualities" [30, p. 324]. The aim is to provide "perceptions of central, enduring and distinctive features of the organization by providing them with legitimate and consistent narratives that allow them to construct a collective sense of self" as stated in Czarniawska's [8] and Whetten and Mackey's [9] studies. These identity claims that are organizational self-definitions proposed by organizational leaders and spokesmen are nominated as the *sense-giving* function of the organization as stated in Albert and Whetten's [10] and Whetten's [11] studies. Haslam et al. see "the social actor perspective as relating to the central and enduring attributes of an organization that indicate what category (of organizations) it belongs to" [30, p. 324]. Furthermore, the proponents of the social actor perspective, like Whetten and Mackey, in their study back in 2002 [9] "observe how deeply held beliefs, embodied in formal claims, tend to change only rarely and never easily" [7] (as human beings are adverse to change [26]). Thus, external events that challenge an organization's claim of identity often result in a counter-action in order to preserve personal and external representations of what the organization is or stands for [7]. Organizational identity is seen as binding, central, enduring, and distinctive [30], which may be linked to dysfunctional processes with "damaging consequences for both individuals and collectives" [30, p. 324].

2.2 Social Constructionist Perspective (E.g. Open-Minded Companies)

On the other hand, "realities are socially constructed" [29, p. 458], and multiple scholars' studies like Corley and Gioia's study in [12], Dutton and Dukerich's study in [13], Fiol's study in 2002 [14] and Gioia and Thomas's study in 1996 [15] point out that the members' beliefs about the identity of their organization may indeed change when confronted with internal and external stimulus. Human beings are very susceptible to suggestion and to outside events and may change their views, in fact, very quickly, often without knowingly doing so. Therefore, in this view there is a collective understanding of what the identity of the organization is and, thus, as concluded in Gioia's study in 1998 [16] the organizational identities reside in the shared perspective schemes that members collectively construct in order to provide meaning to their experience. In constructionist and postmodern assumptions, [identity] change is "understood as a continuous, emergent, self-organized, and socially constructed process [...] based on the perspective that change could be created, planned, and managed against ideal models" [29, p. 458]. Ashforth and Mael's study in 1996 [17] adds that these schemes may or may not correspond to their organization's official narrative. These shared understandings are the results of *sensemaking* processes where the members interrogate themselves on central and distinctive features of their organization. This means that, as pointed out by these scholars, "the shared beliefs are subjected to periodic revision, as organizational members modify their interpretations in light of environmental changes" [7]. Therefore, "emerging "new OD [Organizational Development] practices" (Marshak and Grant 2008), in contrast, [are] more dialogical or conversational in the sense that they [provide] spaces for new social agreements to be created" [29, p. 458].

2.3   Identity Changes in View of Environmental Changes

Albert and Whetten's seminal article from 1985 [10] advocates that external pressures (such as those felt today, including increased competitiveness and technological change [27]) increase the likelihood of organizational members engaging in reflection on identity issues - "in terms of their core assumptions and organizing principles" [30, p. 320]. The line of argument was extended by Huff (cited in [7]) by giving more importance to the interpretation of said pressures by organizational members as a source of stress. Furthermore, studies such as Bouchikhi and Kimberly's study in 2003 [18] and Brunninge's study in 2004 [19] claim that substantial environmental changes (e.g. technological development, including those which affect jobs, by way of automating traditional functions) may threaten the sustainability of organizational identity (in the case of layoffs, for example). A study conducted by Humphreys and Brown, in 2002 [20], claimed that members are likely to reject new conceptualizations that they perceive as incoherent with organizational history, tradition, and their sense of self, along with the changes they are expected to promote. Studies like these fore-shadow a relationship between organizational identity (which may be a personal perception [30]) and culture (e.g. shared norms [30]).

2.4   Organizational Identity and Culture

In the article Three Bell Curves, Rosauer claims that the best definition of organizational culture is the one developed by professors and organizational performance experts Deal and Kennedy, and states the following: "Culture is the way things get done around here" [21]. Although there are multiple definitions for this concept, [7] defines it in a more scientific way as "a set of shared mental assumptions that guide interpretation and action in organizations by defining appropriate behavior for various situations". The studies of Schein in 1992 [22] and Trice and Beyer in 1984 [23] further claim that these largely tacit assumptions and beliefs are expressed and manifested in a web of formal and informal practices and of visual, verbal, and material artifacts, which represent the most visible, tangible, and audible elements of the culture of an organization. One model of organization culture that stood out among scholars was proposed by Edgar Schein (a pioneer in the organizational culture field, who stated further that founders and leaders greatly influence culture [26]): The Iceberg Model [24]. Schein believed that there are three levels in an organization culture:

- **Artifacts and Symbols (easier to access [26]):** The first level is the "above the water level". These are the things that are clearly visible, that can be heard as well as felt by individuals and which are collectively known as artifacts [22, 25]. "The dress code of the employees [e.g. formal or casual], office furniture, facilities [e.g. open space, promoting sharing and proximity, or closed offices, which promote individualism and privacy], behavior of the employees [e.g. punctual, dedicated, and respectful of superiors, rather than gossipy, laid back and where firm policies are not followed], mission and vision of the organization all come under artifacts and go a long way in deciding the culture of the workplace" [24].
- **Values (harder to access [26]):** The second level is the "below the water level", and it is not as visible as the artifacts. According to Schein, the mindset of the

individual in any given organization (e.g. where there is individual and personal responsibility, and a mindset where individuals want to do the right thing, even if that means disagreeing with superiors and being insubordinate, as one will try to change the boss's mind), will influence the culture of the working environment, thus, what the individuals actually think, what they value and how they carry themselves (their attitudes) play an essential role in the defining of the organizational culture [22].

– **Shared Assumptions (deeper down and the hardest to access [26]):** The third and last level is the "deep water level". This level accounts for the employees' shared assumptions (e.g. that ideas come from individuals, who are capable of doing the right thing, and where no single individual is smarter than the group [22]), assumed values, beliefs and facts that are hard to measure, see and cognitively identify in everyday interaction between organizational members. "Organizations where female workers dominate their male counterparts do not believe in late sittings as females are not very comfortable with such a kind of culture. Male employees on the other hand would be more aggressive and would not have any problems with late sittings" [25].

The major difference between organization identity and culture is defined in [7] as: "while organizational culture tends to be mostly tacit and autonomous and rooted in shared [and mostly automatic] practices, organizational identity is inherently relational (in that it requires external terms of comparison) and consciously self-reflexive". Organizational identity is the interpretation of the organizational members, who are trying to satisfy their search for meaning, and this interpretation is deeply influenced by their activities and beliefs, which are grounded in and interpreted using cultural assumptions and values.

3   Altice & Altice Labs

Altice is a Dutch multinational that provides services of telecommunication, contents, media, entertainment and publicity in several countries such as Portugal, France, Belgium, United States of America and Israel. Owned by Patrick Drahi, Altice was founded in 2001 and by 2017 it was worth 50 billion euros. In 2015, Altice publicly announced the acquisition of Portugal Telecom (debt free) bought from the Oi group for the value of 7.4 billion euros showing once again Altice's strategy for growth by making price-disciplined acquisitions [5]. Portugal Telecom was once a case of great success in the European telecommunication market. Unfortunately, Portugal Telecom faced serious problems due to corruption and debt accumulation before being bought out by Altice. In 2016 the former Portugal Telecom, and now Altice Portugal, announced the transformation of the former PT Inovação into Altice Labs, which would become responsible for the innovation and development of new solutions, technologies and trends in the telecommunications area of all Altice group. Altice Labs is now the company of the Altice Group which core business is the development of new IT solution that add value for companies in the group as well as competing in the internal and international market promoting processes of innovation of services, technologies and operations [6]. To this point, Altice Labs already won several awards such as the Dell EMC Award in 2018, the international Technology Leadership Award in 2017 and the NGON Award in 2017, proving the quality of it's R&D department. Altice Labs is in a privilege position that only a few companies are: by being part of a bigger and successful telecommunication group, it has a direct access to the international market, doesn't have to worry as much about the company survival (because it is cherish by the group), has a truly faithful wallet of customers (all the other companies in the group) and has the support and means to preform high tech R&D to support numerous services in different countries. All of above facts conjugated with the access to cheap and qualified work force make Altice Labs a unique interesting company in today's market.

4 Methodology

After researching on the topic, the authors conducted a semi-structured interview with an expert on the area. For this purpose an interview script was made with questions that would lead to the obtainment of information about the following key topics: *The effects of technological advances on companies*; *The way cultural identity and the method of work is changing within companies*; *Where companies are investing in order to stay competitive in the future.* Jorge Sousa, the Head of M2M/IoT & API Management & Virtual Assistant - Network and Service Solutions in Altice Labs, with a bachelor degree in Automation and a Master's in Economics and Innovation management, from Faculdade de Economia do Porto, was the go-to person in order to obtain information about the case study, since his team's work orbits, among other areas, around AI, bots/virtual assistants and its ramifications. His work path started in 2001 in Altice Labs, previously known as PT Inovação, and was the former Division manager - Head of SmartCities and Development Team Leader before getting to his current position in the company. The interview, performed on the 18th of October, 2018, was recorded and transcribed in order to retain the full information. The redundant parts were filtered out in order to provide clear answers to the main questions of this study.

5 Interview

This interview was conducted at the Altice Labs facilities during approximately one hour with the interviewee Jorge Sousa. In order to be more legible the questions are subdivided into three subchapters that correspond to the major topics of this study.

5.1 How Are Companies Following These Technological Advances?

Regarding the first topic, which was to raise awareness about how companies are following these new technological advances, the interview provided surprising information about what was to be expected. "These products don't present AI as people think" was the first sentence stated by Jorge Sousa when asked about the services provided by Altice Labs that are known to use AI. The majority of the products do not use the AI concepts so much, using instead AI tactics to allow them to process information more easily when there are a large number of information vectors.

On the other hand, Altice Labs has been working for one year and a half on Virtual Assistants, a purely AI dedicated project, which has the main purpose of facilitating work in companies. "Despite the fact that there are already products like Alexa and Google Home, these are not easily transferred into small and medium-sized enterprises (Portuguese PMEs - *Pequenas e Médias Empresas*, which stands for small or medium size companies). There is difficulty for companies to have access to channels like these." In this way, Altice Labs is adapting the products used in domestic situations to something else that can be used at work and to make a company more productive.

"In companies it is not hard to think that there is going to be a virtual assistant helping humans to program, providing advice, and it is to be expected that in the future, workers will have a more pleasant experience, by decreasing time spent on bureaucracy and allowing people to be more focused on creative content." These advances in AI technologies are promoting a more evolved future, with more commodities and where people are going to work more in what they really appreciate. "From my point of view AI is going to be a technological transformation that can even create a new economic cycle due to the impact it is having" stated the interviewee, who firmly believes that all companies will start to introduce more and more intelligence into their products and services that will potentiate a new hype in technology.

5.2   How Is the Cultural Identity and Method of Work Changing Within Companies?

"Like when electricity was created, and like the first wave of industrialization, the first impact will be on the more traditional jobs" stated Jorge Sousa when asked what he thought about AI constituting a threat to the *status-quo* of the labor force, since AI will start by performing these same services. "Alongside with these jobs, the mechanical and procedural workers will have to leave their comfort zones and improve their skills or take the risk of losing their jobs." Although this view sounds pessimistic, he believes that there is going to be a large number of opportunities appearing. "This has been happening in every new cycle, society has always more yield than before and there were still jobs for people." About a somewhat fearsome idea, namely AI quality of service outperforming humans, the main idea of the interviewee is that, unlike a human, who has highs and lows, a machine will always have the same specified throughput. "From the moment that a machine repeats the same work N times, does not send text messages to a friend and is never distracted, the quality level will be the one it was programmed with." On the other hand, the fact that Humans are not consistent, being more productive in some phases and less in others, is what allows them to be so innovative, and is what has brought humans so far. "The major asset of a company is and will always be the people working on it." This is what distinguishes every single company, so Humans will always be on top of the equation even in an era of intelligent machines.

5.3   Where Are Companies Investing in Order to Prepare for the Upcoming Competitive Future?

"No one sells AI just for itself. We see Microsoft and Amazon working on machine learning but not much more." Unlike these companies, the market space for Altice Labs is in its products to evolve and integrate into themselves AI capabilities to be competitive with the market because all the other companies are doing the same. It ceases to be a matter of developing more value for the company and becomes a compelling need in order to keep up with the competitiveness among each other. Slowing down in this moment, when the AI development is starting, could mean losing the race and in the near future having products which are just not right for the current market. Leading the topic into universities, not only is "the university training in Portugal one of the best" but also "the labor market in Portugal is very competitive" which ends up leading companies to have a great offer in terms of students with specific knowledge. Because of the competitiveness of the work market in Portugal and the lack of brand equity, which is much more present abroad, the resource costs are more affordable than in other developed countries, which leads to more investment from the part of companies like Altice Labs into the Portuguese Market. This noticeable increase on AI education branches has been seen with good eyes from the companies, since the current workers did not have the opportunity to study AI during their superior education neither possess a background on these topics. This is a reason for Altice Labs to invest in strategic seminars and projects alongside universities. "This has two main purposes, the first is to show what is being done inside Altice Labs, promoting the brand and motivate the future workers to apply for a job at Altice. And the second is to try to influence the universities path, showing the importance of Machine Learning and other new areas, and the importance of investing in those areas."

6   Discussion of the Results Achieved

Despite the huge interest in AI and the development of new products and services that differentiate a company from their competition, companies that are joining the "AI race" are focused on expanding the features of their existing services and/or products by reinventing and transforming what the consumer already knows to exist, instead of trying to create something new, from scratch, using AI.

However, these new features are also being used to adapt and improve their workplaces, in an attempt to outperform their competition. Pure AI products include virtual assistants and bots that would work in parallel with human employees and after interviewing the Altice Labs' Head of Virtual Assistants we can affirm that the future is arriving sooner: the predicted shift in the *status quo* of the labor force is already happening, due to the introduction of these assistants. Open-minded companies, i.e., that follow a *Social Constructionist Perspective*, are already incorporating these assistants in their working methods, in an attempt to maximize the strategic advantages enabled by these solutions, due to the fear of being outperformed by the competition, in an era of accelerated technological change [27].

While these changes look promising, there is already the awareness among top executives that AI will replace several jobs and force layoffs, ultimately completely realigning company visions and forcing a new workplace culture and working methods. However, this impact can be seen through an optimistic lens, when comparing it to all the industrial revolutions of the past. Therefore, we must not fear it [26], but embrace it, accepting the transformational force that it brings to the workplace and to the quotidian activities of a business [31, 32]. Mundane tasks, such as writing code or analyzing data, can be easily automated and the interviewee sees this as a way for "(…) workers to have a more pleasant experience, decreasing the time spent on bureaucracy and allowing people to be more focused on creative content". The future workplace will praise creativity, innovation and content creation [28], instead of repetitive tasks, bureaucracy and secretariat.

While some companies are already embracing this change, many others will start to follow, provoking a radical shift in company values and culture. Nevertheless, companies that follow a *Social Actor Perspective* of their own organizational identity will have a hard time adapting themselves, since they will most likely see the adoption of AI tools in the workplace as a threat to their established work methods, job security and organizational culture, fearing layoffs due to the automation of several tasks.

Staying true to their core values and working methods, in an era of massive changes both in company culture and working methods, will dictate a company's downfall in this new industrial era. In the opinion of Jorge Sousa, failing to keep up with this race for the inclusion of AI in services, products and in the workplace will be a major setback from which large and especially well-established companies will have a very hard time recovering from, if they can recover at all. By the time that AI development starts to be exponential, large companies will fail to adapt and innovate, undermining their possibility to deliver desirable products to the newly created future market.

Furthermore, not only companies are being competitive, but also universities are competing amongst themselves. This "race for AI" is also acknowledged by universities, while undertaking an attempt to establish education programs for the future workplace [33] and also to train new students with skills oriented for AI and machine learning, strategically aligning them with companies' needs, such as is the case of Altice Labs and the University of Aveiro. In this growing market, Portugal presents a unique opportunity due to the high quality of Portuguese higher education and the competitiveness of the Portuguese market, generating the conditions for affordable and highly qualified technical personnel which companies can take advantage of to stay competitive and innovative. This is one of the main reasons for Altice Labs' continuous investment in R&D in Aveiro and for the development of closer relations between Altice and the University of Aveiro.

7   Conclusions and Suggestions for Future Research

In this case study a particular view is given on a very mediatic topic: artificial intelligence and the new industrial revolution. However, instead of focusing on the advantages, market impact and advances that will be possible, this case study presents a bold approach on the transformative force AI brings to the workplace, by seeing it as a new employee. This new approach to this topic unveils the possibility to discuss pressing matters on the subject, such as the impact of AI on the business culture and workplace organization.

Let us state that "it is important to accept that the construction of any social reality is marked by constant tensions between forces for stability (traditions of truth and "taken for granted" assumptions) and forces for transformations (new shared meanings)" [29, p. 471].

External events that challenge the *status quo* of the labour force are already happening in market-leading companies, changing the shared beliefs of the work force. These shared beliefs are deconstructing the organizations' official narrative, rebuilding the company culture for the future workplace, where several jobs will be completely automated. After interviewing the Altice Labs' Head of Virtual Assistants, we can conclude that the increasing interest in AI and in its capabilities, despite all the impacts on services and products, is having a tremendous role in rewriting the company working methods, accelerating the inevitable revolution of the *status quo* of the labour force and company culture. The tendency for the future workplace is to set aside repetitive or bureaucratic work, leaving AI in charge of those tasks while cooperating alongside with human employees. The awareness that AI will replace several jobs and force layoffs is also forcing a change on the job profiles of the future, rewarding more creative, innovative and content creation capabilities.

As has happened in the past, an industrial revolution has the possibility to change all of society for the better. While the fear associated with machines outperforming humans, people losing their jobs, impersonalization and alienation of the working environment, Jorge Sousa's final message is reassuring: "the major asset of a company is and always will be the people working in it". Despite all the tools, capital, market penetration and resources, there are the people and their ideas that truly represent the biggest asset of any company. Following this last premise, for future research it would be interesting to evaluate how employees react to the possibility of working with virtual assistants side-by-side, in order to further develop this study on the impact of AI on the organizational culture of a company. Are workers able to relate to AI colleagues and to feel empathy and companionship with them? Human beings are also a source of conflict and thus of added stress and so do AI workers present benefits in so far as they are not a source of conflict? The relations between human coworkers may change due to AI. Will human employees in the future complain to AI about a colleague who is always late to meetings, or who does not do his or her work properly, for example? A longitudinal study at a firm such as Altice Labs to measure these effects is suggested.

References

1. LinkedIn by the Numbers: Stats, Demographics & Fun Facts (n.d.). Accessed 24 Oct 2018. https://www.omnicoreagency.com/linkedin-statistics/
2. Lets Talk About AI Ethics; Were on a Deadline (n.d.). Accessed 24 Oct 2018. https://www.forbes.com/sites/tomvanderark/2018/09/13/ethics-on-a-deadline/#72aeddf42e21

3. The Guardian View on the Ethics of AI: It's About Dr Frankenstein, Not his Monster — Editorial — Opinion — The Guardian (n.d.). Accessed 24 Oct 2018. https://www.theguardian.com/commentisfree/2018/jun/12/the-guardian-view-on-the-ethics-of-ai-its-about-dr-frankenstein-not-his-monster
4. Artificial Intelligence - The New York Times (n.d.). Accessed 24 Oct 2018. https://www.nytimes.com/topic/subject/artificial-intelligence
5. Altice's Business Model Is Not Sustainable - ALTICE S A ADR (OTCMKTS:ATCEY) — Seeking Alpha. (2017). Accessed 25 Oct 2018. https://seekingalpha.com/article/4123284-altices-business-model-sustainable
6. História: A evolução histórica da empresa — Altice Portugal (n.d.). Accessed 27 Oct 2018. https://www.telecom.pt/pt-pt/a-pt/Paginas/historia.aspx
7. Ravasi, D., Schultz, M.: Responding to organizational identity threats: exploring the role of organizational culture. Acad. Manag. J. **49**(3), 433–458 (2006). https://doi.org/10.5465/AMJ.2006.21794663
8. Czarniawska, B.: Narrating the Organization: Dramas of Institutional Identity. Bibliovault OAI Repository. The University of Chicago Press, Chicago (1997)
9. Whetten, D.A., Mackey, A.: A social actor conception of organizational identity and its implications for the study of organizational reputation. Bus. Soc. **41**, 393–414 (2002)
10. Albert, S., Whetten, D.A.: Organizational identity. In: Cummings, L.L., Staw, B.M. (eds.) Research in Organizational Behavior, vol. 7, pp. 263–295. JAI Press, Greenwich (1985)
11. Whetten, D.A.: A Social Actor Conception of Organizational Identity. Brigham Young University, Provo (2003, unpublished manuscript)
12. Corley, K.G., Gioia, D.A.: Identity ambiguity and change in the wake of a corporate spin-off. Adm. Sci. Q. **49**, 173–208 (2004)
13. Dutton, J., Dukerich, J.: Keeping an eye on the mirror: image and identity in organizational adaptation. Acad. Manag. J. **34**, 517–554 (1991)
14. Fiol, M.C.: Capitalizing on paradox: the role of language in transforming organizational identities. Organ. Sci. **13**, 653–666 (2002)
15. Gioia, D.A., Thomas, J.B.: Identity, image and issue interpretation: sensemaking during strategic change in academia. Adm. Sci. Q. **41**, 370–403 (1996)
16. Gioia, D.A.: From individual to organizational identity. In: Whetten, D.A., Godfrey, P.C. (eds.) Identity in Organizations: Developing Theory Through Conversations. Sage, Thousand Oaks (1998)
17. Ashforth, B.E., Mael, F.A.: Organizational identity and strategy as a context for the individual. In: Baum, J.A.C., Dutton, J.E. (eds.) Advances in Strategic Management, vol. 13, pp. 19–64. JAI Press, Greenwich (1996)
18. Bouchiki, H., Kimberly, J.R.: Escaping the identity trap. Sloan Manage. Rev. **44**(3), 20–26 (2003)
19. Brunninge, O.: Translating strategic change in companies with strong identities: the dynamics of identity and strategic change at Scania and Handelsbanken. Paper presented at the 20th Egos Colloquium, Ljubljana, Slovenia, 13 July 2004
20. Humphreys, M., Brown, A.D.: Narratives of organizational identity and identification: a case study of hegemony and resistance. Organ. Stud. **23**, 421–447 (2002)
21. Rosauer, B.: Three Bell Curves, January 2013. http://www.threebellcurves.com/THREEBELLCURVESCust1st41415.pdf
22. Schein, E.H.: Organizational Culture and Leadership. Jossey-Bass, San Francisco (1992)
23. Trice, H., Beyer, J.: Studying organizational cultures through rites and ceremonies. Acad. Manag. Rev. **9**, 653–669 (1984)

24. Schein's Model of Organizational Culture - apppm (n.d.). Accessed 29 Oct 2018. http://apppm.man.dtu.dk/index.php/Schein\%27smodeloforganizationalculture#TheIcebergModel
25. MSG.: Edgar Schein Model of Organization Culture. https://www.managementstudyguide.com/edgar-schein-model.htm. Accessed 30 Nov 2018
26. King, D., Lawley, S.: Organizational Behavior, 2nd edn. Oxford University Press, Oxford (2016)
27. Moreira, F., Au-Yong-Oliveira, M., Gonçalves, R., Costa, C. (eds.) Transformação digital – Oportunidades e ameaças para uma competitividade mais inteligente [Digital transformation – Opportunities and threats for a more intelligent competitiveness]. Sílabas & Desafios, Faro, Portugal (2017)
28. Au-Yong-Oliveira, M., Moutinho, R., Ferreira, J.J.P., Ramos, A.L.: Present and future languages – how innovation has changed us. J. Technol. Manag. Innov. **10**(2), 166–182 (2015)
29. Aguiar, A.C., Tonelli, M.J.: Dialogic organization development and subject– object dualism: a social constructionist perspective on dialogic methods in an organizational context. J. Appl. Behav. Sci. **54**(4), 457–476 (2018)
30. Haslam, S.A., Cornelissen, J.P., Werner, M.D.: Metatheories and metaphors of organizational identity: integrating social constructionist, social identity, and social actor perspectives within a social interactionist model. Int. J. Manag. Rev. **19**, 318–336 (2017)
31. Branco, F., Gonçalves, R., Martins, J., Cota, M.: Decision support system for the agri-food sector–the sousacamp group case. In: New Contributions in Information Systems and Technologies, pp. 553–563. Springer (2015)
32. Martins, J., Goncalves, R., Pereira, J., Oliveira, T., Cota, M.: Social networks sites adoption at firm level: a literature review. In: 2014 9th Iberian Conference on Information Systems and Technologies (CISTI), pp. 1–6. IEEE (2014)
33. Au-Yong-Oliveira, M., Gonçalves, R., Martins, J., Branco, F.: The social impact of technology on millennials and consequences for higher education and leadership. Telematics Inform. **35**, 954–963 (2018)

The Role of AI and Automation on the Future of Jobs and the Opportunity to Change Society

Manuel Au-Yong-Oliveira[1,2(✉)], Diogo Canastro[1], Joana Oliveira[1],
João Tomás[1], Sofia Amorim[1], and Fernando Moreira[3,4]

[1] Department of Economics, Management, Industrial Engineering and Tourism,
University of Aveiro, 3810-193 Aveiro, Portugal
{mao,diogocanastro,jrfo,joaotomas.va,
soamorim.ferreira}@ua.pt
[2] GOVCOPP, Aveiro, Portugal
[3] IJP, REMIT, Universidade Portucalense, Porto, Portugal
fmoreira@upt.pt
[4] IEETA, Universidade de Aveiro, Aveiro, Portugal

Abstract. In today's world, technology is an indispensable part of our daily
life. More and more products are produced to satisfy the needs of a growing
population and the Internet is creating new services every day. However, to be
able to keep up with the growing demand, new technologies needed to be
invented to increase the pace of production and to lower costs. Automating tasks
was the solution and for many years machines did repetitive tasks, replaced
people and created new and better jobs to substitute the old ones. Nowadays
automation is reaching incredible levels and it is not creating enough jobs to
replace the old ones. Will unemployment increase in the upcoming years or will
humanity be able to adapt to a different job market? Our study focuses on the
types of unemployment caused by automation and on the possible solutions that
society needs to implement. The research also points out how people from
different social classes face the changing job market and how all can benefit
from it. A new form of governance of society may be needed, in view of
previous failed forms including fascism, communism, and, more recently, liberalism. In the future there will be fewer and fewer jobs that cannot be replaced
by a robot. With this, the probability of mass unemployment is very high. An
ideology promoting a Universal Basic Income could be a solution to combat the
massive unemployment that automation may cause in the future, especially in
medium-skilled jobs.

Keywords: Automation · Industry · Unemployment · Artificial Intelligence

1 Introduction

In the last 100 years, Automation has been applied in several kinds of fields such as
services, industry or personal devices. Many inventions were created to replace human
labor and others to improve jobs that already existed. In factories, robots were
developed to do repetitive handiwork without making human errors and increasing the

© Springer Nature Switzerland AG 2019
Á. Rocha et al. (Eds.): WorldCIST'19 2019, AISC 932, pp. 348–357, 2019.
https://doi.org/10.1007/978-3-030-16187-3_34

pace of production. In our homes, we can watch a movie without the need to find one. With the simple pressing of a button, Netflix or another streaming platform will advise us as to what movie we should watch, using its powerful algorithm. And while we are shopping at the supermarket, we can scan the bar code of the products we want and pay at the exit without the need to get the products out of the shopping cart – putting them on the cash register mat, waiting for the employee to scan the bar codes and putting the products in the shopping cart again. In sum, automation benefits our life in multiple ways, making our tasks easier, faster and more convenient.

But talking about automation, we cannot forget about an important part of this subject: Artificial Intelligence (AI). We can define AI as a part of computer science that develops systems capable to perform tasks that required human intelligence and are able to adapt to different scenarios [1]. Although a big portion of people do not know how to define what artificial intelligence is, they come into contact with it every day and it is not a strange thing to them.

Automation and AI have changed our way of life and are more present than ever before in our daily tasks. Human kind is getting more and more dependent on new technologies that are able to do tasks with a minimum of interaction with people. Machines have taken over tasks that were entirely done by people. Although this is a very good improvement in our daily life that can help us get more time to do other things, from a job point of view it is becoming a big problematic discussed by numerous specialists.

Automation appeared with a single purpose: to let machines perform repetitive and monotonous tasks. This gave the opportunity for people to transit from low skill jobs in the industry to medium and high skill jobs in services and even in the same industry.

But adding Artificial Intelligence to the equation makes this subject even more complex. Machines have become so smart that they can mimic the more complex tasks that many years ago most people did not think was possible. Various specialists have different opinions on what will happen in the future. The majority think that unemployment will reach a new high and others think this will offer better jobs to everyone.

For example, according to a study done by Boston University, since the first implementation of the Automated Teller Machine (ATM), the number of human tellers has grown considerably [2]. These numbers raise an important question. Why automation in this sector has not eliminated or, at least decreased, the number of employers? And although the first ATM was installed in 1967, can we compare it with today's and future automation innovations?

In this article, the authors will explain this battle and present their own perspective.

2   Literature Review

As was already explained, there is a conflict between automation and general employment. If the machines can do our work, why are we still working? And will this happen the same way in the future?

Every company needs workers with a multiplicity of skills. With only automated systems, it is not possible to achieve this goal. It emphasizes the need for human work. Since there is a dependency on the control of machines, the value of specialized people increases.

According to David Autor, we can use two economic principles to explain this fact. The first principle is the O-ring principle. It analyses human creativity and capability [3]. As an example, we can think about the implementation of the first Automated Teller Machines (ATM). It had two different consequences on employment. As we can imagine, in the beginning, there was a lot of hand work being replaced by ATMs. These machines handle cash tasks faster.

With this transformation, banks started to open new branches because of the lower costs. More branches meant more tellers but with different work. They started a new business and adapted their jobs. Instead of doing the cash-handling tasks, they bet on a relationship with customers, by choosing a more cognitively demanding job and introducing new products (for example, investments or credit cards) [2].

By analyzing the teller's reaction, it is possible to comprehend human economic value. Since machines were not able to relate to the customer, the importance of human work and the capability to adjust to market changes was clear (problem-solving and judgment skills).

This recognition of the importance of human work is present in many jobs. A professor can be substituted by online/digital classes, but it will never compensate the relationship between a student and a teacher. It will not draw the student into the passion of learning and the exchange of knowledge.

The evolution of automatization magnifies the importance of human labor by selecting the type of work that can only be done by people. When only the first principle is used, the importance of human work is easily known, but it is not possible to know how many people are needed to do the job. Here is when the second principle is needed, the Never-get-enough principle. It analyses human insatiability [3].

Most of today's work only started to exist in the last century. But why? Every day, there is a constant need to look for new ideas, new products, new technologies or even new services. People always want more and are willing to invest time and money to find or create this [4, 5]. This greed supports innovation in companies and creates more types of careers. Not just the ones that control the machines and the automatization but also the creative and marketing jobs. They need to be the first to capture their consumer's attention with innovative products.

So, if we accept these two principles, does it mean that there is an employment problem? No. We need to take into consideration the evolution of automatization and innovation.

Alex Williams, in his article called "Will Robots Take Our Children's Jobs?" wonders if Artificial Intelligence could make countless professions obsolete in a very near future [6]. Professions like radiologists, lawyers and surgeons are being threatened by intelligent machines. Arterys, a start-up company, has created software that can perform a magnetic-resonance imaging analysis of blood flowing through a heart in 15 s while a human needs 45 min to perform this [7]. And are professions like journalism safe in The Information Age? According to another article the answer is no. The Associated Press already used a program to produce a copy covering Wall Street earnings and some college sports [7]. A study done in 2013 by the Department of Science Engineering of the University of Oxford estimated that 47% of current jobs will be victims of automation, in a period of 20 years [8]. By 2033, new jobs certainly will have been created, but will they be enough to compensate for the jobs that were eliminated?

Because of this changing environment, there is also the problem that society does not know what are the skills that we should teach to children because the skills that are needed today, are not the same as those that will be needed in 20 or 30 years from now [9–11].

3   Methodology

Initially, a work study was conducted in order to analyze the current situation in the job market caused by automation and to identify the main causes of unemployment.

To arrive at conclusions about automation in the industry field, a questionnaire was created, having as a target audience the lower working-class in industry. The authors aimed to study the knowledge of the workers about the famous concept "Industry 4.0", that is, if they were aware of what tasks automation could do and the impact that it could have on employability in the upcoming years.

For this to be able to happen, companies were contacted, and the authors asked their permission to distribute the questionnaires to their employees. Six companies in the center/north region of Portugal were contacted, which asked to remain anonymous, but didn't give the permission needed. One of the persons who provided an explanation referred that by asking those questions to the operators it could have a negative effect on their perception of new automated technologies that the company would implement in the future and cause some discomfort in the work environment.

With this, it was realized that asking this type of questions could easily have a big impact on the satisfaction of factory workers and that those people might suspect of future investments of their company in automated jobs.

Since it was not possible to make the questionnaire, the authors chose to analyze more scientific articles and more case studies with real examples from different companies.

The authors started to analyze which jobs are more at risk of being substituted by machines. After this, it was necessary to investigate what people think about new technologies in their work space and how they adapt to a changing environment. Various future possible scenarios were already thought of by numerous experts and the ones analyzed were those that the authors think are the most probable. A possible solution was also discussed.

4   A Theoretical Model

There are three types of employment. The first is high-wage jobs with high-education. It represents work like that of engineers, sales and marketing managers, programmers or doctors. The second is middle-class jobs like operative positions. The low-skill is the last type and it represents low-education jobs (for example, cleaning service or home health aides).

According to David Autor, at TEDex Cambridge in 2017, the high-education and high-wage jobs are very likely to not be replaced by machines and even increase the number of jobs available. Regarding medium-skill jobs, the number available is decreasing and making people with not so much education getting jobs that don't require as much skills as they have. Those middle-skills jobs are getting substituted by machines and algorithms that understand the procedures needed to get the job done [3].

The best jobs are becoming more and more specialized. The first robots were created to substitute repetitive operations and predictable tasks. With the evolution of automation, this is no longer true. In 2017, Google's AlphaGo Software defeated a nineteen-year-old Chinese master at Go, considered the most complicated board game in the world that not even the best players can explain what they are doing [10]. This means that machines are capable of doing more types of jobs, and people need to invest more in their education and specialization if not to be replaced my machines.

According to Frey and Osborne [8], that distinguish jobs between high, medium and low risk occupations, came up with the following graphic to explain what is going to be probability of computerization, regarding the kind of jobs that were mentioned below.

In sum, Frey and Osborne [8] show that the high skill jobs like Management, Business, Finance, Computers, Engineering, Science and Services are amongst the sectors with lower probability of computerization. On the other hand, for the low skill jobs like Transportation, Material Moving, Production, Sales, Office and Administrative Support the probability of computerization is higher, since they are related to simple, monotonous and easy tasks that can easily be replaced by automation.

And what about Portugal? According to an article from Bruegel, and cited by the World Economic Forum, around 60% of jobs are at risk of suffering computerization or automatization. This is a high percentage compared to less than 47,5% in the UK and Sweden [13].

To try to solve this problem that will affect all countries in the world even if they are underdeveloped, the authors found various articles that talk about what are the opinions of people about the implementation of new automated machines in their work space.

According to Melonee Wise, CEO of Fetch Robotics, when a company presents a new machine or new software to their employees, there are usually 5 stages of acceptance [14]:

1. The first stage is Fear. This happens because people think that those robots are there to substitute them, but most of the time they are there to make their jobs better.
2. The second stage is apprehension. The employees see that the robots will not replace them but think that they will not be qualified to interact with them.
3. The third stage is curiosity. After the employees start to work with them, they wonder what tasks the robots can do.
4. The fourth stage is tolerance. In this stage, the productivity increases because employees are more used to the new machine.

5. And last stage is satisfaction. Usually workers like the new job experience and it facilitates their job.

This might be true in this situation, but in other situations the robots are there to substitute workers and not to help them. Look at the example of Amazon Go. Amazon launched three new supermarkets without any employees that works completely autonomously and eliminates the need for hiring workers [15]. In this case automation might be good for employers and clients but it is not good for workers.

According to Viktor Weber, Founder & Director of the Future Real Estate Institute, there are 4 possible scenarios for this problematic [12]:

1. Automation might automate only some tasks and in other cases human labor would do the jobs. This could happen if human labor is cheaper than investing in new machines.
2. Automation would still evolve, but with limitations. Society would limit the autonomous machines in areas were human labor is an important part and it is more valuable.
3. Humans would focus on an "experience-based economy" where tasks like cooking or handicraft are more valued in society instead of engineers or economists.
4. Automation will not have limitations and will automate every task possible. Only one small part of people would profit with this and a big part of society would be unemployed. Social inequality would increase dramatically.

5   Discussion

With the First Industrial Revolution (1820–1840), human work got easier and productivity rose. Services and products could be produced in a shorter time while using the same number of workers. While this innovation eliminated a lot of jobs, at the same time a lot were created. It should be mentioned that those jobs were better than the previous ones due to their complexity. Workers put down manual agricultural tools to start handling big machines that could do the exact same job much faster and in a lighter way.

In sum, innovation led to higher productivity, fewer old jobs and many new and often better jobs. Overall, this worked well for most people and living standards improved. Meanwhile, humans shifted into service jobs and only a few decades ago in human history, the Information Age happened. Suddenly, the World changed its rules and our jobs are now taken over by machines much faster than they were in the past.

5.1   Innovation

Nowadays, there are new information industries appearing and growing. But, comparing to the past, they are creating less new jobs. To comprehend this problem, we analysed the car industry. One hundred years ago, cars transformed the way of life and everything around us. Everybody could find a job either directly or indirectly to this industry. Over time, there was a lot of very considerable investment until it became largely complete and nowadays, cars don't create as many jobs as was supposed, even if looking at electric cars that are a recent innovation. By comparison, in 2004, Blockbuster made $6 billion US Dollars a year and had 84.000 workers, while Netflix with 4.500 employees made $9 billion US Dollars in 2017 [16].

5.2 New Machines and New Jobs

Over thousands of years, human jobs became more and more specialized. While even the smartest machines are bad at doing complicated jobs, they are extremely good at doing defined and predictable tasks. That's the reason why so many jobs are disappearing.

Machines are becoming so good at breaking down complex jobs into simple ones, that for a lot of people, there will be no further room to specialize.

Machines teach themselves because we make this possible by giving a computer a lot of data about something people want to become better at. As an example: if a person shows to a machine all the things she bought online, it will slowly learn what to recommend to her, so she buys more things. What humans created by accident, is a huge library that machines can use to learn how humans do things and learn to do them better, by giving them information and records about everything that happens. Because of this, digital machines might be the biggest problem of all and destroy the most jobs.

It is not enough to substitute old jobs with new ones. There is a need to generate new jobs constantly because the world population is growing, and in the past, mankind has solved this through innovation. People were sure that with rising productivity, more and better jobs would be created.

This time, the nature of innovation in the Information Age is different from everything that was encountered before. Self-driving cars have been created. Self-working robots have been created. Speaking robots have been created. And so on. It seems like innovation is going too far.

5.3 A Solution that Can Change Society

In the future there will be fewer and fewer jobs that cannot be replaced by a robot. With this, the probability of mass unemployment is very high. And what can one do about it? It is a massive social challenge that humankind is not prepared to face.

Various experts have already suggested a solution that we consider that is a great solution for the massive unemployment that automation may cause in the future. That solution is called Universal Basic Income (UBI).

UBI is considered the most ambitious social policy of our times. According to the Basic Income European Network (BIEN), UBI is a way to improve our social structure. Each person would receive an amount of money periodically without the need of work or of any other form of payment. That amount of money would be enough to cover all basic expenses needed for the person to survive above the poverty line [17].

This brings many questions and there are many skeptics about it that say that people would stop working and spend that amount of money on unnecessary items. Tests run in Canada in the 1970s showed that around 1% of the recipients stopped working, mostly to take care of their kids. On average, people reduced their working hours by less than 10% and the extra time was used to achieve goals like going back to school or looking to get better jobs [18].

The easiest way to pay for a UBI is to end all welfare and use the free funds to finance it. Although this can be a good solution, it could also leave many people worse off than before. The second way to make this possible is higher taxes especially for the very wealthy. Nowadays the gap between the rich and the poor is rapidly increasing and UBI would maybe solve this problem [19].

UBI would be a solution to combat the massive unemployment that automation may cause in the future, especially in medium-skilled jobs. However, if everyone receives a basic income, would people still want to have low-skilled jobs that are often bad and unpopular? Some say that UBI might give them enough leverage to demand better pay and working conditions. There would still be poor and rich people, but the conditions of living of the poor would be substantively better.

So, is UBI a good idea? The honest answer is that no one knows. There needs to be a lot more research and experiments to know the right answer. Society needs to think about what kind of UBI it wants and what it is prepared to give up to pay for it. The potential is huge. It might be the most optimistic model to sustainably eliminate poverty. It might seriously help humanity to be much less stressed out.

6   Conclusions and Suggestions for Future Research

In a world where digital technologies are so present in people's daily lives, it is difficult to imagine a world without them, although not many years ago, most people lived in rural areas and their livelihood was from agriculture and handicrafts. In the last 100 years, the world has changed dramatically, and the pace of innovation will increase even more in the upcoming years.

Most people when they think about technology think about the products and services that they interact with in their daily lives. Most people do not worry about the technologies that are being implemented in factories and services that automate completely tasks that before were done by people.

In our view, we perceive that automation will create a high percentage of unemployment that will affect mainly people with an education. People with a high- and low-level education will not be affected as much as people with a medium-level education. Low skill jobs that do not require any specialization and usually are the worst paid jobs will increase and the differences between them and high-skill jobs will also increase causing a massive inequality between those two social classes. This will happen because those jobs are cheap and implementing machines or automating their tasks will not be economically worth it.

In our point of view, there is no doubt that automation will change the job market even more than it has already changed it. Although this will implicate serious problems, society should not give up on this. Automation is a good thing. Humankind just needs to find a way to adapt.

The Information Age could be a huge opportunity to solve poverty and the biggest problems facing society. Mankind just needs to think in a faster and different way. And that is one of the problems of Automation: although there is already a notion about what automation can do as regards unemployment, there is no major concern about it. People think this is a subject that will happen in many years from now, but they are wrong. This future is already here.

In our point of view, UBI would be a solution that could help solve many problems and not only the mass unemployment caused by automation. Tests are already ongoing in Finland and many developed countries, like Iceland, are already discussing it. However, that is not enough. There should be more tests and analysis about this solution and in an even faster manner. Although the mass unemployment caused by automation and the solution that is UBI are already discussed throughout the world, they are not discussed at the pace that they should be to accompany the transformation in industry and services.

A new form of governance of society may be needed, in view of previous failed forms including fascism (which did not endure the II World War), communism (which failed in the 1980s), and, more recently, liberalism (which has shown signs of failing, with Brexit and the rise of Donald Trump, in the USA) [20]. Globalization and liberalism have benefited only small elites, leaving the masses at the mercy of new waves of technology, which threaten their professions and livelihood [20]. An ideology promoting a UBI could be a solution. Capitalism thus allied to a form of widespread social responsibility may be a way forward for mankind.

Will Automation continue to benefit society and will UBI be the right solution to solve the problems caused by it? No one knows. But we hope that with our discussion we have provided information necessary for future research and showed that solutions need to be found faster.

References

1. Artificial intelligence | Definition of Artificial Intelligence in English by Oxford Dictionaries. https://en.oxforddictionaries.com/definition/artificial_intelligence. Accessed 27 Oct 2018
2. Bessen, J.E.: How computer automation affects occupations: technology, jobs, and skills. Boston University School of Law, Law & Economics Working Paper No. 15–49 (2016)
3. Autor, D.: Why are there still so many jobs? The history and future of workplace automation. J. Econ. Perspect. **29**(3), 3–30 (2015)
4. Martins, J., Gonçalves, R., Pereira, J., Cota, M.: Iberia 2.0: a way to leverage Web 2.0 in organizations. In: 2012 7th Iberian Conference on Information Systems and Technologies (CISTI), pp. 1–7. IEEE (2012)
5. Branco, F., Gonçalves, R., Martins, J., Cota, M.: Decision support system for the agri-food sector–the Sousacamp group case. In: New Contributions in Information Systems and Technologies, pp. 553–563. Springer, Cham (2015)
6. The New York Times: Will Robots Take Our Children's Jobs? https://www.nytimes.com/2017/12/11/style/robots-jobs-children.html. Accessed 21 Oct 2018
7. Arterys Company Profile, MedTech Innovator. https://medtechinnovator.org/company/arterys/. Accessed 06 Nov 2018
8. Frey, C.B., Osborne, M.A.: The future of employment: how susceptible are jobs to computerisation? Oxford Martin Programme on Technology and Employment, Oxford Martin School, University of Oxford Working Paper, September 2013

9. Harari, Y.N.: The Rise of the Useless Class (2017). https://ideas.ted.com/the-rise-of-the-useless-class/. Accessed 16 Oct 2018
10. Au-Yong-Oliveira, M., Gonçalves, R., Martins, J., Branco, F.: The social impact of technology on millennials and consequences for higher education and leadership. Telematics Inform. **35**, 954–963 (2018)
11. Martins, J., Branco, F., Gonçalves, R., Au-Yong-Oliveira, M., Oliveira, T., Naranjo-Zolotov, M., Cruz-Jesus, F.: Assessing the success behind the use of education management information systems in higher education. Telematics and Inform. (2018, in press)
12. The New York Times: Google's AlphaGo Defeats Chinese Go Master in Win for A.I. (2017). https://www.nytimes.com/2017/05/23/business/google-deepmind-alphago-go-champion-defeat.html. Accessed 19 Oct 2018
13. Mesnard, X.: What happens when robots take our jobs? World Economic Forum (2016). https://www.weforum.org/agenda/2016/01/what-happens-when-robots-take-our-jobs/. Accessed 21 Oct 2018
14. Wise, M.: The 5 Stages of Acceptance as Robots Enter the Workforce, World Economic Forum (2018). https://www.weforum.org/agenda/2018/10/robots-are-coming-to-your-workplace-here-s-how-to-get-along-with-them/. Accessed 19 Oct 2018
15. Wingfield, N., Mozur, P., Corkery, M.: Retailers Race Against Amazon to Automate Stores, The New York Times (2018)
16. Feldman, D.: Netflix has Record-Breaking Fourth Quarter in 2017, Exceeds $11B in Revenue, Forbes (2018)
17. Basic Income Earth Network: What is Basic Income? https://basicincome.org/basic-income/. Accessed 23 Oct 2018
18. Flowers, A.: What Would Happen if We Just Gave People Money? FiveThirtyEight (2016). https://fivethirtyeight.com/features/universal-basic-income/. Accessed 20 Oct 2018
19. Straubhaar, T.: On the economics of a universal basic income. Interecon. Rev. Eur. Econ. Policy **52**(2), 74–80 (2017)
20. Harari, Y.N.: 21 Lessons for the 21st Century. Jonathan Cape, London (2018)

Smart Cities and Smart Tourism: What Future Do They Bring?

Ana Matos[1], Bruna Pinto[1], Fábio Barros[2], Sérgio Martins[2],
José Martins[3,4], and Manuel Au-Yong-Oliveira[1,5(✉)]

[1] Department of Economics, Management, Industrial Engineering and Tourism,
University of Aveiro, 3810-193 Aveiro, Portugal
{anaclaudiamatos, pintobruna, mao}@ua.pt
[2] Department of Electronics, Telecommunication and Informatics,
University of Aveiro, Aveiro, Portugal
{fabiodaniel, sergiomartins8}@ua.pt
[3] INESC TEC and University of Trás-os-Montes e Alto Douro,
Vila Real, Portugal
jmartins@utad.pt
[4] Polytechnic Institute of Bragança - EsACT, Mirandela, Portugal
[5] GOVCOPP, Aveiro, Portugal

Abstract. We have sought to understand the current state of the art on smart tourism and on smart cities. Furthermore, we have sought to understand community awareness and the will to embrace innovation, as they are decisive factors to acquire base knowledge and overcome barriers in (soon to be) overpopulated cities and for those who are looking for a limited time culture experience - known as tourists. We live in an age where technology is increasingly present in our lives and provides us solutions to societal problems. Problems such as traffic, infrastructure and natural resources management, or even increasing citizens' participation in governance, bringing them closer to decision-making. The objective is to understand the current level of people's knowledge about the impact that technologies have on the society in which we live and their perception of the usefulness in solving these same problems. Therefore, an anonymous questionnaire was carried out (176 valid answers were received), as well as a focus group with two experts on the Smart Cities subject. What future is brought by those who live and breathe technology? Are people willing to accept a paradigm shift?

Keywords: Smart cities · Smart tourism · Smart guides · Technology · Sustainability

1 Introduction

Smart technologies reach into nearly every aspect of life nowadays, although they are often unnoticed by users and taken for granted most of the time. Cities and tourism are not an exception.

Today, cities are experiencing a severe resource crisis since the number of people living in a city far outnumber the rural population. Thus, it is not possible to continue

© Springer Nature Switzerland AG 2019
Á. Rocha et al. (Eds.): WorldCIST'19 2019, AISC 932, pp. 358–370, 2019.
https://doi.org/10.1007/978-3-030-16187-3_35

managing cities as they were managed in the last century. It is therefore necessary to find medium- to long-term alternatives to this new reality. As such, the concept of "smart city" arises, which is defined by the use of technology to improve urban infrastructure and make urban areas more efficient while improving the quality of life for their inhabitants. Studies [8, 10] consider the existence of six essential pillars in a smart city; smart economy, smart mobility, smart environment, smart people, smart governance and smart living [10].

Smart tourism is one sub category of the smart cities' pillars. It is included within smart living and described as:

"It responds to the requirements of the present global and mobile elites by facilitating access to tourism and hospitality products, services, spaces and experiences through Information and Communication Technology based tools and where a healthy social and cultural environment can be found through a focus on the city's social and human capital. On the other hand, it also implements innovative and entrepreneurial businesses and fosters the interconnectedness of businesses" [16].

This concept is particularly relevant in the Portuguese economy, since tourism is one of the most important sectors of the country, represents more than 5% of the national Gross Domestic Product [3] and is one of the strategic activities of a developing country. Today's tourist is increasingly dependent on information and communication technologies, growing more demanding and informed, looking forward to new emotions and a new variety of experiences and creating new trends of consumption.

The increase in mobile computing and the popularization of mobile devices has enabled the development of innovative applications in several areas such as smart guides for tourism. Being a sub-category of smart tourism, these suggest personal and more relevant feedback according to the preferences and location of their users. Thus, it aims to replicate a human tourist guide.

To better understand the awareness level and the kind of challenges to be faced in the Portuguese community, on the topics of both smart cities and smart tourism, a questionnaire was publicly launched online, and a focus group meeting was performed with experienced people in the technology field. This allowed for the development and structuring of the current article, to provide relevant insights on what the future holds.

2 Literature Review

2.1 Smart Cities

According to data provided by World Urbanization Prospects [11, 13, 14], the majority of people currently reside in cities, especially since 2010 [14]. The trend is to increase, with cities being prepared for this new reality. As mentioned by Mitchell [15], the infrastructure of cities faces difficulties due to the fact that more than half of the world population lives in cities, thus it is necessary to find solid long-term solutions; in addition, cities cannot continue to be managed as they used to. According to the United Nations (*UN*), urban populations will grow by more than 2 billion people over the next 40 years [15] and may become less attractive if there is excessive, untidy construction and no spatial planning, which will lead to social, environmental, energy efficiency, transport, and public and private service problems [15]. Besides that, cities only represent 2% of the Earth's surface, but its inhabitants consume more than ¾ of the world's natural resources and produce more than 60% of all carbon dioxide [24]. Hence, there is a need for changes in infrastructure, in the distribution of goods and services, as well as regarding people's overall concern for environmental sustainability.

Accordingly, there comes the concept of "smart city" as a solution to these problems. This concept involves sustainability, energy, waste and traffic management, integration between public systems, and provision of services to the entire population, such as education, health, transportation and energy [8, 10]. All these aspects must be managed in an integrated and intelligent way in order to create synergies aimed at improving the quality of life of the residents while creating efficiency in the services of a given city. Moreover, smart cities also focus on the future, emphasizing the importance of resource and sustainability application for future generations [8].

According to Hollands [11, p. 10], there are four factors that have dominated the different designs and projects of smart cities in implementation around the world, and they are:

- "Focus on information and communication technologies and network infrastructures";
- "Market-induced urban development, with cities shaped by large multinational corporations";
- "Emphasis on technology intensive industries";
- "Concern about environmental sustainability, including energy issues".

The use of technologies is a means for the management of intelligent cities. However, it is not only the adoption of intelligent systems and other technologies that transform a city into a smart city; it is necessary to maintain a certain level of cooperation between technologies, management and improvement of a social economic system, according to Blanco [15]. Information and Communication Technologies (*ICT*) should be used as a new central nervous system potentiator for the management of smart cities, state Gupta and Toppeta [15]. It is also necessary for citizens to interact with governance in order to find solutions to the challenges large urban areas face, improving the quality of life and respecting the environment.

The report of the University of Technology in Vienna, the University of Ljubljana and the University of Technology in Delft, focused on the development of a ranking of medium-sized smart cities [8, 10], considers the existence of six essential pillars in a smart city (as illustrated in Fig. 1): Smart Economy, Smart Mobility, Smart Environment, Smart People, Smart Governance and Smart Living [10].

Smart economy refers to the economic competitiveness of cities, integrating issues associated with innovation and the emergence of new services that lead to the growth and development of the economy. The local accessibility of cities and the *ICT* network are the main components of smart mobility, such as traffic reduction and the adoption of electric transport with benefits related to CO_2 emissions. The smart environment is defined by the attractiveness of natural conditions, environmental protection and resource management, such as waste, water and energy management in order to achieve more efficiency [10]. Smart people relate to the degree of qualification of human resources, with the openness and level of social interactions [10]. Well-informed and

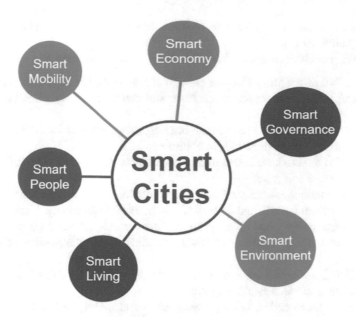

Fig. 1. The six pillars of Smart Cities (based on [8]).

prepared people to deal with the city in an "intelligent" way are essential to achieve the desired efficiency in the implementation of those systems. Smart governance, on the other hand, encompasses aspects related to public participation, services provided to citizens, the functioning of the public administration and greater interaction between citizens and governance. Finally, the intelligent way of life integrates several issues related to the quality of life, such as culture, health, safety, tourism and housing [10].

2.2 Smart Cities in Europe

A study was performed on smart cities in Europe, carried out by the Policy Department A: Economic and Scientific Policy of European Parliament, whose analysis is based on the alignment between the objectives and characteristics of each city's project portfolio and the Europe 2020 objectives [9].

Initially, the 468 EU-28 cities with more than 100,000 inhabitants were considered. Through a research process, the level of smart city activity present in each selected city was studied. Based on this initial analysis, 240 EU-28 cities with verifiable smart city activity were identified and from this group, a sample of 50 smart city initiatives were examined in 37 cities. Within this sample, stakeholders, funding and scalability of the initiatives were observed. To explore the relationship between intelligent cities and Europe 2020, relevant evidence was gathered in a structured panel, but the sample of cities used was restricted to 20, due to data limitations. Afterwards, a quantitative survey was carried out to analyse the alignment between smart city initiatives within the sample of 20 cities and Goals of the Europe 2020 Strategy. Finally, a restraint was made on a range of innovative implementation strategies in the six major cities that run the smartest city projects to identify cross-cutting themes and potentially replicable smart city solutions [9].

The main conclusions drawn from this study were as follows [9]:

- In 2011, 240 of the 468 EU-28 cities with at least 100,000 inhabitants (51% of the total) had at least one smart city feature and can therefore be classified as smart cities;
- There are more small smart cities than big ones; also, smart cities can be found in all size categories and in most EU-28 countries;
- Larger cities tend to have better-structured projects, with at least one fully launched or implemented initiative;
- The most common of the six features are intelligent environment and intelligent mobility, present in 33% and 21% of the initiatives, respectively. Each of the other four characteristics is addressed in approximately 10% of smart cities;
- The size of the city is clearly positively correlated with the number of smart city initiatives;
- Smart living initiatives are found across the EU-28, while initiatives focused on other features are less evenly distributed;
- Smart governance and mobility projects are found mainly in Northern Europe.

2.3   Portugal and Smart Cities

Portugal, although not one of the most advanced countries in Europe in this area, already integrates several projects of smart cities.

The latest project, to be implemented in the city of Aveiro, is a smart bus which, in addition to driving itself, also allows access to the Ultra TV content management platform, developed by Altice Labs in partnership with the University of Aveiro [5]. This project also counts on the participation of Ericsson, CarMedia and EasyMile [5]. The bus accommodates six people and uses 5G technology in conjunction with artificial intelligence. The future objective is for passengers to be able to engage in leisure or professional activities during their travels [5].

Another project that is under implementation in Portugal within the scope of smart cities is the Sharing Cities programme [6, 7], which has three pilot cities in Europe, Lisbon being one of these cities. The project seeks to develop affordable solutions that result in commercial provisions with high market potential for smart cities. It bles citizen involvement and cooperation at a local level, enhancing trust between cities and citizens. The changes taking place in the capital of Portugal are to the extent of building rehabilitation, shared electric mobility services, energy management systems, smart lighting poles and an urban sharing platform involving citizens. The goal is that by the end of the project, the changes will extend to the rest of Europe [6].

2.4   Smart Tourism

Smart tourism is one of the characteristics of smart cities and an important part of the six pillars lying at the base of this concept; considering that tourism is an important sector in the economy of many countries, even of primary importance in some cases, smart tourism can support sustainable development tourism and has the potential to be of impact in tourist destinations [21, 25].

"A smart destination is an innovative tourist destination, consolidated on a state-of-the-art technological infrastructure that guarantees the sustainable development of the tourist territory, accessible to all, which facilitates interaction and integration of the visitor with the environment and improves the quality of their destination experience" Stated by Segittur [22, p. 1].

Gordon Philips defined smart tourism as *"simply taking a holistic, longer term and sustainable approach to planning, developing, operating and marketing tourism products and businesses"*. Additionally, smart tourism is modelled by two main techniques [12]:

- Smart searching and usability. To manage research and access content;
- Smart marketing techniques. Used to segment clients to deliver personal and customized feedback.

Another concept of smart tourism emerged from the Organization for Smart Tourism in the United Kingdom and was formulated by Jannie Germann Molz, who identified a relation between smart tourism that links digital mobility to create more intelligent and sustainable connections between tourists and their destination [12].

Mobile technology seems to have accelerated this trend further through the pursuit of information, communication, social networks, as well as mobility-related features to help travellers wherever they might go [21].

2.5   Smart Guides for Smart Tourism

Smart guides for tourism are applications that, using the tourist's location and preferences, provide services that are customized and appropriated to the environment in which they are located, such as nearby places and locations that suit certain areas of interest - monuments, historical and cultural areas. These guides aim to mimic the human tourist guide through building relationships between knowledge-based systems, looking to provide a professional service that best meets customers' needs and the desire of gaining sufficient information and an objective understanding of the places visited together, adding value to the tour and raising the level of satisfaction regarding the overall experience [1].

Ergo, it may be stated that today's tourism is close to the concept of "custom-tailored" activities, being personalized rather than a general tourism offer. Increasingly, customers create their own experiences and personalize the places they wish to visit. Tourists are prosumers - responsible for many of their own goods and services [2]; this phenomenon is strongly boosted by the technological development we are witnessing today.

This type of applications targets the tourist and not the resident, although both benefit from the process. Meaning, it fits both the tourist and resident, changing according to the need, but serving everyone; always working towards the satisfaction of tourists and improving the quality of life of residents.

These smart guides exemplify one of 2018's trends, according to TrendWatching - assisted development, since they provide its users with behavioural guidelines and accessibility of information. An example of an application that produces travel guides in a mobile format with content culturally adapted to the tourist is the JiTT.travel. It offers a customized high-quality experience, transforming users' smartphones into smart tourist guides, giving inputs based on their desired journey according to current location and available time frame to be spent in each location. It is absolutely free and comes with a huge feature: it works offline! It was nominated by World Tourism Association as the "best mobile application designed to enrich tourists' experience" [4], working alongside the tourism development market and taking advantage of the growing need to serve tourists.

3 Methodology

To evaluate the general awareness on smart cities and smart tourism as contemporary topics, a questionnaire and a focus group session were conducted. The questionnaire, public and anonymous, contained around twenty questions, split into two different sections (see Table 1). The first section was related to personal data, with closed questions, and the second section was referring to the analysed theme.

Table 1. Questions and answers of the questionnaire.

Question	Summary of the answers
Do you agree on applying new technologies in tourism?	• 94.9% - Agree • 5.1% - Do not agree
Why?	Innovation, less third-party dependencies, better experience, easier to access and accessible at any time
Do you know any smart tourism applications?	• 17.7% - Yes • 82% - No
Have you ever used any of the applications you know?	• → out of the 17.7% that know any application: • 20.4% - Yes • 79.6% - No
Have you ever heard that tourism may be replaced with virtual reality?	• 51.9% - Yes • 48.1% - No
Which one do you prefer?	• 20.9% - Virtual reality guidance • 79.1% - Human guidance
Why (Virtual reality guidance)?	→ out of 20.9% that prefer virtual reality guidance: Cost efficiency, privacy, no third-party dependencies
Why (Human guidance)?	→ out of 79.1% that prefer human guidance: Employment stability, human interaction, cultural experience

(*continued*)

Table 1. (*continued*)

Question	Summary of the answers
In your personal opinion, does virtual reality guidance impact tourism positively or negatively? (*)	• 39.2% - Negatively • 60.8% - Positively
Why (negatively)?	→ out of 39.2% that said it negatively impacts: Unemployment, may not be qualified to personal requirements, no human/cultural interaction
Why (positively)?	→ out of 60.8% that said it positively impacts: Availability, cost efficiency, multi-language, the guidance given may support profile customization
Would you use virtual guidance if recommended by any friend/family?	• 25.4% - I would • 10.2% - I would not • 64.3% - Probably
Do you know any location using smart tourism technologies?	• 9.5% - Yes • 90.5% - No

(*) - Yes or no question to gather information on which side people would support more (positive or negative impact). Giving no neutral option to further discuss choices made.

Distinguishing between gender, age and education qualifications was crucial in order to identify common patterns. That was the first part of the study. Being more generalistic, composed by both closed and open questions, the second section provided some insight on subjects such as common knowledge on the smart tourism topic and community willingness to embrace new technologies in tourism.

3.1   Summary of the Field Work

Table 1 shows a summary of the questionnaire questions and answers (176 valid answers were received). We can see that technology and tourism are closely linked, in the perspectives of the respondents, and in effect there has been a very significant boom in tourism due also to technology. Albeit, most respondents (82%) do not know any smart tourism applications. Furthermore, the sample is divided on whether "tourism may be replaced with virtual reality?" Human guidance and human interaction are much preferred over virtual reality (79.15%), though virtual reality guidance is seen to impact tourism positively (60.8%). Virtual reality is seen to lead to unemployment, a negative effect. As concerns the question "Would you use virtual guidance if recommended by any friend/family?", 25.4% would, and 64.3% probably would – which shows the receptivity to this sort of technology.

Relatively to the focus group (see Table 2), we had the pleasure to set up a session with João Costa and Diogo Correia, who welcomed us at Ubiwhere.

Table 2. Information about the interviewees in the focus group.

Interviewees in the focus group	Gender and age	Job at Ubiwhere	Academic qualifications
Diogo Correia	Male, 23 years old	Smart Cities Manager (which involves developing business strategies, sales management and providing insight on smart cities solutions)	Master's in Industrial Engineering and Management; Currently a PhD student
João Soares da Costa	Male, 30 years old	Marketing Manager (the goal is to make Ubiwhere more attractive for customers and employees alike, through a simple and focused brand)	Bachelor's degrees in Public Administration and Marketing; MBA in Marketing Management

The focus group was based on the following goals:

- To evaluate the relevance of related projects within organizations;
- To understand more about the user experience concerning smart technologies;
- To retain a personal point of view from people that develop technological solutions and use technology on a daily basis.

The script had six questions, discussed during the focus group meeting:

- Ubiwhere works on smart city-based solutions. Tell us more about them.
- Which cities are currently using these solutions? And where specifically are they most used (concrete locations/places)?
- Are any of the solutions connected to smart tourism?
- Do you have any thoughts on smart tourism agents?
- Pros and cons of applying technologies around cities and the challenges faced while working on such solutions?
- Opinion on "Tourism may be replaced by virtual reality".

4   Discussion of the Fieldwork

4.1   Questionnaire

The questionnaire had a clear input on how people in general are likely to embrace new technologies around their cities. Despite it being noticeable how new this topic is in the community, the majority tend to accept and support its positive impacts.

The 45+ year-old public were the only ones with a full acceptance rate (100%) in the questionnaire, accepting new technologies in their cities and in tourism itself with more ease. Furthermore, every one of them stated that they did not know any smart tourism applications, nor have they used any, which led us to retain that people with the most willingness to accept and acknowledge existing applications are not the ones using them the most - there was no direct relationship.

Generation X, which is now between 30 and 45 years old, tends to be less willing to embrace new changes, even though this trend is also changing [23]. Having been born and having grown up along major technological development, generation X is known for its human interaction and multi-tasking qualifications. Living for work, family and friends, this generation embraces with more ease applications on related subjects.

Since gender was not as revealing as age, qualifications were also taken into consideration for further analysis. Unfortunately, no particular patterns were found.

It is clear how the community understands the statement "Have you ever heard that tourism may be replaced with virtual reality?", being totally divided (nearly 50%/50%). Furthermore, it is unanimous that a real person is preferred to serve as a guide rather than having virtual guides. People still need to relate to other people. Empathy, culture and specifically human contact along with a noticeable dismay by an increase in unemployment were mentioned.

While trying to understand if people, even those with none or minimal knowledge on the subject, were more likely to start using smart applications, if suggested by friends and family, it was noticed that even when suggested by the closest friends or relatives, there is still a "maybe" on the lips of the majority, affirming just how new and fresh the concept still is.

4.2   Focus Group

Considered a "Bleeding Edge Technologies" [17] organization and founded in 2007, Ubiwhere has been growing ever since. Their headquarters are currently in Aveiro, having moved there in 2017. They also may be found in Porto, Coimbra and even in Lisbon. Ubiwhere mainly provides smart cities and telecommunications solutions.

Related to smart cities, they are currently working on areas such as waste management, smart guides for wineries and wine exhibitions, and museums and parking traffic management. A wide variety of solutions are available for locations such as Guimarães, Aveiro and Porto. Note that one of their solutions for waste management, implemented in Aveiro, named PAYT (Pay as you Throw) [18], tends to support reducing taxes around communities that promote waste recycling. The more a user recycles, the less waste tax he will be paying, improving the environment and reducing individual expenses by doing good.

João and Diogo share a common opinion that smart cities solutions, being related to the improvement of the community experience while living in the city, and smart tourism solutions, aiming to improve the community experience going in and out of a city, are strongly connected. Both concepts are related, supporting each other directly or indirectly.

It was also impressive how professionals, who improve, use and work with technologies on a daily basis state that "Human interaction cannot be replaced". Both support technology usage and acceptance worldwide, but both think that culture is still of major value to the users' experience, not believing in the statement that "Tourism may be replaced by virtual reality". The simple act of taking a photo, human interaction and even the use of Google maps while connecting with people during a trip seem irreplaceable in the near future, and this can only strengthen the fact that tourism experiences and experiences in general should be "personal experiences" and not generalized. However, benefits such as cost efficiency and bringing tourism to everyone – the poor, the rich, the able and the disabled – were also mentioned.

As referred to and concluded at the smart cities congress during Techdays Aveiro 2018 [19], the biggest obstacles remain as being financing and political support. Having both, the last challenge will be obtaining community support, which is inevitable since such solutions will bring major benefits to the city. João, who was present at the congress, also mentioned that they themselves share these concerns, but consider – both being very positive individuals – that it is only "a matter of time" until they are overcome.

5   Conclusions and Suggestions for Future Research

With today's technological advancement, the population growth of cities (as illustrated in Graphic 1) and the subsequent natural resource shortage have given rise to the need to restructure some processes in order to efficiently manage available resources.

The smart cities concept presents itself as a viable answer aiming to support the restructuring of these processes, focusing on smart mobility, smart economy, smart people, smart governance, smart environment and smart living (as illustrated in Fig. 1). Being a relatively new concept and still rather less-known, cities are investing more often in related solutions, Sharing Cities [6, 7] and Aveiro Steam City [20] holding €24 million and €6.1 million in investments, respectively. Obtaining funds is still the major obstacle to be overcome, since such investments do not cover every focus point of the smart cities concept. Although just in the phase of initial investment, it will generate additional mid- and long-term funds resulting in higher sustainability levels, governance and living quality around cities.

An application of smart cities is smart tourism, which is subtracted from the smart living concept (as illustrated in Fig. 1). As smart living is related to the experience of living in the city, smart tourism, on the other hand, is related to the experience of going in and out of the city. This can cause the community to become weary, mainly due to the lack of knowledge and exposure to the subject. As well as smart guides, smart tourism aims to respond to customer needs by crossing data around personal preferences and current location. People in general are not using smart tourism applications to enrich themselves but as a way to smooth their tourism experience, meaning that people still prefer human contact and the whole cultural experience. "… culture is not replaceable. Pictures are not enough." – interviewee João Costa, October 2018, affirming that virtual reality is not going to replace tourism in a near future.

In conclusion, smart city solutions will be going ever further and they will overcome barriers such as financing and lack of political support. Although communities have a relatively low level of knowledge on the subject, exposure will be increasing, and people will embrace smart city technologies with more ease.

For future research, we suggest connecting investment in smart city solutions to tourism growth, to gather information on tourism preferences in cities using more technology. Furthermore, we need to understand smart city exposure issues and how to improve them, as well as community apprehension on technological paradigm embracement and the implication of data privacy around new solutions in smart cities.

Acknowledgements. We would like to thank João Costa and Diogo Correia for having shared their opinions, experiences and current projects with us. Furthermore, a thank-you is due to everyone else involved in the questionnaires who contributed to a deepening of our knowledge on the subject.

This work is financed by the ERDF – European Regional Development Fund through the Operational Programme for Competitiveness and Internationalisation - COMPETE 2020 Programme and by National Funds through the Portuguese funding agency, FCT - Fundação para a Ciência e a Tecnologia within project POCI-01-0145-FEDER-031309 entitled "PromoTourVR - Promoting Tourism Destinations with Multisensory Immersive Media".

References

1. Owaied, H.H., Farhan, H.F., Al-Hawamdeh, N., Al-Okialy, N.: A model for intelligent tourism guide system. J. Appl. Sci. **11**, 342–347 (2011)
2. Toffler, A.: The Third Wave. William Morrow and Company, New York (1980)
3. Pena, J.: O Turismo em Portugal: Oportunidades e Desafios. Lisbon: Seminar AESE, 27 May 2002
4. JiTT.travel. https://jitt.travel/pt/. Accessed 27 Oct 2018
5. Notícias de Aveiro: 5G 'made in Aveiro' promovido com mini bus autónomo. https://www.noticiasdeaveiro.pt/5g-made-in-aveiro-promovido-com-mini-bus-autonomo/. Accessed 27 Oct 2018
6. UE – Lisboa é uma das Smart Cities que vai guiar a Europa. http://www.lisbonne-idee.pt/p4132-lisboa-uma-das-smart-cities-que-vao-guiar-europa.html. Accessed 28 Oct 2018
7. Sharing Cities. http://www.sharingcities.eu/sharingcities/city-profiles/lisbon. Accessed 27 Oct 2018
8. Al Nuaimi, E., Al Neyadi, H., Mohamed, N., Al-Jaroodi, J.: Applications of big data to smart cities. J. Internet Serv. Appl. **6**(1), 1–15 (2015). A SpringerOpen Journal
9. Manville, C., Cochrane G., Cave, J., Millard, J., Pederson J.K., Thaarup, R.K., Kottering, B.: Mapping Smart Cities in the EU. Policy Department A: Economic and Scientific Policy, European Parliament (2014)
10. Selada, C., Silva, C.: Smart Cities in the European Agenda: Opportunities for Portugal. II Conferência de PRU, VII ENPLAN e Wokshop APDR: "Europa 2020: retórica, discursos, política e prática"
11. INTELI: Índice de Cidades Inteligentes - Portugal. INTELI (2012)
12. Li, Y., Hu, C., Huang, C., Duan, L.: The concept of smart tourism in the context of tourism information services. Tour. Manag. **58**, 293–300 (2017)
13. Caragliu, A., Del Bo, C., Nijkamp, P.: Smart cities in Europe. J. Urban Technol. **18**(2), 65–82 (2011)
14. World Urbanization Prospects - Population Division - United Nations. https://population.un.org/wup/. Accessed 06 Oct 2018
15. Guardia, S., Guardia, M.: An essay on smart tourist destinations. Revista Turismo Desenvolvimento **27**(28), 1305–1314 (2017)
16. Calisto, L., Gonçalves, A.: Smart citizens, wise decisions: sustainability-driven tourism entrepreneurs. In: Carvalho, L. (ed.) Handbook of Research on Entrepreneurial Development and Innovation within Smart Cities, pp. 20–43. IGI Global, Hershey (2017)
17. Ubiwhere | Bleeding Edge Technologies with Custom Research and Development. https://www.ubiwhere.com/. Accessed 16 Oct 2018
18. PAYT - Portugal. http://www.payt-portugal.com/. Accessed 20 Oct 2018

19. Techdays. Building our Future. https://www.techdays.pt/pt/techdays. Accessed 30 Oct 2018
20. Aveiro futurista. Dos moliceiros elétricos ao 5G, a cidade já quer ser mais que a 'Veneza do Portugal' - Tecnologia - SAPO 24. https://24.sapo.pt/tecnologia/artigos/a-aveiro-futurista-dos-moliceiros-eletricos-ao-5g-a-cidade-ja-quer-ser-mais-que-a-veneza-de-portugal. Accessed 30 Oct 2018
21. Gretzel, U., Koo, C., Sigala, M., Xiang, Z.: Special issue on smart tourism: convergence of information technologies, experiences, and theories. Electron. Mark. **25**(3), 175–177 (2015)
22. SEBRAE: Destinos Turísticos Inteligentes – Tecnologias de Informação e Desenvolvimento Sustentável. SEBRAE (2016)
23. The New York Times: Generation X More Addicted to Social Media Than Millennials, Report Finds. https://www.nytimes.com/2017/01/27/technology/millennial-social-media-usage.html. Accessed 29 Oct 2018
24. Climate Change – UN-Habitat for a Better Urban Future. https://unhabitat.org/urban-themes/climate-change/. Accessed 06 Oct 2018
25. Martins, J., Gonçalves, R., Branco, F., Barbosa, L., Melo, M., Bessa, M.: A multisensory virtual experience model for thematic tourism: a Port wine tourism application proposal. J. Destin. Mark. Manag. **6**, 103–109 (2017)

Personal Data Broker Instead of Blockchain for Students' Data Privacy Assurance

Daniel Amo[1], David Fonseca[1](✉), Marc Alier[2],
Francisco José García-Peñalvo[3], and María José Casañ[2]

[1] La Salle, Universitat Ramón Llull, Barcelona, Spain
{daniel.amo,fonsi}@salle.url.edu
[2] Universitat Politècnica de Catalunya, Barcelona, Spain
marc.alier@upc.edu, mjcasany@essi.upc.edu
[3] Universidad de Salamanca, Salamanca, Spain
fgarcia@usal.es

Abstract. Data logs about learning activities are being recorded at a growing pace due to the adoption and evolution of educational technologies (Edtech). Data analytics has entered the field of education under the name of learning analytics. Data analytics can provide insights that can be used to enhance learning activities for educational stakeholders, as well as helping online learning applications providers to enhance their services. However, despite the goodwill in the use of Edtech, some service providers use it as a means to collect private data about the students for their own interests and benefits. This is showcased in recent cases seen in media of bad use of students' personal information. This growth in cases is due to the recent tightening in data privacy regulations, especially in the EU. The students or their parents should be the owners of the information about them and their learning activities online. Thus they should have the right tools to control how their information is accessed and for what purposes. Currently, there is no technological solution to prevent leaks or the misuse of data about the students or their activity. It seems appropriate to try to solve it from an automation technology perspective. In this paper, we consider the use of Blockchain technologies as a possible basis for a solution to this problem. Our analysis indicates that the Blockchain is not a suitable solution. Finally, we propose a cloud-based solution with a central personal point of management that we have called Personal Data Broker.

Keywords: Blockchain · Smart contracts · Learning analytics · Educational data mining · Academic analytics · Data privacy · Digital identity · Moodle

1 Introduction

Learning Analytics has become a key tool for assessment [1]. The students' interactions within online learning environments are collected, processed by statistical models, and finally presented to teachers and other stakeholders. These results are growing in detail and complexity. The student's personal data is generated from interactions in Learning Management Systems (LMS) [2, 3]. That data is analyzed by algorithms and afterwards

© Springer Nature Switzerland AG 2019
Á. Rocha et al. (Eds.): WorldCIST'19 2019, AISC 932, pp. 371–380, 2019.
https://doi.org/10.1007/978-3-030-16187-3_36

visualized by stakeholders in dashboards [4]. This helps to provide insights about behavior and learning needs [5]. These enhanced results enable teachers to better understand the progress of students and other related educational context aspects [6], such as validity of resources or even assignments. Hence, analytics have become essential to understand the students, get actionable recommendations [7] or even predict patterns of learning [8].

This paper is structured in four further sections. In Sect. 2, we introduce the objectives, the research question and the methodology used. In Sect. 3 we introduce the fundamentals and argument that Blockchain is a novel technology which still has some security flaws that makes it an unreliable protocol in terms of data privacy. In Sect. 4, we introduce an alternative solution to ensure students' data privacy and its architecture in the cloud, which we call Personal Data Broker. Section 5 concludes and closes the paper and describes future works.

2 Methodology

In the Learning Analytics process, different tools are used to collect private, personal and highly sensitive data from students, even from minors. This data collection generates privacy issues related to data leakages and misuses. The collected data can be stored in unknown servers, making it possible for administrators without legal permissions to access them. Moreover, even if the logs could be stored in the same students' institution, the data inside the logs cannot be trusted due to possible alterations by system administrators. This context generates fears against the use of educational analytical approximations. Some data misuses such as inBloom schools case [9] highlight the importance of respecting students' privacy. They also manifest the urgency of finding a definitive and global solution to student data protection.

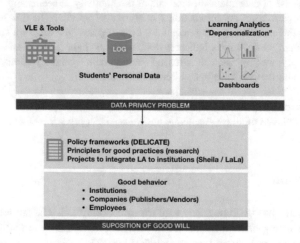

Fig. 1. Data privacy problem treated by law under supposition of good will.

Current solutions are delivered as policies, such as General Data Protection Regulation (GDPR) laws and regulations [10]. There are also research results in form of frameworks, good practice guides or principles [11–14]. Policies, regulations or guidelines do not correct the problem from its root because they do not prevent data leakages or misuses in real time. They are only applied when the problem has occurred and someone has reported it. Moreover, policies assume that the intentions of each of the stakeholder are in the interests of the students –see Fig. 1-, but reality shows that this is not always the case [5, 9]. Laws, regulations, and principles are not enough to assure data privacy in any kind of educational data log created by Learning Analytics, Academic Analytics, Educational Data Mining -or even other no analytical tools-. Therefore, there is a fragility in educational data privacy that needs to be addressed.

This situation can be solved by support the legislation with a layer of automation technology to enforce their policies and regulations in real time. This new approach could safeguard the privacy of students' data in its current and future uses. Some emerging technologies such as Blockchain seem to be strong candidates to ensure privacy and secure sensible data of students. The use of Smart Contracts inside Blockchain allows the automation of legal actions. They could be executed as soon as irregularities are detected in the use or collection of data. However, there is an ongoing academic debate and growing research on the security and privacy of the Blockchain [15–18]. The Blockchain approximation might not be ready to ensure data privacy in educational contexts.

Our hypothesis is that Blockchain is not the best technology to protect and manage safe access to personal data in the educational context. Consequently, the main purpose of this manuscript is double. On the one hand, we show the limitations of Blockchain as a possible solution to the problem posed. On the other hand, we present an alternative solution to Blockchain.

To achieve the objectives and validate the hypothesis, we propose to expose the weaknesses of Blockchain related to the privacy and data security dimensions in the educational context. As a guide to formulating a suitable methodology, we launch the research question: Is the Blockchain a suitable technology to ensure data privacy for students? We find publications that are motivated by the upward trend of Blockchain's use, advocate the use of it in education [19]. However, there are studies based on the very foundations of Blockchain that question it as a private and secure technology [20–24]. All of them repeat a series of security flaws that can be included in different categories: Identity filtering, Lack of transactional privacy, Private Key insecurity, Vulnerabilities in Smart Contracts, Discovery of user identities and Quantum computing.

3   Blockchain Fundamentals

Nakamoto [25] announced a cryptographic solution to a number of game theory problems. This would also allow the creation of a peer-to-peer electronic cash system. After that, Nakamoto open sourced the implementation of the solution for the cryptocurrency Bitcoin introducing Blockchain as a cryptographic networked platform. Since then, Blockchain has attracted different opinions in education. This platform implemented a distributed ledger managed by consensus with a four core characteristics [26, 27]. Such characteristics are both the strengths and weaknesses of Blockchain:

1. **Immutability:** When a block is added, it cannot be altered. This data persistence and irreversibility creates trust in the transaction ledger.
2. **Decentralized:** Each user of the network has a copy of all the transactions. All the data is distributed through the network and decentralized.
3. **Consensus Driven:** Data is validated through cryptographic proof-of-work. Hence, no central authority is needed nor trusted.
4. **Transparent:** All transactions are public and history records are accessed by anyone in the network. Hence, users' data is public.

3.1   Consensus and Permission-Less

Blockchain is permission-less by default. No one needs a special permission, nor a central authority, to access the Blockchain users' network. In a context with untrusted users, where data is decentralized and no central authority is required, a solution was needed to assure data integrity. Blockchain solved this problem. Each transaction in the Blockchain must be validated by consensus [25]. A network of peers that perform cryptographic calculations achieves this consensus when each peer validates the transaction by proof-of-work. The proof-of-work consists in solving a cryptographic algorithm using intensive computation power. If the majority of the network that has completed the proof of work says a transaction is valid, then it is. The peers validating transactions are called "miners" because they get monetary rewards depending on the computing power they commit – linked to the energy they consume – and randomness.

Although Blockchain seems to be a secure and private technology, there are some factors that support the criticism and skepticism it has received in education that make it unsuitable as a form of safeguarding data privacy. Blockchain is adequate for data and transaction trustworthiness, certification and validation by consensus. However, data privacy in Blockchain is not fully assured by design. Blockchain, as stated by Nakamoto, is only anonymous until the public key is shared. Therefore, Blockchain is pseudonymous and arises some insecurities about data privacy in it.

Considering that it is possible to link the public key to a real person, any user in Blockchain could discover real identities. Although users' data can be encrypted in Blockchain, users' real identity can be discovered through different attack techniques [28–30]. Hence, Blockchain is a novel technology that needs a new design approach and definition to assure users' data privacy in Blockchain.

3.2   Limitations in Permission-Less Architecture

Blockchain has a permission-less design. Such designs presents some limitations [28]:

1. **Sequential execution:** Blockchain platforms, such as Ethereum, can execute Smart Contracts. For each transaction, each Smart Contract is executed sequentially in each and every node of the network. In some cases, it can provoke a Denial of Service when the execution takes too long.

2. **Non-deterministic execution:** Non-deterministic execution does not assure the same results. In a Blockchain network, this can lead to ambiguous results.
3. **Execution on all nodes:** This is at odds with confidentiality, given that each and every node has to execute the smart contracts. These should only be executed in the granted contracts.
4. **Privacy-invasive:** Blockchain transactions and data are public. The ledger is distributed to all nodes, so all users have the same data. This situation is clearly privacy-invasive for many use-cases. Although data in blocks is encrypted, each and every user of the Blockchain network has the same data. This data can be exported and decrypted in the future where decrypting conditions could be optimal.
5. **Hard-coded consensus:** The consensus protocol is hard-coded to any Blockchain service. Thus, is very difficult, if not impossible, to change without any recoding or code refactoring.

These limitations could affect the performance and efficiency of possible automatic legal enforcements in the Blockchain. This automation will be executed in every node by smart contracts, which could result in a potential bottleneck. Moreover, the hard-coded consensus cannot be reprogrammed quickly enough to assure data trustworthiness. This could lead to a situation where data in Blockchain is not reliable. Hence, Blockchain is not suitable to automatically enforcing legal controls, principles and good practices for data privacy agreements.

3.3 Data Immutability

Data immutability in Blockchain is a problem due to:

1. **Privacy-invasion:** If each user of the Blockchain network has the same data and this data can be exported, better decryption conditions can release real data and expose it to non-permitted users.
2. **Bureaucracy in itself:** The users' network of Blockchain is the central bureaucracy in itself. The Blockchain exists if there are enough users to validate data, as in the networks of centered authorities.
3. **Need of a database:** If the word "Blockchain" can be substituted by "database" inside a text without losing any sense, maybe what is needed is a database instead of a Blockchain. Future needs may arise to change the data inside the Blockchain. These could be solved using links to data storage services such as Drobox or Google drive. However, links to data may change in the future. New services could appear with new and stronger encryption algorithms or current outdated data storage services y disappear.

Blockchain is about data immutability and unbreakable encryption algorithms. What will happen when ultrahigh computing systems are able to break encryption of such encrypted data? Standard cryptographic systems are known to be vulnerable [31]. Eventually, it will be impossible to ensure data privacy. Such a consequence will be unacceptable for any Blockchain user. The data will be immutable as long as we live in a computational power era in equilibrium with the hard-coded consensus protocol. Hence, data privacy is not assured.

4 Solution and Architecture

In the area of learning analytics, both digital artifacts and real devices are involved. Digital artifacts involve all virtual learning environments. Real devices involve all those that allow to take biometric data or measure body actions such as face detection, sweating, body position or even "graphological". Both could potentially collect students' private data. All collected data is stored in logs. These logs may be physically stored in the educational institution in which they are generated. However, depending on the platforms used, the data may be stored on the servers of the companies that provide them. These storages keep personal data of students, including minors. Therefore, data privacy in learning analytics is a sensitive issue –the same for other contexts where educational data is traced, collected, stored and used in some manners.

The fragility in data privacy is a problem within learning analytics. It generates fears and feelings against its use. In this sense, different proposals have been developed to ensure the privacy of data and ensure the digital identity of students. We find policy frameworks such as DELICATE, principles and rules of good practice such as Sheila Project or LALA Community, or even European laws such as the General Regulation of Data Protection and other specific ones depending on the country such as the Organic Law on the Protection of Personal Data in Spain. These privacy policies and agreements are contingent upon good-doing. Consequently, the approach of ensuring the privacy of data bylaws or policies is dysfunctional, since it assumes that institutions, companies or workers will behave in a correct manner. Reality shows that there are misuses of personal data collected in processes of learning analytics. The most dramatic case was the closure of inBloom schools, which shared personal data of students to third parties without parental consent. Even in other areas, there are data leaks such as the Cambridge Analytics from Facebook.

It is necessary to develop an approach that addresses the problem from its design and creation. It is necessary for laws to be applied before or at the time of data misuse. Consequently, the study and generation of a technological framework of architecture and concrete rules to allow the automation of policies, laws, principles and good practices in a safe and interoperable environment is required. The solution has to be:

1. **Centrally managed:** Distributed data in a public ledger is not suitable to assure data privacy and real anonymity. A central platform required to enable students to manage their data.
2. **Data Link based:** Private and personal data do not have to be stored in the solution. The solution has to be a gate to the data where students can use any platform.
3. **Interoperable:** Has to be able to communicate with different platforms to collect data and perform CRUD (Create, Read, Update, and Delete) actions.
4. **Secure and reliable:** The architecture has to be reliable and the data behind it have to be secured and not accessible without any consent. The slightest failure can raise the same fears and angst as those that appeared in Learning Analytics.
5. **Scalable:** Private and personal data of so many students from all connected platforms could be considered as Big Data. The architecture of the solution has to provide mechanisms to scale in time with interoperability with different platforms.

6. **Automated agreements:** The data privacy agreements between students and entities have to be enforced automatically to provide real time checking. In addition, the laws, the regulations and all the policies involved have to be checked automatically in real time.

Hence, the solution has to be a technological mechanism to automatically enforce policy and private data agreements.

4.1 Personal Data Broker

The idea of the Personal Data Broker (PDB) is to mediate between entities -such as students and universities-. This Fig. 2 activates different abilities for the students: (A) Store the data wherever they want (logs, portfolios, certificates, qualifications …) and in whichever technology they want (relational databases, xAPI Learning Record Stores, cloud storage services, etc.); (B) Automate privacy agreements between entities; (C) Decide the CRUD actions to be taken and how long to activate them for certain entities; (D) Allow to read encrypted or open data. Our narrative of the PDB mechanism reads: "An entity that links to student data which incorporates different privacy agreements related to the relationships students have with different entities. It is able to allow the access, modification or storage of their data. These actions are determined by the privacy agreements."

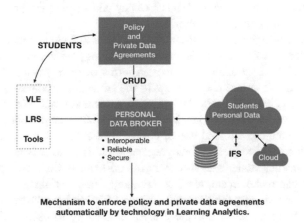

Fig. 2. Personal Data Broker scheme.

4.2 Smart Contracts for Automation and Blockchain for Trustable Context

Digital automation technologies such as Smart Contracts are candidates for study. Their use can make the interoperable and secure framework of automation of policies, laws, and principles of good practices a reality. Smart Contracts automate business rules between two entities. This technology is based on short programs whose activation depends on specific conditions. These conditions are found in the privacy agreements or policies between entities, such as agreements between universities and students. Hence, Smart Contracts are applicable in this solution.

In educational contexts, smart contracts can be used to enforce regulations, laws, principles and good practices to ensure students' data privacy. The use of Blockchain as a certification mechanism between entities could prove interesting as it could generate trust in the educational context while preserving private and personal data with PDB at the same time. In no case would Blockchain be used to protect the privacy of students.

5   Conclusions

Analytics in education is a growing aspect in educational technology. The results are very useful for stakeholders and tools providers and be used to enhance education and learning processes in many different ways, as other studies have demonstrated [32–34]. Despite of improvements, students' personal and private data is being exposed, shared and traded by the same service and tools providers. Regulations, ethics, laws, principles and good practices are not enough to stop data leaks and misuse. The personal and private data collected from students also includes data from minors. Therefore, it is a sensitive issue. There is an urgent need to develop a solution that can leverage the technology to automate those regulations and principles to enable real-time detection. The solution has to assure students' data privacy and respect the privacy agreements between students and educational entities.

Emergent technologies such as Blockchain seemed to be a potential solution. In this paper, we argued that Blockchain is a novel technology with some security flaws by design that turns it an unreliable protocol in terms of data privacy. Hence, is not a suitable technology to assure data privacy in the educational context.

Smart contracts can automate data privacy agreements between entities, such as students and educational institutions. We propose an alternate solution to Blockchain that uses smart contracts, is cloud based, enables law automation and assures the data privacy of students and respects privacy agreements. We call it Personal Data Broker (PDB), which enables students to take control and manage their own data and decide who and when can make create, read, update, and delete actions.

We are now working on the implementation of PDB in Moodle to save logs outside the LMS, in real time, and to provide a layer of security to users' private data. The results of this ongoing and other future work will be published in article format.

Acknowledgment. To the support of the Secretaria d'Universitats i Recerca of the Department of Business and Knowledge of the Generalitat de Catalunya for the help regarding 2017 SGR 934. This work has been partially funded by the Spanish Government Ministry of Economy and Competitiveness throughout the DEFINES project (Ref. TIN2016-80172-R) and the Ministry of Education of the Junta de Castilla y León (Spain) throughout the T-CUIDA project (Ref. SA061P17).

References

1. Filvà, D.A., Forment, M.A., García-Peñalvo, F.J., Escudero, D.F., Casañ, M.J.: Clickstream for learning analytics to assess students' behavior with Scratch. Futur. Gener. Comput. Syst. **93**, 673–686 (2019)
2. Gros, B., García-Peñalvo, F.J.: Future trends in the design strategies and technological affordances of e-learning. In: Learning, Design, and Technology, pp. 1–23. Springer, Cham (2016)
3. Conde, M.Á., García-Peñalvo, F.J., Rodríguez-Conde, M.J., Alier, M., Casany, M.J., Piguillem, J.: An evolving Learning Management System for new educational environments using 2.0 tools. Interact. Learn. Environ. **22**, 188–204 (2014)
4. Amo, D., Alier, M., Casañ, M.J.: The student's progress snapshot a hybrid text and visual learning analytics dashboard. Int. J. Eng. Educ. **34–3**, 990–1000 (2018)
5. Lupton, D., Williamson, B.: The datafied child: the dataveillance of children and implications for their rights. New Media Soc. **19**, 780–794 (2017)
6. Conde, M.Á., Hérnandez-García, Á., García-Peñalvo, F.J., Séin-Echaluce, M.L.: Exploring student interactions: learning analytics tools for student tracking (2015)
7. Chatti, M., Dyckhoff, A., Schroeder, U.: A reference model for learning analytics. Int. J. Technol. Enhanc. Learn. **4,** 318–331 (2013)
8. Papamitsiou, Z., Economides, A.A.: Learning analytics and educational data mining in practice: a systematic literature review of empirical evidence. J. Educ. Technol. Soc. **17**(4), 49–64 (2014)
9. Herold, B.: inBloom to Shut Down Amid Growing Data-Privacy Concerns - Digital Education - Education Week. http://blogs.edweek.org/edweek/DigitalEducation/2014/04/inbloom_to_shut_down_amid_growing_data_privacy_concerns.html
10. Hoel, T., Chen, W.: Implications of the European data protection regulations for learning analytics design (2016)
11. Drachsler, H., Greller, W.: Privacy and analytics: it's a DELICATE issue a checklist for trusted learning analytics. In: Proceedings of the Sixth International Conference on Learning Analytics & Knowledge, pp. 89–98 (2016)
12. Tsai, Y.-S., Moreno-Marcos, P.M., Tammets, K., Kollom, K., Gašević, D.: SHEILA policy framework: informing institutional strategies and policy processes of learning analytics. In: Proceedings of the 8th International Conference on Learning Analytics and Knowledge - LAK 2018, pp. 320–329. ACM Press, New York (2018)
13. Sclater, N., Biley, P.: Code of practice for learning analytics | Jisc. https://www.jisc.ac.uk/guides/code-of-practice-for-learning-analytics
14. Pardo, A., Siemens, G.: Ethical and privacy principles for learning analytics. Br. J. Educ. Technol. **45**, 438–450 (2014)
15. Forment, M.A., Filvà, D.A., García-Peñalvo, F.J., Escudero, D.F., Casañ, M.J.: Learning analytics' privacy on the blockchain. In: Proceedings of the Sixth International Conference on Technological Ecosystems for Enhancing Multiculturality – TEEM 2018, pp. 294–298. ACM Press, New York (2018)
16. Filvà, D.A., García-Peñalvo, F.J., Forment, M.A., Escudero, D.F., Casañ, M.J.: Privacy and identity management in Learning Analytics processes with Blockchain. In: Proceedings of the Sixth International Conference on Technological Ecosystems for Enhancing Multiculturality – TEEM 2018, pp. 997–1003. ACM Press, New York (2018)
17. Bartolomé Pina, A.R., Bellver Torlà, C., Castañeda Quintero, L., Adell Segura, J.: Blockchain en Educación: introducción y crítica al estado de la cuestión. Edutec. Rev. Electrónica Tecnol. Educ. **0**, 363 (2017)

18. Grech, A., Camilleri, A.F.: Blockchain in Education. JRC Sci. Policy Rep. (2017)
19. Sun, H., Wang, X., Wang, X.: Application of blockchain technology in online education. Int. J. Emerg. Technol. Learn. **13**, 252 (2018)
20. Henry, R., Herzberg, A., Kate, A.: Blockchain access privacy: challenges and directions. IEEE Secur. Priv. **16**, 38–45 (2018)
21. Karame, G., Capkun, S.: Blockchain security and privacy. IEEE Secur. Priv. **16**, 11–12 (2018)
22. Casino, F., Dasaklis, T.K., Patsakis, C.: A systematic literature review of blockchain-based applications: current status, classification and open issues. Telemat. Inform. **36**, 55–81 (2019)
23. Feng, Q., He, D., Zeadally, S., Khan, M.K., Kumar, N.: A survey on privacy protection in blockchain system. J. Netw. Comput. Appl. **126**, 45–58 (2019)
24. Li, X., Jiang, P., Chen, T., Luo, X., Wen, Q.: A survey on the security of blockchain systems. Future Gener. Comput. Syst. (2017)
25. Nakamoto, S.: Bitcoin: A Peer-to-Peer Electronic Cash System (2008)
26. Sultan, K., Ruhi, U., Lakhani, R.: Conceptualizing blockchains: characteristics & applications (2018)
27. Puthal, D., Malik, N., Mohanty, S.P., Kougianos, E., Das, G.: Everything you wanted to know about the blockchain: its promise, components, processes, and problems. IEEE Consum. Electron. Mag. **7**, 6–14 (2018)
28. Vukolić, M., Marko: Rethinking permissioned blockchains. In: Proceedings of the ACM Workshop on Blockchain, Cryptocurrencies and Contracts - BCC 2017, pp. 3–7. ACM Press, New York (2017)
29. Wan, Z., Lo, D., Xia, X., Cai, L.: Bug characteristics in blockchain systems: a large-scale empirical study. In: 2017 IEEE/ACM 14th International Conference on Mining Software Repositories (MSR), pp. 413–424. IEEE (2017)
30. Halpin, H., Piekarska, M.: Introduction to security and privacy on the blockchain. In: 2017 IEEE European Symposium on Security and Privacy Workshops (EuroS&PW), pp. 1–3. IEEE (2017)
31. Ikeda, K.: Security and privacy of blockchain and quantum computation. Adv. Comput. **111**, 199–228 (2018)
32. Pinto, M., Rodrigues, A., Varajão, J., Gonçalves, R.: Model of functionalities for the development of B2B e-commerce solutions. In: Cruz-Cunha, M.M., Varajão, J. (eds.) Innovations in SMEs and Conducting E-Business: Technologies, Trends and Solutions, p. 26. IGI-Global (2011)
33. Pereira, J., Martins, J., Santos, V., Gonçalves, R.: CRUDi framework proposal: financial industry application (2014)
34. Pires, J.A., Gonçalves, R.: Constrains associated to e-business evolution. In: E-business Issues, Challenges and Opportunities for SMEs: Driving Competitiveness, pp. 335–349. IGI-Global (2011)

The Strategy of *Josefinas*: Building a Well-Known Brand Through Social Media

Ana Inês[1], Patrícia Pires[1], André Carvalho[2], Luís Santos[2],
Frederico Branco[3], and Manuel Au-Yong-Oliveira[1,4(✉)]

[1] Department of Economics, Management, Industrial Engineering and Tourism,
University of Aveiro, 3810-193 Aveiro, Portugal
{anapereiraines,p.pires,mao}@ua.pt
[2] Department of Electronics, Telecommunications and Informatics,
University of Aveiro, 3810-193 Aveiro, Portugal
{andrecajus,luisfsantos}@ua.pt
[3] INESC TEC and University of Trás-os-Montes e Alto Douro,
Vila Real, Portugal
fbranco@utad.pt
[4] GOVCOPP, Aveiro, Portugal

Abstract. The Internet has changed the way people and businesses communicate. Social media is now seen as a marketing tool, as it influences the promotion of products and helps to raise brand awareness and recognition. Indeed, companies now want to build long-term relationships with clients through online channels, as they are more convenient, faster and reach a greater public. Our article emphasizes the role of social media as a driving force to promote online businesses and build a well-known brand. However, to build and manage customers' relationships and loyalty, brands need to develop an online communication strategy that will help to promote the brand. Our case study focuses on the success story of *Josefinas*, a Portuguese luxury shoe brand. It was created exclusively online and became a well-known brand due to social media and digital influencers. However, is it enough for businesses to be online? To answer this question and to analyse the major role social media played in building *Josefinas*, our study involved two interviews and the analysis of secondary data.

Keywords: Social media · *Josefinas* · Digital marketing · Digital influencers ·
Storytelling · Flagship store · Pop-up store · Online business

1 Introduction

Undoubtedly, the Internet has changed the way people and businesses communicate, leading to a more globalized world. With the widespread use of the Internet and its fast evolution, social media now assumes a central role in people's lives. It is also used by businesses to raise brand awareness, build customer relationships and increase brand loyalty [1, 2], changing completely the communication strategies between brands and customers [3]. Indeed, the emergence of new information technologies has led to disruptive changes, both on an individual and on a business level. Due to globalization, potential clients can be anywhere around the globe, thus the need for brands to master

© Springer Nature Switzerland AG 2019
Á. Rocha et al. (Eds.): WorldCIST'19 2019, AISC 932, pp. 381–391, 2019.
https://doi.org/10.1007/978-3-030-16187-3_37

social media. Their channels are used as a marketing strategy tool, allowing companies to reach a wide range of customers [4, 5]. Besides, consumers now rely much more on information found on social media, than on traditional media (television, newspapers). Nowadays, the biggest technological trend is telecommunications technology, as they allow a faster transmission of information [6]. Plus, social media is also believed to be the best tool to build customer relationships [1], as it is a way of being part of people's daily lives. However, this technological evolution has led to a "communication overload" [7], and customers are increasingly more selective. They choose brands that have a communication strategy and that support certain values [8]. When clients are satisfied, they tend to share their positive experience with other customers, giving them positive feedback. Word-of-mouth is, thus, very important to online businesses. Positive feedback can also be considered free advertising, as customers are "selling" the brand to other potential clients [9]: that is why digital influencers are increasingly more important to brands. They act as intermediaries between brands and customers and their opinions influence greatly the attitudes and behaviours of their large network of trusted followers [10]. *Josefinas'* brand, created exclusively online, was chosen as a case study for this article, as two of the co-authors have followed the brand since its beginning, on Facebook and on Instagram. It is thus of great interest to study how social media had an impact on building this brand and what is *Josefinas'* online strategy. According to Hitt, Ireland and Hoskisson, strategy is a "coordinated set of commitments and actions designed to exploit core competencies and gain a competitive advantage" [6]. In *Josefinas'* case, its strategy relies on the use of social media, not only to advertise their products, but to raise awareness to women's rights, as it is a brand designed specifically for women. *Josefinas'* strategy is not duplicated by competitors, as there is no other shoe brand that focuses so much on women and feminism, while relying on digital influencers that share the same values to promote the brand. The way the brand advertises their products, through storytelling, is also a strategy; as well as its need to create temporary physical stores to raise recognition. *Josefinas* was also chosen as it is an "online-based fashion brand" and fashion is one of the areas that has increased its online presence [3].

The article does a brief presentation of *Josefinas*, followed by the literature review. Some key concepts related to the impact of social media when building a well-known brand are discussed. The qualitative methodology used in our case study comprises a secondary data analysis and the interviews with *Josefinas'* CEO and a former employee of the company.

The research questions of this paper are the following:

- Is it enough for businesses to be online?
- What was the major role that social media played in building *Josefinas'* brand?

2 Brand Presentation

Josefinas is a Portuguese luxury shoe brand designed specifically for women. Figure 1 shows its logo. It belongs to Bloomers SA Company, headquartered in Braga, northern Portugal. It is a small company with six employees. *Josefinas* takes its name after the grandmother of one of its co-founders, Filipa Júlio, as she used to take her to ballet classes [11]. Filipa Júlio had the idea of starting a brand that would sell ballet flats. She then met Maria Cunha in an idea contest and, in 2013, "they co-founded the company together with a third partner, Sofia Oliveira" [12], launching ballet flats in six colours. *Josefinas* is a medium-luxury brand, whose shoes retail between 149€ and 3369€. However, *Josefinas'* accessories, such as shoelaces or stickers, are available at a cheaper price [11]. Its mission is to "help women lift themselves and follow their dreams", and each product "is synonymous with female empowerment" [12]. This idea is conveyed daily through social media.

Fig. 1. Josefina's logo [11]

The brand was created exclusively online. Its website is available in four languages: English, Portuguese, Spanish and Japanese and the products are shipped everywhere. Besides the website, which functions as a blog and as an online store, the brand can also be found on a number of social media channels: Twitter, Pinterest, Tumblr, Snapchat, Google Plus, WhatsApp, Facebook and Instagram [11]. *Josefinas* has 288,563 likes on its Facebook page and 86,3 k followers on Instagram (on October 25th 2018). Its main communication channels are these online platforms, word-of-mouth and the partnerships with digital influencers [11]. *Josefinas* "has managed to build a fast national and international recognition over the last several years through well-known trendsetters and has already made an appearance in international magazines, such as Vogue" [3]. Through social media, *Josefinas* communicates with its clients, creating emotional bonds. It was, indeed, the digital environment that has led to *Josefinas'* growth and its use of social media is one of the brand's main strengths, being able to create competitive advantage over its competitors [13]. *Josefinas'* online communication conveys a very strong message that boils down to its motto "Proud to Be A Woman".

At first, *Josefinas* offered handmade ballet flats known for their all-day wearing comfort. However, to expand its market, the brand followed a diversification strategy. *Josefinas* now offers a wide range of flat shoes (sneakers, boots, mules, sandals, brogues) and accessories. All products and their packages are handmade in northern Portugal by artisans who pay very close attention to all of the details. In fact, a single pair of shoes can take up to 16 h to make [12]. Thus, craftsmanship is thus one of the values that *Josefinas* cherishes. Furthermore, the brand establishes a connection between tradition - offering the best of Portuguese craftsmanship, known abroad for their art of shoemaking - and our modern world, due to the brand's online presence. The brand operates globally, with the USA being their second market, followed by Southern Korea and Japan [13].

There have been two temporary physical stores: one flagship store in New York (from 2016 to 2017) and one pop-up store in Lisbon (for two days in December 2017). The flagship store in New York was a true milestone for *Josefinas*, as it was the first "woman-led brand that opened a store in New York" [14]. Being a feminist brand, focusing on women empowerment and the advocacy of women's rights, *Josefinas* supports social causes related to violence against women. There is a strong component of corporate social responsibility as the brand contributes for the NGO Women for Women International, that helps women in war-torn countries. For that purpose, the brand has invited Gloria Steinem, "the mother of feminism" [12], to design a special and limited edition. For every ten pairs of shoes sold, *Josefinas* helps a woman at risk. The brand has also partnered with APAV (Portuguese Association for Victim Support), a non-profit organization that helps victims of domestic violence. For this purpose, the brand has created a collection named "You can leave", giving the NGOs a sales percentage of each pair sold [11]. The corporate responsibility of *Josefinas* is very important as "luxury customers are increasingly interested in the social (…) impact of the luxury goods they buy" [15].

3   Literature Review

3.1   Social Media and Its Impact on Brands' Success

Social media consists of different online platforms that enable people to create and share content; and to communicate with each other. This is the biggest trend of the 21st century, being used both for private and for businesses purposes. For example, Facebook is one of the most popular platforms, as users can find a lot of functionalities in the same place. It can be used to interact with friends, but also to search for brands, finding more information about them and communicating with them as well [16]. We have been noticing the growing influence of social media, while in comparison to traditional media. More than ever, customers search for brands online. According to a DEI worldwide study on the impact of social media platforms on purchase decisions, "70% surveyed consumers visited social media platforms to get information on a brand" [3]. With globalization and the democratization of the Internet, businesses understood that social media allowed a higher level of convenience and efficiency. Social media is now a key element of businesses' marketing strategy, being able to influence purchasing behaviour [17]. Social media also makes communication easier, allowing a quicker exchange between entities. Therefore, it is a required element for any company that wants to improve their sales, increase brand awareness, reduce marketing costs and be known worldwide. Due to a "communication overflow" [7] brands must create coherent content on their social media platforms, being responsive and actively engaging with customers. As these platforms allow brands to be constantly present in people's lives, they are an effective tool to influence customers' behaviours. Thus, social media might be the key to success, as it builds relationships, increases brand awareness and loyalty. They revolutionized the way businesses interact with customers, being able to influence their purchases, mostly through digital influencers [18]. Digital influencers behave as intermediaries, publicizing the brand and shaping opinions. The messages they share reach several people very quickly. Therefore, it is important that companies create strategic partnerships with them [10].

3.2 Digital Marketing

Companies' marketing strategies are activities and processes used to communicate with customers. Therefore, digital marketing communicates through digital platforms to create value. Digital marketing is becoming an important tool for companies as it ensures that the necessary information is available online for everyone. This type of marketing can happen through videos, tweets, photos or any kind of sales strategy that implies the use of interactive communication [19]. According to the Digital Marketing Institute, digital marketing is "the use of digital technologies to create an integrated, targeted and measurable communication which helps to acquire and retain customers while building deeper relationships with them" (Smith 2007) [20]. If a good marketing strategy implies that companies can deal with both market and consumer changes, it is imperative in the 21st century that companies rely on digital technologies to succeed. In fact, customers rely increasingly more on the Internet, thus the major key role of digital marketing in the brand's promotion [21, 22]. Besides, digital marketing allows brands to easily communicate with a broader group of customers. However, Kotler believes that digital marketing and traditional marketing should coexist, since both are relevant when creating an emotional relationship with customers and understanding customer's characteristics and behaviours. Customers are now co-creators, as they share their opinions and experiences through digital platforms. If companies create value for customers, customers will be loyal to the brand [21].

3.3 Flagship Stores and Pop-up Stores

Josefinas had two temporary physical stores: a flagship and a pop-up store. Flagship stores "represent a market entry method" [15] and thanks to the flagship store in New York *Josefinas* gained increased visibility. For instance, in July 2017, before the store's first anniversary, the brand was recognized by British Airways in their segment "Inside New York: Finding Hidden Gems" [11]. A flagship store strategy is the "organization of in-store events", such as the *Josefinas'* monthly #ProudToBeAWoman talks in 2017, where successful women were invited to share their stories to inspire other women [11]. Flagship stores are very important in the luxury market, as they lead to brand reinforcement, through the creation of a place that spreads the brand's identity and values. Therefore, *Josefinas'* was a place where customers could experience the brand. The "in-store communication" is thus crucial [15], therefore, in flagship stores the staff members' mission is to present the brand, its history and its values. *Josefinas'* flagship store was designed by Christian Lahoud, a well-known architect in the luxury market, who has also designed Tiffany&Co stores. Flagship stores distinguish themselves as they create a "luxury experience for the customers", mostly due to "store architecture" and "added value services" [15]. *Josefinas'* pop-up store for two days in a hotel in Lisbon was also part of the brand's strategy. Clients could book an hour with one of *Josefinas'* consultants to try the shoes [11]. Pop-up stores bring "web-only" products to an "in-person" environment, being a way to capture new customers and to increase customer loyalty. They are an opportunity for consumers to experience the product and the brand, serving basic marketing needs, such as introducing the brand and providing direct contact with customers [23]. Pop-up stores also draw attention to the product, being adopted by several digital brands as a way of testing the waters and raising brand awareness. Food52, Organic Valley coffee shop and REVEAL are some examples of successful pop-up stores [23]. Both flagship and pop-up stores aim to offer a unique experience, thus making the customer feel special, focusing on customer experience and allowing the brand to get feedback on the spot. Having temporary stores may be a strategy for online businesses to reach new markets and raise brand awareness.

3.4   Storytelling

Storytelling, which can also be named narrative advertising, is a way of communicating a brand, product, or a service through a story [24]. In a globalized world, all brands try to have an online presence. However, when building a brand through social media, it is important to create a strategy to gain competitive advantage through differentiation. Brands need, simply said, to find their own voice to build emotional connections with customers around the world. This is one of the reasons why this paper focuses on *Josefinas*, as they use the strategy of "storytelling" to advertise and to communicate with their customers [7]. Every post the brand makes, whether it is on Facebook or on Instagram, has a story and so it is easier for the customer to understand the brand positioning and its mission. According to Filipa Júlio, the brand has come up with a strategy that is applied to all products, each one "has an identity, a concept, and a story" and every story "includes the values of the brand – meaning, dreaming, femininity, strength, happiness, proactivity and the value of hand-made work and craftsmanship". Customers identify themselves with the story, thus creating a connection to the brand [7]. Storytelling is therefore part of *Josefinas'* strategy, allowing the brand to differentiate itself from other footwear companies.

4   Methodology

This paper's main purpose was to identify the major role social media can play in building a well-known brand, focusing on *Josefinas'* case study. Two main research questions, as stated in the introduction, were determined. To answer them, a qualitative methodology was followed, as it is believed that a case study design favours qualitative methods, especially interviews [25]. We conducted two semi-structured interviews, in October 2018, with *Josefinas'* CEO, Maria Cunha, and with a former employee, Olga Kassian. The interview with *Josefinas'* CEO, Maria Cunha, occurred via e-mail, the interview script having been sent directly to the CEO of *Josefinas*. The interview with Olga Kassian occurred in person, and took 20 min; it was recorded in audio and transcribed in full. Secondary data, which includes "any data that is examined to answer a question other than the question(s) for which the data were initially collected" [26], about *Josefinas* was also analysed. This study's main research objectives were to collect more information about *Josefinas*, as well as understanding how social media has helped to raise brand awareness and recognition, and if that is the only tool the brand relies on.

5   Discussion on the Fieldwork

5.1   Analysis of Secondary Data

After its first collection launch, *Josefinas* partnered with Portuguese fashion bloggers, as a way of boosting its online presence. The first Portuguese blogger to work with the brand was Maria Guedes [11], which now has 84,5 k followers on Instagram (on October 29th 2018). In May 2013, Maria Guedes announced on her personal blog, *Stylista*, the online launch of *Josefinas*. After that, *Josefinas* started to make partnerships with many international bloggers and digital influencers, making it possible for the brand to appear in fashion magazines [7]. This relationship was planned, since famous bloggers can influence customers' opinions. Some influencers buy *Josefinas* and share their opinion online; others are given the products by the brand itself. These influencers are carefully selected [3]. For instance, Chiara Ferragni, known as one of the biggest digital influencers worldwide, shared pictures with a pair of *Josefinas* on her Instagram account. She has about 15 million followers on Instagram (on October 29th 2018), which allowed *Josefinas* to reach more customers. Last year Chiara was considered one of the Forbes' Top Influencers in the Fashion category. Due to these partnerships, *Josefinas* is recognized worldwide and recommended by the fashion and business press [28]. By analysing the work of Hermanaviciute and Marques [3], one may conclude the following:

- Its strong presence in all these social media platforms allows *Josefinas'* team to reach and target different markets and countries, and communicate with its clients, who give their feedback and ask questions;
- *Josefinas'* website is the only sales platform the company has, since it doesn't have any physical stores;
- The way the company communicates with customers creates a competitive advantage for *Josefinas*, since there is special treatment for each client;
- *Josefinas'* team makes an effort to give their customers a positive experience while buying their products through their website. By using specific approaches, clients will enjoy the process and become loyal;
- All the information on *Josefinas'* website is related to its target market, the type of product sold and charitable causes.

5.2   Interviews

Two interviews were conducted. One with *Josefinas'* co-founder and CEO, Maria Cunha, graduated in management, and with an MBA in Marketing; and the other with a former employee of *Josefinas*, a Master's student in Industrial Engineering and Management at the University of Aveiro, Olga Kassian. Olga worked full-time, from July 2016 to February 2017, at *Josefinas'* store in New York. She had three main responsibilities: store sales, NY public relations and brand ambassadorship. As an ambassador, she attended fashion and feminist events to promote the brand. Olga's mission was to offer a personalized service aiming at establishing long-term relationships with customers. She also met with local influencers to establish local partnerships.

Both interviews revealed that *Josefinas* was born to be an online brand, being able to communicate with their customers, regardless of where they are. It is, indeed, as Maria stated, "a brand made in Portugal for the world". Nevertheless, the physical dimension is still important. In fact, clients frequently ask the brand to open a store as they wish to try *Josefinas'* shoes. Therefore, there have been two temporary physical stores: one flagship store in New York from 2016 to 2017; and a temporary pop-up store in Lisbon, for two days in December 2017. According to Maria, New York was strategically chosen, as it is known as the "world's fashion centre", but also because the American market is their second most relevant in terms of sales. As Olga mentioned, the retail staff assistants in New York acted as brand ambassadors. Indeed, her main task was to present *Josefinas* to customers, thus developing brand attachment and generating positive brand experiences. Therefore, the flagship store in New York allowed *Josefinas* to raise and increase brand awareness.

Concerning social media, both interviewees agreed on how important these channels are as a tool to build relationships. For instance, Olga referred that the brand uses several techniques when talking with clients, like always referring the client's name when answering them, establishing a personal connection. Maria mentioned that to reach their target audience – "independent modern women, tech savvy, that use social media, more precisely Instagram", *Josefinas* establishes connections with digital influencers, who wear the brand because they like it and believe in its mission. Olga highlighted the wider range of Instagram in comparison to other social media. Olga and Maria mentioned that the brand contacts the digital influencers, trying to find out if they have any interest and then they send the product. The digital influencer only posts a picture on social media if she likes the product.

Maria mentioned that the brand positioned itself as a flat shoes brand, whose mission is to inspire women. The brand strategy, according to Maria, relies on Public Relations, social media and social media pay. The team always knew that their strategy would rely on social media, as *Josefinas* "was born in this new digital era". Maria stated that as an online brand, social media are its main communication channel, bringing different kinds of turnover. Maria mentioned that not only do they increase sales, but they also allow clients' feedback and quicker problem solving. She highlighted the importance of interacting with clients and when asked if the brand became well-known due to social media, *Josefinas'* CEO answer was "Yes, absolutely". The interview aimed to quantify the *Josefinas'* annual sales figure and profit margin, however the CEO stated that she "won't disclose the exact number". It was only revealed that the product profit margin was about 67%. Another important topic discussed was *Josefinas'* partnership with NGOs such as APAV (Portuguese Association for Victim Support) or Women for Women International. Maria Cunha stated that Josefinas regards social media as a tool not only to advertise its campaigns and products, but also to raise awareness to social causes, especially those related to women's rights. Supporting these causes, according to Maria, is an important part of *Josefinas'* strategy, as they are aligned with its values. In fact, Maria told us that if she were to describe *Josefinas* to someone who has never heard of it, she would say that the brand "doesn't believe just in beautiful shoes, but in a fair world and in each other's inner beauty". As Olga said, what sets *Josefinas* apart from other brands is their online presence and their symbolism, as everything the brands works for is shaped by its mission.

6 Conclusion and Suggestions for Future Research

Due to the fast evolution of the Internet, social media has become an important medium of communication, allowing different kinds of exchanges between individuals and between customers and businesses. Businesses realized the need to use social media as a communication channel to reach their clients and to promote their brands. To create an online competitive advantage, brands should communicate having in mind their brands' values as well as their clients'. Social media's growth has also led to the emergence of digital influencers that are able to shape the opinions and the purchase behaviours of millions of people. Thus, another strategy to promote a brand is by creating partnerships with these influencers. Our case study revealed *Josefinas'* main strategy relies on using social media as a communication channel and as a way of promoting the brand and its values. It is obvious that *Josefinas'* success happened really quickly, since the company was founded in 2013 and is now a well-known brand. *Josefinas*, as a feminist brand, has engaged socially with APAV and Women for Women International to support women's rights. By advertising these causes, the brand warns people about real problems that affect women and proves that it really is a brand made by and for women. Opening a physical store in New York was a huge step for the brand since the American market is one of *Josefinas'* target markets. It provided the opportunity to create new partnerships and to improve its visibility. In conclusion, social media channels are an important tool, but *Josefinas* cannot rely only on them, as they have also opened two physical stores until now. They were a strategic decision, as they fulfilled the clients' wish of having an in-store experience.

Like *Josefinas*, other companies became well-known due to their online presence, relying highly on the digital market. *Farfetch* is a good example of this also. This fashion company, founded by a Portuguese entrepreneur named José Neves, in 2008, created an online store selling luxury products. Not having any physical store was a strategic decision [27]. For future research, we would like to suggest a comparison between both companies. The way companies use social media platforms can be different so it would be interesting to understand how each company reaches its success through them. On the other hand, it would also be interesting for future research to establish a comparison between *Josefinas* and other shoe brands that do not rely on social media to communicate their products and that have physical stores. In that way, one would be able to understand the role of social media in building a brand.

Acknowledgements. The authors would like to thank the interviewees Maria Cunha and Olga Kassian for their time. We are grateful to *Josefinas'* CEO for having verified this article and for having authorized its publication.

References

1. Martins, J., Goncalves, R., Pereira, J., Oliveira, T., Cota, M.: Social networks sites adoption at firm level: a literature review. In: 2014 9th Iberian Conference on Information Systems and Technologies (CISTI), pp. 1–6. IEEE (2014)
2. Riyanto, A., Renaldi, F.A.: Effect of social media on e-commerce business. In: IOP Conference Series: Materials Science and Engineering (2018)
3. Hermanaviciute, G., Marques, A.: Importance of digital brand presence in the portuguese footwear companies with different types of market retails. In: Congresso Internacional de Negócios da Moda (2016)
4. Martins, J., Gonçalves, R., Pereira, J., Cota, M.: Iberia 2.0: a way to leverage Web 2.0 in organizations. In: 2012 7th Iberian Conference on Information Systems and Technologies (CISTI), pp. 1–7. IEEE (2012)
5. Pires, J., Gonçalves, R.: Constrains associated to e-business evolution. E-business issues, challenges and opportunities for SMEs: driving Competitiveness, pp. 335–349. IGI Global (2011)
6. Hitt, M., Ireland, R., Hoskisson, R.: Strategic Management: Concepts. Competitiveness and Globalization, 9th edn. South-Western, Mason (2011)
7. Dias, L., Dias, P.: Beyond Advertising Narratives: Josefinas and their storytelling products. Anàlisi Quaderns de Comunicació i Cultura **58**, 47–62 (2018)
8. Thomkaew, J., Homhual, P., Chairat, S., Khumhaeng, S.: Social media with e-marketing channels of new entrepreneurs. In: AIP Conference Proceedings (2018)
9. Hajli, N., Shanmugam, M., Papagiannidis, S., Zahay, D., Richard, M.O.: Branding co-creation with members of online brand communities. J. Bus. Res. **70**, 136–144 (2017)
10. Uzunoglu, E., Kip, S.M.: Brand communication through digital influencers: leveraging blogger engagement. Int. J. Inf. Manag. **34**, 592–602 (2014)
11. Josefinas Portugal. https://josefinas.com. Accessed 27 Oct 2018
12. Spanggaard, A.: The Scandi Female Code: An empowering education tool to help girls and women unlock their potential, pp. 70–77. BoD - Books on Demand (2018)
13. Abreu, J.: Empresariato - Casos de Sucesso Empresarial, Branding, pp. 129–150. Idioteque (2018)
14. Bloomidea. https://bloomidea.com/en/projects/josefinas. Accessed 27 Oct 2018
15. Arrigo, E.: The flagship stores as sustainability communication channels for luxury fashion retailers. J. Retail. Consum. Serv. **44**, 170–177 (2018)
16. Kaplan, A.M., Haenlein, M.: Users of the world, unite! The challenges and opportunities of Social Media. Bus. Horiz. **53**(1), 59–68 (2010)
17. Felix, R., Rauschnabel, P.A., Hinsch, C.: Elements of strategic social media marketing: a holistic framework. J. Bus. Res. **70**, 118–126 (2017)
18. Swani, K., Milne, G.R., Brown, A.B., Assaf, A.G., Donthu, N.: What messages to post? Evaluating the popularity of social media communications in business versus consumer markets. Ind. Mark. Manag. **62**, 77–87 (2017)
19. Kannan, P.K., Li, H.A.: Digital marketing: a framework review and agenda. Int. J. Res. Mark. **34**(1), 22–45 (2017)
20. Wymbs, C.: Digital marketing: the time for a new academic major has arrived. J. Mark. Educ. **33**(1), 93–106 (2011)
21. Kotler, P., Kartajaya, H., Setiawan, I.: Marketing 4.0: Moving from Traditional to Digital. Wiley, Hoboken (2017)

22. Gonçalves, R., Martins, J., Branco, F., Perez-Cota, M., Au-Yong-Oliveira, M.: Increasing the reach of enterprises through electronic commerce: a focus group study aimed at the cases of Portugal and Spain. Comput. Sci. Inf. Syst. **13**, 927–955 (2016)

23. AMA - The Magic of Pop-Up Shop Marketing. https://www.ama.org/publications/MarketingNews/Pages/magic-of-pop-up-shop-marketing.aspx. Accessed 27 Oct 2018

24. Laurence, D.: Do ads that tell a story always perform better? The role of character identification and character type in storytelling ads. Int. J. Res. Mark. **35**, 289–304 (2018)

25. Bryman, A., Bell, E., Mills, A.J., Yue, A.R.: Business Research Methods, Canadian edn., pp. 93–102. Oxford University Press, Oxford (2011)

26. Vartanian, T.P.: Secondary Data Analysis. Oxford, New York (2011)

27. About Farfetch - Company Information & Press Releases. https://aboutfarfetch.com. Accessed 21 Oct 2018

28. Hazlehurst, B.: This Portuguese brand is reviving the early 2000s ballet flat. Paper. Fashion, 11 July. http://www.papermag.com/josefinas-ballet-flat-back-2583252036.html. Accessed 26 Jan 2019

Validating Game Mechanics and Gamification Parameters with Card Sorting Methods

Eva Villegas[1], Emiliano Labrador[1], David Fonseca[1(✉)],
and Sara Fernández-Guinea[2]

[1] La Salle – Universitat Ramon Llull, 08022 Barcelona, Spain
{eva.villegas,emiliano.labrador,fonsi}@salle.url.edu
[2] Universidad Complutense de Madrid, Madrid, Spain
sguinea@psi.ucm.es

Abstract. The article is based on the application of a Card Sorting to establish a criteria in accordance with the most relevant game mechanics at present. The evaluation has been carried out with experts in the gamification sector who are currently active. The study presented is based on the application of user experience techniques: interviews and card sorting system, which have allowed the first definition of the concepts applicable to gamified systems and describe, compare, eliminate, group and hierarchies three types of game mechanics, the most used, starting with the first definition criteria of game mechanics and what is not. The results are founded on a first approximation that indicates parameters to be taken into account in the possible application in gamified systems contributing with both strong and weak points of the three systems and allowing to establish a starting point for more possible closed evaluations.

Keywords: Game mechanics · Card sorting method · Gamification

1 Introduction

The game concept has been introduced in marketing, banking, health and education [1–3]. In order to do this, a new system called gamification is created although, there is no consensus on its definition or on how the system is applied. The authors Hunicke, Leblanc and Zubek (2004) clarify in their work "MDA: A Formal Approach to Game Design and Game Research" [4], that the game can be diversified into three elements M: Mechanics, as rules that make up the game system, D: Dynamics, such as the interaction established between the system and the users, and finally A: Aesthetics of the users before the application. This is the data is taken into account to establish a game system in an environment that is not game-friendly. It is worked on the hypothesis that collecting its meaning and contrasting it with consultants in gamification has a certain lack of consensus. The study presented below, is part of an arising need to create a new methodology named "Methodology I'M IN" [5]. Both disciplines, user experience and gamification, are mixed and connected using emotional behavior [6–8].

To incorporate gamification in the method, the MDA framework is used as the starting point. Once the method is created, the concepts to validate the approximation

© Springer Nature Switzerland AG 2019
Á. Rocha et al. (Eds.): WorldCIST'19 2019, AISC 932, pp. 392–401, 2019.
https://doi.org/10.1007/978-3-030-16187-3_38

are evaluated with experts [6]. A specific study has been carried out with senior gamification consultants with the aim of proposing a first approach and establishing a conjoint definition, a classification of concepts and a hierarchy of content.

2 Literature Review

The gamification [9] is a widespread practice, proven by the many scientific publications that exist today. Despite the multiple definitions, it is only worked on with those that are based on the design of the system [10, 11], that is those which connect the implementation of elements/game components with the requirements or objectives to bear in mind with the system [12–14]. The definition of the MDA method is taken into consideration [4], (M) Mechanics, (D) Dynamics and (A) Aesthetics, (M) Mechanics, (D) Dynamics and (A) Aesthetics, as it is mentioned earlier.

To make a first approximation and consensus of what is considered game mechanics and what is not, a system called Card Sorting is used [15]. Card Sorting is an empirical system that helps obtain information from the mental model of users [16, 17] individually, regarding the knowledge of a concept. It is a standard method of research within the discipline of user experience and is used mainly in the area of information architecture research [18]. It helps to assess concepts related to terminology and the organization of concepts according to the emotional experience of users [19–21]. In the method, users receive cards in order to group, organize and priorities those with all the information already analyzed. The application of card sorting can be open or closed. In the case of an open card sorting, users receive the cards and can freely create the categories [22] they find necessary. As it concerns the closed card sorting, the user already has the name of the categories and should only place them where it is appropriated. Regarding this study, a Hybrid Card Sorting was carried out since it seeks to obtain a first approach to the categorization of predefined (closed) cards and freedom of grouping and creating new elements (open) [23].

3 Application

The technique used for the analysis is based on an initial interview and the completion of a hybrid Card Sorting, once the test is finished, the session closes with possible arisen doubts.

The study is carried out individually and in person with 5 participants [24]. All of them are senior consultants in gamification, between 5 and 14 years of experience. Regarding gamification, in full swing since 2004, the consultants have high knowledge of the discipline and its application with an average age of 45 years and a standard deviation of 3.08. The average duration of the test is one and a half hours. The method [25] carried out is based on:

- Initial closed questions as a starting point for the basic definitions of concepts: gamification, mechanics, dynamics and aesthetics, according to the experts' perspective.

- Application of the card sorting method of the most valued game mechanics in the sector.
  - Actionable gamification, Octalysis of Yu-Kai-Chou [26]. Based on 8 factors: meaning, empowerment, social influence, unpredictability, avoidance, scarcity, ownership, accomplishment and with a total of 76 game mechanics. Unlike the creation of the first version of cards, the total sub classifications of the 8 indicated parameters have been considered.
  - Gamification Model Canvas 2.0 of Sergio Jiménez [27]. Based on: platforms, mechanics, dynamics, components, aesthetics, behaviors, players, simplicity, costs, re-venues and with a total of 59 cards distributed among the different mentioned parameters.
  - Gamification Inspiration Cards by Andrej Marczewski [28], with a total of 54 cards including options of mechanics and options of profiles.

To create the groups, it is suggested that the succeeding process is followed:

1. Step 1: The three decks of cards are delivered, each of them is created according to the game mechanics mentioned above, keeping in mind that all the experts are aware of what is involved.
2. Step 2: A first classification of the given cards is made, distinguishing between those that are considered game mechanics and those that are not.
3. Step 3: A second concept classification is done regarding their affinity.
4. Step 4: For each of the groups, a title is defined.

Since the first step, the consultants can discard the cards that they do not consider to be part of the game mechanics, modify the name of the card in the case that it was not considered optimal and create new cards for a new addition considered mechanical.

From the beginning, qualitative data is collected helping to understand the reason for the entire structure created and once the session is finished, the quantitative data is collected to be subsequently codified. Below are two images that reflect the classifications made in two of the sessions (see Fig. 1).

Fig. 1. Image of the classification made by user 1 and user 4.

4   Results

As indicated in the previous section, the qualitative part is collected during the whole test and is based on the perceptions, behavior and comments of the users and the quantitative part is based on the analysis of the results of the final categorization performed.

4.1   Qualitative Results

The results of the initial interview are shown below:

- Definition of gamification: As a definition of gamification, all experts agree that these are game concepts applied to environments that are not intended to play, mainly in education and in human resources departments. Comments such a "achieve the goal more comfortably", "improvement of the user experience", "to gamify is to create experiences" were made.
- Definition of mechanics: As a single definition it is understood as the strategy to follow to apply the game concept. It received comments such as: "are the rules of the game", "is the mission", "is the basic element of the game system", "everything you can do", "is everything that a user does to get closer to the goal".
- Definition of dynamics: When talking about the definition of dynamics there is not as much consensus as the mechanics, as some consultants speak of the term as: "I treat them as motivators," others define them as: "relationship between the user and the mechanics "," interactions that occur with the group ". Or as: "it has to do with hidden roles".
- Definition of aesthetics: In the definition of aesthetics, it is mainly spoken of: "relationship between the mechanics and the dynamics of the person", "things that influence in creating experiences", "it is the narrative", "it is the perception of the user", or well "is the visual component that helps immersion".

4.2   Quantitative Results

Result of the First Classification. The results of the classification between the frameworks of cards delivered. Bellow you can find the percentages for each of the options (Table 1).

Table 1. Sample of the average of distinction of mechanics of game according to the type of mechanics analyzed.

Average	No mechanics	Mechanics	Doubt
Octalysis Yu-Kai-Chou	19%	36%	45%
Gamification Inspiration Cards Andrej Marczewski	26%	40%	34%
Gamification Model Canvas Sergio Jiménez	90%	10%	0%

During the first classification, all the elements of the three analyzed options are taken into account, which initially were not all considered mechanical by the authors themselves. This is the main reason why Sergio Jiménez's Gamification Model Canvas option has a non-mechanical percentage of 90%. However, in the case of Gamification Inspiration Cards by Andrej Marc-Zewiski it has a percentage of 26% considered non-mechanics and 40% considered mechanics. The cards classified as doubt, 45% in Octalysis by Yu-Kai-Chou and 34% by Andej Marczewski refer to cards whose meaning and application is not understandable to the evaluated experts and, therefore, have been discarded in order to continue with the test (Figs. 2, 3 and 4).

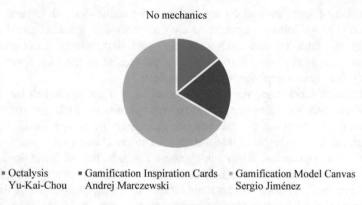

Fig. 2. Distribution of the percentage of non-mechanical according to the three options presented.

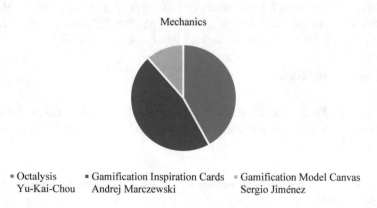

Fig. 3. Distribution of the percentage of mechanics according to the three options presented.

Doubt

▪ Octalysis   ▪ Gamification Inspiration Cards ▪ Gamification Model Canvas
    Yu-Kai-Chou   Andrej Marczewski   Sergio Jiménez

Fig. 4. Distribution of the percentage of doubts according to the three options presented.

Result of the Second Classification. Result of the second classification in which the users organize by affinity the cards selected as game mechanics and the cards selected as non-mechanical. Below is the data according to the classifications created from a first classification determined as game mechanics (Table 2).

Table 2. List of the classifications of the mechanics according to the coincidence among the users of the test.

Classification of mechanics	Co-occurrence users
Altruistic // Anonymous // Challenges // Content of the product //	1
Creativity // Economy // Go back in the system // Help	1
Learning, competence growth /Increase your abilities // Time	2
Me // Motivation // Random // Share // Warnings, Handicaps, Disruptions	1
Narrative	2
Progress, how do you assume the purposes? Progression, verification	3
Reward / Awards / Gift	4
Social world / Social	3
No group	1
Total number of different classifications	20

The results indicate a classification of 20 different elements from which three of the elements: "skills", "reward", "social world" have different nomenclature but a different meaning. The following table shows the data according to the classifications created from a first classification determined as non-mechanical game (Table 3).

The results of 5 different classifications show the dynamics with 4 coincidences and the game elements as the main concepts that are worked on as non-mechanical. The narrative is considered essential in non-mechanical and even gamification, for this reason it does not get classified. The results of the mechanics that close the second

Table 3. List of the classifications of the mechanical NOs according to the coincidence between the users of the test.

Classification of NO mechanics	Co-occurrence users
Betting	1
Dynamics	4
Game elements	3
Profiles	2
Tools	1
Total number of different classifications	5
Others (they are always used, they are not an option)	1

classification, are shown by the following global results according to the grouping of the cards, whose graph indicates the number of times that each mechanic is placed in the same group. The cards of which there were more than one for the same name have been joined, such as: leaderboards or badges since they refer to the same concept.

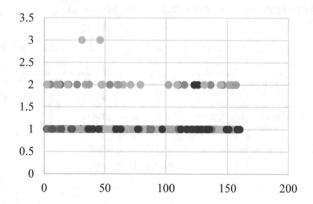

Fig. 5. Representation of the degree of dispersion in the grouping of the cards.

In the session graph (see Fig. 5) you can see all the groupings in the 161 selected cards distributed based on the number of consultants that have included it in the same group, the degree of dispersion is very high since only 3 consultants coincide in the group of Real prizes within "Rewards" and "to win to Reward", within Reward.

5   Conclusions

The aim of the study is to identify the first impressions regarding the categorization of elements used to gamify, and therefore, the consensus in its use by all the experts in the sector. The bases are defined by qualitative results, but in gamification there is a consensus that these are game elements applied in non-ludic environments. However, it is differentiated as it always has a narrative that serves as an immersive element, or with the fact that the narrative is a mechanic of game more and that, therefore, its use can be chosen. The rest of the worked concepts are defined differently but their meaning is unique. The dynamics is always worked as a connection between the user and the mechanics. The mechanics is treated as that as it is applied so that the system is gamified and the aesthetics is unified as the perception of the users before the application is made.

From the quantitative results, it can be determine that the main findings are based on the diversity of groupings created due to differences in criteria. The classification has been made based on the use of the cards not on their meaning. The cards based on the meaning, have been rejected as none of the authors has tried to find the meaning to enter to value it: if it is not understood, it is not useful. The study shows consensus mainly due to the self-taught learning which bases the application on the experience acquired by the expert in his professional career, as shown in the literals named in the results section. The interpretation of the elements is based on intuition, not on the rigor of reviewing the author's intention, the proof of this is the diversity of classifications and the high number of doubts that are collected as rejected cards. The next step is to expand the number of users analyzed in order to obtain more information from this first approach. Even so, and seeing the results, it is intended to perform a closed Card Sorting from the definition of the cards classified as non-mechanical to achieve a more precise selection of cards. The key to the study is determining a unified meaning of game mechanics, unifying the definitions of each of them and defining a single classification of mechanics and applicable criteria, following previous works previously validated [29, 30].

Acknowledgment. To the support of the Secretaria d'Universitats i Recerca of the Department of Business and Knowledge of the Generalitat de Catalunya for the help regarding 2017 SGR 934.

References

1. Villagrasa, S., Fonseca, D., Redondo, E., Duran, J.: Teaching case of gamification and visual technologies for education. Cases Inf. Technol. **16**(4), 38–57 (2014). https://doi.org/10.4018/jcit.2014100104
2. Fonseca, D., Navarro, I., Villagrasa, S., Redondo, E., Valls, F.: Sistemas de Visualización Gamificados para la mejora de la Motivación Intrínseca en Estudiantes de Arquitectura. In: CINAIC 2017 - IV Congreso Internacional sobre Aprendizaje, Innovación y Competitividad, pp. 209–214 (2017). https://doi.org/10.26754/cinaic.2017.000001_043
3. Calvo, X., Fonseca, D., Sánchez-Sepúlveda, M., Amo, D., Llorca, J., Redondo, E.: Programming Virtual Interactions for Gamified Educational Proposes of Urban Spaces. In: Lecture Notes in Computer Science, vol. 10925 LNCS, pp. 128–140 (2018). https://doi.org/10.1007/978-3-319-91152-6_10
4. Hunicke, R., LeBlanc, M., Zubek, R.: MDA: A Formal Approach to Game Design and Game Research. In: Proceedings of the Association for the Advancement of Artificial Intelligence Workshop on Challenges in Game AI (AAAI 2004) (2004). doi:10.1.1.79.4561

5. Villegas, E., Labrador, E., Fonseca, D., Fernández-Guinea, S: Mejora de las metodologías de experiencia de usuario mediante la aplicación de gamificación. Metodología I'm In. In: 13th Iberian Conference on Information Systems and Technologies (CISTI), pp. 1–6. Cáceres (2018). e-ISBN: 978-989-98434-8-6, Print-ISBN:978. https://doi.org/10.23919/cisti.2018. 8399386

6. Villegas, E., Labrador, E., Fonseca, D., Fernández-Guinea, S.: Methodology I'M IN applied to Workshop. Successful educational practice for consultants in user experience with gamification fields. Universal Access in the Information Society (2019)

7. Fitz-Walter, Z., Johnson, D., Wyeth, P., Tjondronegoro, D., Scott-Parker, B.: Driven to drive? Investigating the effect of gamification on learner driver behavior, perceived motivation and user experience. Computers in Human Behavior (2017). https://doi.org/10. 1016/j.chb.2016.08.050

8. Fonseca, D., Villagrasa, S., Navarro, I., Redondo, E., Valls, F., Llorca, J., Gómez-Zevallos, M., Ferrer, Á., Calvo, X.: Student motivation assessment using and learning virtual and gamified urban environments. In: Proceedings of the 5th International Conference on Technological Ecosystems for Enhancing Multiculturality - TEEM 2017, pp. 1–7 (2017). https://doi.org/10.1145/3144826.3145422

9. Attali, Y., Arieli-Attali, M.: Gamification in assessment: Do points affect test performance? Comput. Educ. **83**, 57–63 (2015). https://doi.org/10.1016/j.compedu.2014.12.012

10. Mekler, E.D., Brühlmann, F., Tuch, A.N., Opwis, K.: Towards understanding the effects of individual gamification elements on intrinsic motivation and performance. Comput. Hum. Behav. (2017). https://doi.org/10.1016/j.chb.2015.08.048

11. Dicheva, D., Dichev, C., Agre, G., Angelova, G.:. Gamification in education: A systematic mapping study. Educ. Technol. Soc. (2015). https://doi.org/10.1109/educon.2014.6826129

12. Prowting, F. (n.d.). Gamification: engaging your workforce

13. Amo, D., Valls, A., Alier, M., Canaleta, X., García-Peñalvo, F.J., Fonseca, D., Redondo, E.: Using web analytics tools to improve the quality of educational resources and the learning process of students in a gamified situation. In: Proceedings of 12th Annual International Technology, Education and Development Conference, p. 5 (2018). https://doi.org/10.21125/ inted.2018.1384

14. Sanchez-Sepulveda, M., Fonseca, D., Franquesa, J., Redondo, E.: Virtual interactive innovations applied for digital urban transformations. Mixed approach. Future Gener. Comput. Syst. **91**, 371–381 (2019). https://doi.org/10.1016/j.future.2018.08.016

15. Spencer, D.: Card Sorting: Designing Usable Categories. In: Card Sorting: Designing Usable Categories (2009)

16. Schmettow, M., Sommer, J.: Linking card sorting to browsing performance – are congruent municipal websites more efficient to use? Behav. Inf. Technol. (2016). https://doi.org/10. 1080/0144929x.2016.1157207

17. Zimmerman, D.E., Akerelrea, C.: A group card sorting methodology for developing informational Web sites. In: IEEE International Professional Communication Conference (2002). https://doi.org/10.1109/ipcc.2002.1049127

18. Tullis, T., Albert, B.: Measuring the User Experience: Collecting, Analyzing, and Presenting Usability Metrics, 2nd edn. (2013). https://doi.org/10.1016/c2011-0-00016-9

19. Nawaz, A.: A Comparison of card-sorting analysis methods. In: Proceedings of the 10th Asia Pacific Conference on Computer-Human Interaction (2012). https://doi.org/10.1590/s1806- 83242011000200014

20. Petrie, H., Power, C., Cairns, P., Seneler, C.: Using card sorts for understanding website information architectures: technological, methodological and cultural issues. In: Lecture Notes in Computer Science (including subseries Lecture Notes in Artificial Intelligence and Lecture Notes in Bioinformatics) (2011). https://doi.org/10.1007/978-3-642-23768-3_26

21. Sakai, R., Aerts, J.: Card sorting techniques for domain characterization in problem-driven visualization research. In: Eurographics Conference on Visualization (EuroVis) - Short Papers (2015). https://doi.org/10.2312/eurovisshort.20151136
22. Blanchard, S.J., Banerji, I.: Evidence-based recommendations for designing free-sorting experiments. Behav. Res. Methods (2016). https://doi.org/10.3758/s13428-015-0644-6
23. Barnum, C.: Usability Testing Essentials. (2011). https://doi.org/10.1016/c2009-0-20478-8
24. Nielsen, J.: Why You Only Need to Test with 5 Users. Nielsens, Jakob (2000). http://www.useit.com/alertbox/20000319.html
25. Jorgensen, D.L.: Participant observation: a method for human studies. Participant Obs. (1989). https://doi.org/10.4135/9781412985376.n1
26. Chou, Y.-K.: Octalysis: Complete Gamification framework. Website (2015). https://doi.org/10.1016/j.jallcom.2006.04.035
27. Jiménez, S. (n.d.). Gamification Model Canvas | Game Marketing (2013)
28. Marczewski, A.: 48 Gamification elements, mechanics and ideas. Gamified UK (2015)
29. Fonseca, B., Morgado, L., Paredes, H., Martins, P., Gonçalves, R., Neves, P., Nunes, R., Lima, J., Varajão, J., Pereira, Â., Sanders, R., Soraci, A.: PLAYER-a European project and a game to foster entrepreneurship education for young people. J. Univ. Comput. Sci. **18**(1), 86–105 (2012)
30. Martins, J., Gonçalves, R., Branco, F., Barbosa, L., Melo, M., Bessa, M.: A multisensory virtual experience model for thematic tourism: A Port wine tourism application proposal. J. Destination Mark. Manag. **6**(2), 103–109 (2017). https://doi.org/10.1016/j.jdmm.2017.02.002

What Is the Effect of New Technologies on People with Ages Between 45 and 75?

Ana Fontoura[1], Fábio Fonseca[1], Maria Del Mar Piñuel[1],
Maria João Canelas[1], Ramiro Gonçalves[2],
and Manuel Au-Yong-Oliveira[3(✉)]

[1] Department of Languages and Cultures, University of Aveiro,
3810-193 Aveiro, Portugal
{anafontoura, fabiof, mariapinuel,
mariajoaocanelas}@ua.pt
[2] INESC TEC and University of Trás-os-Montes e Alto Douro,
Vila Real, Portugal
ramiro@utad.pt
[3] GOVCOPP, Department of Economics, Management,
Industrial Engineering and Tourism, University of Aveiro,
3810-193 Aveiro, Portugal
mao@ua.pt

Abstract. When we look around us, we see people of all ages using all kinds of technology, more specifically, devices like cellphones are used by younglings, adults and older people. There is no escaping it. Consequently, we decided to analyse how technologies affect the lives of these men and women, focusing on an older age group, mainly because of the lack of studies that focus on people over 65. If most of them are able to speak, focus, and learn, how many of them have given in to the new waves of evolution that have feasted on the world for the past few years? That is exactly what we are trying to find out: How older people react to new technology (more specifically to information and communication technologies or ICT), how often and how they use it. For this study, we decided to focus on the littoral north of Portugal, given that the author-group's members live along the coast line. A survey was performed (with 56 valid responses) as were eight interviews. Regarding the view that the older people we have interviewed have of millennials, most of our respondents show some concern. Notably, they see the Internet as an addiction of the younger elements in society, contributing to them getting into trouble, losing interest in school, meeting up with strangers and only being concerned with being popular on social media. As concerns ICT usage by elders, the psychological component of the "I can't" does not help them.

Keywords: Old people · Technology · Ageing · North and littoral · Seniors · Smartphones · ICT

Á. Rocha et al. (Eds.): WorldCIST'19 2019, AISC 932, pp. 402–414, 2019.
https://doi.org/10.1007/978-3-030-16187-3_39

1 Introduction

Portugal is not a country without history, its impact on the world is known and it has fought through time, conquering land and defeating a large number of many enemies. In sum: Portugal is an ancient nation, but contrary to what might be expected, it is not a country filled with younglings. It is quite the opposite in reality. As shown by the latest data retrieved from INE (the National Institute of Statistics of Portugal), the proportion of the resident population aged 65 years and more by place of residence, in Portugal (according to the data of the last census undertaken, in 2011) represents a percentage of 19.03% [8]. As such, Portugal is an old country filled with old people. However, these people are not invalids, they are capable people that retain certain abilities, such as sight and memory.

Technology is here to stay, and it brought with it innovation, new devices and the Internet – one of the biggest steps in human history. After the Internet entered the homes of the Portuguese people, it brought the ones that were away, at a distance, one step closer to each other. But have the elderly, whose mentality is already "set in stone", embraced these new technologies? Have they accepted this great unknown?

The aim of this study is to understand how people between 45 and 75 years of age feel about this huge wave of new technology, how it has influenced their lives and if they are in on the new waves, especially regarding smartphones.

The number of researchers' focusing on observing the benefits of the Internet in elderly people is now growing. The web has benefits for this group in order to increase communication, escape isolation and loneliness and to make them age dynamically [14].

The goal is to study an age group that has been rather overlooked, people of another generation with different livelihoods. Will there be acceptance, understanding and a good usage by these individuals, or do they only see the new technological evolution wave as a big untamed monster that has come to corrupt young people?

The article continues with a review of some of the literature on the topic. Then there is a methodology section, followed by a discussion of the results of the field work (questionnaires and interviews). We then conclude and suggest avenues for future research.

2 A Review of Some of the Literature

2.1 Technology

Nowadays, technology is something intrinsic to our daily lives, something so common that, contrary to ancient times, it is considered odd when someone abstains from using it. From the moment we wake up until we go to bed at night, we see ourselves surrounded by resources that aim to make our lives easier, that aim to ease our daily activities.

Technology has left a huge footprint in society, so big that it is required as a qualification for most jobs, to engage in certain activities, to socialize and to find the new information that swallows the new world of today. As such, a basic set of skills is required to access most devices and informational forums, and we therefore witness the arising of "New conceptions regarding information systems, such as technological ecosystems" [15, p. 240].

These technological ecosystems are constantly evolving, and such is their growth that we now can interact with whoever we want from any device and from any given location, "that is to say, we are approaching a ubiquitous and mobile environment" [15, p. 241]. It is important to note that people who dominate the necessary skills to navigate such equipment have a greater understanding of the ecosystem they are inserted in. Therefore, a person who effectively knows how to deal with a phone or computer will have a greater use of the services provided to aid in such tasks, demanding even more services and "greater flexibility of access, thus creating a new potential gap in technology for those users with a reduced number of digital competences" [15, p. 241]. Consequently, development and the increased technological insurgence creates a barrier that divides the people who possess the required abilities and those who do not and "between those who are able to reach their full potential within the ecosystems that shape our social and professional activities and those who are just getting by, and perhaps unable to keep pace with the rapid evolution of the technology that surrounds us" [15, p. 241].

Thus, technology usage occupies a great percentage of the activities done at home, in companies or at institutions, and that demands a bigger follow-up effort by individuals to know what is new [9]. In this light, we can see a growing reliance on the usage of electronic resources.

New technologies of information and communication are boosting the economy, with job creation, innovation and lower costs, providing services in education, health and social cohesion. It is something usual in our lives, but not all people use it equally, and the best example is seen in elderly people [13].

However, a question arises: Does the general population have equal access to these resources? Or is there a sample of this population that does not utilize such tools? The common belief is that younger people, who usually are seen as being more versatile, adapt easily and have an easier time using technological devices that are aimed for the general population, but does that assumption make any sense? Is there any other factor worthy of being taken into consideration besides age? This was the motto that led to the development of the present study.

In order to achieve the intended outcome, a sample of people between 45 and 75 years of age was selected. This age group belongs to an older generation that does not englobe the people writing this paper, who are millennials. Several factors were considered, including residency (given that the majority of people interviewed live in the littoral north of Portugal), gender, scholarly habilitations and professional situation.

The final goal of this paper is to understand how people utilize the technological means they possess, what use they give to their devices, including cellphones and computers, and if they use social networks and related apps, as well as the reason behind all of these questions.

It is also important to highlight that the impact of new technologies of communication has not affected all sectors in the same way, as this depends on the social structure in which one is integrated and depends on how people react and adapt to change [16, 17]. This technological evolution has been affecting a significant number of individuals, and among them it is possible to notice a lack of social integration of those that do not possess any or only a few technological abilities. As such, the digital inclusion process is an option that helps to fight this tendency. At times, in the groups where there is more social exclusion due to the lack of knowledge of technology, the cause resides not only in not knowing, but also in not having the means to be able to afford and access it.

In a society as old as the one in Portugal, it is important to try and understand how the older generations see and deal with the technology that has taken over the planet. And even though the Portuguese population is older, contrary to common belief (and knowing that younglings have greater ease in learning new things), most people that are older do not complain about cognitive ability loss. Skills such as memory, vision or the ability to understand others is not an issue. Therefore, taking into account that the older population continues to rise in number as do the ways they deal with the growing availability of technology, "new areas of study are being explored, such as the ones Coulson (2000) calls Gerontechnology" [3, p. 2].

2.2 Gerontechnology

This scientific area seeks to research, develop and implement certain kinds of technology for an elderly section of society. As such, this generational technology tries to include older people so that they do not feel left out, so that they do not feel the advance of the world without them accompanying it. As Herman Bouma [12] states, the technological environment that we create should be adaptive to the changing requirements of ageing processes [12, p. 2].

With the growing impact of technology [4] comes a constant accompanying of new trends, an adaptation of the virtual and artificial world and, obviously, it brings with it a new perception of the world, as well as a change of habits (nowadays, heating food in a microwave is way much more practical than to heat it up on a stove). Therefore, it is important to illustrate the urgency of adaptation and change that technology brings with it.

Adaptation to change is always easier for the younger generations, because they are born and raised surrounded by different kinds of sensorial stimulation. The fact that they grow up in this environment and that they adapt easily helps them to better understand and use new types of tools. Having a good education and access to adequate resources is fundamental for an effective adaptation to technology [4].

These adaptations, however, become complicated through the course of time, and because we get older, the familiarity of that with what we grow up with, opposes directly to the things that are outlandish in our eyes. Besides, new technologies demand new abilities, be it to use them or even to understand them.

As [12] states: "This is called 'structural lag' between individuals and society" [12, p. 94]. Given such a fact, it is impossible not to agree with the [12] when he mentions the difficulty in "unlearning" abilities that were gained through life. Additionally, maybe the problem does not belong to learning in old age, perhaps the problem is present in handling new and old skills.

Following along this point of view, given that the elderly are distinct to the present separation, their needs are too, as [12] explains, and technology should serve them differently, and it is from that need that Gerontechnology is born [12, p. 94].

Gerontechnology does not just worry about providing a better life to older people that find themselves debilitated somehow, rather, it worries about their integration and the integration of those who lead an independent carefree life, and those who wish for more information but have no way of accessing it. Gerontechnology aims to respond to the needs of its target audience (informational needs, among others, of various groups, be them in their 20 s, 30 s, 40 s or 70 s).

2.3   The Ageing Population

Old age is the last stage in human life and ageing, simply put, is a process that is originated at birth and ends with death [5]. Ageing is a complex time process where the choices we make in life weigh later on in the consequences they produce and, therefore, ageing produces different results on different individuals [5]. Due to the large adjustments people experience in life, it is usual for the physical part of the human being to become weaker. The social and psychological maturing of the mind has also different rhythms, and each person finds a balance between growth and decline in his/her own ways.

Considering The WHO (2002, 2012) the concept of active ageing is defined as a growing "process of optimizing opportunities for health, participation, and security, in order to enhance quality of life as people age, facilitating active ageing in better conditions" [14, p. 38].

Having as a base the census created with the population information in 2011 [8], the elderly population, people older than 65, rose to 19% since the last study (in 2001 they represented 16% of the population). The percentage of younger population (people under 14 years of age), remains at 15%. This big discrepancy attributes to Portugal the title of one of the most ageing countries in the world, with an ageing index of 128.

Analyzing such a country with a high rate of old people, it is only natural that we observe a high number of new technologies being utilized by older people. According to the national statistical institute, 74% of people between 16 and 74 use the Internet, and each day the number of people older than 45 using the Internet increases.

It is important to talk about one of the most relevant and bigger problems that affect the society of today, regarding elderly people, that is social isolation. Loneliness and isolation are two of the biggest negative impactors on the lives of the elderly.

2.4   Home Access to the Internet

Nowadays, it takes only a glance around us to understand how surrounded by technology we are, how encircled by it we find ourselves and that is because, "We live in a highly digitalized society, with continuous access to direct and indirect technology regardless of age, education, or profession" [15, p. 240]. In a way, if you really want to be tangled in the new way of social life, you must possess your own device with a connection to the world wide web, as a means of connecting with the modern set of mind, as such "modern society's technological bias makes learning necessary for all groups to get a job, learn an activity, or simply communicate with other people and be informed" [15, p. 240].

Using data collected by the "Inquiry to the use of information and communication technologies by families" [2] in 2016, we can observe that 74% of the Portuguese population has access to the Internet at home and, in the center and north of Portugal, the study area of this essay, the percentage is at 72% and 70% respectively.

Also based on that same study, it is possible to understand that men are more frequent users of the Internet, as opposed to women (the percentages are 72% for men and 69% for women). The Internet is more frequently used, however, by employed people and by students. And our empirical study intends to show us why.

Finally, families with children (94%) and those living in Metropolitan Lisbon (82%) have the best broad band access to the Internet [2].

2.5 Ageing and New Technologies

Older people have been embracing new technologies over time, and even though the number of people older than 65 using the World Wide Web is still small, it has grown in comparison with previous years.

According to the study written by Azevedo [3] cellphones and the Internet are a positive influence in the social relationships of people over 61, for they allow more social interaction to happen, keeping connections alive and breaking geographic barriers, bringing them closer to other people and preventing isolation from society [3].

Gracia and Herrero (2008), cited by [14], found that aged Internet users had better physical health, experienced less problems related to mental disorder and showed higher levels of integration and involvement than non-users.

"Technology has been shown to be beneficial to older people" [11, p. 1], and according to the article of Chen and Chan, of the University of Hong Kong, old people are neglected when it comes to studies about new technologies. It is important to know what the relationship with technology of this generation really is in today's society. According to their study, adults "have a positive attitude towards technology", however, they show little interest in using it [11, p. 1].

Studies have been and are being done so that we can understand in what way technology can be used to fight loneliness and thus get more adults to participate in society's daily life. In a way, it is in this area that Gerontechnology inserts itself, because its goal is to eradicate the loneliness issue, a problem that haunts thousands of elderly people. It is not so much a physically debilitating problem as it is a psychologically worrying one. All of us, in a way, as members of a functioning society, should try and fight this serious problem.

Considering the study presented in [14] and regarding the population segment they have considered for the experiment in the paper (it focused on the elderly) they concluded that "older people use the computer and the Internet less as they advance in age" [14, p. 44], and that "differences emerge between rural and urban areas in favour of the latter" [14, p. 44] given that "the majority of older people wish to learn how to use computers and how to navigate on the Internet, as they consider that these are useful skills to acquire knowledge, stay up-to-date, and to participate in leisure activities" [14, p. 45].

Certain technological advances have benefitted the elderly, such as touch-screen technology and regarding, for example, Australian citizens. "To say that times have changed for this group of Australians [aged 65 and over] would be an understatement. By way of context… in 1952, the only screens most Australians could access were found in cinemas. The launch of mainstream television was still four years away, and this cohort of Australians would be in their 40 s or older before consumer access to dial-up Internet gained momentum in the mid-1990s" [1]. The elderly are each day that passes more connected to technology, and to the information on how it is used, because it reaches them more easily.

New technologies help to increase the social interaction of older people, allowing them to keep contact with their family and friends, improving their quality of life. However, the devices need to be adapted to these people and their needs [13].

Social networks are also on a rising wave, having increased on a large scale, especially Facebook usage, "increasing from around 1 million monthly active users in 2004 to over 1.2 billion monthly by 2013" and "social media use is increasingly becoming a part of older adults' lives" [8].

One of the biggest technological devices present today is the mobile phone, and the smartphone. "A high percentage of older people in developed countries owned one of these devices, however they only use mobile phones for very limited purposes, such as for calling or texting in emergencies" [13, p. 118]. The is because it is not adaptable to their needs.

A study suggests the creation of an App repository just for older people [13] to satisfy their needs. With flexible apps for the common operating systems, with adaptations that would make older peoples' lives easier, such as TalkBack. The repository and the apps would be evaluated according to their usability. The layout of the repository would be suitable to older people, simple and with aiding tools. All of this repository would be able to help and make older people's lives easier when they are using new technologies [13].

3 Methodology

For the writing of this paper, about the effects of new technologies on mature adults, it was decided to limit the target age early in the development of the paper. It was established that the focus would be ages ranging from 45 to 75 years of age. Following that decision, in order to collect the best data and new information, research for the literature review began, using previously published articles by other authors. So that legitimacy could be attributed to the investigation, empirical data was collected. Eight interviews were thus performed with three women and five men, of different ages, different genders and different professional situations (Table 1). The interviews had a duration of about ten minutes each, except for one, that lasted for 49 min. All of the interviews were audio recorded with permission. The sample of interviewees (interviewed from the 18th to the 22nd of October 2018) was made up by people who actively participate in our lives, people we knew who dealt (or did not deal) with technology. A questionnaire was also created so that we could research which group has the most difficulty using new technologies. This questionnaire, which was placed online and divulged through social media and through the authors' personal network

Table 1. Profile of the interviewees

Name	Gender	Age (years)	Profession
Interviewee 1	Male	63	Painter
Interviewee 2	Female	59	Housekeeper
Interviewee 3	Male	75	Retired
Interviewee 4	Male	53	Electrician
Interviewee 5	Male	49	Professor
Interviewee 6	Male	60	Security guard
Interviewee 7	Female	50	Worker in a factory
Interviewee 8	Female	47	Shoemaker in a factory

(including the authors' as well as the authors' colleagues' parents and grandparents), was answered by 66 people (a total of 56 valid responses were received).

Thus, the paper is divided into two kinds of analysis. In the quantitative study data was retrieved from the questionnaires made available online through social media, and in the qualitative analysis interviews (with eight people) were performed with the goal of obtaining answers and results and in order to come up with a reasonable and realistic conclusion.

4   Discussion of the Results

4.1   The Questionnaires

Of all the respondents to the questionnaires 30.4% are men and 69.6% are women. The questionnaire, structured according to the study undertaken beforehand, focuses on the littoral north of our country, covering the Aveiro and Greater Porto areas, in order to better understand the reality of this Portuguese region.

Although there were people between 21 and 76 years of age answering the questionnaire, we will only acknowledge the answers that were given by people who are between 45 and 75 years old, for that is our target demographic. That being the case, out of the 66 answers we will only consider 56 as valid responses. We received more responses from people belonging to the group age of 45 and 54 and less from the age group varying from 65 to 75 years old. This may suggest that people with an advanced age have little means to access our survey. Out of all the ages, the one with the biggest percentage of answers was the age of 49 with 7 answers, that is 12.5% of the total.

In fact, most of the answers present in our survey are from within the 40 to 50 age group and we would even argue that, after this observation, we may conclude that unlike people in their 60 s and 70 s, people between 40 and 59 are the largest users of technology within our target demographic. We claim this because, after the age of 49, it is the age of 53 that answered the survey the most, with a total of 6 answers and a percentage of 10.7% of the total sample.

It is also important to highlight the fact that the bigger percentage of answers come from women, with 69.6%, again showing the tendency for women to be more active in academic pursuits in Portugal.

Regarding academic qualifications, we have obtained 20 responses from people who have completed the 6th grade, a percentage of 35.7%. Following the people who have finished the 6th grade, come the ones who have completed the 4th year of school, with a percentage of 17.9%, a clear contrast to those who have obtained a PhD (5.4%) or a Master's degree (1.8%), which present the lowest percentages of the response sample. In a deadlock, with the same percentage of 12.5%, we observe the ones who have completed their 9th grade and the Bachelor's degree level.

Concerning employment, most people who have answered the questionnaire have a job, a percentage of 71.4%, whilst 10.7% are currently unemployed. The retired percentage is at 17.9%, with 10 people.

A total of 78.6% of the people who have answered the questionnaire have a computer at home, and 82.1% of the respondents claim to have a smartphone, with only one small percentage using older cellular devices, the so-called "dumbphone". It is possible to observe a significant fraction for cellphone usage (having been classified from 0 to 5 in which 0 is no use and 5 is very frequent use), 26.8% of people voted 5, while 5.4% voted one, a clear difference as regards phone use. In terms of specific use given to the phone, the consensus is that almost all of the participants use the phone for making calls (100%) and for messaging (85.2%). In terms of social network use, only 46.3% of the sample use it for that purpose, which matches less than half of the participants, and within those who use it, it is important to enhance that Facebook is the most frequently used, with 69.6% of older people using it, followed by WhatsApp (26.8%). However, the biggest percentage that rivals the high percentage of Facebook is the option "none", showing us that there is a strong use of technological equipment without accessing digital social platforms.

Somehow opposing the fact that many of the participants do not have an account on social networks, most participants (67.9%), refer that they do not have any issues in using their devices. But is that the factual truth? Or is there any sort of reluctance in admitting to the lack of skills as regards using phones and computers?

4.2   The Interviews

In a more qualitative analysis, we will review the interviews we undertook and study them one by one, to find common aspects and answers we consider important to present and detain.

The opinions about old cellphones, the ones with physical keys instead of a touchscreen, are very similar. Generally, the equipment was described as being a decent piece of technology, with a long-lasting battery that wouldn't immediately shut down. In fact, this would only happen after many hours of usage. They were also described as being resistant pieces of equipment, that when they fell would not break, with few functionalities and consequently rarely blocking or "crashing".

On the other hand, the use of smartphones is very imbedded in our society, and that is noticeable when we observed that most of the subjects interviewed, nowadays, use the device to contact other people. Besides its use for calls, it also works as an agenda, to take notes and to make lists. "All the things that were done on paper before I have now on my cellphone" (interviewee 5). Smartphones also double as mechanisms to access social networks, e-mails, Google maps and even meteorology. Smartphones are practical, they fit in our pockets and are portable, allowing us to use them wherever we want to. This mobility that new technologies allow us to have is an asset, especially for those who have a busy schedule, as they help us to organize our time and allow us to work anywhere, anytime.

"The phone is like a soldier's gun: without it many tasks would be impossible. It is a fundamental piece in the daily work of individuals, without it they are unreachable. Furthermore, work is ever present via the smartphone as well as the contacts of other people with whom we need to work with. Losing our smartphone means wasting a lot of time because it is necessary to restore all the old information, if it is available at all, in recent backups. Without smartphones, people would not be able to work as efficiently, the same way that a soldier would not be able to defend himself without his gun" (interviewee 5).

In most of our interviews, people told us that they use online finance-related apps that allow them to gain more time, since they do not need to go to the physical bank and wait in line for hours. It is simpler and even more economical to use apps, given that you can do it comfortably from home.

Regarding the view that the older people we have interviewed have of millennials, most of our respondents show some concern. They see the Internet as an addiction of the younger elements in society, contributing to them getting into trouble, losing interest in school, meeting up with strangers and only being concerned with being popular on social media.

One of the questions asked in the interviews concerned the choice between travel agencies and online traveling services. We realized that most people still prefer to use travel agencies whenever they want to go on holiday, since they provide more security in case anything goes wrong.

As regards the interviews, the last question inquired about how people feel in relation to all the evolution that has flooded the market. The answers, ranging from people in their forties to people in their sixties, were very diverse, as shown below:

- "I do not feel like I am retarded, I have evolved along with technology, I am not that far behind, I have managed to keep up with evolution. It is a good and a bad evolution. There are easier things, we do not have to go to some places, and it is also bad for a lot of the youths, they learn a lot they should not learn, they get lost in social networks, there are young people who cannot live without it. I can live without my phone, I am just not reachable, and that is not an issue." (interviewee 4).
- "Rotten, they only learn bad things, it only creates issues, without the Internet there would not be so much tragedy" (interviewee 2).
- "It is reasonable, it is evolution, it is easier to learn new information and it is more practical" (interviewee 1).
- "Important for evolution, it avoids time loss, it is more economical. Regarding online shopping though, it is expensive, we do not really know what we are buying, we buy "a pig on a poke", the image [often] has nothing to do with the product. I am

very unhappy with young people, they go where they should not and cause a lot of tragic events. And indecent ads are bad " (interviewee 3).

- "So, technology has changed everything. Before, car windows would have to be manipulated by hand, the radio had to be synchronized, it was not automatic. So much has changed… I got lost so often before, now I do not. No one does. Not being able to access the Internet is a tragedy." (interviewee 5).
- "Well, it has facilitated a lot of processes where waiting in line and transportation were needed. I can see the TVI channel easily, online and on TV. Seeing the news online has also helped save a lot of money, since we do not spend it on physical newspapers anymore (the ones made of paper, they pollute a lot) and we are now able to read the news for free and at the distance of a click" (interviewee 6).
- "Happy. It has made it possible for me to meet with people I have not seen in a while" (interviewee 6).

5   Conclusion and Suggestions for Further Research

One of the most important conclusions to retrieve from this paper has, for a basis, our reduced number of people that answered our online questionnaire, which may have not been the best choice to produce this project. We have had a difficult time collecting answers, given that most of the people that are in our target demographic had limited resources to access our survey.

It is also important to reference that we have overlooked one of the most important methods required, which is observation [6, 7, 9, 10]. As such, we give the suggestion for further research, in that one should observe the behaviors of people that allege that they have no problems handling a smartphone, with one in their hands. We suggest doing so because, in our survey, one of the most common answers given was to claim that no problems in using a cellphone exist, but is that so? Or is the lack of problems in dealing with smartphones simply reduced to a shutting down of the device, without even solving the issue?

Of course, with this lack of a notion about how electronic devices really work, comes the realization that, as Bouma has previously written, it is extremely difficult to combine older and newer knowledge, and that an even bigger challenge resides in the resistance created around giving in to the new wave of technology, because, just like one of the people interviewed has stated "…I think we should evolve, I just didn't really feel the need to try it, I'm good as I am right now. I appreciate the familiar" (interviewee 8), and such opinions make the inclusion of the elderly in the present wave of technological growth more difficult.

There are also people who vehemently refuse the use of new technology, smartphones, and instead choose what has been around longer, just like landline phones. We have not considered that variable in this paper, however, one of the people that was questioned by us states that she only possesses a landline phone: "I only own a landline phone. I have no cellphone, nor do I need one" (interviewee 2).

On the other hand, there is also the psychological side, many seniors are retired, and as one of the elders interviewed said "I'd like to know how to use financial services, but it's hard to use and because it's not something we use every day, it doesn't stick in my memory" (interviewee 3). So, even though people want to use it, the psychological component of the "I can't" doesn't help them. Because they're not used to using these technologies since a young age, it is hard for them to realize how to use it and that they should. This negative attitude and the lack of effort (due to resistance to change) is what makes them stick to the very basics. That being the case, we have concluded that new technologies have brought a new life and vision to people. That vision can be either positive or negative. On the one hand, we have those who look at the Internet and to the new smartphone devices as being used for evil, where young people get lost and end up ruining their own lives, be it meeting people who hurt them or becoming the perpetrators of evil deeds. However, not all of it is bad, and there are older people who view this innovational wave that has reached our homes with a more joyous gaze. Many people see the smartphone as a life savior, a time saver, a path finder, an information keeper and a gasoline warden. All that in the old days was done personally, by hand and that required extra work is now literally in our hands, all that is needed is a Wi-Fi connection or mobile data. New technologies have come to make the lives of many elderly easier and, as such, some of them embrace it, learn from it and let them become a part of their daily lives.

Acknowledgements. We would like to thank the interviewees and questionnaire respondents who offered their time and experience to help us in our study.

References

1. ACMA: Digital lives of older Australians (2016). https://www.acma.gov.au/theACMA/engage-blogs/engage-blogs/Research-snapshots/Digital-lives-of-older-Australians. Accessed 13 Dec 2018
2. INE: Inquiry to the use of technology of information and communication by families (2016). https://www.ine.pt/xportal/xmain?xpid=INE&xpgid=ine_destaques&DESTAQUESdest_boui=250254698&DESTAQUESmodo=2. Accessed 13 Dec 2018
3. Azevedo, C.: Tecnologias e pessoas mais velhas: Importância do uso e apropriação das novas tecnologias de informação e comunicação para as relações sociais de pessoas mais velhas em Portugal. Master's Disssertation. Universidade Nova de Lisboa (2013)
4. Bez, M.R., Pasqualotti, P.R., Passerino, L.M.: Inclusão digital da terceira idade no Centro Universitário Feevale. In: Brazilian Symposium on Computers in Education (XVII Simpósio Brasileiro de Informática na Educação - SBIE), Vol. 1, No. 1, pp. 61–70 (2006)
5. Dias, I.: O uso das tecnologias digitais entre os seniores: Motivações e interesses. Sociologia, Problemas e Práticas **68**, 51–77 (2012)
6. Gonçalves, R., Oliveira, M.A.: Interacting with technology in an ever more complex world: designing for an all-inclusive society. In: Wagner, C.G. (ed.) Strategies and Technologies for a Sustainable Future, pp. 257–268. World Future Society, Boston (2010)
7. Au-Yong Oliveira, M., Gonçalves, R.: Restless millennials in higher education – A new perspective on knowledge management and its dissemination using IT in academia. WORLDCIST, Porto Santo, 11–13 April. In: Rocha, A., Correia, A.M., Adeli, H., Reis, L. P., Costanzo, S. (eds.), Recent advances in information systems and technologies, vol. 2., Advances in Intelligent Systems and Computing (Book of the AISC series), vol. 570. Springer, pp. 908–920 (2017)

8. INE. Censos 2011
9. Au-Yong-Oliveira, M., Gonçalves, R., Martins, J., Branco, F.: The social impact of technology on millennials and consequences for higher education and leadership. Telematics Inform. **35**(4), 954–963 (2018)
10. Martins, J., Branco, F., Gonçalves, R., Au-Yong-Oliveira, M., Oliveira, T., Naranjo-Zolotov, M., Cruz-Jesus, F.: Assessing the success behind the use of education management information systems in higher education. Telematics and Informatics, pp. 1–12 (2018)
11. Chen, K., Chan, A.H.S.: A review of technology acceptance by older adults. Gerontechnology **10**(1), 1–12 (2011)
12. Bouma, H.: Gerontechnology: emerging technologies and their impact on aging in society. In: Graafmans, J., Taipale, V., Charness, N. (eds.), Gerontechnology: a sustainable investment in the future, pp. 93–104, IOSpress (1998)
13. García-Peñalvo, F.J., Conde, M.Á., Matellán-Olivera, V.: Mobile apps for older users – the development of a mobile apps repository for older people. Learning and Collaboration Technologies. In: Technology-Rich Environments for Learning and Collaboration, pp. 117–126. Springer International Publishing, Switzerland (2014)
14. Casado-Muñoz, R., Lezcano-Barbero, F., Rodríguez-Conde, M.J.: Active ageing and access to technology: an evolving empirical study. Comunicar **23**(45), 37–46 (2015)
15. Fonseca Escudero, D., Conde-González, M.Á., García-Peñalvo, F.J.: Improving the information society skills: is knowledge accessible for all? Univ. Access Inf. Soc. **17**(2), 229–245 (2018)
16. Martins, J., Gonçalves, R., Branco, F.: A full scope web accessibility evaluation procedure proposal based on Iberian eHealth accessibility compliance. Comput. Hum. Behav. **73**, 676–684 (2017)
17. Gonçalves, R., Martins, J., Branco, F.: A review on the portuguese enterprises web accessibility levels – a website accessibility high level improvement proposal. Procedia Comput. Sci. **27**, 176–185 (2014)

Economic Relations Modification
During the Digital Transformation of Business

Sergei Smirnov[1]([envelope]) [iD], Eugen Cheberko[1] [iD],
Igor Arenkov[1] [iD], and Iana Salikhova[2] [iD]

[1] St. Petersburg State University,
7/9 Universitetskaya nab., St. Petersburg 199034, Russia
sergej-smir@yandex.ru
[2] St. Petersburg State University of Economics,
21 Sadovaya str., Saint-Petersburg 191023, Russia

Abstract. This paper is devoted to economic relations transformation under the influence of digital transformation, the recent socio-economic trends and the overall crisis of the free market system. Recently researchers have shown an increased interest in changes of commodity-money relations, expansion of pseudo-market relations, de-commercialization and, vice versa, hyper-commercialization, that spreads out to cover new markets globally. By authors opinion economic relations are in the process of transformation that ultimately will lead to sufficient decrease of for-profit organizations and period of non–profit organizations flourishing. Digital transformation even now creates major challenges for current commodity-money relations. Excessive abundance of available information distort the market functioning and create obstacles for the productive forces development. Pseudo-market relations have become increasingly common especially in the spheres of culture, art and sport. Under these circumstances during the forthcoming long period of digital transformation the increasing importance has to be given to the moral aspects of economic relations, and these problems could not be any more ignored by governments.

Keywords: De-commercialization · Sociocultural factors ·
Commodity-money relations · Digital transformation

1 De-Commercialization of the Contemporary Economics

Among the profound changes taking place in the system of social relations' system the particular attention is drowned to the processes that often referred as the de-commercialization. In the economic literature, this phenomenon is described as a significant growth of nonprofit organizations with strive to penetrate in traditional "for profit" markets. These dramatic changes are large scaled and unevenly spread throughout the global economy. In developed countries, non-profit sector employment is expected to reach 50% of the total employment in the economy. These processes were reflected in the number of theoretical concepts and usually referred to the de-commercialization paradigm [1, 2]. The ambivalent nature of commodity production finds its ultimate expression in the paradigm of commercialization. Generally speaking,

© Springer Nature Switzerland AG 2019
Á. Rocha et al. (Eds.): WorldCIST'19 2019, AISC 932, pp. 415–421, 2019.
https://doi.org/10.1007/978-3-030-16187-3_40

every business entity has immediate and the definite goals. A definite goal is the consumer value, it is a satisfaction received by a customer from getting products or services. An immediate or actual objective of the business is determined by the owner. In case of a commercial organization the immediate objective is a profit. In a subsistence economy both goals are equal. The emergence of private property rights and labor division has led to the disconnection of above-mentioned business objectives.

In a simple commodity production the main need was to maintain exchange equivalence but not a profit maximization that later became dominant. Aristotle was the first scholar who paid attention to the contradictory character of commodity production. He determined two types of human activities: creating favorable living conditions and accumulating monetary wealth. The science about the first type of activities he named as economics and the second one as chrematistics. At the same time the strengthening of commercial component of production was observed, so Aristotle had a negative attitude towards this trend.

Speaking about contemporary de-commercialization phenomenon we mean movement from the concept of profit maximization to better concept where needs and aspirations of society meet. Presently we do not speak about rejection from the entrepreneurial spirit, but we mean the establishment of a reasonable balance between final and immediate production goals. Herewith we should keep in mind that an entrepreneurship is based on economic activity but not every economic activity is an entrepreneurship. Such widely spread sameness of entrepreneurial and economic activity is not so harmless as it could be thought from theoretically and practically points of view. In the past the absolutization of one economic activity aspect led to serious negative consequences. It would be enough to remind the collapse of the Soviet experiment when there was an attempt of total rejection of entrepreneurial spirit from the economic activity.

During the digital transformation a new level of proximity to the consumer occurs. As the customer takes part in product or service design he begins to feel as co-creator who participates in the supply chain process and who is responsible for key product or services characteristics. It means that the customer performs some function of vendor and even entrepreneur and respectively wants to know more about economic aspects of the production and to modify it to his own advantage.

Moor close social interactions between people through the digitalization of social life also contributes to this. As a result, there are created further prerequisites for the joint consumption non-profit models development, which will arise in more and more new industries.

2   Pseudo-Market Relations

Development of new social relations is proceeded not by pre-designed scenario but with trials and errors. Not all changes could be classified as positive [3]. Among all the variety of options, we can see the stalemate and unfavorable ones. Commodity-money relations are one of the forms of social relations, which originated from high antiquity but not at the moment when human civilization appears. In the very long term it will disappear as useless when the need of equivalent exchange disappears as well as the need to measure and compare consumer value in monetary terms. In course of time the sector where commodity-money relations exist only formally but not substantively will expand. During the USSR period there were such relations when trade resources took the form of product and exchange was occurred in monetary form, but there were no full-valued market relations. Can we reasonably discuss the reality of commodity-money relations in the spheres of culture and sport? Or we can easily recall unrealistic high prices on paintings of famous artists.

Very often premium pricing in luxury markets is not correlated to their quality. Also, we can not skip the examples in sports. The wages of football players are inadequate enormous and have no correlation with their sports clubs' revenues. We can conclude that commodity-money relations in culture have artificial nature enforced from the outside [4]. Culture in the narrow sense of this word (contemporary arts, and all that is connected with the satisfaction of emotional needs) has significantly become the place of money laundering, gained in the shadow economy by big capital. Moreover, commodity-money relations in this sphere are formal and have no real value base. The trade product has to be valued and useful but it is not enough as the monetary value should be adequate to its usefulness.

3 The Material Base of Changes in the Market Relation and Appearance of New Economic Subculture

Nowadays pivotal changes in the nature and fundamental principles of market activity have occurred under pressure of huge shifts in manufactural forces emerging from the digital transformation and the transition to 6-th technological mode. According to another approach it is the advent of MANBRIC-technologies (a complex that includes medical, additive, nano and biotechnology, robotics, information and cognitive technologies) [5]. There is no fundamental difference between the above mentioned approaches, but the second is more specific. Let us consider how digital transformation and new technologies will affect the process of de-commercialization. A number of authors discuss that the main reason for capitalism's collapse will be information technologies, which destroy market mechanisms. First of all over-accessible and excessive information hinder the ability of the markets to form the prices correctly. Even more dangerous to the economy can affect a distorted signal which the manufacturer receives from the market.

The serious consequences of this situation are warned by Fituni and Abramova. They draw attention to the fact that persons often act under influence of not real market signals, but subjective information with suggestive and manipulative features (ratings, unreasonable forecasts, pressure from lobbyists, ideological labels, and stamps, economic boycotts, sanctions, misinterpreted public good, etc.) [6]. The notions "market", "market system" are changed under influence of information technologies, digitization and digital transformation, not only in terms of social production dissemination, but the mechanism of their action transform.

The classical market was based on the mechanism of price signals, but nowadays we receive that it could be easy to make mistakes when deciphering these signals with all contemporary distortions. In order not to be mistaken we should have professional skills and institutions that ensure the information search and its correct interpretation. Therefore, when we speak about the market in general with its efficient mechanisms, we mean the situation that no longer exists in real life, since the mechanism of self-regulation does not automatically work anymore. Large changes in the productive forces and economic relations are associated with the advent of additive technologies. On the one hand, it is already obvious that they will strengthen the level of the producer's isolation. In many areas factory organization of production will come to naught. Certainly final products can be obtained from the manufacturer, who works at home and use additive technologies. Of course, there are a lot of technical difficulties (high energy intensity of production, its high harmfulness, security, etc.). But this road cannot go away. On the one hand, this will lead to increased isolation, atomization of production, and consequently to the strengthening of market principles. On the other hand, this should lead to an intensification of economic regulation on a fundamentally new basis. Most likely it is a question of a new quality of market network interaction between producers. Already nowadays new forms of social relations lead to the necessity of revision of the established ideas about the economic system functioning.

It was assumed that the strengthening of the social role of production would be accompanied by an increase in the role of the state, or rather its function – to ensure sustainability. And the process of raising the level of production socializing will lead to a diminishing role and shrink the size of the commodity-money relations. Why? As useless. The mandatory condition of commodity-money relations is an existence of a large number of independent owners who are destined to disappear in direct manufacturing era. Market mechanisms do not disappear but are transformed. On one hand, an excessive abundance of available information reduces efficiency and distorts the price mechanism of market regulation. On the other hand, it strengthens the public component of the economic reproduction process by compelling the private owner to obey decisions handed down to them through information channels from outside. In the context of globalization and regardless of the capitalist's will, socialization takes place under the pressure of the growing volumes of network information. In the era of digital transformation the entrepreneurs' actions are more and more predictable and "planned", coordinated or conformed to the colleagues' and competitors' actions, and are based on the proposed model solutions, etc." [6]. Information coming from markets became a significant force for the promotion of labor socialization process.

The first condition for commodity-money relations existence is a fundamental need of exchange between different producers. The need for exchange is the basis for any economic activity. But nowadays the cost value basis of exchange processes begins to erode. Instead and in addition to it, the form of exchange using non-monetary values, free time, mutual services without compensation is emerging. The new network structures of production chains appear without traditional hierarchy relationships. Incoming manufacturers do this voluntarily and they also can leave it without significant costs. Are these relations equivalent? Yes, but they are not based on monetary value. Integrating into the new networks, their members receive from cooperation exactly as much as they want, because there is no coercion and there is no decision center in the traditional sense. If this is not a new generation of commodity-money relations in fact, then at least in their place new social relations are emerging. In recent years the new business subculture has emerged, which is called the "sharing economy", the key concepts of which are categories of "common" and "peer-production".

There is a spontaneous growth of peer production of goods and services outside of both market relations and any management hierarchy which relies on self-organizing communities of individuals. The motor of capitalism transformation is the mass collaboration (or social enterprises emerging), i.e. local economic entities that are not incorporated into the capitalist economy and carry out their activities in particular through the exchange of time (time banks), the practice of gifts and care for others. Paul Mason, the economic editor of the BBC in the article "The end of capitalism has begun", writes that it is enough just to look more closely and you will see this new economy [7].

Digitization of socio-economic relations simplifies the information exchange and leads to an increase in the intensity of ground-level socio-economic activity.

When non-governmental organizations in Greece examined the country's food cooperatives, alternative producers, parallel currencies and local exchange systems, they found more than 70 significant projects and hundreds of small initiatives - from squats and carpules to free kindergartens. Such things can hardly be qualified as economic activity - but that's the thing. They exist because they provide the customer value, sometimes inconsistently and inefficiently, using the currency of post-capitalism: free time, network activity and free things and services. It may seem that it is impossible to create a suitable alternative to the global economic system based on these beginnings, but in the era of Edward III, they treated by the same way money and credit [8].

Socialization of production proceeds as transformation and de-commercialization of capitalist commodity relations. This will become absolutely clear when under the influence of the productive forces development the need to earn a living by working will disappear. Already today it becomes clear that in the future society the measure of wealth will be free time. Obviously, the opportunities provided by the digital transformation of society and business will be used not only by traditional entrepreneurs, but also by the activists and non-governmental organizations in the creation of large non-commercial platforms of goods and services exchange, and later for the production of new goods and services.

4 Sociocultural Factors of De-Commercialization

The formation of a new mentality may not be a decisive factor in the process of the emergence of new economic relations, but it is very important and requires careful study of cultural influence on the economy. Human capital has come to the forefront as the main factor of economic development and now it defines the welfare and well-being of society. Already now, digital talents are a key resource of digital transformation projects, they are distinguished by digital thinking, the ability to see the enterprise activities in the form of algorithms [9]. But at the same time, bare algorithmization of business processes is impossible without disregarding the moral side of dramatic changes.

It is probably early to declare that among all characteristics of a modern manufacturer, the level of moral development and other similar aspects of his personality are decisive. Meanwhile, the first places among them are occupied by education, creativity, the level of professional training. The transformation of the economy from the material category into a category of moral one is not in the nearest future.

The difficulties in analyzing this phenomenon increase motivation for further development of institutional-economic models. That transforms the idea of the "economic man" and considers the agent's motivations in the broader context of adaptation to institutionalized rules and traditions. In the evolutionary paradigm of the economy a new step has been taken that requires taking into account the evolution of economic agents, studying the socio-economic genotypes of their population, taking into account the interaction of acquired factors and factors of "social heredity" [10].

The beginning of new social relations which are difficult, and sometimes impossible to explain, relying on the modern methodology of research may be experienced in different parts of our economic life. Thereby, it is promising to look at the economy using the methodology and categorical apparatus of the cultural studies, since it provides a methodology for research contemporary political–economic relations.

5 Conclusions

Digital business transformation leads to a change in the nature of the contemporary economic system. In this context, the de-commercialization paradigm attracts special attention. It implies not the divorce from the market economic principles, but rather the redressing the imbalance induced by hyper-commercialization phenomenon. It may be premature to introduce the issue of decreasing the scope of entrepreneurial activity as an obligatory imperative for social relations development.

The spirit and nature of the phenomena of entrepreneurship are changing under the influence of overall digital transformation. This is a subject for further research.

It is equally dangerous for the economy both to disqualify commercial approach and to turn it into a single form of economic development. Market principles of economy in their pure form ensure social justice, but by their nature, they can not and should not ensure social equality. At the same time, justice can be violated when there are pseudo-market relations occurred. When under the guise of market relations, their surrogates are used with only one purpose to create conditions for carrying out anti-social activity. The emergence of pseudo-market relations with a negative public result means that hyper-commercialization has gotten out of control. Digital transformation and digitalization create additional distortions by providing technical means to the above mentioned negative aspects. The pursuit of profit, by all means, has led to the emergence of commercial market relations where they are unacceptable. At the same time, it should be stipulated that it is not necessary to resist against market profit-oriented relations but to resist those which led to hyper-commercialization of economic activity. That is when the remedy turns into a goal.

The process of economic relations transformation occurs under the influence of profound changes in the productive forces that provide a material basis, the prerequisites for changing the essential characteristics of commodity-money relations in the modern society. In such a manner digital transformation of businesses and society destroy market mechanisms by the means of information and communication technologies.

References

1. Kleyner, G.B.: Paradigma dekommertsializatsii: globalnyie imperativyi i natsionalnyie interesyi. Sovremennyie globalnyie vyizovyi i natsionalnyie interesyi: XV Mezhdunarodnyie Lihachevskie nauchnyie chteniya, SPb.: SPbGUP. (2015). (in Russian)
2. Kleyner, G.B.: Dekommertsializatsiya ekonomiki kak kulturnyiy proekt (na puti k sozdaniyu kulturologicheskoy teorii ekonomik. Gumanitariy yuga Rossii, № 2. (2015). (in Russian)
3. Harcourt, W.: The Future of Capitalism: A Consideration of Alternatives, vol. 38, pp. 1307–1328. Cambridge j. of economics, Cambridge (2014)
4. Bogomazov, G.G., Davidova, D.A.: Sphere of culture as object of study economic science. St Petersburg Univ. J. Econ. Stud. **33**(3), 415–432 (2017). https://doi.org/10.21638/11701/spbu05.2017.304
5. Grinin, A.L.: Kiberneticheskaya revolyutsiya i istoricheskiy progsess (tehnologii buduschego v svete teorii proizvodstvennyih revolyutsiy, Filosofiya i obschestvo. № 1 (2015). (in Russian)
6. Fituni, L.: Zakonomernosti formirovaniya i smenyi modeley mirovogo ekonomicheskogo razvitiya (in Russian), Mirovaya ekonomika i mezhdunarodnyie otnosheniya (in Russian). № 7 (2017)
7. Mason, P.: The end of capitalism has begun http://www.theguardian.com/books/2015/jul/17/postcapitalism-begun-end-of. Accessed 30 Dec 2017
8. Mason, P.: Postcapitalism: A Guide to Our Future. Paul Mason, Allen Lane (2015)
9. Smirnov, S., Arenkov, I., Yaburova, D.: On the way to the organization of the future: practice, patterns, and prospects. In: Proceedings of the 32th International Business Information Management Association Conference (IBIMA), pp. 2928–2937, At Seville, Spain, ISBN: 978-0-9998551-1-9
10. Kleyner, G.B.: Evolyutsiya institutsionalnyih sistem. Nauka, Moscow (2004). (in Russian)

A Model to Define an eHealth Technological Ecosystem for Caregivers

Alicia García-Holgado$^{(\boxtimes)}$ ⓘ, Samuel Marcos-Pablos ⓘ,
and Francisco J. García-Peñalvo ⓘ

GRIAL Research Group, Computer Science Department,
Research Institute for Educational Sciences, University of Salamanca,
Salamanca, Spain
{aliciagh, samuelmp, fgarcia}@usal.es

Abstract. The ageing of world population has a direct impact on the health and care systems, as it means an increase in the number of people needing care which leads to higher care costs and the need for more resources. In this context, informal caregivers play an important role as they enable dependent persons to stay at home and thus reduce care costs. However, long-term continuous care provision has also an impact in the physical and mental health of the caregivers. Moreover, geographical barriers make it difficult for caregivers to accessing psychoeducation as a way to alleviate their problems. To support caregivers in their needs and provide specialized training, technology plays a fundamental role. The present work provides the theoretical basis for the development of a technological ecosystem focused on learning and knowledge management processes to develop and enhance the caregiving competences of formal and informal caregivers, both at home and in care environments. In particular, a platform-specific model to support the definition of the ecosystem based on Open Source software components is presented, along with a Business Model Canvas to define the business structure as part of the human elements of the technological ecosystem.

Keywords: Model Driven Development · eHealth · Software ecosystems · Technological systems · Software engineering · Business Model Canvas

1   Introduction

It is a fact that world population is aging. According to the United Nations world population prospects, the number of older persons—those aged 60 years or over—is expected to be more than double by 2050 than it is today, and growing faster than all younger age groups [1]. A common characteristic of the elderly is the frequent occurrence of both cognitive and physical impairments, which results in an increase in the cost of care and resources needed for this population.

In this sense, informal care plays a very important role within the care systems of many countries [2], as it prevents the institutionalisation of the dependent persons enabling them to stay at home and thus reducing care costs. However, the average age of the caregivers is also rising, and providing long-term continuous care entails a high

© Springer Nature Switzerland AG 2019
Á. Rocha et al. (Eds.): WorldCIST'19 2019, AISC 932, pp. 422–432, 2019.
https://doi.org/10.1007/978-3-030-16187-3_41

physical and mental health impact for the caregivers [3], who suffer problems such as work overload, depression or anxiety, significantly reducing their quality of life and increasing their social isolation.

The caregivers need, therefore, a way to obtain answers to the questions that daily arise during their care duties, psychological support to help fulfil their tasks, information, advice and guidance, as well as an access to a community of equals and experts that can help them. But only when they can all get it easily, in low cost formats and when the information is adapted for them and easy to understand.

Psychoeducation [4] can be an alternative solution to alleviate the aforementioned. Psychoeducation involves providing information in a coherent, simple, accurate and objective way, both for the dependent people and their caregivers. However, access to psychoeducation, which is mostly performed through face-to-face interventions, is usually difficult for caregivers as they cannot leave alone the person they are attending, or because of geographical barriers especially for those who live in the rural environment.

In this sense, IT solutions can allow (in)formal caregivers and dependent persons to receive support in their needs. These technological solutions can make it possible to design and develop personalized attention services, provide remote teaching-learning environments as well as social networks for social inclusion and contact with experts regardless of their situation and context.

The present paper aims to providing a technological ecosystem [5, 6] that allows the caregivers to develop and enhance their caregiving competences both at home and in care environments, and also to share that knowledge with the persons they take care of. In order to support the definition of this ecosystem, a platform-specific model and a Business Model Canvas were developed.

The rest of the paper is organized as follows. Section 2 describes the methodology used to define the technological ecosystem. Section 3 presents the model to define eHealth ecosystems for caregivers. Section 4 depicts the business structure using a Business Model Canvas. Finally, Sect. 5 summarizes the main conclusions of this work.

2   Methodology

The definition and development of technological ecosystems has greater complexity than traditional information systems. The problems inherent to software engineering, such as the interoperability between components or the evolution of the ecosystem, are combined with the difficulty of managing complex knowledge and the diversity of people involved [7]. In order to improve the definition and development of this kind of technological solutions, two metamodels were defined and validated in previous works [8–10]. First, a platform-independent metamodel (PIM) that identifies the main concepts of a technological ecosystem and the relations among them (https://doi.org/10.5281/zenodo.1066369). On the other hand, a platform-specific metamodel (PSM) to define ecosystems based on Open Source software (http://doi.org/10.5281/zenodo.1284567). Both metamodels are instances of Ecore [11], a simple meta-metamodel based on Meta Object Facility (MOF), the standard provided by the Object Management Group (OMG) to support Model-Driven Architecture (MDA).

In order to provide a technological ecosystem for formal and informal caregivers, the ecosystem metamodels were used as a starting point. Figure 1 shows the different models and transformations in the four–layer metamodel architecture of MDA. First, the eHealth ecosystem model was instantiated from the platform-independent ecosystem metamodel. Later, the model was transformed in a platform-specific metamodel using a set of rules defined with ATL [12].

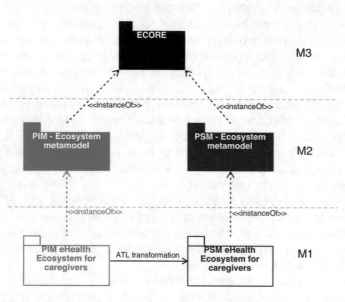

Fig. 1. Models and transformation to get the platform-specific model of the eHealth ecosystem for formal and informal caregivers

To complete the definition of the eHealth ecosystem for caregivers, the platform-specific model is completed with the definition of the business structure through the Business Model Canvas (BMC) [13]. This structure is part of the human components of the ecosystem. One of the main differences between technological ecosystems and other approaches such as software ecosystems (SECO) [14], is the human factor, although it has presence in any kind of technological solution, in the case of technological ecosystems the human factor has the same relevance than the software components. People are not only end-users but also an important component of a technological ecosystem [15].

The BMC is a template that depicts a methodology to design and describe a business model. Business models are defined by [13] as a description of how an organization creates, delivers, and captures value, where 'value' is the benefit an actor gets from the ecosystem in the form of need satisfaction or problem solution [16]. The BMC consists in nine different blocks that describe the business infrastructure, value propositions, customers and finances.

The obtained platform-specific model and the Business Model Canvas serve as a guide to later develop the corresponding eHealth technological ecosystem.

3   eHealth Ecosystem for Caregivers

The platform-independent ecosystem metamodel provides the basis to define different kinds of technological ecosystems focused on knowledge management, consisting of a set of software components, human elements and the relationships between them, without a core software system that provides the basic functionality. Regarding the platform-specific ecosystem metamodel, it provides the basis to get the guidelines to define those technological ecosystems using Open Source software components.

The aim of the eHealth technological ecosystem for caregivers is to support the learning and knowledge management processes to develop and enhance the caregiving competences both at home and in care environments of formal and informal caregivers.

As we described in the methodology section, the model of the eHealth techno-logical ecosystem was instantiated from the ecosystem metamodel [10]. The instance is a PIM to define eHealth technological ecosystems for caregivers. Figure 2 shows this process on the right column.

Fig. 2. Instances of the ecosystem metamodel, both platform-independent and platform-specific metamodels

This PIM provides the concepts and the relations between them in order to support the definition of different real ecosystems. To get a model with details about the software used to implement the ecosystem, a set of ATL rules was applied. These rules transform each element in the PIM to an instance of the platform-specific ecosystem metamodel. Figure 2 shows the results of this transformation on the left column.

The platform-specific metamodel has no visual editor associated, the result of the transformation is an XMI file with the description of the model. In order to complete this information, the model was represented using a CASE tool. In particular, the model was represented by three views or packages. The views correspond to the three main parts identified in a technological ecosystem: software components, human elements and the relationship among each other. These views show the elements from the model in black and the elements from which they are instantiated in grey.

First, Fig. 3 shows the view of software components that compose the eHealth technological ecosystem for caregivers. There are two main types of components: infrastructure and tools. The *eHealthEcosystem* represents the ecosystem, it is the main element that contains the other model elements. It is instantiated from the *Ecosystem* class of the metamodel.

The infrastructure is composed of: *MailServer*, to send emails from the different software components, it provides the *SMTPConfig*; *DataAnalysisSupport*, to monitor the activity of the users and their interactions; and a *CentralAuthenticationServer* to centralize the user management, both data and login to the ecosystem.

The tools that provide the user-level services are instances of different Open Source tools represented in the metamodel: *SocialNetwork*, *LearningPlatform* and *Dashboard*. The *SocialNetwork* represents a private social network for patients, caregivers and relatives. The *LearningPlatform* is focused on providing psychoeducation support for formal and informal caregivers. Finally, the *Dashboard* is focused on decision makers and caregivers' managers in order to support the decision-making processes.

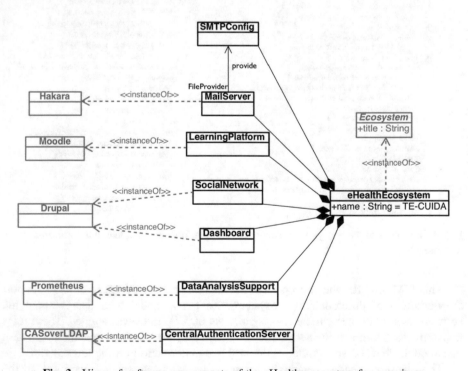

Fig. 3. View of software components of the eHealth ecosystem for caregivers

Figure 4 is focused on the human factor as a key element of the eHealth techno-logical ecosystem. The main users are represented as instances of *Manager*, *Method-ology* and *Management* classes of the platform-specific ecosystem metamodel. In particular, the manager of the hospital or care center is represented by *HospitalMan-ager*, and the manager of the caregivers as *CaregiversManager*. These users establish a set of methodologies (*TrainingPlan* and *MedicalProtocol*) and perform the business model (*BusinessModel*) which is described in detail in the Sect. 4.

Finally, Fig. 5 shows the view of the services which implement the information flows between the different software components of the eHealth technological ecosys-tem. There are three main services that are instances of *RESTfulAPI*: *InteractionData* to get the information of the users in the social network; *TrainingData*, to get information about the activity of the caregivers in the learning platform; and *AnalysisResults*, that provides the data analysis layer to support the information shown in the Dashboard.

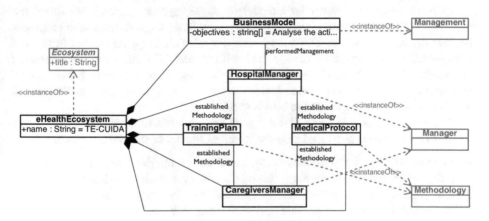

Fig. 4. View of human factor of the eHealth ecosystem for caregivers

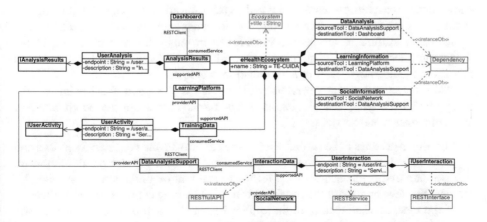

Fig. 5. View of services of the eHealth ecosystem for caregivers

4 Business Structure

In order to design the business structure, we use the BMC approach developed by Osterwalder and Pigneur [13]. Osterwalder's approach has proven to be a valuable tool for describing not only commercial business models, but also in many other contexts including health-related ecosystems [16–18]. From a business perspective, one of the main distinctions of technological ecosystems when compared to other types of business models resides in the involvement of the different ecosystem actors in value co-creation. As stated in [19]: "the motivation of actors to participate and engage within a medical ecosystem arises from the reciprocal benefits, namely the value propositions that variant types of actors within the ecosystem offer and seek". As such, value in ecosystems also emerges through interactions and collaboration between the different ecosystem actors [20], and this collaboration should be fostered by the ecosystem business design. These values are not independent from the other parts of the business model as they influence all other elements and building blocks.

In order to apply Osterwalder's model to our ecosystem proposal, we based the design in a literature review aimed to identify how technological ecosystem proposals address the business component, along with a systematic mapping study of the most relevant EU funded projects related to the health domain [21]. As a result, we obtained an insight of the main trends, lacks and opportunities, involved actors (stakeholders, end users, etc.), employed technologies, projects' investment, etc. related to the development of ecosystems in the health sector.

The BMC is available on http://doi.org/10.5281/zenodo.2273646. The outcomes of these reviews were applied to define the different blocks of the BMC as follows:

- In terms of the **Key Partners**, health and care authorities are needed as main information providers regarding the care delivery models and legal and privacy requirements that may vary across different regions, as well as being potential ecosystem investors and provide access to other stakeholders.

  Hospitals and care institutions partnership offers real scenarios and personnel involved in the provision of professional care. They also act as providers of the knowledge and expertise in care giving and psychoeducation, helping to build a trusted environment for the informal caregivers, the dependent persons and their relatives.

  Research institutions are needed in order to maintain the state of the art of the ecosystem's technological components, as well as for providing the latest trends in eLearning methodologies and psychoeducation protocols.

  Platform suppliers and operators will guarantee the good functioning of the ecosystem's underlying technologies, as well as provide means for the incorporation of new components and services.

- The **Key Activities** to be carried out in order to provide the considered services and generate value for the ecosystem stakeholders include: the design, development, operation and maintenance of the technological platform components and ontologies; the development of the psychoeducation eLearning contents; the identification of the most relevant data from the application of care and psychological activities within the ecosystem that could be relevant for the health and care stakeholders; and the identification of the platform's key performance indicators (KPIs) that allow monitoring its performance and the extraction and analysis of data (usage, trends, most demanded services) that could be valuable for current or potential involved ecosystem actors.

- The **Key Resources** needed in order to develop the proposed business model are the ecosystem components: social network, eLearning platform, mailing system and dashboard (which could evolve based on the observed ecosystem activity and actor needs); the platform hosting infrastructure; the technological expertise required to develop the platform components and ontologies, data mining, analysis and visualization; and finally the medical and educational expertise for providing adequate online psychoeducation contents.

- From a global perspective, the three main benefits that the ecosystem offers to its community of users are: the improvement of the quality of life of patients and (in)formal caregivers, a better quality of care and, as a consequence, a reduction of care provision costs. **Value propositions** include and are built around these three main values, based on the other model blocks (partners, resources, activities, etc.) to translate these benefits into concrete services. As stated before, these values greatly depend on achieving the involvement of the different ecosystem actors and not just the key partners (patients, relatives, third party care organizations and providers, etc.).

It has to be noted that value propositions are interrelated as, for example, the access to the social network or the psychoeducation learning program enhances the quality of care, but also improves the quality of life of patients and carers (better care planning that results in more spare time, contact with other users with the same needs and specialized personnel that provides support, advice and guidance etc.), and all these values will, in turn, result in a reduction of care costs. Also, it has to be taken into account that value propositions in an ecosystem are also subject (and meant) to evolution as the ecosystem itself evolves, since the incorporation of new actors or services can cause new functionalities to appear.

Thus, value propositions include a personalized learning itinerary for (in)formal caregivers by accessible psychosocial programs; a network that includes care professionals for social communication and support; valuable data for: the analysis of the different care and psychological action protocols, monitoring of the psychological disorder evolution and trends analysis (personal, geographical, etc.), analysis of ecosystem usage, user needs and trends; and the possibility of promoting third parties care solutions, incorporate new services and/or modify existing ones.

- The first of the **Customer Segments** refers to the patients and relatives as informal carers, as they are the main target consumers of the ecosystem. On the other hand, health and care providers shall be considered either as direct customers or act as solution intermediaries providing the proposed services to their carers in a business-to-business-to-consumer (B2B2C) approach. Also, health and care related service providers could act as ecosystem complementors or third-party providers who could be interested in promoting their services through the platform (e.g., care institutions, travel agencies, pharmaceutics, etc.), employing the platform available channels (social networks, mail), and using the trends data obtained from the ecosystem usage. Public health authorities are also a potential customer segment because they can act as funders (especially in countries with public health systems). In addition, they may be interested in making use of the ecosystem valuable data for the analysis of health evolution trends to enhance the health provision system.

- The **Customer Relationships** that must be maintained with each customer segment are the psychoeducation training program, the provision of professional medical and care contacts, training on platform usage, allow updates of platform contents and services and the promotion of third-parties care-related services.

- Considered **Channels** to reach the customer segments include the different direct contacts attained through the key partners that would include other regional/national health authorities and care related networks, and the promotion through social networks and web marketing.

- The **Cost Structure** that determines the costs of the business model is directly related to the tasks of development and deployment of the ecosystem, and includes: the platform development and maintenance, it's hosting, the communications infrastructure, and the IT and health personnel needed to develop and provide the considered services and functionalities.

In terms of the **Revenue Streams**, proposed values to the customer segments may generate different revenues at different levels. The considered revenue streams include: customers using ecosystem services; providing training to clients for the usage of the ecosystem; the IT support and maintenance; the ecosystem related data and its analysis tools; and the publishing of third parties care-related services.

5 Conclusions

The ecosystem metamodel allows to define models to support learning and knowledge management processes in heterogeneous contexts. The platform-independent and the platform-specific ecosystem metamodels were validated in our previous work, and their quality was also validated using the metamodel quality framework proposed by López-Fernández, Guerra and de Lara [22].

The platform-specific metamodel to define an eHealth technological for caregivers is based on a robust Model-Driven Development (MDD) solution. This model provides a guide to develop a real ecosystem to support learning and knowledge management processes to develop and enhance the caregiving competences of formal and informal caregivers, both at home and in care environments.

Taking into account the business structure is a fundamental aspect in the design and development of technological ecosystems, as it allows to reduce the gap between the technological structure and the actors' real interactions after ecosystem deployment. If these characteristics are omitted from the ecosystem design, they will surely affect the ecosystem health and performance once it has been deployed in real world scenarios. However, taking into account the business structure during the technological platform conception is not an easy task. To ease this process, we have modelled the business structure following the Business Model Canvas (BMC). On this respect, the research contributions have been within the identification, from the existing literature and EU funded developed projects, of the care ecosystem human components and their inter-actions, their value propositions and needs, and the key resources, costs and revenue streams associated with the care-related services provision chain.

The results of this work, both the platform-specific model and the BMC, have a number of important implications for future studies. In particular, it would be inter-esting to develop several case studies in order to test the proposal in the eHealth context.

Acknowledgments. This work has been partially funded by the Spanish Government Ministry of Economy and Competitiveness throughout the DEFINES project (Ref. TIN2016-80172-R) and the Ministry of Education of the Junta de Castilla y León (Spain) throughout the T-CUIDA project (Ref. SA061P17).

References

1. World Population Prospects: The 2017 Revision, Key Findings and Advance Tables. ESA/P/WP/248. United Nations, Department of Economic and Social Affairs, Population Division (2017)
2. OECD/EU: Health at a Glance: Europe 2018: State of Health in the EU Cycle. OECD Publishing, Paris (2018)
3. Verbakel, E.: How to understand informal caregiving patterns in Europe? The role of formal long-term care provisions and family care norms. Scand. J. Public Health **46**, 436–447 (2018)
4. Bäuml, J., Froböse, T., Kraemer, S., Rentrop, M., Pitschel-Walz, G.: Psychoeducation: a basic psychotherapeutic intervention for patients with schizophrenia and their families. Schizophr. Bull. **32**, S1–S9 (2006)
5. García-Peñalvo, F.J., García-Holgado, A. (eds.): Open Source Solutions for Knowledge Management and Technological Ecosystems. IGI Global, Hershey (2017)
6. García-Holgado, A., García-Peñalvo, F.J.: The evolution of the technological ecosystems: an architectural proposal to enhancing learning processes. In: Proceedings of the First International Conference on Technological Ecosystem for Enhancing Multiculturality (TEEM 2013), Salamanca, Spain, 14–15 November 2013, pp. 565–571. ACM, New York (2013)
7. García-Holgado, A.: Análisis de integración de soluciones basadas en software como servicio para la implantación de ecosistemas tecnológicos educativos. Programa de Doctorado en Formación en la Sociedad del Conocimiento. University of Salamanca, Salamanca, Spain (2018)
8. García-Holgado, A., García-Peñalvo, F.J.: A metamodel proposal for developing learning ecosystems. In: Zaphiris, P., Ioannou, A. (eds.) Learning and Collaboration Technologies. Novel Learning Ecosystems, LCT 2017, Part I, vol. 10295, pp. 100–109. Springer, Cham (2017)
9. García-Holgado, A., García-Peñalvo, F.J.: Learning ecosystem metamodel quality assurance. In: Rocha, Á., Correia, A., Adeli, H., Reis, L., Costanzo, S. (eds.) Trends and Advances in Information Systems and Technologies, WorldCIST 2018. Advances in Intelligent Systems and Computing, vol. 745, pp. 787–796. Springer, Cham (2018)
10. García-Holgado, A., García-Peñalvo, F.J.: Validation of the learning ecosystem metamodel using transformation rules. Future Gener. Comput. Syst. **91**, 300–310 (2019)
11. Eclipse Foundation. http://bit.ly/2KBR9AR

12. García-Holgado, A., García-Peñalvo, F.J.: Code repository that supports the PhD thesis "Integration analysis of solutions based on software as a service to implement Educational Technological Ecosystems" (2018)
13. Osterwalder, A., Pigneur, Y.: Business Model Generation: A Handbook for Visionaries, Game Changers, and Challengers. Wiley, Hoboken (2010)
14. Manikas, K., Hansen, K.M.: Software ecosystems – a systematic literature review. J. Syst. Softw. **86**, 1294–1306 (2013)
15. García-Holgado, A., García-Peñalvo, F.J.: Human interaction in learning ecosystems based on Open Source solutions. In: Zaphiris, P., Ioannou, A. (eds.) Learning and Collaboration Technologies. Design, Development and Technological Innovation, LCT 2018, vol. 10924, pp. 218–232. Springer, Cham (2018)
16. Christensen, H.B., Hansen, K.M., Kyng, M., Manikas, K.: Analysis and design of software ecosystem architectures - towards the 4S telemedicine ecosystem. Inf. Softw. Technol. **56**, 1476–1492 (2014)
17. León, M.C., Nieto-Hipólito, J.I., Garibaldi-Beltrán, J., Amaya-Parra, G., Luque-Morales, P., Magaña-Espinoza, P., Aguilar-Velazco, J.: Designing a model of a digital ecosystem for healthcare and wellness using the Business Model Canvas. J. Med. Syst. **40**, 144 (2016)
18. Locatelli, P., Cirilli, F., Gastaldi, L., Solvi, S., Pistorio, A.: Progressively developing a business model to assist elderly patients with cognitive impairment through a digital ecosystem: a methodological approach. In: Proceedings of the International Conference on E-Health, EH 2017, pp. 19–26. Curran Associates, Red Hook (2017)
19. Litovuo, L., Makkonen, H., Aarikka-Stenroos, L., Luhtala, L., Makinen, S.: Ecosystem approach on medical game development: the relevant actors, value propositions and innovation barriers. In: Proceedings of the 21st International Academic Mindtrek Conference, pp. 35–44. ACM, New York (2017)
20. Akaka, M.A., Vargo, S.L., Lusch, R.F.: The complexity of context: a service ecosystems approach for international marketing. J. Int. Mark. **21**, 1–20 (2013)
21. Marcos-Pablos, S., García-Holgado, A., García-Peñalvo, F.J.: Trends in European research projects focused on technological ecosystems in the health sector. In: García-Peñalvo, F. J. (ed.) Proceedings of the 6th International Conference on Technological Ecosystems for Enhancing Multiculturality (TEEM 2018). ACM, New York (2018)
22. López-Fernández, J.J., Guerra, E., de Lara, J.: Assessing the quality of meta-models. In: Boulanger, F., Famelis, M., Ratiu, D. (eds.) MoDeVVa, vol. 1235, pp. 3–22. CEUR Workshop Proceedings, Valencia (2014)

Internet and Local Television in the Context of Digital Transformation

Kruzkaya Ordóñez[1]([✉]), Ana Isabel Rodríguez Vázquez[2]([✉]), and Abel Suing[1]([✉])

[1] Investigadores Grupo de Comunicación y Cultura audiovisual,
Universidad Técnica Particular de Loja, Loja, Ecuador
{kordonez, arsuing}@utpl.edu.ec
[2] Profesora de Ciencias de la Comunicación,
Universidad Santiago de Compostela, Santiago de Compostela, Spain
anaisabel.rodriguez.vazquez@usc.es

Abstract. The local televisions they are not oblivious to digital convergence, its competition is wide and extensive. The production they generate cannot remain for a specific audience. Local televisions can take advantage of technologies as a local-global potential, to cover new audiences that are outside the coverage space. This article presents a compilation of arguments described by experts, through the application of semi-structured interviews, these constitute a fundamental methodological instrument of qualitative studies. Among the conclusions it is highlighted that, local television in the field of digital transformation is of the utmost importance, these means allow the articulation of more specialized communities and the connection with the closest problems.

Keywords: Internet · Local television · Digital transformation · Convergence

1 Introduction

Nowadays, the Internet has become a tool for basic information and social interaction, what Castells calls the network society. "The Internet is the heart of a new sociotechnical paradigm that constitutes in reality the material basis of our lives and of our forms of relationship, work and communication" [1]. "From the technological point of view, the Internet is a network of neutral transport with respect to the data it transmits. It has a capacity to access huge amounts of content, in a personalized way, that is, at the user's full choice" [2].

The rapid development of the Internet means that the media, including television, can take advantage of the many advantages offered by this technology. Internet translates into "new multimedia languages, the establishment of a new hypertext grammar and, above all, a new way of transmitting the image and corporate identity" [3].

Both the use of the Internet and the number of connections that are increasingly increasing allow us to define the strategies that in this particular case of study, local television, I should adopt in order to achieve the audience loyalty.

In this context, the Internet as a tool for the integration of societies and cultures, through the web and the apps, have been part of the business policies adopted by some

© Springer Nature Switzerland AG 2019
Á. Rocha et al. (Eds.): WorldCIST'19 2019, AISC 932, pp. 433–439, 2019.
https://doi.org/10.1007/978-3-030-16187-3_42

television channels, with different ownership, national, regional, local, cable or satellite. With the aim that:

"People access haven to the programming of the chain, and Attract attention the new users who find attractives content. Viewers consume content across multiple platforms, so that the Internet not only does not cannibalize television, but complements and enhances the synergistic and joint offer" [2].

Web pages have a leadership increasingly pronounced. This is the case of television stations that have opted for a space on the Internet, such as: the pages associated with the programming of the channel, in some cases live newscasts and in others as information sites; the promotion of their programs, their talents and advertisers.

"The presence of TV channels television in the net supposed creating a space in the eyes of any cybernaut, since the Internet transcends the geographical and cultural borders of the countries gathering together all the television products offered within the audiovisual panorama. The television channels sare aware, of the importance that their space on the Internet has as a communication tool with the different interest groups, spectators, competitors, suppliers or even international platforms" [3].

There are new forms of communication that are obtained on the Internet. These allow television channels to usually present their programming grid, with detailed information on each television product and links to sites that may interest the user. Go beyond what is offered on the screen. "More and more television channels and production companies are restructuring their contents to offer them [...] them through short excerpts, which appear combined with written information, photographs and audio fragments" [4].

In addition, the television transferred to the web, attend the needs of the new generations, who use the virtual space more every day through the computer and various devices. The audiences gradually leave television consumption for open signal. In this sense, Pérez Dasilva and Santos Díez [5] warn that audiovisual media "must adapt to new forms of information consumption if they do not want to lose audience". Therefore, televisions in the web environment:

"they must combine in their content rigor, amenity, plurality, graphic/visual richness, personalization of information, interactivity, easy accessibility and diversity of linguistic offer. [...] We can not fill new media with old content. Therefore, imposes the need to handle novel keys for describe topicality the differently" [6].

Local televisions can take advantage of technologies as a local-global potential tencompass new audiences that are found outside the coverage space. "The territory as a geographical boundary ceases to be a disadvantage and the small local media that jump to the Web have before a potential audience that has nothing to envy to that of the big journalistic companies" [5]. Internet for the local television could configure as an ally for the exhibition of its contents, cross borders, reaching the proximity audiences inside or outside its territory of reference. In addition, they become a viable option before the presence of multiple content distribution platforms offered by the audiovisual market.

According to Statista [7], until the closing of 2018 the most consumed medium in Latin America continues to be television. However, the Internet presents gradual growth with sights a remarkable penetration in the media market. The technological reconfiguration to which the media and in particular television are exposed, incorporates multiplatform production proposals with interaction results through social networks and other tools provided by the Web. 2.0.

The Internet and television relationship allows a hybridization of contents. Action that becomes effective across the programming of information, entertainment, promotion and, even, live transmission through the websites. In these spaces that are opened for local television, the fact of proximity must be considered because it is the local media that can in some way with their content "defend the signs of local identity, thanks to its representative function and its capacity to show an ethnographic vision of the facts, that is, from the inside out" [2].

Identity, undoubtedly, is an element that is not easily erased and even less with technological development, more its consolidation is needed. "Identity is a narrated construction we exist as an identity to the extent that we can tell others that I do not exist for myself but for the other who recognizes that I am different" [8].

The local televisions they are not oblivious to digital convergence, its competition is wide and extensive. The production they generate cannot remain for a specific audience. The technological changes that allow the transmission of contents in high definition and the Internet with its multiple possibilities (for example, television on demand), allow the incorporation of production dynamics among them the calls 360, theses "contents are exploited through diverse screens." [9].

2   Methodology

The methodology used in the research is qualitative through the application of semi-structured interviews, these were conducted on the basis of a questionnaire (Table 1), designed with topics related to digital convergence, the incidence of Digital Terrestrial Television and Internet, Digital technologies and audience fragmentation.

Table 1. Expert interview questionnaire

1. What is the value of local television in the current panorama of digital convergence? Label any of these values (for example, proximity, connection, etc.).
2. Which is the incidence of DTT in the evolution of proximity television. Is an opportunity? A threat to the survival of these channels? Or that the need to face this technological adaptation should not condition the current situation or future of these TV channels?
3. How can local televisions use the possibilities of personalization, interactivity or diffusion, through multiple screens that offer technologies such as DTT or Internet?
4. In the current scenario, in which digital technologies provoke an evident fragmentation of audiences how do local televisions compete?

Source: Own elaboration

The interviews were conducted from distance (via Skype) and face-to-face in two periods: July and November 2015 and the month of July 2016. The following Table 2 details the place and date of the interviews with the experts:

Table 2. Reference data the interviews with experts

Experts	Place and date	Modality
Ángel Badillo Matos	Salamanca, july 2016	From distance/on Skype
Miguel de Moragas	Santiago de Compostela, october de 2015.	Face to face
Francisco Campos	Loja, july de 2015	Face to face
Xosé López García	Santiago de Compostela, october de 2015	Face to face
Omar Rincón	Colombia, november de 2015	From distance/on Skype

Source: Own elaboration

3   Results

Local communication has the value to articulate the community and the closest levels in a media system. The transformation of communication, digital conversion, the gigantic transformation of communication in which we are immersed, allows the articulation of much more specialized communities and therefore in levels even lower than local communication [10].

The local television and the local audiovisual have an extraordinary importance in the articulation of the communities. Therefore, one of the main labels that define and give value to local TV is the participation in the construction of the communities to which it is directed, its identity the connection with realities and the closest problems.

Proximity in the ambit of digital transformation is relevant whenever it is imply and talk about the complexity between both transmitter and receiver. Proximity is that which occurs directly in the community. In everyday life, only what is local, what is close, has media repercussion [11]. What defines the local is not the transmission system, it is how it is implied in the community [12].

In this sense, for the media, radio, press, television, the transmission of the future goes to the Internet and in this sense, they do not require large economic infrastructures. All they need are human resources, innovation and adaptability. Strategies must be designed to take advantage of these spaces and redesign their content for the Internet, although dissemination through this platform is not a business, it will be. Therefore, with the Internet local television will have the opportunity to reach the people of your community scattered in various territories. Local broadcasters in Internet are part of the globalization without restriction of any communication politics [13].

In what refers to the fragmentation of the hearings, in the event that the public politics of a given territory choose to increase the offer and evidently, if there is this increase, this fragmentation will occur. In this case, to compete, operators will have to sign up, optimize, targeting targets specific and differentiated audiences [10].

Table 3. Descriptores convergencia digital

Digital convergence	TDT	Digital technologies
Local TV participates in the construction of communities connected to the closest problems	It is related to publics politics and the interest of operators	Fragmentation occurs in the event that policies determine an increase in offer and local specialization is consolidated
The local TV its local contents have media repercussion, it is the television of proximity	The irruption of DTT has meant a loss of initiatives and broadcasters	It is a paradigm shift, technologies are fragmented by the intervention of more actors
The local is not the transmission system, it is how the community is involved	Local TVs are in a situation of deep transition which will depend on factors such as consumption, trends, bandwidth and the Internet	Adoption of digitalization strategies that help the reconversion and empowerment of TV
With the Internet, local television will reach people in your community scattered in different territories	It is an opportunity the TDT. The local media should create collaborative exchange and joint investment organizations	Fragmentation affects all media, but there is room for the local
	It puts in doubt the insertion of the local TV by the industry and the quality technology	Local TVs have to return to the classic criterion of community, which makes sense when it generates connecting link, in so far as it is solidary

Source: Own elaboration based on the opinion of the experts

The DTT entails increasing the number of channels among which the local television stations would be, these, in turn, find on the Internet an opportunity for expansion and connection with the public. Therefore, in territories of countries or regions that opt for an increase in supply and decide to empower local television the fact of orienting the content to specific local audiences is an opportunity for that sector.

Any local television must aspire, if he is capable, to emit for supports all media in which it can distribute its signal and, therefore, try to be in all areas. And of course, one of the basic aspects in the communities and in the proximity is to involve users through the network, through initiatives of all kinds, from the most traditional to the most current with a semi-automated system on the Internet. Fragmentation affects everyone, but there is room for the local [12].

4 Conclusions

Who competes with the TV local are the new forms of local communication that are being articulated in the virtual space. For social networks, multimedia, transmedia and crossmedia contents are built in a new technological environment, in which the receiver access all the time through different devices. That is the main niche of competition and the main challenge that local TV have.

The main problem that is detected not only affects local television, but the whole of the traditional audiovisual, it is the competence of the communication in the networks and before that competition, there is the need to restore the means of financing, the business models of the traditional audiovisual.

The television systems of the countries are in a moment of transformation in which their main challenge is to know how they will survive the competition of the new media, how they will be transformed within the new digital environment, with multi-screens with cell phones, and that they are providing permanent and instantaneous, personalized information, within communities much more specific, even, than the local one.

The impact and irruption of Digital Terrestrial Television can mean a threat in the local television sector, since economic limitations hinder the sustainability of these means. In this sense, governments must seek formulas that make viable a friendly technological change that goes against a local television decrease.

Local television is a means of present and future, will have prominence in the communication ecosystem in the measure that it can respond to the demands of information and be involved in these local communities and especially in those where it gets the community involved. What it has to demonstrate is its usefulness to the community regardless of whether they are, more or less, commercial.

Finally, local television stations must be prepared to broadcast their signal on various screens, including the Internet, as a subsistence alternative with the opening of the spectrum and the digital transformation in which the media plays a remarkable role with the information they generate.

References

1. Castells, M.: Internet y la sociedad red. La Factoría, No. 14–15 (2001). https://goo.gl/RKEsvr
2. García Avilés, J., García Martínez, A.: Nuevos retos de la televisión ante la convergencia digital. Dadun Depósito Académico Digital, pp. 275–286. Universidad de Navarra (2010). https://goo.gl/sS2bZM
3. González Oñate, C.: Nuevas estrategias de televisión. El desafío digital. Identidad, marca y continuidad televisiva. Editions de las Ciencias Sociales, Madrid (2008)
4. León, B., García, J.: La información audiovisual interactiva en el entorno de la convergencia digital: desarrollo y rasgos distintivos. Comun. Soc. **13**, 141–179 (2000)
5. Pérez Dasilva, J., Santos Díez, T.: Las televisiones locales del País Vasco en Internet. Revista Latina de Comunicación **64**, 192–202 (2009). https://doi.org/10.4185/rlcs-64-2009-816-192-202. https://goo.gl/kGhiRP

6. López, X., Neira, X.: Los medios tradicionales, en los nuevos escenarios de la comunicación. Revista Latina de Comunicación Social **32** (2000). La Laguna- España. https://goo.gl/0cR3Fi
7. Statista: Los mercados líderes de la TV pagan on line en América Latina (2018). https://goo.gl/BQCW1Z
8. Martín Barbero, J.: Diversidad cultural y convergencia digital. Revista Científica de Comunicación e Información **5**, 12–25 (2008). https://goo.gl/nsoFPR. Recuperado al 23 de julio de 2015
9. Godoy, S.: Televisión digital en Chile: aspectos regulatorios y modelos de negocio. Mesas Redonda Universidad Católica de Chile, 24 de mayo 2007. https://goo.gl/aZQ7En
10. Badillo, Á.: Televisión local. Regulación. (K. Ordóñez entrevistadora) Salamanca, España, 4 de julio de 2016
11. De Moragas, M.: Televisión local de proximidad. (K. Ordóñez entrevistadora) Santiago de Compostela, España, 21 de octubre de 2015
12. López García, X.: Proximidad, identidad local. (K. Ordóñez entrevistadora) Santiago de Compostela, España, 25 de octubre de 2015
13. Campos, F.: Industrias culturales. Televisión local. TDT. (K. Ordóñez entrevistadora) Loja, Ecuador, 10 de junio de 2015

Microtransactions in the Company's and the Player's Perspective: A Manual and Automatic Analysis

Pedro Gusmão[1], Tiago Almeida[1], Francisco Lopes[1], Yuriy Muryn[1],
José Martins[2,3], and Manuel Au-Yong-Oliveira[4(✉)]

[1] Department of Electronics, Telecommunications and Informatics,
University of Aveiro, Aveiro, Portugal
{pedrogusmao,tiagomeloalmeida,lopes.francisco,
murynyuriy}@ua.pt
[2] INESC TEC and University of Trás-os-Montes e Alto Douro,
Vila Real, Portugal
jmartins@utad.pt
[3] Polytechnic Institute of Bragança - EsACT, Mirandela, Portugal
[4] GOVCOPP, Department of Economics, Management,
Industrial Engineering and Tourism, University of Aveiro,
3810-193 Aveiro, Portugal
mao@ua.pt

Abstract. Microtransactions dominate today's video game industry and will continue to do so for the foreseeable future, despite all the controversy it brings. To approach this problem, we created a survey, shared it on several gaming forums (a total of 1661 answers were obtained), then we designed a theoretical model and based on that, an automatic analysis was performed to understand what microtransactions are adequate to certain types of videogames. In parallel, we also performed a manual analysis that helped us gain insights into player preferences. Through the manual analysis we can conclude that players show a greater tendency to spend on microtransactions in mobile games. On average, respondents spend more on microtransactions than on purchasing videogames per month; with this, we can understand why the market of microtransactions has been growing greatly in recent years. Players that have jobs spend more on time savers microtransactions, and this probably happens because of the lack of time these players have comparing to the rest and the fact that they have an income to spend. Players aged 25 and above have shown to be more inclined to spend money to remove advertisements from games; however, players under the age of 25 are more inclined to spend money on general microtransactions in contrast to their older counterparts. It is also noticeable the negative sentiment towards players that spend money on advantageous items.

Keywords: Microtransactions · Video games · Pay-to-play · Free-to-play · Gaming platforms · Cosmetics · Neural network

Á. Rocha et al. (Eds.): WorldCIST'19 2019, AISC 932, pp. 440–451, 2019.
https://doi.org/10.1007/978-3-030-16187-3_43

1   Introduction

The digital and online world, and technology in general, have changed the face of many industries in recent years – including education, the finance industry and how commerce takes place [14–18]. Leisure activities have also been transformed. The first video game was created in the mid-1900s, but it was not until the 1970s, when Atari got involved with arcade games, that the video game industry increased significantly; but only in the 1980s video games became a popular hobby. At the beginning, publishers adopted a very basic approach to make profit out of their video games, in which they would sell the entire game, at full price, to a customer (pay-to-play (P2P) concept). Although this concept seems appropriate to make a profit, the excess of video games as well as the diverse preferences of the market motivated publishers to try a different approach, and they soon began to sell partial content instead of the entire version of the game. This approach would make the base game more accessible, or even make the game free, while all additional content or extensions would be charged through microtransactions (games that can be played for free are called Free-to-play (F2P)).

There is not a specific definition for microtransactions that perfectly encapsulates and represents the term. In general, a microtransaction is anything you pay extra in a video game, outside of the initial purchase. The purchases may unlock specific features or content. Microtransactions that unlock new content are more commonly known has DLC (Downloadable Content) and provide new ways to experience an already existing game through new extensive storylines, levels, challenges, game modes, etc., in order to expand the game's life span. Microtransactions that unlock specific features are more focused on consumable and cosmetic items. A cosmetic item is simply for aesthetic purposes and will only change the appearance of a character, weapon or any other item of a game. Consumable microtransactions have different types that serve different purposes:

– Time savers, that can boost experience to level up or simply skip levels, for example, in puzzle games like Candy Crush Saga;
– Advantageous items, where you can purchase stronger weapons or useful items that give the player an edge against his/her opponents. Games with this type of microtransaction are often considered as Pay-to-Win (P2W), as you are almost required to spend money on the game to play it competitively and win;
– Loot boxes, where a player can receive a random item, based on a collection, of different qualities (or rareness);
– In-game virtual currency, which is the currency a game may have so players can purchase virtual items. In most cases, virtual currency can be earned by playing the game, although at a slow rate.

There is not a concrete definition of a microtransactions. Whether it is good or bad, there is always controversy [1]. However, we can find examples considered good and bad from which we can learn from.

The F2P game Warframe is one of the best examples of a microtransaction model that's considered good [2]. The developers interact often with the community discussing if the prices are fair, how much play time is needed to get certain items, and when the community dislikes an update on the model an immediate response is given [3] and the revenue generated is then used to further develop updates, that are given for free, to offer players more content.

Examples of microtransaction models considered bad are everywhere, since this theme generates a lot of controversy. Star Wars Battlefront II is the most recent case. There were many aspects of the game that outraged players [4], many of which involved loot box microtransactions. A user estimated that, in order to unlock all the base content for the game, one would have to spend 4,500 h playing the game, or hand out $2,100. This is certainly not a feature a player is expecting from a game that retails for $60.

With these described cases we can affirm that not every model is successful, and we try to understand, from a set of defined variables, which of those are important and have a good impact on the success of a microtransactions model. Based on this set of variables, we also propose a theoretical model that suggests the best microtransaction model to be used in a hypothetical video game. This model would have a positive effect on players and help support the development/maintenance costs of the game. These ideas, however, are in the company's perspective and it is also important to understand what influences players to spend money on microtransactions, such as financial status, time that they spend playing or social aspects (pressure from other people or friends).

We can describe our main research questions as follows:

– Company perspective:
  • What variables are important to the success of a microtransactions model?
– Player perspective:
  • What influences or not the player to spend money on microtransactions?
  • How does a microtransactions model affect the sentiment of a player towards the game or other players?

2   Literature Review

Videogame microtransactions is a highly controversial topic that has been studied by many authors. By analyzing some of the works, we can conclude that:

1. Mobile games rely heavily on microtransactions because it is not possible for them to have the same price as PC games [5].
2. In multiplayer games, players want to be distinguished from others and so the additional content (microtransactions) that allows that is more easily accepted [8].
3. F2P games with microtransactions have much less problems related to piracy compared to P2P games [9].
4. Many players only accept cosmetic microtransactions as they don't create imbalance in the game [5, 6].
5. P2W microtransactions are especially unpopular in the gaming community. However, opinions on microtransactions that allow users to progress or get some items faster than a player that did not buy them are mixed. Some authors [5] state that these are accepted by the gaming community while others [6] declare that these have the same negative reaction as P2W microtransactions.

6. DLCs are a great source of extra income because developers can reuse the already created assets of the original game. In this way it is possible to charge almost the same amounts as for the original game and without requiring nearly as much work [10].
7. F2P games are accompanied by less negative publicity in case of lower quality content than P2P games [5], meaning that there's a lower probability that players are turned away from the game.
8. In general, the gaming community does not like it when premium games add microtransactions [5]. They feel like paying the full price for the game should be enough and that publishers are being greedy.
9. In-game currencies are used to create confusion and dematerialize payments [11]. In the end, most of the time users don't have a clear idea of the cost of the microtransaction that they want to purchase, this only making the said decision much easier.
10. Despite the negative attitude towards some microtransactions, players "become tempted to spend money on microtransactions themselves if they are confronted with other players who use them" [6].
11. Loot boxes have been considered a form of gambling, and it can lead to players overspending money because of an addiction to gambling [7, 12]. Some countries have already deemed these types of microtransactions as illegal [13].

3 Methodology

As observed in the previous section, the reviewed articles have a greater focus on the player's point of view, however, these lack research on the company's perspective of what variables influence positively or negatively a microtransaction model. In order to cover this, to answer the main questions that were mentioned in the Introduction (Sect. 1) and to gather sufficient data, we ended up, with the use of the Google Forms platform, creating an online survey. The survey was distributed, in the first instance, among the authors' network, and after that it was published in several subreddits (Reddit) and Facebook groups related with gaming. We tried to reach groups of all gaming platforms (Console, PC, Mobile) to avoid gathering biased data. The survey was made available in two languages, Portuguese and English, and was divided in five main sections:

- **Personal Information** such as gender, age, country of residence and if the person is a student and/or has a job;
- **Gamer Information** such as the favourite gaming platform, average hours a week spent playing video games, how much money a year they spent on video games and favourite game genre;
- **Buying Microtransactions**; here we asked if the player spends money or not on microtransactions, which buying methods are preferred, what type of monetization that games follow is preferred (free-to-play, buy-to-play or pay-to-play). We also placed some questions regarding the types of microtransactions, other video game characteristics and gaming platforms.

- **Social Aspect of Microtransactions**; here we try to understand if friends and other players pressure them into the purchase of microtransactions.
- **Real World Cases**, in this section we gave some freedom to the survey taker to comment on actual models and on the future of microtransactions.

The survey was available for two weeks and in this time, we were able to obtain 1661 answers. As mentioned in the introduction, a set of variables that influence microtransactions models were defined; these are based on the questions in the section Gamer Information. The variables are comprised of the genres and platforms that are included in the following questions:

1. "What's your favourite genre?"
2. "What's your favourite gaming platform?"

A few examples of genres included are: MMORPG, Fighting and MOBA. The platforms are PC, console and mobile; added up there are 15 variables (12 genres and 3 platforms) in total. Other variables could have been used however these were the ones that seemed, intuitively, to show more potential in demonstrating correlations with each microtransaction type.

4   Theoretical Model

With the variables that were defined in the section above, we propose a theoretical model that describes an ideal microtransaction model; this is a model that is, in theory, more successful in terms of the company's perspective and its results try to not disrupt the will to play that players have - in other words, it doesn't make players quit or feel that it jeopardizes their game.

To develop this theoretical model, we started with an abstraction; that is, from the set of variables that describe some of the properties of the game, the model will tell what kind of microtransactions should be present. We can observe a representation in Fig. 1. On the left, it's possible to observe the set of variables, which are the model input. In the middle box there's our proposed theoretical model, and in the right results a value that represents the affinity the set of variables has with a type of microtransaction, from the types of microtransactions that were discussed in the Introduction (Sect. 1).

Fig. 1. The abstraction of the proposed theoretical model.

We define our theoretical model (middle box) as a function that calculates the affinity value for each type of microtransaction

$$F(x; \theta) : R^d \rightarrow R \tag{1}$$

Here, function 1 is parameterized by θ, which can be represented as a matrix of parameters, these are inferred during data analysis, d is the total number of variables. The output is a real numeric value belonging to a $]0, 1[$ interval; the closer this number to 1 the higher the affinity will be. This function has to be able to extract nonlinear relations.

Since we defined our theoretical model as a function we can also find the maximum values of it; this corresponds to finding the variables (game characteristics) that would theoretically have more success in each type of microtransaction.

$$x = x - \lambda \nabla J(\theta) \tag{2}$$

Equation 2 defines an iterative process, since manual analysis is impracticable due to the function being multidimensional, to find which variables x could give us the maximum affinity value related to each type of microtransactions. Where λ is the chosen learning rate constant value, $J(\theta)$ is the loss function of Eq. 1, which represents the error of our theoretical model; lastly $\nabla J(\theta)$ corresponds to the gradients - these basically show us how to modify the values of our variables x so that we can find a local or optimal maximum of $F(x; \theta)$.

Now, the crucial part is to find the parameters θ and b for Eq. 1, and for that we propose an approach using artificial intelligence, more precisely, a machine learning algorithm that will leverage the data gathered in the survey to approximate a function.

5   Discussion on the Fieldwork

We opted to divide the analysis of the data in two parts:

- The manual analysis that gets conclusions directly from the survey;
- The automatic analysis in which the theoretical model is applied.

We decided to introduce the last one, due to the impracticality that it is to manually find relations with such a great amount of variables that can influence the affinity result; this is because there's the need to consider the combinations of all the different variables (32768 combinations) and also some relationships have been shown to be nonlinear or non-trivial. Finally, we compared these results with the conclusions from the other works presented in the Literature Review.

5.1   Manual Analysis

To give some context on the results we will present later on, basic statistics about the survey takers are shown in the following statements, these are **direct results** from the survey:

– A total of 1661 answers were obtained;
– 91.1% of the survey takers are male;
– 75.6% of the survey takers are between 18–34 years old;
– 52.2% of the survey takers are from the U.S.A., 8.3% are from Canada and 5.7% are from the U.K.; we had answers from 66 different countries;
– 71.4% of the survey takers have full-time or part-time jobs;
– 57.7% of the survey takers prefer credit card as a payment method followed by e-wallet (e.g. PayPal) with 30.3%.

More importantly:

– Only 2.7% of the survey takers spend less than 1 h per week playing video games;
– 86.5% of the survey takers have bought microtransactions and the main reasons for the remaining 13.5% who haven't, were because they're too expensive (54.5%) and, specifically talking about advantageous/time saving microtransactions, because they don't want to make their gameplay easier by taking shortcuts (44.6%). We also gave freedom to the survey taker to give other answers and we noticed quite a few of them said they don't want to support the practice of microtransactions;
– 38% would rather buy microtransactions in F2P games than P2P, because they didn't have to pay for the game, while 43.5% of them are indifferent;
– 36.6% would be more inclined to buy microtransactions in mobile games rather than with consoles (5.7%) or PCs (16.2%).
– Players don't like and even hate players that spend money on P2W microtransactions;
– In general, players don't feel pressured by their friends to spend money on microtransactions; however they do feel inclined to buy cosmetics when they run into strangers that bought them (38.3%).

In Table 1, we can observe that players do spend more money on microtransactions than buying video games.

Table 1. Comparison between money spent monthly on microtransactions and the purchase of video games.

	Microtransactions ($)	Videogames ($)
Worst case	18.16	12.78
Average	39.80	13.88
Best case	61.44	20.61

In Fig. 2, each graphic represents a different type of microtransaction and a histogram in which the scale goes from one to five (one represents that the player will quit because of the model and five represents that the player will buy microtransactions that are from the model). At first glance, it's noticeable how advantageous items are not accepted, as a great amount of people will quit because of those microtransactions or simply have negative sentiments towards them. In contrast, cosmetics are much more accepted compared to all the other types.

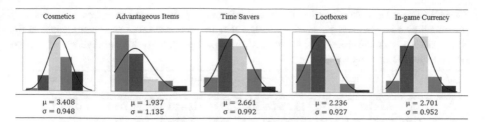

Cosmetics	Advantageous Items	Time Savers	Lootboxes	In-game Currency
$\mu = 3.408$ $\sigma = 0.948$	$\mu = 1.937$ $\sigma = 1.135$	$\mu = 2.661$ $\sigma = 0.992$	$\mu = 2.236$ $\sigma = 0.927$	$\mu = 2.701$ $\sigma = 0.952$

Fig. 2. Distribution of players' preferences regarding the types of microtransactions. The mean and standard deviation are presented below the graphics.

With the provided answers we were able to draw some conclusions and some possible deductions by crossing the given data from different questions; in other words, we can reach some certain indirect results by usually combining two specific questions from the survey. Next, a few examples of these **indirect results** are given:

- 73.4% that purchase microtransactions have a job (full-time or part-time);
- 79.6% that work would buy time savers and 74.4% like this type of microtransaction; we can deduce that people who work have less time to play, so they tend to buy this type of microtransaction;
- Adults (age 25+) are more likely to spend money to remove ads (49%) than younger people (ages 12–24), where only 35% would do so.

5.2   Automatic Analysis

Moved by the idea described in Sect. 4 of finding good microtransaction models and what variables influence them, we implemented the function 1 as a simple FeedForward Neural Network (NN) composed by 4 layers, an analytic representation of this is presented in Eq. 3, with the objective of predicting qualification (from 1 to 5, where 1 as the lesser and 5 as the higher affinity with the type of microtransaction), given the variables.

$$F(x; \theta, l) = \begin{cases} \sigma(\theta_l x + b_l), & if\ l = 0 \\ F(\gamma(\theta_l x + b_l), l - 1), & otherwise \end{cases} \tag{3}$$

This is a recursive function that calculates in each layer l, from a total of L layers, a linear combination $\theta_l x + b$ followed by a nonlinear function γ, this enables the function to extract nonlinear relations. In the last layer, which is the output layer $l = 0$, a sigmoid σ function is used to transform the values to the mentioned (Sect. 4)]0, 1[interval.

After the definition of the NN, we proceeded to the training phase that corresponds to finding the parameters θ and b, for each layer, of Eq. 3 using the answers from our survey, and after this phase we have a nonlinear function that maps the variables to the qualifications of different types of microtransactions.

In Table 2, the F1 score and accuracy of the function were registered, and we also created two baseline models, the first one, "truly random" only returns with a

Table 2. Recorded results for the f1 scores followed by the accuracy scores in parenthesis performed on 431 samples, the remaining 1032 samples were used for the learning phase of the NN; it is also worth mentioning that we used the stratified strategy for the division of the test samples and the training samples.

	Cosmetic	Advantageous	Time savers	Loot boxes	In-game currency
Truly random	17.7% (20.6%)	15.8% (19.5%)	15.4% (18.4%)	16.3% (19.7%)	18.2% (20.4%)
Cumulative random	19.1% (28.7%)	21.6% (32.2%)	16.6% (25.5%)	17.9% (28.8%)	17.7% (30.2%)
Our function	**29.7% (35.7%)**	**32.8% (36.2%)**	**27.7% (33.2%)**	**30.4% (36.7%)**	**30.7% (36.6%)**

probability of 20% each type of qualification (1 to 5) and the second one "accumulative random" uses the number of occurrences to predict qualification. The f1 score metric should be given more importance due to the distribution of the data shown in Fig. 2. We can also see that our approximation function in general has a 10% better score than the baseline scores, which gives us the perception that there is some relation between the variables and the qualification of each type of microtransaction.

From this function we proceed to the next step, where we compute, in an iterative way, equation number 2 with the purpose of calculating a plausible local maximum that will tell us what kind of variables contribute for a better qualification of that type of microtransaction. The results are presented in Table 3, and for some cases it wasn't possible to find a valid maximum; this is related with the lack of data for that type of qualification. In general, we agree with the results; for example, we can understand why "Console, PC, MOBA, Shooter, Sports" are extremely related with the qualification (5: "I would buy it") for the type "cosmetics"; however, we are not sure why; for example, single player has a great influence on the qualification 5 for type "in-game currency", and this could be associated with the error of our function, indeed. Therefore, it is worth mentioning we don't produce a good approximation function as

Table 3. Results of what kind of variables X are more important to the different types of microtransaction model; the qualification (1) is associated with the idea of "will make me quit" and qualification (5) with the idea of "I would buy it". This qualification is derived from the survey data.

	Qualitative 1 (will make me quit)	Qualitative 5 (I would buy it)
Cosmetic	Not found	Console, PC, MOBA, Shooter, Sports
Advantageous	MMORPG, Simulation, Single player, Strategy	Puzzle
Time savers	Single player, Strategy	Not found
Loot boxes	Console, Racing, Strategy	PC, Simulation, Sports
In-game currency	MMORPG	Cards, Single player

seen in 2; this could be related with the limited and incoherent data. A better neural network could also be used, but this was not the focus of the paper[1]. Also, other variables can be used so that the function produces better predictions.

5.3   Results Comparison with the Literature Review

Through both analyses, we compared our results with some of the points drawn from the Literature Review (Sect. 2). Work [8] stated that players like to distinguish themselves from others, so content that provides that is more easily accepted; we can reinforce this statement since, through our automatic analysis, we observed that online games such as shooters and MOBAs are usually successful with these types of microtransactions. Other works [5, 6] also stated that players tend to accept more these types of microtransactions due to them not creating imbalance in the gameplay; we can support this through the results obtained by the manual analysis that can be observed in Fig. 2. Lastly, as mentioned in the paper [6], players do feel pressured or inclined to buy microtransactions when others do so; through the answers obtained in the survey we can conclude that players don't feel pressured by their friends but do feel inclined to buy cosmetics when strangers show them what they bought for themselves (38.3% of the players feel this).

6   Conclusions and Suggestions for Future Research

Since a lot of mistakes happened throughout the years regarding microtransactions (Battlefront II [4] for example), one of the main objectives of this paper was to find how companies could decide on a model of microtransactions for their hypothetical created video game. With the proposed abstraction in Fig. 1 an artificial intelligence based automatic procedure for analysing the data of the survey was developed. This made it possible to reach conclusions that otherwise, with only a manual analysis, it would not be possible to reach. Through the automatic procedure it was possible to correlate gaming platforms and video game genres with each type of the defined microtransactions. The consequent results were shown to be consistent, that is, the variables that influence the types of microtransactions appear to be plausible in Table 3, despite the metrics being poor; this is due to data limitations. Through the manual analysis we can conclude that players show a greater tendency to spend on micro-transactions in mobile Games. On average, respondents spend more on microtransac-tions than on purchasing videogames per month; with this, we can understand why the market of microtransactions has been growing greatly in recent years. From the types of microtransactions, advantageous items and loot boxes are considered the least accepted types (loot boxes probably due to their gambling aspect, among other negative factors that can be pointed out), on the other hand - cosmetics are the most accepted as these don't create imbalances in the games - this can be seen in Fig. 2. Players that have jobs

---

[1] In some of the tests we pushed the neural network to 48% of accuracy (the F1 score was not implemented yet); this gives us an idea that there is space for improvement.

spend more on time Savers microtransactions; this probably happens because of the lack of time these players have comparing to the rest and the fact that they have an income to spend. Players aged 25 years old and above have shown to be more inclined to spend money to remove advertisements from games; however, players that are under the age of 25 are more inclined to spend money on general microtransactions in contrast to their older counterparts. It is also noticeable the negative sentiment towards players that spend money on advantageous items. For future work, more precisely in the manual analysis, we left a wide area to be explored. Also, with a bigger and more coherent data set the automatic analysis can show a significant improvement in results due to its consistence.

References

1. Anderton, K.: Controversy on microtransactions. forbes.com/sites/kevinanderton/2018/03/07/the-on-going-controversy-of-microtransactions-in-gaming-infographic/#282ecc4c1d9c. Accessed 7 Nov 2018
2. Brightman, J.: Microtransactions on warframe. gamesindustry.biz/articles/2017-12-04-warframe-dev-industry-must-get-better-at-giving-players-a-choice-and-a-voice. Accessed 7 Nov 2018
3. Gach, E.: Warframe microtransaction removal. https://kotaku.com/warframe-removed-a-microtransaction-because-a-player-us-1824002323. Accessed 7 Nov 2018
4. Looper. https://www.looper.com/96597/really-went-wrong-star-wars-battlefront-ii/. Accessed 7 Nov 2018
5. Tomić, N.: Effects of micro transactions on video games industry. Mega-trend revija **14**(3), 239–258 (2017). https://doi.org/10.5937/megrev1703239t
6. Evers, E.R.K., van de Ven, N., Weeda, D.: The hidden cost of microtransactions: buying in-game advantages in online games decreases a player's status. Int. J. Internet Sci. **10**(1), 20–36 (2015)
7. King, D.L., Delfabbro, P.H.: Predatory monetization features in video games (e.g., 'loot boxes') and Internet gaming disorder. Addiction **113**(11) (2018). https://doi.org/10.1111/add.14286
8. Rosenberg, D.: MMORPGs, microtransactions, and user experience. CNet, 16 July 2009
9. Nickinson, P.: How high is 'unbelievably high' piracy? Dead Trigger dev's not saying, Android Central, 23 July 2012
10. Campbell, C.: This is why paid DLC is here to stay, Polygon, 27 January 2015
11. Gilardoni, P., Ringland, E., Angela, H.: In-game currencies: in the line of fire?, 19 August 2014. Lexology.com
12. Lum, P.: Addictive effect of Lootboxes. https://www.theguardian.com/games/2018/aug/17/video-game-loot-boxes-addictive-and-a-form-of-simulated-gambling-senate-inquiry-told. Accessed 7 Nov 2018
13. Gerken, T.: Lootboxes are Illegal in Belgium. https://www.bbc.com/news/technology/43906306. Accessed 7 Nov 2018
14. Gonçalves, R., Martins, J., Pereira, J., Cota, M., Branco, F.: Promoting e-commerce software platforms adoption as a means to overcome domestic crises: the cases of Portugal and Spain approached from a focus-group perspective. In: Mejia, J., Munoz, M., Rocha, Á., Calvo-Manzano, J. (eds.) Trends and Applications in Software Engineering, pp. 259–269. Springer, Cham (2016)

15. Pereira, J., Martins, J., Santos, V., Gonçalves, R.: CRUDi framework proposal: financial industry application. Behav. Inf. Technol. **33**, 1093–1110 (2014)
16. Pinto, M., Rodrigues, A., Varajão, J., Gonçalves, R.: Model of funcionalities for the development of B2B E-Commerce solutions. In: Innovations in SMEs and Conducting E-Business: Technologies, Trends and Solutions, pp. 35–60. IGI Global (2011)
17. Au-Yong-Oliveira, M., Gonçalves, R., Martins, J., Branco, F.: The social impact of technology on millennials and consequences for higher education and leadership. Telematics Inform. **35**, 954–963 (2018)
18. Martins, J., Branco, F., Gonçalves, R., Au-Yong-Oliveira, M., Oliveira, T., Naranjo-Zolotov, M., Cruz-Jesus, F.: Assessing the success behind the use of education management information systems in higher education. Telematics Inform. (2018, in press)

Mobile Applications and Their Use in Language Learning

Cátia Silva[1], David Melo[1], Filipe Barros[1], Joana Conceição[1],
Ramiro Gonçalves[2], and Manuel Au-Yong-Oliveira[3(✉)]

[1] Department of Languages and Cultures,
University of Aveiro, 3810-193 Aveiro, Portugal
{mlsc, davidrafael.melo, barros.filipe, jconceicao}@ua.pt
[2] INESC TEC and University of Trás-os-Montes e Alto Douro,
Vila Real, Portugal
ramiro@utad.pt
[3] GOVCOPP, Department of Economics, Management, Industrial Engineering
and Tourism, University of Aveiro, 3810-193 Aveiro, Portugal
mao@ua.pt

Abstract. It is undeniable that the world is taking a turn for the digital. Banking, shopping, working, education, entertainment, everything has its own digital equivalent or means thereof. This includes language learning. Nowadays, there is a myriad of (mobile and digital) possibilities to study, learn and improve a person's knowledge of virtually any language. Thus, there should be potential benefits in their use in academic level (foreign) language learning, but also in an individual "self-learning" process. In this study, we choose to tackle mobile apps and their impact in the teaching/learning of languages, the perception of their usefulness and their impact (positive or otherwise) on the subject. To achieve this, a literature review on the subject was conducted, in addition to twenty interviews performed with students of the University of Aveiro. Supported by our literature review, we can safely state that language learning is just one of the many segments of both life and academia that can only be positively affected by digitalisation, and that we will continue seeing a rising trend towards that goal.

Keywords: Mobile · Apps · Digital · Language · Learning · Teaching ·
M-learning · Mobile-assisted · Digitalization

1 Introduction

Every day we rely more and more on digital media. Through our smartphones, tablets and computers, we have the world at our fingertips. More businesses and activities rely more deeply and more often on digital media [16]. Through this study, we hope to be able to evaluate and contextualize the current perception and state-of-the-art of applications in the field of language learning. We have conducted a thorough literature review to attempt to ascertain an academic consensus from our findings, while compiling together the different studies and data-gathering that has been previously elaborated on the subject.

© Springer Nature Switzerland AG 2019
Á. Rocha et al. (Eds.): WorldCIST'19 2019, AISC 932, pp. 452–462, 2019.
https://doi.org/10.1007/978-3-030-16187-3_44

This document is thus intended to become a contextualizing piece: We plan on compiling and revising the current state-of-the-art of research in the field, to see if an academic consensus can be achieved and then seek to compare it to our own ideas and findings, to finally be able to elaborate an educated perspective on the subject. Through this work, we plan on creating a solid piece of literature to lay the groundwork for any further research into the topic. Interest in this topic was further heightened as the authors are all language students themselves, and thus the idea of digitalizing or otherwise applying modern, digital techniques to language teaching and learning is of particular personal and academic interest.

In effect, we have been seeing a number of applications spring up that cover more and more activities: shopping apps, banking apps, taxi service apps, food delivery apps, translation apps, learning and self-review apps; all sorts of apps. Thus, the authors felt it relevant to inquire further into the topic of apps in language learning: to try and ascertain if they were being well-received by the academic community, if they were being accepted and used, if they had solid advantages and/or disadvantages compared to more traditional teaching methods, and thus to find out if there could be a future for m-learning (mobile learning) apps in formal education, as opposed to being mostly a self-review aid; as well as checking and comparing user demographics, prejudices and expectations regarding m-learning apps. In fact, even the very concept of "m-learning" is a very recent thing, as the literature explains [14, 15].

Therefore, we are motivated by wanting to find the trend for the development of m-learning and its possibilities in the future as a complement or, perhaps, a replacement of traditional teaching methods on the subject of language learning.

In addition to the literature review, a number of interviews were conducted amongst students of the University of Aveiro. A summary of our findings, conclusions and afterthoughts will be presented later.

2 A Look at the Literature

As an emerging tendency, language learning through mobile applications can make an interesting and most valuable turn in the teaching in classrooms approach. The authors henceforth cited will shine a light on the different prospects, equally the positive and negative points of view relating to this issue.

"The integration of mobile devices and vocabulary learning needs to take into consideration portability, social interactivity, context sensitivity, connectivity and individuality" [3, p. 44]. Additionally, according to [11, abstract] "studies into the use of mobile applications (apps) for language learning have found positive results on language improvement and learner engagement", which leaves one optimistic as regards the future of the digitalisation of society.

[7] aim to track the technologies used by language learners to support their language learning in an Australian University in 2011 and compare those findings with two previous studies on students' use of technology in the UK and Canada in 2006. The scholars ranked the top three technologies they believe are the most beneficial to their language learning process.

The results demonstrated that students in 2011 used at least 8 technologies for their language studies, both inside and outside of class. Online dictionaries and web-based translators, YouTube social networking and mobile applications were the most used in practice [7].

When reviewing the data, some subtle changes are noticeable in the years of 2006 through 2011. The traditional language tools such as dictionaries, verb conjugators and electronic translators have been transformed and adapted to new technologies, to strengthen their capabilities [7].

The development of mobile apps was also reflected on by [7], namely:

"Nearly 55% of our students use mobile apps to support their language learning and 134 ranked them in their top three beneficial technologies. The popularity and uptake of mobile apps represents a significant emerging technology that has impacted on language learning over the five years, to 2011" [7].

[7] conclude that students are accessing their own technologies in language learning more by the day, both inside and outside of the classroom, and are similarly becoming more autonomous and independent in their learning process. The study also evidences that learners find the use of technology tools more beneficial to acquire the basics of language learning, such as vocabulary.

[4] refers to the increasing preference in the use of the mobile phone or smartphone over the traditional computer. Furthermore, this study refers to the interchangeability and versatility of the mobile phone or smartphone as positive factors to favour this option. The recent adaptation of language learning methods to the technological era brings new notions such as Mobile Assisted Language Learning (MALL) or mobile learning (m-learning), which change the idea that learning a language could only be accomplished through teaching in a classroom. Now students can study any language, anywhere and in a complete informal and autonomous way, which allows them to be more comfortable with making mistakes.

The two main differences between MALL and the traditional face-to-face language learning are mobility and connectivity. Learning with mobile applications makes it feasible to have easy access to learning materials, individual place and time of study, instant feedback and self-testing. The learning process is more attractive and dynamic, which encourages students to study [5].

[2] also argues that it is important to create content that connects to students' lives while bringing those experiences along to the classroom. The use of a smartphone that students relate to is an opportunity to enhance digital devices and encourage life-long learning, autonomy and critical digital literacy.

The fast-paced development of mobile devices and their ownership is having an impact on language education, as well as on other learning contexts. Learning environments are happening more in mobile devices, enabling independent language learning [8].

The research developed by [4] was mainly focused on the use of language learning apps on mobile phones and smartphones and their limitations as well as strengths, with special attention to the teaching of the English language. The discoveries accomplished by the author show that many studies have already been conducted regarding apps with this purpose. The studies found are issued from a great number of countries across the globe but all of them show the same properties confirming relevant focus on linguistic aspects such as vocabulary, and all of them indicate positive effects of the use of apps on mobile phones and/or smartphones for the teaching of English as a foreign language.

[4] gathers further positive effects from the different studies showing that the learning of a language through apps is beneficial given that students can store more information because they learn at their own pace. In relation to the precedent, it's also mentioned that they can assess themselves as well as their progress, and subsequently take time to check for specific areas of their learning that might require attention. These are all factors that are not paid attention to in a classroom, given the fact that teachers can't focus on the needs of a specific student for them to improve, and since everyone retains information at their pace, it becomes easier for someone to learn on their own rather than in a classroom.

In [5] the researchers studied two populations: people who have decided to use a language-learning application of their own volition, and high school students of foreign languages who were asked to use a mobile application to complement their language learning.

The study establishes the concept of "CALL", or Computer Assisted Language Learning, to refer to any and all instances a computer was used to improve their language learning; and of "MALL", Mobile Assisted Language Learning, pertaining to the usage of mobile apparatus such as phones and tablets. The authors distinguish CALL from MALL via the platform of use (computer vs phone) and via the mobility and connectivity that MALL offers in comparison to regular language learning.

The authors [5] then go on to postulate on the perceived advantages and disadvantages of MALL as a whole, as understood from their preliminary research. The main disadvantage listed is the dependence on Internet access, and the small screen and keypad which can cause severe accessibility issues to learners with related physical handicaps.

The positive findings were thus summarized:

- The users were unanimously satisfied with the freedom and flexibility of the MALL m-learning app, allowing them to use it at their own pace, on their own time, at their own level. It is interesting to notice such a finding is confirmed by [4].
- The users were unanimously happy with the degree of self-testing, immediate feedback and adaptation of the materials that the MALL m-learning app provided them with.
- The users were generally satisfied with the "game" aspect of the application, which encouraged them to return to it more often than they previously assumed they would, to resume their learning process.

As for the negative findings:

- The necessity of Internet availability. The authors interestingly note that, when questioned about it, the voluntary user group mentioned that despite the Internet being cheaper now, more readily available and of better quality than it was some years back, relying on it is still perceived as a disadvantage.
- Noisy environments. In a day and age where the usage of mobile devices is unrestricted by the environment, their findings indicate that people have problems with using MALL m-learning methods in noisy or otherwise disturbing environments.

- The final disadvantage was the mobile device itself, as most of the participants also view their mobile device as a tool for socializing and entertainment, and therefore it provides too many easily-available distractions and temptations, potentially damaging their learning concentration and pace.

In general, the authors [5] conclude that the usage of MALL and m-learning apps was a very positive experience for all the participants, and thus postulate that it is indeed an area worthy of continued investment and study, as well as offering suggestions to enhance the gamification of the application and the fostering of a competitive atmosphere between the students using the application.

"It wasn't that long ago that the most exciting thing you could do with your new mobile phone was to download a ringtone. Today, new iPhone or Android phone users face the quandary of which of the hundreds of thousands of apps they should choose" [1].

Along with the rise of PDAs (personal digital assistants), language dictionaries, e-books and flashcard programs also made their breakthrough. However, the operating systems for phones and PDAs at the time weren't good enough for mobile language learning, due to the lack of features such as high-resolution screens and unlimited storage. The development of better apps, adding improved functionalities, started with the arrival of brands of smartphones in the market such as the iPhone (in existence since 2007), and Android and Windows phones. These devices have all the features and functionalities that PDAs lacked, such as larger data storage as well as memory, improved interfaces and responsiveness to touch screens, solving the problems that early MALL projects had [1].

[9] sought to investigate the contribution of mobile devices to language learning. A meta-analysis of 44 peer-reviewed journal articles and doctoral dissertations was conducted (from the period 1993–2013). According to the authors, the study had three purposes: the first was to "provide an overview of the use of mobile devices in language learning, including the target population, types of hardware and software, teaching/learning methods, the setting in which the teaching/learning, language skills were implemented (…)" [9]. The second was to "determine the overall effectiveness of employing mobile technology in education on the language achievement by students" [9]. The third was to "investigate whether moderator variables influence the effects of mobile devices on language learning" [9].

Ultimately, [9] concluded that their research "provides concrete evidence regarding the overall effects of using mobile devices in language education". Statistically, according to [9], "around 70% of students in the experimental groups who were learning with the aid of mobile devices would outperform their counterparts who learned languages without such devices" [9]. Furthermore, some interesting patterns were found about mobile language learning:

- Adults and school children had similar beneficial effects from MALL;
- The functionalities of mobile devices in multiple learning settings generated more marked effects/learning achievement than the more restricted settings in classrooms or outdoors;

- Integrating mobiles with multiple teaching/learning strategies produced better effects/learning achievement than with only lectures, or inquiry-oriented or cooperative learning;
- Using mobile devices for vocabulary or mixed language skills produced better effects/learning achievement than for single skills such as listening and reading [9].

[1] argues that the new technological advances in the smartphone industry, specifically in the evolution of the well-known brand Apple, have significantly changed the way education is seen.

"The Apple-inspired touchscreen smartphone is not just another technological innovation, but rather a device that has ushered in a new era in the human–machine relationship and that, thereby, it has the potential (not yet realized) of fundamentally disrupting teaching and learning, including L1 and L2 literacies and learning." [1].

In his ensuing work, [2] states that learners will continue to use apps to socialise or incidentally improve their language learning through gaming activities, for example, whilst taking advantage of the utility of apps for translation and dictionary look-ups. However, it is not an easy task to incorporate computing, communications and collaboration capabilities in smartphones for serious learning.

The different functionalities of the modern smartphones are additionally discussed by [1]. The versatility of today's devices and the importance that they are given by creating dependent users of this type of technology can also be seen as a condition to which smartphones should be perceived as of foremost importance in language learning nowadays. Some positive factors are also pointed out, such as the ease of access to Wi-Fi networks on public and private sites. This interdependence is taken with great relevance by the author in one of his subsequent works when he alludes to the potential of the smartphone as a "game-changer in education" [2].

Supporting the preceding point of view, [6] developed a study aiming to give an insight on the application of technological tools such as WhatsApp and Duolingo on a French class in Brazil, through an evolution period of two years with the same students. The intent is for students to be capable of learning the language autonomously out of the class environment, resorting to these mobile apps.

[6] believe that with the evolution of mobile systems and the dedication to improve apps on mobiles, language learning through this channel is a more efficient way for students to be able to learn foreign languages. Given that a great majority of students are already big consumers of the mobile industry, the use of apps for learning purposes would not only be beneficial, as it would also be logical. Since little or no investment at all would have to be made to have this teaching system employed in practice, the authors defend this idea with the present study.

Two simultaneously linked experiments were developed by [6]. The primary one was through the language learning app Duolingo, and its results were assessed through the social app WhatsApp in a more informal and less direct way in a second phase of the investigation. To compliment the experiment, some questionnaires and interviews were also applied to the students.

At first, it was asked of the participants to use the app Duolingo daily, to acquire better knowledge of the specificities of the language. The results were then further analysed by the teacher during the experiment. Concurrently, the social app WhatsApp was thought of, to better assess and complement the research. A group was created on the app with the students who showed interest in taking part in the project, and the teacher would try to endear them to write, send audio or songs and images in French, in a less formal way. It was thought that by using this strategy they would feel more at ease and be keener to use the target language.

Overall, [6] conclude that the experiment was successful. The language learning app Duolingo and WhatsApp were a good complement for the research, and between one another, they allowed students to have a more serious and linguistic-related component, while being able to learn on a less formal platform at the same time.

Formerly, [5] had already conducted a research study on what perception students have by learning through the mobile app Duolingo and which attributes they find useful or disadvantageous.

What students thought would be a hindrance changed after experiencing the use of the m-learning application: they thought mobile device distractions would pose a conflict to their learning, but found to make it less boring instead. In addition, they also thought the absence of human feedback would make a difference in their learning process but in the end found it unnecessary.

[5] conclude that most students found in Duolingo an enrichment in their learning progress. The factors pointed out were the ease of use, ubiquity and gamification, giving special attention to the latter and stating that:

"Gamification of m-learning applications can promote and encourage the use of these applications." [5].

[1] also takes MALL into consideration, but unlike other authors' points of view, this is a less optimistic approach, where it is stated that some features are lacking attention for this type of learning system to improve. An additional downside indicated relates to the fact that the greater part of studies conducted around this issue take a broad variety of devices into consideration, which doesn't allow for a deeper focus of improvement on certain features on smartphones that could enhance the performance of MALL projects. Another important asset [1] finds to be lacking in most studies is the fact that they only analyse the institutional use of these devices, disregarding the possibilities of research that could be accomplished if subjects were to be studied on all the informal daily activities that require contact with another language.

This setback is further taken into consideration a few years ahead where [2] states that early MALL projects such as using SMS for vocabulary learning is a teacher-centred approach, very limited pedagogically compared with recent features in mobile apps such as Memrise or Anki.

While language learning might be a given entitlement to most of us, [2] refers to the importance of second language learning to migrants and refugees, to whom the several apps developed by some countries such as InfoAid in Hungary, or Gherbtna, an app developed by Turkey aimed for newly arrived Syrians or even Ankomme, an app developed by the Goethe Institute and federal agencies responsible for immigration and employment, constitute a form of adaptation and integration to a new culture.

[2] further on adds that the development of powerful and versatile apps is increasing, also believing that more users will make them a primary focus, and language educators shouldn't ignore this trend.

3 Methodology

To complement the previously reviewed literature, we have conducted a series of twenty personal interviews, based on an interview script, amongst language students in the University of Aveiro, hailing from various courses in the Department of Languages and Culture (a qualitative research method, where each interviewee was individually interviewed, to better access his/her opinions and thoughts).

The interviewees were approached in person and sorted via the following questions:

1: Are you a student of a foreign language in this Department, be it as a Free Course or as part of a degree?

   1.1: If so, would you be interested in answering a short list of questions regarding mobile apps and their impact on language learning?

If both of these questions were met with positive replies, we advanced to the main interview in a designated room with the interviewee and two members of our author group. The interviews lasted an average of 15 to 20 min and notes were written down.

The interview script was comprised of the following questions:

Preliminary: Could you please tell us your name, gender, age and what language(s) you are learning/dedicating most of your time to learning in the University?

1: Have you ever used, do you regularly use or do you plan on using, a mobile application (Duolingo Mobile) to assist you in your language learning? (Yes – Have used in the past/Yes – Regularly use/No – Has not used).

   1.1: If yes, did you use/plan on using the said application as a complement/aid to your formal learning classes, or as a substitute for them? (As a complement/As a substitute).

   1.2: If you use them as a complement/aid, do you do so mostly to assist you in your classwork or in your homework? (Both/Mostly one).

2: Before you started using the mobile application, what were your main fears, prejudices or issues regarding the said application? (Open answer).

3: After using the said mobile application, did any of the issues you listed before manifest themselves? Did you find different, new issues? (Open answer).

4: What language did you use the application to assist you with the most? (Open answer).

5: Are you overall pleased with the usage of the application and its impact on your learning? Do you have any personal complaints about it? (Very pleased/Pleased/Displeased; Open answer).

Thus, the questions were aimed at understanding whether or not the students had ever used, regularly used or planned on using language learning applications (the example we provided was Duolingo Mobile) and if positive, whether they used it as a complement, substitute or aid to their own formal learning classes; as well as if they had any prejudices or *a priori* concerns before using the apps, and a brief summary of their final experiences with it.

Of the twenty interviewees, five were male and fifteen were female. The respondents were aged from 22 to 25 years old.

The analysis of the data gathered was achieved by listening to the voices in the data, creatively and with interaction, while considering the complexities of the context; much as [13] suggests how interpretive research should be done.

4 Field Work

Interestingly enough, despite the wildly different cultural context, our findings were similar to those of [5] in that the majority (17) of the students inquired admitted to using a mobile app to assist them in their language classes, as well as those same students claiming they used it both to assist them with their homework and classwork. Among their *a priori* issues, the most commonly stated were, similarly to the afore-mentioned study, concerns about battery life of their devices when using the application. However, all of the students who named such a concern stated that it turned out to not have been an issue at all, after having used the app for some time. The students were all native Portuguese speakers and used the application to assist them in their German, English, Spanish and French lessons. All but one of them claimed they were pleased with their usage of the mobile applications, with the detractor stating that he still preferred traditional teaching methods as he found it difficult to learn from a screen as opposed to from a teacher or a book. However, none of the other interviewees saw any issue or distinction between these methods.

In this particular study, the annoyances of keypads or small screens were generally not mentioned by the interviewee participants. We would be led to believe that people have simply become used to them to a point that it is no longer worth mentioning. The student group also found worthy of note the comparison of their progress, of their quiz responses and of their achievements via the application.

Regarding m-learning, one male interviewee stated that he was "Very pleased with no complaints." Another female respondent stated she was "Very pleased, particularly with the quality of the application." Yet another female respondent stated that she was "Very pleased with the quality of the information and its versatility of use." She stated further that "I expected the application to be a bit "janky" like Google Translate is at times, and I was pleasantly surprised when that wasn't the case".

Despite all the disadvantages, in the interviews most students responded that the issues with noisy environments and smartphone temptations that they had *a priori* imagined turned out to not be such a major factor after all. Furthermore, as [10, p. 24] stated, "enjoyment plays a significant role in mobile services acceptance" and language learning is no exception, with m-learning providing for an enjoyable experience, according to the interviewees. Finally, the interviewees generally communicated how mobile language learning applications have potentially transformed language-learning. In agreement with [12], our primary data supports the evolution of such technology to "adapt to suit the skill sets of individual learners [while] offering explanatory corrective feedback to learners [and incorporating] more contextualized language".

5   Conclusions and Afterthoughts

Through the compilation of previous research on the subject and our own data gathering, we have found most of our preconceived ideas about apps in language learning to have been vindicated: we did not believe that there would be any major, user-end issues regarding their usage – and both the literature review and our own research have come to prove us right on that assumption, as most if not all of the *a priori* issues turned out to have become non-issues after using the applications (such as battery life, screen size, distractions, and Internet connectivity).

As young students ourselves, we have come to believe that digital is the future, and the research we have gathered in this study seems to corroborate our thoughts on this matter: m-learning apps have been received with almost unanimous, overwhelmingly positive acclaim and impact both as an aide to formal teaching and as a complement to it. Thus, we're led to believe that, as [2] postulated, the development of new and better apps will only continue, as well as their positive reception and implementation by schools and universities in their classes; not to mention voluntary usage on the students' part.

In summary, this manuscript has compiled the current state of research and data gathering on the subject of m-learning (mostly geared towards students, admittedly) and compared it with our own findings and preconceptions.

Supported by our literature review, we can safely state that language learning is just one of the many segments of both life and academia that can only be positively affected by digitalization, and that we will continue to see a rising trend towards that goal.

As a suggestion for further research, we find it would be interesting to inquire into the possibility of language learning through exclusively digital means, as a complete replacement for traditional/formal teaching in the near future (as the quality of apps tends to exponentially improve) as well as the impact of the "human factor" in language learning (cultural immersion, jargon, lingo, human interaction).

We sincerely hope this study can serve as a basis for anyone seeking to learn more about the current state-of-the-art, preconceptions and perceptions of m-learning in the 21st century and the projections for its future development.

References

1. Godwin-Jones, R.: Emerging technologies mobile apps for language learning. Lang. Learn. Technol. **15**(2), 2–11 (2011)
2. Godwin-Jones, R.: Smartphones and language learning. Lang. Learn. Technol. **21**(2), 3–17 (2017)
3. Hu, Z.: Emerging vocabulary learning: from a perspective of activities facilitated by mobile devices. Engl. Lang. Teach. **6**(5), 44–54 (2013)
4. Klímová, B.: Mobile phones and/or smartphones and their apps for teaching English as a foreign language. Educ. Inf. Technol. **23**(3), 1091–1099 (2018)
5. Gafni, R., Achituv, D., Rachmani, G.: Learning foreign languages using mobile applications. J. Inf. Technol. Educ. Res. **16**, 301–371 (2017)

6. Souza, T., Mourão, M.I.: Teaching French by mobile devices: an experience with Duolingo and Whatsapp. Texto Livre Linguagem e Tecnologia **10**(2), 206–219 (2017)
7. Steel, C.H., Levy, M.: Language students and their technologies: charting the evolution 2006–2011. ReCALL **25**(3), 306–320 (2013)
8. Yang, J.: Mobile assisted language learning: review of the recent applications of emerging mobile technologies. Engl. Lang. Teach. **6**(7), 19–25 (2013)
9. Sung, Y.T., Chang, K.E., Yang, J.M.: How effective are mobile devices for language learning? A meta-analysis. Educ. Res. Rev. **16**, 68–84 (2015)
10. Ovčjak, B., Heričko, M., Polančič, G.: Factors impacting the acceptance of mobile data services–a systematic literature review. Comput. Hum. Behav. **53**(C), 24–47 (2015)
11. Rosell-Aguilar, F.: Autonomous language learning through a mobile application: a user evaluation of the busuu app. Comput. Assist. Lang. Learn. (Early Access) (2019)
12. Regina Heil, C., Wu, J., Lee, J., Schmidt, T.: A review of mobile language learning applications: trends, challenges, and opportunities. EuroCALL Rev. **24**(2), 32–50 (2016)
13. Elharidy, A.M., Nicholson, B., Scapens, R.W.: Using grounded theory in interpretive management accounting research. Qual. Res. Acc. Manag. **5**(2), 139–155 (2008)
14. Fonseca, B., Morgado, L., Paredes, H., Martins, P., Gonçalves, R., Neves, P., Nunes, R., Lima, J., Varajão, J., Pereira, Â.: PLAYER–a European project and a game to foster entrepreneurship education for young people. J. Univ. Comput. Sci. **18**, 86–105 (2012)
15. Martins, J., Gonçalves, R., Santos, V., Cota, M., Oliveira, T., Branco, F.: Proposta de um Modelo de e-Learning Social. RISTI-Revista Ibérica de Sistemas e Tecnologias de Informação **16**, 92–107 (2015)
16. Au-Yong-Oliveira, M., Gonçalves, R., Martins, J., Branco, F.: The social impact of technology on millennials and consequences for higher education and leadership. Telematics Inform. **35**, 954–963 (2018)

Digital Bubbles: Living in Accordance with Personalized Seclusions and Their Effect on Critical Thinking

Beatriz Ribeiro[1], Cristiana Gonçalves[1], Francisco Pereira[1],
Gonçalo Pereira[1], Joana Santos[1], Ramiro Gonçalves[2],
and Manuel Au-Yong-Oliveira[3(✉)]

[1] Department of Languages and Cultures,
University of Aveiro, 3810-193 Aveiro, Portugal
{beatrizfribeiro, cris.goncalves, franciscopereira,
gnspereira, joanacsantos}@ua.pt
[2] INESC TEC and University of Trás-os-Montes e Alto Douro,
Vila Real, Portugal
ramiro@utad.pt
[3] GOVCOPP, Department of Economics, Management, Industrial Engineering
and Tourism, University of Aveiro, 3810-193 Aveiro, Portugal
mao@ua.pt

Abstract. Since the emergence of the Global Village, the information flow changed drastically. Digital Technologies changed how people communicate, how they access information and how they share it. It gave people an unlimited exposure to information and knowledge. However, it also seemed to limit it. Recommendation algorithms are used in order to provide a customized experience that captivates users. Although they play an important role in selecting information that is considered relevant to the user, significant information/content may be omitted. Consequently, users end up closed in a bubble of limited information, which affects critical thinking skills and appears to influence and guide personal opinions. Little attention has been given to the negative effects of information bias on people's critical thinking. Thus, it is hoped that this study will at the same time educate and bring awareness to this issue. In a survey we performed (with 117 answers) the majority of the survey sample (approximately 54,7%) revealed discomfort regarding the storage and filtering of data. Interestingly, 29,9% of the participants were found to be indifferent regarding this issue. From these results, the authors can conclude that, although most of the participants feel uncomfortable, they prefer to be passive about this, which reinforces the idea of conformity and the false sense of organization mentioned herein. An interview with an expert in the area drew attention to the fact that social pressure most often leads users to comply and rely on the group's beliefs and attitudes, which facilitates social relationships and avoids confrontation.

Keywords: Technology · Filter bubble · Critical thinking · Social media · Algorithms · Rational behaviour · News feed · Information · Online · Customization

© Springer Nature Switzerland AG 2019
Á. Rocha et al. (Eds.): WorldCIST'19 2019, AISC 932, pp. 463–471, 2019.
https://doi.org/10.1007/978-3-030-16187-3_45

1 Introduction

The advent of new information and communication technologies allowed significant changes not only in interpersonal relationship patterns but also in the ways in which information is received. The flow speed allows greater ease of communication and, consequently, a wide and endless diffusion of news and ideas. Although it may go unnoticed, such an ever-increasing amount of information implies a dissemination of thoughts, conceptions and values that shape the surrounding reality.

Digital technologies increasingly enable a constant customization which changes the flow of information. Much has been discussed regarding the influence of the Internet on users' viewpoints and beliefs. Though the exposure to a wide range of information allows an expansion of knowledge, it is also true that it seems to limit it. Yet, to date, little attention has been dedicated to this tendency that most often users fail to properly acknowledge. Considering this context, the present study attempts to examine the impact of algorithms on users' perspectives as they demonstrate their influence on the formation of a vicious circle of unobtrusive manipulation, as well as on the formation of future inputs [1–3].

The term "filter bubble" introduced by [5] will play an important role in the course of this study as it delineates this closed individual world of information, which supports our point of view regarding the importance of developing judiciousness and a reasoned judgment, either prior to algorithm construction or during their exposure.

In view of the ongoing technological changes in an increasingly technology-dependent society, a careful awareness about the manipulated content will always appear to be lacking, most likely due to the false sense of planning that users experience. However, such passivity presupposes an impact on users' "ownership" of thoughts and opinions. In addition, a lack of critical thinking might compromise the personal, academic and working spheres of a user's life, which draws attention to the importance of learning strategies and reflecting skills. Perception is key since people construct their surrounding reality and perspectives through this process. Thus, in the light of the aforementioned, digital constraints on users' perceptions seem to be a current issue that needs further discussion. By exploring this subject, the current article looks at the following research questions:

- How might critical thinking be affected by algorithms?
- Is the lack of critical thinking the result of algorithm influence or is it the cause of the bias behind their construction?

2 Literature Review

Prior studies have highlighted the interrelationship between the filtering process and the partiality that results from it. Considering the current digital revolution and the continuous development of an information society, the set of built-in instructions – Algorithms – bring about the need to understand how they actually impose themselves on our lives. They have become a means through which a certain power and social control are exercised. However, such power and control encompass different subjective perspectives as algorithms are originally engineered by people. There is thus a lack of objectivity [4]. The precedent bias in their construction, though often overlooked, demonstrates that transparency constitutes an essential part of online navigation, not only in terms of data usage but also in terms of the development of a thoughtful and careful interpretation of inputs and outputs. This paper will mainly focus on recommendation algorithms due to the influence exerted on user perceptions and engagement.

Programmer Pariser [5] is responsible for creating the term "filter bubble" in his first book, "The Filter Bubble: What the Internet is Hiding from You". The author's theory outlines the fact that online navigation is subjected to algorithms which foster the rise of online filters, which play a prominent role in analysing users' preferences and predicting future interests. By reaping data/information and guessing what a user would like to see in the future, filters become controllers whereby bubbles are built, preventing people from properly deciding what to be displayed and hidden [5]. Many argue that users can, however, exert control by manually choosing what to display or hide in online profiles. Nevertheless, a drawback with such an argument is that users do not really seem to perceive that choice is affected by intrinsic perspectives and even possible dissonances, which may have been caused and influenced by social representations, which in turn, are formed by several factors. This emphasises critical thinking as a key element throughout the entire navigation process.

Content diversification seems to have become less important, which leads many to believe that users are not allowed to choose deliberately whether to enter the above-mentioned bubbles or not. According to Pariser's theory [5], algorithms immediately assume users' preferences, preventing them from being exposed to different content that may defy their points of view. Consequently, there might be a consolidation of already existing assumptions. Other observations, on the other hand, seem to indicate that the influential power is actually employed by users as they are the ones who provide the main input [6]. In that sense, both the purpose of an algorithm and people's responses become relevant in explaining the reasoned judgment applied by users and their future contribution to content creation.

An increasing content personalization has become a current matter of concern on social networks regarding voluntary and involuntary manipulation and persuasion. It should be noted, however, that it has been a matter of discussion since the advent of search engines. It is known that, in order to prevent an overload of information, search engines apply a ranking system. While it brings advantages when it comes to the selection of the most significant information, it is also true that users' data are collected in order to satisfy "third parties" demands, i.e. to reach specific target audiences. In this context, both users' data and search contexts are relevant, since both influence search rankings and results. It is important to emphasize the existence of techniques that raise the ranking of a website [7], as people might be exposed to low quality information, thereby not accurately evaluating evidence and angles.

Encouraging critical discussions and connecting people are often considered as desirable features of social networks such as Facebook and Twitter. However, one should take into account that these have become markets that need unceasing profit in order to prosper [7]. In this respect, users' data seems to have become a fundamental source of revenue, not only because of the online advertising market, but also because of the user-generated content. The latter becomes relevant as the phenomenon of modern global village enables democratic communication. By generating content, users play the same social role of mass media, contributing to the propagation of values, ideas and behaviours. They become powerful socializing agents, thereby contributing to the establishment of mindsets and beliefs. Although this may be strategically used as a channel of marketing distribution, it can also be a means to enhance engagement, since the data collected from one user will directly affect the input of another, based on the premise that online behaviour of like-minded people has an impact on the filtering process. Following this line of thought, the fact that users are increasingly becoming content creators might contribute to the emergence of singular and stereotyped visions of the world. The filtering process will just reinforce such misguided views.

The recommendations filter has proven to be a double-edged sword. It provides users with the information they have been regularly searching for, assuming it is what they want and need to see. After information is gathered, a link between the recommendation system and the user is created, facilitating online navigation in many ways. That transforms the personal web experience into an easy activity, whereby the user feels comfortable and pleased while navigating [5]. There is, however, a great deal of debate due to the disadvantages that the bubble filter can bring to Internet users. As above-mentioned, there is a tendency for the user to be exposed to subjective appealing content. Such a directed vision brings negative personal consequences, as well as negative effects on the development of a democratic society. As the tendency is for people to spend more and more time in the virtual world, the gathering of users with the same preferences and opinions will gradually contribute to unilateral visions of the various themes of society [8]. Thus, users must be aware of the inconvenience that recommendation systems bring about, either at a personal or social level.

As aforementioned, an overload of information might impact users' navigation, not only in the way they perceive online environments but also in how they discern the varied subjects therein contained. Prior to the change of paradigm in news media, people would be exposed to unbiased and standardized information. However, the advent of new information and communication technologies changed the journalism industry. Most social media channels seem to have forgotten an important mission: to transmit ethical, quality information with credibility and honesty to a vast and heterogeneous public. Today, information seems to be created and transmitted to a "right" group, considered to be the best target to read it. A customization system is responsible for filtering the information to match users' interests, thereby closing them in a bubble of similar content. Consequently, all the information users receive will be more of the same, echoing what they already think [6]. These "Eco chambers" will limit users' visions of the world. Bearing this in mind, if users' vision is limited, so will their judicious observations be. In other words, if there is a tendency to see, listen and comment on the same things, and thus a tendency to cohere with like-minded people, users' horizons can become narrowed [9, 10]. Over time, such facts may lead users to be increasingly extreme in their beliefs, since no one seems to question them [6]. Therein lies the issue: Fake news that uncontrollably overflow social media seem to be the result not only of specific and purposive algorithms but also of previous bias, based on the premise that the user is a producer of content and value.

Fake news might lead to misunderstandings about the circumstances under which a certain event occurred, given that the constant presentation of the same kind of information might lead to a lack of judiciousness [11]. Although there is a wide range of trending fake news topics, it has become observable that most of them have been increasingly used as political weapons [12]. This is one of the reasons why a considerable political polarization and fragmentation have become increasingly noticeable in many countries [13]. The existence of techniques that enable a website to raise in ranks has already been mentioned. The release of fake news appears to be one of the most used practices, as a fact might promptly be distorted or presented as a sensational one. In view of that, a user's perception and reason will be crucial, since reliable information becomes a key factor for users to weigh up different facts and make informed decisions.

When navigating online, users might experience some contradictions regarding their online behaviours and what they actually believe to be acceptable. There is thus a cognitive dissonance [14]. This theory seems relevant to explain the resulting inconsistencies in users' judgments caused by online navigation and by algorithm influence. Online behaviours may be the result of an intrinsic desire for acceptance. They might thus be adopted according to the reward the user might get: sense of inclusiveness and recognition within a group.

The access to personalized information does not depend on users' visits on social media sites or social networking sites. Whenever this is done, a permission to activate the cookies will appear. Most of the time, users do not really consider the "conditions" behind their activation. The question then arises: Are cookies tracking more than necessary? As abovementioned, third parties attempt to raise ranking positions by using several techniques, which may involve advertising practices. Third-party cookies are mainly employed by analysing and advertising companies that aim at getting involved in users' browsing [17]. A cookie, in other words, a package of data sent by websites is kept in the user's computer hard disk, whereby user's information (IP, idiom, e-mails and respective passwords, shopping cart features) is retrieved. They were created to improve online navigation, increasing the search efficiency without being considered malware. Websites can, thus, offer more content that users are likely to enjoy, omitting what they possibly would not. Website algorithms use cookies as a means to choose the information shown to the user, playing a role in the phenomenon of filter bubble, as an element of information storage. There are, however, some precautions users should consider. Unlike many people think, cookies do not pass on viruses. Nevertheless, the real risk is when users log in with personal emails and social media accounts, since there is a gathering of information that tracks their online activity. It has been argued that personal information may be retrieved at an international level, which is a matter of concern. This decreases the feeling of control and security of users, although it does not seem to alter their perception about personalized content.

3   Methodology

An inter-triangulation method was employed in order to assess users' awareness of the impact exerted by algorithms and better address the two initially stated research questions. A literature review was performed with the purpose of obtaining a better understanding and to provide a theoretical approach of the phenomenon. Subsequently, a survey questionnaire based on the information and findings gathered in the literature review was developed as an attempt to determine users' perceptions about recommendation algorithms, filter bubbles and their possible impact on critical thinking. Considering the importance of "Critical Thinking" for the current study, an informal conversational interview was conducted in order to clarify concepts and conceptualise related information. The usage of both qualitative and quantitative methods was based on the premise that "multi-method studies generate richer data that may be applied to more robust theory building, hypothesis testing, and generalising" [15, 16].

Close-ended questions were encompassed in the survey, which included rating scale, multiple-choice and yes and no questions. The survey questionnaire addressed personal and demographic information, social network usage/engagement and critical thinking indicators. Firstly, the authors collected secondary data as a means to obtain a broader range of information about the domain. A keyword search was conducted in Scopus and Springerlink databases to collect relevant academic journal articles. Although the search for keywords/phrases related to the subject allowed the retrieval of relevant data, significant information about critical thinking was difficult to access. In order to collect specific information for the present study, the authors decided to use primary data. Initially, a survey questionnaire was conducted in order to gather information about people's positions about the subject of study. Thereafter, as secondary research on the value of critical thinking proved insufficient, the authors decided to carry out an informal unstructured interview to gather primary data on the topic and better address its connection to the study. Thus, an attempt of contact was made to Dr Amanda Franco, a Postdoctoral Researcher at CIDTFF (Didactics and Technology in Teachers' Education Research Centre), who is currently working on the project "Critical thinking and college training: Impacts on students and their academic performance". After having arranged a meeting and following an informal conversational interview, the authors were able to get a better insight and understanding of this field of study. Furthermore, Dr Amanda provided some additional reading material, which allowed the authors to further discuss about the importance of this subtheme.

4 Discussion

The results of the questionnaire suggest that most participants are aware that their preferences, likes and clicks are being collected by algorithms/cookies. The sample consisted of 117 people. The majority of the sample (approximately 54,7%) revealed discomfort regarding the storage and filtering of data. Interestingly, 29,9% of the participants were found to be indifferent regarding this issue. From these results, the authors can conclude that, although most of the participants feel uncomfortable, they prefer to be passive about this, which reinforces the idea of conformity and the false sense of organization abovementioned. The majority of the sample declared using Facebook (111) and YouTube (107), which came to be the most relevant and most used social networks. Half of our sample considers the content of their social networks profiles as diverse, which corroborates the fact that users do exert control by providing input. It should be noted, however, that such input might be the effect of previous bias and socially-constructed perspectives.

[5] stated that his Facebook feed shows more posts and comments of his democratic friends than his republican friends, which emphasises a possible polarization in societies, as content is specifically chosen to coincide with users' inclinations. Thus, algorithms are responsible for hiding some content that he might be interested in. The authors decided to ask in the conducted questionnaire if participants felt frequently that their social media friends have the same opinion about certain online subjects. The authors intended to analyse the social pressure and conformity to other's opinions regarding the different information displayed (72,6% stated that sometimes people have the same opinion/20,5% people stated they do not have the same opinion). Regarding reflecting skills, the sample of participants consider having good critical thinking skills. While half of the sample occasionally participates in discussions in social media, the other half prefers not to. In addition, half of the participants consider the content on social media to be neither useful nor useless. On the other hand, an interesting result suggests that social media content is indeed very useful not only in a personal sphere, but also in users' professional and academic life (76,9% - personal life; 47,9% - professional life; 48,7% - academic life).

In order to assess people's critical thinking and reflecting skills, the authors decided to integrate a set of 8 questions related to specific controversial and polemic situations and a set of different contexts. Each question presented an argument and the respondents had to classify them as strong or weak. The results show that people responded to three of the questions based on personal opinions, and not through analysis of the given information. It was possible to infer that there is a tendency to rely on the primary image/perception people have about a certain reality.

The interviewee drew attention to the fact that social pressure most often leads users to comply and rely on the group's beliefs and attitudes, which facilitates social relationships and avoids confrontation. According to Dr Amanda, there is also a tendency to conform in the presence of an authority figure. The fact was also emphasized that when users think, they are subject to a set of cognitive biases. One of these biases is that of self-confirmation. In other words, when people strongly believe in something, they will involuntarily consider the information that confirms it more relevant. In order to be a good critical thinker, a user should always look for reliable, factual information, and compare different sources. Thereafter, he or she must take into account his or her own knowledge and values to think carefully and deliberately about the subject. One should also try to accept other people's arguments as valid.

5   Conclusion and Future Research

The present study sets out to discuss and contribute to the understanding of the increasingly often unnoticed manipulation exerted by algorithms and their behavioural and cognitive effects. Exposure to restrictive and personalized information might prevent users from actually leaving their comfort zone. On the other hand, users' power has gradually increased since they have become the main source of revenue for platforms such as Facebook and Twitter. Nevertheless, one should take into account that personal preconceived notions and partiality can be reflected not only in the construction of algorithms but also in the user's response to them. It is possible to infer that social pressure influences inattentive sharing of information for, depending on the age group, users may feel compelled to follow the same lines of thought as the group. In this sense, it may be easier for users to give in to such pressure, thereby avoiding confrontation. This conformity, although unobserved, shapes future behaviour, which might be reflected in algorithm engineering. Hence, this is a vicious circle.

Given that users are the ones who provide value for online platforms through generated content, transparency regarding the usage of data would be essential. Considering the emergence of a new information society where the filtration of news is recurrent, the development of critical thinking is key. The authors believe that additional research should be done to further investigate the cognitive and emotional impact of digital algorithms. It would be interesting to measure the importance of a filter bubble and its effects on users in the long-term. In the same way, more research should be done in order to study the willingness of an upcoming company/application to adopt recommendation systems, as well as their ethics in their usage. Another interesting topic would be an investigation on how recommendation systems work in networks controlled by the government where censorship prevails (i.e. China). The authors sincerely hope that the provided research will be helpful in a present or near future.

Acknowledgements. The authors would like to thank everyone who took their time to respond to the survey questionnaire. We would also like to gratefully acknowledge the help provided by Dr Amanda Franco, at CIDTFF (Department of Education and Psychology of the University of Aveiro, Portugal). Her sharing of knowledge regarding critical thinking has proven to be fundamental to the authors' approach to the subject.

References

1. Gonçalves, R., Martins, J., Pereira, J., Cota, M., Branco, F.: Promoting e-commerce software platforms adoption as a means to overcome domestic crises: the cases of Portugal and Spain approached from a focus-group perspective. In: Trends and Applications in Software Engineering, pp. 259–269. Springer, Cham (2016)
2. Martins, J., Gonçalves, R., Pereira, J., Cota, M.: Iberia 2.0: a way to leverage Web 2.0 in organizations. In: 2012 7th Iberian Conference on Information Systems and Technologies (CISTI), pp. 1–7. IEEE (2012)
3. Gonçalves, R., Martins, J., Branco, F., Perez-Cota, M., Au-Yong-Oliveira, M.: Increasing the reach of enterprises through electronic commerce: a focus group study aimed at the cases of Portugal and Spain. Comput. Sci. Inf. Syst. **13**, 927–955 (2016)
4. Bozdag, E.: Bias in algorithmic filtering and personalization. Ethics Inf. Technol. **15**(3), 209–227 (2011)
5. Pariser, E.: The Filter Bubble - What is the Internet Hiding from You. Viking, Penguin Books, London (2011)
6. Moeller, J., Helberger, N.: Beyond the filter bubble: concepts, myths, evidence and issues for future debates (2018). https://www.ivir.nl/publicaties/download/Beyond_the_filter_bubble__concepts_myths_evidence_and_issues_for_future_debates.pdf
7. Courtois, C., Slechten, L., Coenen, L.: Challenging Google Search filter bubbles in social and political information: disconforming evidence from a digital methods case study. Telematics Inform. **35**(7), 2006–2015 (2018). https://doi.org/10.1016/j.tele.2018.07.004

8. Bozdag, E., van den Hoven, J.: Breaking the filter bubble: democracy and design. Ethics Inf. Technol. **17**(4), 249–265 (2015). https://doi.org/10.1007/s10676-015-9380-y

9. Garrett, R.K.: The "Echo Chamber" distraction: disinformation campaigns are the problem, not audience fragmentation. J. Appl. Res. Mem. Cogn. **6**(4), 370–376 (2017). https://doi.org/10.1016/j.jarmac.2017.09.011

10. Nechushtai, E., Lewis, S.C.: What kind of news gatekeepers do we want machines to be? Filter bubbles, fragmentation, and the normative dimensions of algorithmic recommendations. Comput. Hum. Behav. **90**, 298–307 (2019). https://doi.org/10.1016/j.chb.2018.07.043

11. Kiszl, P., Fodor, J.: The "Collage Effect" – against filter bubbles: interdisciplinary approaches to combating the pitfalls of information technology. J. Acad. Librarianship (2018). https://doi.org/10.1016/j.acalib.2018.09.020

12. Lazer, D.M.J., Baum, M.A., Benkler, Y., Berinsky, A.J., Greenhill, K.M., Menczer, F., Metzger, M.J., Nyhan, B., Pennycook, G., Rothschild, D., Schudson, M., Sloman, S.A., Sunstein, C.R., Thorson, E.A., Watts, D.J., Zittrain, J.L.: The science of fake news: addressing fake news requires a multidisciplinary effort. Science **359**(6380), 1094–1096 (2018). https://doi.org/10.1126/science.aao2998

13. Geschke, D., Lorenz, J., Holtz, P.: The triple-filter bubble: using agent-based modelling to test a meta-theoretical framework for the emergence of filter bubbles and echo chambers. Br. J. Soc. Psychol., 1–21 (2018). https://doi.org/10.1111/bjso.12286

14. Miller, M.K., Clark, J.D., Jehle, A.: Cognitive Dissonance Theory (Festinger) (2015)

15. McMurray, A.: Research: A Common-Sense Approach. Cengage Learning Australia, Cengage (2004)

16. Loudon, D.L., Della Britta, A.J.: Consumer Behaviour: Concepts and Applications. McGraw Hill, New York (1979)

17. Shuford, E., Kavanaugh T., Ralph, B., Ceesay, E., Watters, P.: Measuring personal privacy breaches using third-party trackers (2018). https://www.researchgate.net/publication/303488658

Could Children Be Influenced by Technology? An Exploratory Study in Portugal

Catarina Martins de Lemos[1], Cláudia Maria Pereira Oliveira[1],
Sofia Andrade Correia Neves[1],
Vanda Maria Mendes Moreira Teixeira[1], Frederico Branco[2],
and Manuel Au-Yong-Oliveira[1,3(✉)]

[1] Department of Economics, Management, Industrial Engineering and Tourism,
University of Aveiro, 3810-193 Aveiro, Portugal
{cmlemos,claudiamaria,sofiaaneves,vandammmt,mao}@ua.pt
[2] INESC TEC and University of Trás-os-Montes e Alto Douro,
Vila Real, Portugal
fbranco@utad.pt
[3] GOVCOPP, Aveiro, Portugal

Abstract. The purpose of this work was to collect information about how technology influences children nowadays, in Portugal, an intermediate technology country in the European Union. The methods used were interviews and questionnaires presented to children (a total of 38 children), parents (15 parents) and teachers (three teachers) in order to find out how they feel about today's technological devices and how they think they can influence the society of the future, and whether they have a negative or positive impact or both. The methods selected were seen to be the best to obtain the desired answers, because public opinion is assuredly a good way to receive feedback about various contemporary products and phenomena. The conclusion was that every child in the sample owns digital devices and this influences them mostly in a positive way, in the opinion of their parents and teachers. Technology can also influence them in a negative way, but only if used so much that it becomes an addiction, affecting the child's attention in school, as well as their daily activities. The motivation to do this research study was mainly that one of the authors has four children and can see the profound changes inflicted in them by technology versus older generations and thus an exploratory study into technological change brought on by devices, gadgets and social media was decided upon.

Keywords: Technology · Children · Influence · Communication · Devices · Inquiries

1 Introduction

Technology appeared a long time ago and it is almost as old as humans. We see technology as being an application of human knowledge to solve a specific problem. Different events throughout history contributed to the appearance and development of much-used and widespread technology, such as the invention of the lamp, by Thomas Edison in 1847, the invention of the telephone, in the XIX century, and the introduction

Á. Rocha et al. (Eds.): WorldCIST'19 2019, AISC 932, pp. 472–481, 2019.
https://doi.org/10.1007/978-3-030-16187-3_46

of the smartphone, in 2007, by Apple, led at the time by Steve Jobs. Technologies such as computers, the Internet, social media and other technologies went through a huge development, and they continue improving even today.

When people talk about technology, telephones, tablets, television (TV), computers and video games are some of the most used devices and applications.

Different age groups, nowadays, from children to seniors are influenced by technologies, but the most affected are the children because they were born in the era of radical changes spurned on by technology [1]. Most children have had contact with tablets, computers or mobile telephones from a young age and a lot of children even have their own such devices. There are many reasons why children spend a lot of time using technology. One of the reasons is how learning in schools currently takes place as the Internet and computers play a significant part in the way classes are conducted. School homework is also aided by technology and is indeed expected to be so by teachers. Another reason is social media channels, such as Facebook, Instagram, and YouTube, which consume much of their time and energy. TV is another way in which children spend their time, watching movies, cartoons and TV shows aimed at younger segments [2].

But why are children obsessed with new technologies? Not so long ago, when technology was not so important, children played with each other outdoors and had other ways to have fun. Nevertheless, with the variety of technologies and advertisements available today, children are motivated to use technological devices and gadgets more often.

Nowadays, children's lives are not the same without technology, which plays an important role. If one wants to punish adolescent children one has only to take away their mobile phones – and the feeling of supposed isolation from their peers may make them desperate to not have such a punishment imposed on them again. Technology can affect children's concentration, health, education, development and even the relationships they form with others, although excessive contact can affect them in both ways, negatively and positively.

For example, communicating is much easier now than it was a few years ago, when people from distant areas had to use letters and the post; with today's progress, talking with people is one click away. Technology improves education and the learning process as well, because children tend to assimilate better in virtual classes, through videos and online books [3, 4].

However, technology also has negative impacts, such as the increase of obesity in children because of less physical activity. Another negative impact is the changes that technology provokes in children- "(…) video games may condition the brain to pay attention to multiple stimuli, they can lead to distraction and decreased memory", as an article from Psychology Today states. Lastly, technology can put children's' safety at risk because they share information online that could become dangerous - "(…) the sex offenders used social networking sites to get information about the victim's preferences" [5].

According to the American Academy of Pediatrics [6], "(…) the average 8 to 10-year-old spends nearly 8 h a day with a variety of different media"; however, "(…) older children and teenagers spend >11 h per day" [6]. Therefore, children spend a lot of hours in front of screens and the parents' concern, and even public's concern in general is the main reason to write this paper.

The objective of this work is to show how technology could change children and influence them. To achieve this purpose, the article is divided in different parts. First is the introduction, where we explain the influence of technology on children. After that, we pursued the exploration of the theme in the literature. A methodology and a brief explanation of our study was made. Finally, conclusions were used to analyse and draw a line about the issue at hand.

2   Literature Review

"The past decade has seen an increase in personal use of electronic technology, with childhood television and video game use similarly increasing. The instant accessibility and portability of mobile devices make them potentially more likely to displace human interactions and other enriching activities" [13].

The fact that these digital devices allow access to information and applications can be both positive and negative in the development of children and youths alike, because skills such as "self-regulation, empathy, social skills, and problem-solving" [13] are primarily learned when children explore the nature around them, interact with other people and play in creative ways. This being said, children's motor and sensory development is being delayed and there are more and more children with disorders (physical, psychological, among others).

Since the use of technology is easily accessible, there is a bigger chance for it to be more present daily, more than necessary. Therefore, they named "technoference" the act of interrupting daily social activities because of the use of technological devices. That kind of interruption may occur during face-to-face conversations, routines where people prefer to stay and play videogames, and many other things.

According to a study realized concerning young children, "most households had television (97%), tablets (83%) and smartphones (77%)", and "almost all children (96,6%) used mobile devices" [12]. Mobile devices are the ones chosen by children to use because of screen size and mobility. Compared to computers, mobile devices are easier to use, and you can use them everywhere. "Most children started using mobile devices in their first year of life" [12] is the reality in our current times, where even since a young age, children use them to play games, watch videos, communicate, take pictures and use apps. Children's ability to use mobile devices is unbelievable and implies independency and exploration. However, "little is known about how children's independent activity on mobile devices affects their cognitive, social, and emotional development" [12]. This is the topic debated in the following topics - the influence of technology on children, the good and the bad side of it, and the consequences it brings.

2.1   Health, Children and Technology

Health, children and technology is a concern and an important topic for society, teachers, specialists and especially for parents. Thus, there are a lot of articles and case studies about it and the relation between these concepts.

Experts say that the excessive use of technology has been studied in terms of Internet addiction. Analysing behaviours regarding cell phones, it is possible to conclude that there are some mental health, depression and anxiety issues associated [8, 9].

Accordingly [13], "The three critical factors for healthy physical and psychological child development are movement, touch and connection to other humans".

The difference between children in the 1990s and children today is evident.

Children, a few years ago, did not have technology; therefore, they played outdoors with other children. Children were more active then than they are now, and "child engagement in rough-and-tumble outdoor play resulted in the achievement of adequate sensory and motor development required for attention and learning" [13].

However, children nowadays play with each other using video games and other technologies. Because of that, most of the experts want to classify playing video games as a mental disorder, and that is because of the time children spend on them; but because of the lack of evidence, such a classification is not possible [8, 9].

Doctors and specialists advise parents to encourage children to be more active and go outside and play with other children as it was done a few years ago because "developing children require 3 to 4 h per day of unstructured, active rough-and-tumble play to achieve adequate stimulation to the vestibular, proprioceptive, and tactile sensory systems" [13].

If someone denies children the possibility to spend time on their electronic devices or even if the parents take them off their hands, children become angry and irritable and this is what we are facing.

Some examples of negative effects due to the use of excessive technology and linked to health diseases are shown below [8–10]:

- Irritability;
- Depression – not only can it generate the previously mentioned behaviours, but disorderly eating habits and academics difficulties as well;
- Aggressive behaviours and consequently suicidal tendencies;
- Risky sexual attitude and substance abuse;
- Lack of social interactions - for example, ignoring phone calls from their friends and isolating themselves to play alone;
- Loss of concentration and diminished attention - the exposure to the radiations generated by the screens could cause that;
- In epileptic children, the flickering of lights may trigger seizures;
- Sedentary life.

2.2   Education and Technology

Nowadays, children use more technology compared to children in 1990, says a survey where the authors conducted a questionnaire to analyse the attitude of 107 schoolchildren regarding the use of gadgets at school. Indeed, to attract and stimulate learning, teachers have started to use gadgets in the classroom because "mastering technologies can transform a classroom into a place where children do not need to mug up or learn by heart, but where they can understand things and think in a different perspective" [7]. Note that, according to another research, there is a good side to the media, such as "learn-to-read" apps and "electronic books", which help children to gain literacy skills because the words are shown in an interesting way through visual design and a touch screen interface, among others. However, the visual design and sound effects can engage them but also distract them, so a balance is needed [11].

"Gadget use in schools has both good and bad results", depending on children. Through the survey mentioned above, the author concluded that [7]:

- 82.2% of participants find E-learning is more effective;
- 92.5% of participants felt technology was an important aspect in their education;
- 68.2% of participants sometimes think that their grades are negatively affected because of the use of gadgets;
- 60.7% of participants also think that their concentration levels are decreasing due to the use of gadgets.

On the other hand, parents and teachers play an irreplaceable role in children's education. Therefore, it is essential for a collaboration between parents and teachers to form, in order to ensure the appropriation and credibility of information searched by children, and to help them to develop their critical thinking skills [10].

2.3   Parent-Children Interactions in the Technological World

Recent studies have suggested that the use of technological devices near kids is a bad influence and this will lead to fewer parent-children interactions, lower responsibility and lower-quality observations made by the parents. Adding to this, studies prove that parents nowadays spend less than 40% of their time with their children, than they did in the 70's.

The biggest problem is that parents play an irreplaceable role since they help children and youths alike to develop an ethical dimension and increase their personal and social awareness. Most of the times, parents try to take their children to psychologists, in order to understand most of their attitudes. Also, there are plenty of activities for all the family to engage in and disconnect from technology, in order to reconnect with each other and try to reverse the "social trends".

Furthermore, both children and youths should take preventive actions to face cybercrime, managing the digital information and devices used. Besides controlling access to information, parents should mainly control digital gaming devices. When it comes to video games, parents should try playing them first, in order to know the content of the game, and then play it with the child if the game is appropriate; at the end they can see if the child is learning something through the game.

To conclude, it is possible to say that the higher the use of technology, in the parents' case, the higher the stress, which can be reflected in the life style of their children.

It is easy to understand that children have different reactions to the use of gadgets in academia and, on one hand, it could have a positive effect because "tech gadgets and gaming might have positive effects on investigating skills, strategic thinking and creativeness" [7]. On the other hand, the addiction to the Internet and social media could cause intolerance, eating disorders, anxiety, and development delay.

The benefits of the virtual world have slowly started to disappear ever since kids started to spend more time on their digital devices instead of spending time with their family, friends or doing school work. Unfortunately, the technological advances are influencing not only the children, but also their parents and the relationship they have with them.

However, when used correctly, digital devices help to develop relationships and create contacts. Therefore, it can empower empathy, cooperation and acceptance of diversity in youths, as well as the development of early literacy skills through educational programming. These devices also help to change thinking skills and the capability to absorb information, since users can access any information they want online, but sometimes the information is not relevant to their needs. Although this is a positive effect of the use of digital devices, it is necessary to keep a balance avoiding prolonged exposure to useless information.

By now, it is possible to conclude that when it comes to children's health, education and relationships with their parents, technology has both positive and negative aspects. However, it is necessary to think about the consequences and keep a balance when it comes to using it, because the negative aspects can have much more serious consequences than the positive aspects.

3 Methodology

The method chosen was to conduct exploratory surveys and interviews in a primary school, in Portugal, while also questioning other kids between 5 and 13 years-old, parents of these kids and their teachers, in order to ask their opinions, analyse information regarding the impact of technology on their lives, as well as the good and the bad side of using it.

The surveys involving the kids analyses were about who already owns certain digital devices, how often do they use them, which technologies are the ones they use the most and what for. The interviews directed towards the teachers, are intended to analyse their opinion on technology and the changes that it can lead to in society nowadays, as well as its positive and negative influence. Lastly, the surveys given to the parents consisted of the same questions in the teachers' interviews.

Our choice was to approach the people with surveys and interviews because these are the most efficient tools to obtain information about the people who are directly affected by the issue at hand, and responses are registered faster and easier.

Thus, the objective is to obtain answers and analyse the percentage of responses, do comparisons, research the changes occurred due to the time period the research was conducted in and if this is something that benefited subjects during that time, or if this made society become so addicted to the virtual world that we forgot that socializing is very important.

To conclude, the age group choice was based on the fact that it is easy to analyse the changes in young people and then understand what technology made of the society, if something good came out of it and if it made people better, kinder and more communicative, or if it made them worse, addicted and violent.

4 Discussion

4.1 Children's Opinions

As said before, a questionnaire was conducted for 38 children aged 5 to 13. By analysing the answers, it is possible to conclude that the lowest percentage of the inquired children was scored by those aged 5 and 11, with 2.6%, while the highest percentage was obtained by the 9 year-olds, with 34.2%; 52.6% of the children are male and 47.4% are female.

About the focus of this questionnaire, 100% of the inquired kids use digital devices daily and the most used devices are the computer, the tablet and the mobile phone. Most of the digital devices used by children belong to their family or to themselves. These digital devices are used, according to this questionnaire, to do researches, to get on social networks and, lastly, to communicate through messages or calls. Still, 100% of the subjects refer that the main reason to use digital devices is to play videogames.

It is important to highlight that 55,3% of the kids inquired use any digital device on a daily basis and 43.5% of them use it less than 1 h per day, whilst 13% use it more than 4 h per day (Graphic 1). Also, 61.1% of the inquired children use digital devices more than 3 times per week, while just 8.3% use it 3 times per week (Graphic 2).

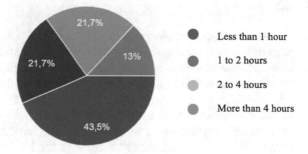

Graphic 1. Hours spent using technologies per day

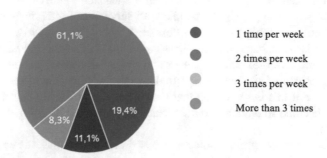

Graphic 2. Time spent using technologies per week

4.2   Parents' Opinions

A total of 15 questionnaires were also answered by parents. In this questionnaire, there are a few points worth mentioning.

First of all, 16.7% of the parents have children aged 5 to 8, 40% have children aged 9 to 11 and 26.7% have children between the ages of 12 and 13.

Parents say that every child in the sample uses electronic devices, which coincides with the children's responses.

Also, 86.7% of the parents consider that technology influences their children positively and negatively at the same time. None of the parents consider that technology influences their children only in a negative way.

About 86.7% of the parents notice that their children are different when using electronic devices.

Every parent sees that technology makes children like learning, which is a positive influence.

The relationship with others and the highly affected attention span are the issues that are highly associated with a negative influence.

Lastly, every respondent feels that children nowadays are different than children from a few years ago.

4.3   Teachers' Opinions

To complete the discussion, three interviews were conducted to analyse the opinion of one group that children are in contact with most of the day, the teachers.

At first sight, two in three teachers say that technology could influence children both negatively and positively, while the other teacher considers that technology only influences students in a positive way. The three teachers observe that when children use technology, they act differently, compared to the times they do not use their devices. Every respondent considers that the relationship with others and school achievements are affected positively by technologies. In addition, two teachers consider that technology influences children's attention negatively. In conclusion, every teacher notices differences between children a few years ago and children nowadays.

5   Conclusions

After reading this article, it is possible to draw much more specific conclusions about the topic "How technology influences children". Everyone knows something about this topic and we are aware that technology is much more present nowadays than it was in the past. Technology is seen everywhere, even while eating at a restaurant, where a child may not be paying attention to his/her meal because he/she is distracted by the cartoons playing on a tablet given to them. It is seen when the parents want to distract a crying child while shopping, so they give them their smartphone, and kids already know how to go on YouTube and search for what they want to see [14]. These are just examples of these days' reality and it is happening with some or most of the children, especially in our sample. However, there are other sides to this topic that are harder to realize, and there are also some good aspects about the technological era that were already discussed and some people might not be aware of.

The information found in scientific articles concludes about how technology has an impact on children's lives. It can bring benefits, such as in the learning process, with more interactive ways of teaching, allowing access to unlimited information and giving us the possibility to keep in touch with family and friends that are far away. Nevertheless, this possibility should not affect the closer relationships kids have, like the one with their parents, although this happens. Also, the health disadvantages can be very serious and alarming, the excessive use of technology leading to irritability, depression, aggressive behaviours and consequently suicidal tendencies, among others explained in the literature review.

Through the questionnaires directed towards children between the ages of 5 and 13, the questionnaires for the parents and the interviews with the three teachers, the conclusions supported what was already stated and gave new information. In fact, technology is very present in our sample of children and in their everyday lives and that has positive aspects as well, such as in the relationship with others, as well as by improving learning capacity and school achievements, as teachers stated. On the other hand, the use of technology has negative aspects, such as lack of attention, the non-existing relationship with others because kids isolate themselves, focusing only on the virtual world, as parents conclude.

In the end, a balance is needed for technology not to be prejudicial for children, because it has positive aspects that they should take advantage of, like access to information, the learning prospects it provides and the possibility to keep in touch with family and friends. However, the main disadvantages are health, social skills and cognitive related issues that can really be harmful for the child and should be avoided by evening out the balance between technology and "real life".

The current younger generation was born surrounded by technology, and the problem (or not) starts with the people around them, because they should give the example of balanced choices when it comes to technology and not introduce them to this world as soon as they are born. They must grow in a healthy manner that helps them develop the right way.

References

1. Au-Yong-Oliveira, M., Gonçalves, R., Martins, J., Branco, F.: The social impact of technology on millennials and consequences for higher education and leadership. Telematics Inform. **35**, 954–963 (2018)
2. Martins, J., Gonçalves, R., Branco, F., Peixoto, C.: Social networks sites adoption for education: a global perspective on the phenomenon through a literature review. In: 2015 10th Iberian Conference on Information Systems and Technologies (CISTI), pp. 1–7. IEEE (2015)
3. Martins, J., Branco, F., Gonçalves, R., Au-Yong-Oliveira, M., Oliveira, T., Naranjo-Zolotov, M., Cruz-Jesus, F.: Assessing the success behind the use of education management information systems in higher education. Telematics Inform. (2018, in press)

4. Ramey, K.: Technology and society - impact of technology on society (2012). https://www.useoftechnology.com/technology-society-impact-technology-society/. Accessed 4 Dec 2018
5. DeLoatch, P.: The four negative sides of technology (2015). http://www.edudemic.com/the-4-negative-side-effects-of-technology/. Accessed 4 Dec 2018
6. American Academy of Pediatrics: Children, adolescents, and the media (2013). http://pediatrics.aappublications.org/content/132/5/958.full#xref-ref-1-1. Accessed 4 Dec 2018
7. Anirudh, B.V.M., Gayathri, R., Vishnu Priya, V.: Attitude of schoolchildren on the use of gadgets on academics. Drug Invention Today **10**(8), 1411–1413 (2018)
8. Singh, M.: Compulsive digital gaming: an emerging mental health disorder in children. Indian J. Pediatr. **86**(2), 171–173 (2019)
9. McDaniel, B.T., Radesky, J.S.: Technoference: parent distraction with technology and association child behavior problems. Child Dev. **89**(1), 100–109 (2018)
10. Huda, M., Jasmi, K.A., Hehsan, A., Mustari, M.I., Shahrill, M., Basiron, B., Gassama, S.K.: Empowering children with adaptive technology skills: careful engagement in the digital information age. Int. Electron. J. Elementary Educ. **9**(3), 693–708 (2017)
11. Radesky, J.S., Schumacher, J., Zuckerman, B.: Mobile and interactive media use by young children: the good, the bad, and the unknown. Pediatrics **135**(1), 1–3 (2015)
12. Kabali, H.K., Irigoyen, M.M., Nunez-Davis, R., Budacki, J.G., Mohanty, S.H., Leister, K.P., Bonner Jr., R.L.: Exposure and use of mobile media devices by young children. Pediatrics **136**(6), 1044–1050 (2015)
13. Rowan, C.: Unplug - don't drug: a critical look at the influence of technology on child behavior with an alternative way of responding other than evaluation and drugging. Ethical Hum. Psychol. Psychiatry **12**(1), 60–68 (2010)
14. Au-Yong-Oliveira, M., Moutinho, R., Ferreira, J.J.P., Ramos, A.L.: Present and future languages – how innovation has changed us. J. Technol. Manage. Innov. **10**(2), 166–182 (2015)

Empirical Studies in the Domain of Social Network Computing

An Empirical Study to Predict the Quality of Wikipedia Articles

Imran Khan[1]([⊠]), Shahid Hussain[2], Hina Gul[3], Muhammad Shahid[4], and Muhammad Jamal[4]

[1] Virtual University of Pakistan, Lahore, Pakistan
`ikniazi786@gmail.com`
[2] COMSATS University, Islamabad, Pakistan
`shussain@comsats.edu.pk`
[3] IQRA National University, Peshawar, Pakistan
`hinaafridi1984@gmail.com`
[4] Govt College No 1, DIKhan, Pakistan
`bluefiber08@gmail.com`, `mustafvimasood@gmail.com`

Abstract. Wikipedia is considered a common way to deliver content in a more effective way as compared to other types of an encyclopedia. However, the quality threat remains an issue regarding the Wikipedia articles. The basic aim of propose research to perform an empirical study to predict the quality of Wikipedia articles. In the proposed methodology, we consider few metrics such as article length (total number of word in an article), number of edits, article age (in the day) and article ranking and perform few statistical tests analyze the quality of Wikipedia articles. Moreover, we observe a significant correlation of proposed metrics with the rating of articles in order to identify their quality.

Keywords: Wikipedia · Correlation · Linear regression · Article length · Number of edits · Article age

1 Introduction

Wikipedia is worldwide most trusted online, open source, the nonprofit organization that owns a large number of articles that almost every topic with a huge viewership (with the total number of 35,147,128 registered users). Wikipedia is one of the socially produced Big Data example. According to the Liu and Ram [8] in September 2017, more than 5,472,000 articles were available on English Wikipedia, now data is produced quicker than ever, and up to now, more than 2.5K Petabyte of data are generated on a day, which brings forth the generally coursed idea of Big Data. By the end of the year 2018, it is more than about 5,763,800 unique articles on English Wikipedia[1]. Wikipedia is a profoundly unique framework; we can change article content as often as possible. Accordingly, the quality of an article is a period subordinate work and a solitary article may contain high and low-quality content in various scopes of its

---

[1] https://en.wikipedia.org/wiki/Special:Statistics

© Springer Nature Switzerland AG 2019
Á. Rocha et al. (Eds.): WorldCIST'19 2019, AISC 932, pp. 485–492, 2019.
https://doi.org/10.1007/978-3-030-16187-3_47

lifetime. The purpose of this research was to analyze the quality of articles on Wikipedia. For instance;

Research Question 1 (RQ1): Is Article Length (AL), highly influence the rating of Wikipedia article?
Research Question 2 (RQ2): Is Number of Edits (NoE), highly influence the rating of Wikipedia article?
Research Question 3 (RQ3): Is Article Age (AA), highly influence the rating of Wikipedia article?

We perform two tests, correlation to analyze the statistic relation between independent variables, to visualize that correlation, to create a scatter plot. The second, to prove our hypotheses use multinomial/linear regression between dependent and independent variables.

2 Related Work

Stvilia [1] and all of them scrutinize the methods in which information quality (IQ) in Wikipedia can be evaluated via proficient methods. Wikipedia article edit history and article features are based on quantitative analysis. Besiki Stvilia and all of them designed a set of seven information quality matrices and nineteen information quality measure that allow us to measure the quality of English Wikipedia articles reasonably and efficiently.

Blumenstock [2] proposed the "word count" metric for measuring the Wikipedia article quality. The Wikipedia article quality can be determined through article length. On the other hand, if our hypothesis does not exist, at that point we can just presume featured articles are long and that long articles are featured.

Qin and Cunningham [3], Propose that commitments are evaluated utilizing alter longevity measures and contributor definitiveness is scored utilizing centrality measurements (metrics) in either the Wikipedia talk or co-creator systems. The outcomes recommend that it is helpful to consider the contributor legitimacy while surveying the data nature of Wikipedia content.

Kittur and Kraut [4] utilizes longitudinal information to inspect the degree to which the quantity of editors contributing to an article in Wikipedia influences its quality and all the more explicitly, the circumstances according to those increasing stakeholder in an article enriches its purity. This represents that adequacy of increasing stakeholder is fundamentally reliant on the degree and stakeholder nature plus the lifespan of an article and the reliance of the assignment engaged with it.

Javanmardi and Lopes [5] design a computerized standard to gauge the purity modifications all through the whole English Wikipedia. Utilizing a numerical approach, pursue the development of substance quality and demonstrate that the portion of time that articles are in a high-quality state has an increasing model subsequent to a few points in time. The author demonstrates to none featured articles will, in general, contain high-quality content 74% of their lifetime and this is 86% for included articles [6–8].

Adler and de Alfaro [9] present a standing method for Wikipedia authors. That mention some of the clearest possibilities, status based content color, Alerting editors,

Restricting edits about low-reputation edits and Provide an incentive for elevated quality contributions [10–13]

Flöck and Acosta [14] develop and envision a framework for organizations of editors communication and called editor-editor organizes. In subsequent study, the authors updated the proposed framework by altering activities on the word-dimension of particular articles. The author also presents the "whoVIS" a work fiction, that is intelligent Website instrument for researching the community-oriented composition procedure a Wikipedia article that joins the majority of the followings highlights, such as (i) a bore down component to figure out how an explicit edge between editors was developed and which words were differ about ("edge setting"), (ii) an interface for intuitively investigating the rising system chart in an explicit article over corrections, improved with meta-data on editors and edges, (iii) mining (re)introduction and erase activities of editors on one another's composed content at word granularity with demonstrated precision, to gather and model supervisor editorial manager contradiction, (iv) some supplementary metrics, and (v) a custom chart drawing strategy for different edges in Wikipedia editing [15–17].

3   Methodology

The aim of the purposed methodology is to empirically investigate the relationship between the variables, (Article length (AL), Article age (AA), Number of edits (NoE) and category of articles (FA, GA, B-class, and C-class)) of an article. The influences of feature articles on Wikipedia and their quality. The layout of the proposed method is shown in figure and description as well.

3.1   Data Collection

Develop crawler to gather a sample of 100 random articles that consist of 25 article of each category i.e. B-class articles, C-class articles, good article and feature articles from the English Wikipedia, December 2018 and that articles gathered from different English Wikipedia points utilizing a stratified testing approach. Feature articles are organized by English Wikipedia into several subcategories including politics, science, sports, and so on. To research the focused territories, we gathered a few specific categories form Wikipedia, applying our required methodology. For this purpose, we decided to begin to collect form feature articles. Wikipedia characterized highlighted articles (i.e. feature article) into multi fundamentally distinct classes including science, technology, health, sports, and so on. The quantity of highlighted articles we arbitrarily have chosen from every class is in the uneven extent to the quantity of included articles in the class. In view of the arrangement of highlighted articles we chose, we at that point arbitrarily chosen a similar number of B-class, C-class articles and good articles from a similar space (Table 1).

Table 1. The dataset variable and its type

Sr. No.	Variables	Type	Description
1	Article Length (AL)	Continuous	Total number of word in an Wikipedia Article
2	Number of Edits (NoE)	Continuous	Total number of edits contain an article
3	Article Age (AA)	Continuous	Total number of days (Article publish to to-do-date)
4	Rating	Category	Article category (Feature Article, Good Article, B-Class or C-Class Articles)
5	Article name	Demographic	Title of an article

3.2 Data Analysis

We categories out data variables into two way independent variable that consists of AL, NoE, and AA and dependent variable Rating (categorized into FA, GA, B-class, and C-class) that is article category. To initialize the four values for the dependent variable, 4 are FA, 3 for GA, 2 for B-class and 1 for C-class.

3.3 Reports

The result will be shown in correlation, linear regression table and scatter plot (Fig. 1).

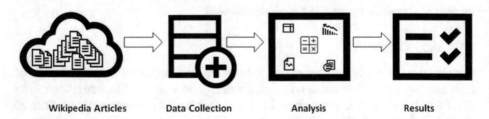

Wikipedia Articles   Data Collection   Analysis   Results

Fig. 1. Workflow of proposed method

4 Experiment Process

We have four variables Article Length (AL), Number of Edits (NoE), Article Age (AA) and Article Category (Feature Article, Good Article, B-Class, C-Class). The AL, NoE and AA are independent/predictor variables to predict the impact on Wikipedia article rating. We initialized the four values for article category, 4 values for the feature article, and 3 for good article, 2 for B-class and 1 for C-class. We perform two tests, first, we perform a Pearson Correlation (because several of these variables are equal) to analyze the statistic relation between independent variables, to visualized that correlation, to create a scatter plot. The correlation between two variables gives us positive, negative or no correlation. If the correlation is positive or negative then find that correlation is the week or strong. The second, dependent variable for our situation is categorical, and consequently, we utilize multinomial/linear regression to test our hypotheses.

4.1   Result and Discussion

We performed certain experiments. In order to respond research questions, the discussion of results is as follows.

Respond to RQ1: In order to response RQ1, first we perform Pearson correlation to analyze the statistic relation between article length (AL) and the number of edits (NoE), and article length (AL) with article age (AA) as well. The output of AL and NoE is 0.716 that show the too strong and positive correlation between them, and the output between AL and AA is 0.436 that show the positive but not strong correlation between them Table 2. To visualize that correlation we create a scatter plot this tells us the correlation between the Al and NoE is strong because the data points are shown in liner/line. But the correlation between AL and AA the data points scatter around, this means the correlation not too much strong but positive Figs. 2 and 3. In linear regression, the significant value of AL is 0.742 Table 3. This value is greater than the 0.05 of our alpha level and that's not supported to relationship AL does not statically fact the article category that indicates that article length is not high influence on the rating of Wikipedia article.

Table 2. Correlation between variables

Correlations		Article length	Number of edits	Article age
Article length	Pearson correlation	1	.716**	.436**
	Sig. (2-tailed)		.000	.000
	N	100	100	100
Number of edits	Pearson correlation	.716**	1	.489**
	Sig. (2-tailed)	.000		.000
	N	100	100	100
Article age	Pearson correlation	.436**	.489**	1
	Sig. (2-tailed)	.000	.000	
	N	100	100	100

** Correlation is significant at the 0.01 level (2-tailed).

Respond to RQ2: In order to response RQ2, first, we perform Pearson correlation to analyze the statistic relation between NoE, AL and number NoE with AA as well. The NoE and AL already discuss in RQ1, and the output between NoE and AA is 0.489 that is approximately 0.5, that shows the positive and strong correlation between them Table 2. To visualize that correlation crate a scatter plot this tells us the correlation between the NoE and AA is strong because the data points are shown in liner/line Fig. 4. In linear regression, the significant value of NoE is 0.124 Table 3. This value is greater than the 0.05 of our alpha level and that's not supported to relationship NoE

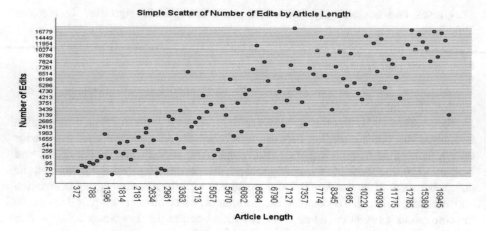

Fig. 2. Scatter plot for number of edits

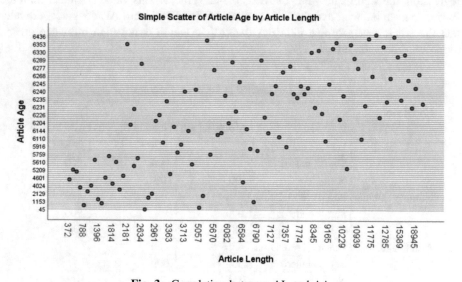

Fig. 3. Correlation between AL and AA

does not statically fact the article category that also indicates that article length is not high influence on the rating of Wikipedia article.

Respond to RQ3: In order to response RQ3, as usual, first we perform Pearson correlation to analyze the statistic relation between AA and AL along with AA with NoE as well. The AA and AL already discuss in RQ1, and the correlation between AA and NoE discussed in RQ2. In linear regression, the significant value of AA is 0.06 Table 3. This value is greater than the 0.05 of our alpha level but not too much and that's supported to relationship AA does statically fact the article category that also indicates that article length is influencing on the rating of Wikipedia article.

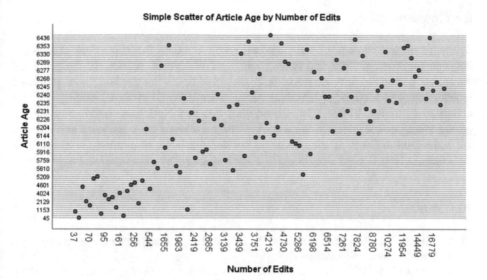

Fig. 4. Correlation between NoE and AA

Table 3. Liner regression between the independent and dependent variable

Coefficients[a]

Model		Unstandardized coefficients		Standardized coefficients		
		B	Std. error	Beta	t	Sig.
1	(Constant)	3.530	.406		8.691	.000
	Article length	−1.115E−5	.000	−.047	−.330	.742
	Number of edits	5.149E−5	.000	.226	1.551	.124
	Article age	.000	.000	−.316	−2.793	.006

a. Dependent Variable: Article Category

5 Threads and Validity

In this proposed study, we observe a few threats. The first threat is related to sample size which is 100 in our case. An increase in sample size might affect the presented finding. The second threat is the use of the number of three metrics. The results regarding the quality prediction of Wikipedia articles might change when the numbers of metrics are increased. Only three feature are to be analyzed (Article Length, Article Age and Number of Edits).

6 Conclusion

The experimental results of the proposed study indicate the significant correlation between proposed metrics and quality rating of Wikipedia articles. In the proposed study, we performed few statistical tests and investigate the correlation of proposed metrics namely Article Age (AA), Number of Editing (NoE), Article Length (AL) with a rating of articles. The results indicate that AL and NoE have influence over the articles rating as compared to the AA. As Wikipedia has a large range of experts who are authorized to check articles, their quality and facts. So, if an article is old enough than ultimately it is dead sure that it has been checked by the authorized experts.

References

1. Stvilia, B., Twidale, M.B., Smith, L.C., Gasser, L.: Assessing information quality of a community-based encyclopedia (2005)
2. Blumenstock, J.E.: Size Matters: Word Count as a Measure of Quality on Wikipedia. School of Information University of California at Berkeley, Beijing, China, 21–25 April 2008
3. Qin, X., Cunningham, P.: Assessing the Quality of Wikipedia Pages Using Edit Longevity and Contributor Centrality. Aalto University and Padraig Cunningham University College Dublin, June 2012
4. Kittur, A., Kraut, R.E.: Harnessing the wisdom of crowds in Wikipedia: quality through coordination (2008)
5. Javanmardi, S., Lopes, C.: Statistical measure of quality in Wikipedia (2010)
6. Hussain, S., Keung, J., Sohail, M.K., Ilahi, M., Khan, A.A.: Automated framework for classification and selection of software design patterns. Appl. Soft Comput. **75**, 1–20 (2019)
7. Hussain, S., Keung, J., Khan, A.A.: Software design patterns classification and selection using text categorization approach. Appl. Soft Comput. **58**, 225–244 (2017)
8. Hussain, S., et al.: Implications of deep learning for the automation of design patterns organization. J. Parallel Distrib. Comput. **117**, 256–266 (2017)
9. Adler, B.T., de Alfaro, L.: A Content-Driven Reputation System for the Wikipedia. University of California, Santa Cruz, November 2006
10. Nasir, J.A., Hussain, S., Dang, H.: Integrated planning approach towards home health care, telehealth and patients group based care. J. Netw. Comput. Appl. **117**, 30–41 (2018)
11. Ahmed, S., Ahmed, B., Iqbal, J., Hussain, S.: Mapping the best practices of XP and project management: well defined approach for project manager. J. Comput. **2**(3) (2010)
12. Hussain, S.: A methodology to predict the instable classes. In: 32nd ACM Symposium on Applied Computing (SAC) Morocco, 4th to 6th April 2017
13. Hussain, S.: Threshold analysis of design metrics to detect design flaws. In: ACM Symposium on Applied Computing (SRC), Pisa, Italy, 4–8 April 2016 (2016)
14. Flöck, F., Acosta, M.: whoVIS: visualizing editor interactions and dynamics in collaborative writing over time (2015)
15. Liu, J., Ram, S.: Using big data and network analysis to understand Wikipedia article quality. Dakota State University the USA, University of Arizona, USA
16. Wilkinson, D.M., Huberman, B.A.: Assessing the value of cooperation in Wikipedia (2007)
17. Hu, M., Lim, E.P., Sun, A., Lauw, H.W., Vuong, B.Q.: Measuring article quality in Wikipedia (2007)

Investigating the Impact of Podcast Learning in Health via Social Network Analysis

Fouzia Tabassum[1]([⊠]), Fazia Raza[2], Fahd Aziz Khan[3],
Syed Bilal Akbar[4], Sheema Rasool Bangash[5],
and Muhammad Shahid[6]

[1] Social Security Hospital, Peshawar, Pakistan
docfzee@gmail.com
[2] Rehman Medical College, Peshawar, Pakistan
faziaraza@yahoo.com
[3] PMU, Peshawar, Pakistan
fah.aziz@gmail.com
[4] Southern Punjab University, Multan, Pakistan
khan_j786@yahoo.com
[5] Institute of Management Science, Peshawar, Pakistan
sheemahizbullah@gmail.com
[6] Govt College No 1, DIKhan, Pakistan
bluefiber08@gmail.com

Abstract. In the context of medical education, research community is focusing on the implications of podcast learning in education. Though, in numerous studies, the impact of podcast technology is investigated and reported. However, to the best of our knowledge, there is no study which has empirically investigated the podcast learning on the patient health outcomes. In order to address this issue, we propose a method to compute the user's perceptions regarding podcast learning and its impact through social network analysis. Subsequently, we performed a case study comprise on 74086 public tweets extracted from the Twitter social network. In this paper, our main contributions are related to the understanding the public opinion and sentiments regarding podcast learning and its impact on the patient's outcomes. We observe 44.4% positive, 40% neutral and 15.6% negative comments to through sentiment analysis. The low value of negative comments indicates the agreement level of public regarding the correlation of podcast learning and its impact on patient's outcomes.

1 Introduction

The use of audio recordings for medical education has been documented in the literature as far back as 1968, when they were used for asynchronous learning in histology classwork. Podcasting in its current form, as a method of distributing audio content and digital recording which played on the digital media player. Usually, MP3 file are used for audio recording but other formats can also be used. Several devices are used to play these files such PC (Personal Computers), laptops and so on. Besides several sort of media players such as Window media players, iTunes and so on [1].

© Springer Nature Switzerland AG 2019
Á. Rocha et al. (Eds.): WorldCIST'19 2019, AISC 932, pp. 493–499, 2019.
https://doi.org/10.1007/978-3-030-16187-3_48

Recently, rapid increase in the podcast users is reported due to variety in devices used to play the audio recording files. Such devices allow audio and video for listening and watching without any time or place constraints. Moreover, in the context of education, same devices can be used to delivered relevant material. Like other domains, in medical education, telemedicine or in telehealth, effective use of podcast has been reported. The interest of professional communities and medical journals are producing numerous podcasts. However, due to blurred images and noises several podcast are immature and demand for high quality cameras and microphones. Research community has also empirically investigated that mostly podcasts are not designed to replace the traditional way of lectures. However, but these podcast can help student to provide online lectures which can be learn anywhere and any tine with any pace. As compared to traditional learning, learning via podcast is more effective because it can enable a learner to understand in-class taught material by revising it several times. Several practice points of podcast are reported such as (1) Its use for medical students, (2) It is useful for exam revision, (3) its impact on patient's outcome and (4) It is available in certain formats [2].

In the context of medical education, podcast are considered as an emerging area for institution, doctors and patient. However, there is lack of study to show the public perception regarding the implementation of podcast and its impact on the patients outcomes. We address this issue and propose a methodology. The aim of propose methodology is to conduct an empirical study to compute the public perception regarding the use of podcast in medical education and its impact on the patient's outcome. In order to achieve our propose objective, we formulate the following two research questions.

Research Question 1 (RQ 1): What is the public perception and opinion regarding the use of podcasts in medical education?
Research Question 2 (RQ 2): What are the public concerns about podcasting in patient's outcomes?

The aim of RQ 1 is to investigate that how public is perceiving Podcasts in medical education and what are their opinion. Similarly, the aim of RQ 2 is to investigate the impact of Podcast on patients' outcome via public perception. In this paper, our main contribution are:

  i. To design and implement a data crawler for twitter Application Programming Interfaces (API).
 ii. To compute the public perception regarding podcast learning.

The rest of the paper is organized as follows: In Sect. 2, we discussed some related work on the topic modeling and sentiment analysis. In Sect. 3, we describe the proposed method. In Sect. 4, we describe the experimental procedure. In Sect. 5, we present the results. Finally, in Sects. 6 and 7, we discuss the implications of the proposed method and conclusion respectively.

2   Related Work

Barnes and Block perform [2] a systematic study of 42 articles and out of which 25 were related to podcasting health education. Authors conclude the two shortcomings of podcasts such as (1) Podcasts cannot replace the traditional lectures, and (2) Podcast have no effect on the class attendance. However, podcasts can be considered as a supplemented material and enable students to study anywhere and anytime.

Vogt et al. [3] use podcasting as an emerging technology of teaching in higher education and indicate its implication with positive satisfaction of students. In this study, the authors select the two junior baccalaureate nursing classes and investigate their earned scores in exams. The sample size of first class was 63 and evaluates them through traditional lectures, while the sample size was 57 and used podcast. The authors observe no significant difference. However, exam scores of second class (who used podcasts) were seem better than another class.

Mostyn et al. [4] perform a case study to explore the impact of biology podcasts on the student's learning. In this study, the authors provide the 9 biological podcasts as a learning tool to 189 first year nursing students. After completion of first year, students were asked to complete a survey. The aim of survey was to frequent use of podcasts tools, reasons why they are using and its impact of on their learning. Finally, the authors concluded that the availability of biology podcast can be more effective for student's learning.

In this paper [5], the authors conduct as study at University of Minho, Portugal to investigate the use of podcast in blended learning. Moreover, 6 lecturers create the podcasts for student's learning (318 students). The aims of podcasts was to support the undergraduate and graduate levels students. Finally, the authors concluded the effectiveness of use of podcast in learning [6–8].

Friedel and Treagust [9] introduce a curriculum inquiry framework. The aim of proposed framework was to investigate the perceptions of 184 students and tutors in the context of nurse education to design the nursing curriculum. Finally, the authors conclude that nursing curriculum are not fulfilled and besides teachings it can be improve. However, the nurses should be the credible members of multi-disciplinary team [16–19].

In the context of healthcare, Dollinger [14] introduce several podcasts in order to facilitate the healthcare professional. The author introduced new podcast every week in the healthcare landscape.

3   Proposed Methodology

The overview of proposed method is shown in Fig. 1. The Fig. 1 illustrates the three phases of proposed approach. We performed linguistic and statistical analysis to tweets regarding podcast learning and its impact on patients outcomes. The three main phases of the proposed approach are (1) data collection and pre-processing, (2) Sentiment Analysis, and (3) Data Visualization.

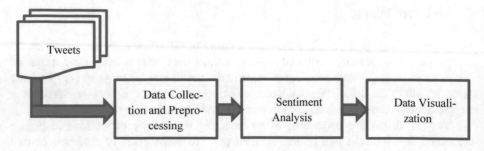

Fig. 1. Overview of proposed approach

3.1   Data Acquisition and Pre-processing

The first phase of proposed approach is responsible to crawl data for some queries. Data is crawled from twitter Application-programming interfaces (API) [15]. Different queries are used to get maximum public tweets regarding the keywords. Table 1 shows the keywords besides the obtained tweets and their percentage. Table 1 Keywords and No. of relevant crawled tweets.

Table 1. Descriptive statistics of Tweets

Keywords	Queries	No. of Tweets
Podcast and patient's outcomes	(#podcasts OR #podcasting) AND (#Education, #Medical) (#podcasts OR #podcasting) AND (#patients, #treatment)	74086

The retrieved data is unstructured and in the form like text, images, special characters and videos. Consequently, it needs some pre-processing steps to get the required format for further processing. We performed and recommend a set of activities in order to present data in structured form. The list of activities is as follows.

Step-1. We convert all tweets into lower case in order to make consistent text.
Step-2. We remove the stop words such as the, a and so on.
Step-3. We apply word stemming using Poster's algorithms to obtain stem word to reduce text noising.
Step-4. We also removed emoticons and special symbols used in tweets.
Step-5. We removed the URLs as they are also not helpful for getting the required results.
Step-6. We extract the hashtag '#' data, as this is a method to spread opinion and information all over the network publicly.
Finally, all processed data is stored in the database in structured form.

3.2   Sentiment Analysis

The second phase of the proposed approach is aid to perform the main roles. Firstly, we used the pre-processed data for sentiment analysis to get the overall public perception about the keywords. Secondly, we also analyzed the data distribution and investigated the nature of data using some statistical analysis. For sentimental analysis of podcast learning and health the cleansed data are retrieved from database. Tags are assigned to each tweet based on their polarity. Tags can be positive, negative, or neutral. The positive tag means that the tweet is giving a positive opinion about the searched trend. The neutral tag means the tweet is not either expressing positive or negative opinion about the given trend.

3.3   Data Visualization

The aim of this phase is to present the analyzed results. Sentimental analysis results are visualized using Pie graphs and bar graphs to show the overall public perception in some particular keyword. The graphs show different aspects of public opinion.

4   Experimental Procedure

The main steps of our experiment are.

Step-1. We design a crawler to retrieve data from multiple clients at the same time. We implement it in Python language and collect data for podcast learning. The preprocessed data is store in NoSQL database in JSON format for fast processing.
Step-2. We performed sentiment analysis to identify the polarity of tweets. We used lexicon based sentimental approach to analyze the tweets polarity. WordNet is used for scoring the opinion words that how is the tweet positive, negative or neutral about the topic. Matplotlib package is used for visualizing all the graphs in python.
Step-3. We used pie and bar charts to describe the results of sentiment analysis.

5   Results and Discussion

Response to RQ-1: Sentiment analysis is performed in order to get the overall user perception on the podcast learning in medical education. Figure 2 shows the retrieved user perception results. That is the distribution of public opinion regarding their sentiments towards the podcasting on patient's outcome. The results present an interesting fact that almost equal percentage of the population has positive and neutral sentiments towards podcasting. This shows that a huge portion of the population is ignorant about the technology. They have not used or they never listened about this technology. That's why they remain neutral about it.

Response to RQ-2: In order to get the public concerns about the use of podcasting in medical is achieved using topic modeling. Topic modeling is applied to get the significant terms occurred in tweets corpus. Cloud representation is used to show the topics along with corresponding frequencies. In this figure (Fig. 3), each some of the topic having fewer frequencies comparatively are missing kids, fake, explosive and failed etc. However, the positive topics or interests about podcasting are bleacher report, recover, and future etc.

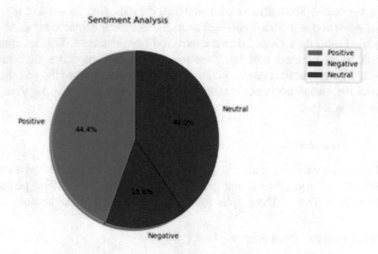

Fig. 2. Sentiment analysis of Podcast learning

Fig. 3. Word cloud extracted from Podcast learning Tweets

6 Conclusion

The public perception regarding new technologies can aid a decision maker of a domain expert to set the effective guidelines for the improvement. Recently, in terms of podcast learning, researchers have reported its implications on students learning and patient's outcomes. In this paper, we propose a approach via social network to compute user perception regarding podcast learning. We retrieved 74086 public tweets from Twitter regarding podcast learning. Subsequently, we perform sentiment analysis.

The results 44.4% positive, 40% neutral and 15.6% negative comments indicate that huge portion of the population is ignorant about the technology. They have not used or they never listened about this technology. That's why they remain neutral about it.

References

1. Sandars, J.: Twelve tips for using podcasts in medical education. Med. Teach. **31**, 387–389 (2011)
2. Barnes, F.M., Block, R.M.: Podcasting in medical education: a literature review. Podcasting in Medical Education (2014)
3. Vogt, M., Schaffner, B., Ribar, A., Chavez, R.: The impact of podcasting on the learning and satisfaction of undergraduate nursing students. Nurse Educ. Pract. **10**, 38–42 (2010)
4. Mostyn, A., Jenkinson, C.M., Mccormick, D., Meade, O., Lymn, J.: An exploration of student experiences of using biology podcasts in nursing training. BMC Med. Educ. **13**(1), 12 (2013)
5. Carvalho, A.A., Aguiar, C., Santos, H., Oliviera, L., Marques, A., Maciel, R.: Podcasts in higher education: students' and lecturers' perspectives. In: WCCE 2009, IFIP AICT, vol. 302, pp. 417–426 (2009)
6. Hussain, S., Keung, J., Sohail, M.K., Ilahi, M., Khan, A.A.: Automated framework for classification and selection of software design patterns. Appl. Soft Comput. **75**, 1–20 (2019)
7. Hussain, S., Keung, J., Khan, A.A.: Software design patterns classification and selection using text categorization approach. Appl. Soft Comput. **58**, 225–244 (2017)
8. Hussain, S., et al.: Implications of deep learning for the automation of design patterns organization. J. Parallel Distrib. Comput. **117**, 256–266 (2018)
9. Friedel, J.M., Treagust, D.: Learning bioscience in nursing education: perceptions of the intended and the prescribed curriculum. Learn. Health Soc. Care **4**(4), 203–216 (2005)
10. Nasir, J.A., Hussain, S., Dang, H.: Integrated planning approach towards home health care, telehealth and patients group based care. J. Netw. Comput. Appl. **117**, 30–41 (2018)
11. Ahmed, S., Ahmed, B., Iqbal, J., Hussain, S.: Mapping the best practices of XP and project management: well defined approach for project manager. J. Comput. **2**(3) (2010)
12. Hussain, S.: A methodology to predict the instable classes. In: 32nd ACM Symposium on Applied Computing (SAC) Morocco, 4th to 6th April 2017
13. Hussain, S.: Threshold analysis of design metrics to detect design flaws. In: ACM Symposium on Applied Computing (SRC), Pisa, Italy, 4–8 April 2016 (2016)
14. Dollinger, N.: The Future of Healthcare Podcast (2018). https://wearethefutureofhealthcare.com/podcast-2
15. Ramage, D., Dumais, S., Liebling, D.: ICWSM, and undefined 2010, Characterizing microblogs with topic models. aaai.org
16. Narula, N., Ahmed, L., Rudkowski, J.: An evaluation of the '5 Minute Medicine' video podcast series compared to conventional medical resources for the internal medicine clerkship. Med. Teach. **34**(11), e751–e755 (2012)
17. O'Neill, E., Power, A., Stevens, N., Humphreys, H.: Effectiveness of podcasts as an adjunct learning strategy in teaching clinical microbiology among medical students. J. Hosp. Infect. **75**(1), 83–84 (2010)
18. Walmsley, A.D., Lambe, C.S., Perryer, D.G., Hill, K.B.: Podcasts–an adjunct to the teaching of dentistry. Br. Dent. J. **206**(3), 157–160 (2009)
19. Bensalem-Owen, M., Chau, D.F., Sardam, S.C., Fahy, B.G.: Education research: evaluating the use of podcasting for residents during EEG instruction: a pilot study. Neurology **77**(8), e42–e44 (2011)

A Methodology to Characterize and Compute Public Perception via Social Networks

Shaista Bibi[1], Shahid Hussain[1], Mansoor Ahmed[1(✉)],
and Muhammad Shahid Zeb[2]

[1] COMSATS University, Islamabad, Pakistan
spl7-rcs-022@student.comsats.edu.pk,
{shussain,mansoor}@comsats.edu.pk
[2] Gomal University, Dera Ismail Khan, Pakistan
bluefiber08@gmail.com

Abstract. Literature shows that the business experts and the data scientist always look at new technologies and its impact on their data centers. Nowadays, in terms of green computing, Internet of Things (IoT), Artificial Intelligence (AI), and Virtual/Augmented Reality (V/AR) are considered and adapted as new technologies. Though, in numerous studies, individual impact of each technology is investigated and reported. However, to the best of our knowledge, there exists no such study that describes the public sentiments and perception regarding V/AR, IoT, and AI; that are required to understand the public demand and improve the business process. In this paper, we propose a computation method for the public perception of new technology(y/ies) in various aspects. Topic modeling, sentiment analysis, and statistical techniques are applied to make the proposed method functional. Subsequently, we have performed a case study, that comprises on 147 million public tweets extracted from the Twitter social network. Moreover, our main contributions are related to the understanding of, (1) Distribution, (2) public perception, and (3) correlation of IoT, AI, and V/AR. The main outcome of the proposed study are; (1) More tweets on AI (51.51%) rather than V/AR (18.37%) and IoT (30.11%), (2) positive comments for IoT and negative comments are identified via sentiment analysis. (3) Some of the noteworthy terms found are blockchain, futurist, user-experience, users-demand, bonus, and presale. These all have been identified as sub-topics for each keyword describing the mutual relationship among the topics. This study is easy to replicate in terms of adaptation of new technology (y/ies) for sustainability and evolution of business process and data centers on the basis of public perception.

Keywords: Green computing · Topic modeling · Sentiment analysis ·
Internet of Things · Virtual Reality · Artificial Intelligence

1 Introduction

In the era of internet and digital communication, latest technologies; being evolved, hold a tremendous impact on social life in certain domains such as healthcare, business, crowd sensing, and digital marketing. On the same way, the upcoming years would witness significant trends in green computing and its implications for business

Á. Rocha et al. (Eds.): WorldCIST'19 2019, AISC 932, pp. 500–510, 2019.
https://doi.org/10.1007/978-3-030-16187-3_49

analytics, Virtual/Augmented Reality (V/AR), Internet of Things (IoT), and Artificial Intelligence (AI). These trends will facilitate Information Technology (IT) leaders and data scientists uplifting their business and data centers as more sustainable[1] than the others technologies could. In spite of differences in their infrastructure and implications, the global use of these trends in certain human fields has been reported. For example, these trends are widely adopted in health [1], education and entertainment [2, 3]. These technologies having an immense effect on our society can revolutionize the world. For example, V/AR have generated USD revenue of 7.7 billion by the end of 2017 and it is expected to break down the industry with 75 billion USD by 2021 [4]. The literature shows that several studies have addressed, how these technologies and their implication have been adopted for the evolution and sustainability of multi-domain systems. However, to the best of our knowledge, there exists no empirical study describing the public perception about these technologies. It is a myth that user (or public) perception and feedback regarding the adaptation of new technologies might help the decision makers to take an effective decision regarding the evolution and sustainability of their system. In this regard, social networks; Facebook and Twitter, are considered as vital sources of information (i.e. Public perceptions) by the research community. Using these platforms, users share their opinions about the emerging trends and their feasibilities in different perspectives. For example, the impact of V/AR, IoT, and AI on the social life, expenses, and revenues with its pros and cons in a multi-dimensional business perspective. In the proposed study, we observe at the capacity of public perception of three emerging trends V/AR, IoT, and AI on Twitter data [5]. We consider Twitter data more reliable because it has been used by the research community in certain domains such as health issues, sentimental, and semantic analysis to extract knowledge of particular topics. The aim of Twitters text mining explores the underlying information patterns, such as in health [1] and economics [2]. Subsequently, topic modeling [7] is also considered a powerful way to identify the hidden patterns from the huge amount of data. The working procedure of the proposed approach is shown in Fig. 1, while described in Sect. 3. In order to achieve our research objective, we formulate the following three research questions.

Research Question 1 (RQ-1): What is the data distribution nature retrieved for VR/AR, IoT, and AI from Twitter?

Research Question 2 (RQ-2): What is the public perception and opinion regarding the aforementioned trends?

Research Question 3 (RQ-3): What is the correlation between the aforementioned trends and how it can play a role in the future business market?

In this paper, our main contribution is:

- To design and implement a data crawler for twitter Application Programming Interfaces (API).
- To leverage and implement a Linear Discriminant Analysis (LDA) to extract topics for the keywords VR/AR, IoT, and AI under study.

---

[1] https://www.mediabuzz.com.sg/technologies-and-products-jan-2017/the-force-of-ai-iot-vr-or-ar-on-digital-marketing-in-2017.

- To compare the data distribution of VR/AR, IoT, and AI in terms of their frequencies.
- To compute the public perception regarding VR/AR, IoT, and AI technologies via statistical tests.

To compute the correlation between VR/AR, IoT, and AI trends

The rest of the paper is organized as follows: In Sect. 2, we have discussed some related work on the topic modeling and sentiment analysis. In Sect. 3, we have described the proposed method. In Sect. 4, we have elaborated the experimental procedure. In Sect. 5, we have presented the experimental results. Concludingly, in Sects. 6 and 7, we have discussed the implications of the proposed method and conclusion respectively.

2 Related Work

The context of the proposed study has been described with the implication of social networks for public perception, opinion mining, and topic modeling. In this regard, we have summarized the related work. From the last decade, the research community has mined and analyzed social network data from various social websites and smart-phone applications. Subsequently, the research community has led to sensors to get real-time data such as diseases' influence in a particular location, detecting traffic congestions from users' messages and natural disasters. Predictions based on user response and sentimental analyses of the data are the key research areas. In recent studies, authors have reported that microblogging websites can effectively resolve the various issues. Twitter data is considered more prominent than any of other social networks. Quantitative analysis of Twitter along with characterization of different aspects was common in early research. For example, Java et al. [8] have shown high degree correlation among twitter' users and their mutual acquaintances. Subsequently, users' geographical classification based on their behaviors has been shown by Krishnamurthy et al. [9].

Arnold and Speier [10] have mined the patients' data under the clinical structures and showed some prediction based on the previous records. Similarly, Karami et al. [1] have investigated the relationship among various health-queries; diabetes, exercise, and obesity, using through topic-modeling of social website Sina Weibo data. Accordingly, few other researchers have also worked in the domain prediction of user-behavior for drug-addiction, smoking, exercise and mental health based on the social media unstructured data [11]. Bollen et al. [12] have targeted the stock exchange and election system via topic modeling techniques. These previous studies show the extraction of common topics regarding some query and discussed the solution using the retrieved results. However, this study has used an innovative approach by analyzing the same data used by other researchers in different contexts. Like these researchers, we have also focused on the characterization of Twitter text on the basis of new emerging trends VR/AR, IoT, and AI to get public responses and analyzing these responses in different perspectives [13–15, 18, 19].

3   Overview of Proposed Method

In this section, we present an overview of the proposed method shown in Fig. 1. Figure 1 illustrates the three-tier architecture of the proposed method. We have used linguistic and statistical analysis to tweets regarding V/AR, IoT, and AI.

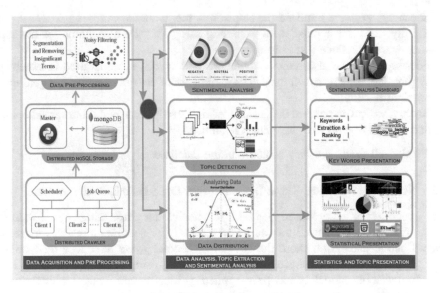

Fig. 1. System architecture

There are three main phases of the proposed method namely data acquisition and pre-processing, topic detection, and analysis of results.

Table 1. Keywords and number of crawled tweets

Keywords	Queries	No. of Tweets	Percentage
VR/AR	#Virtual reality/#Augmented reality OR VR/AR	27176091	18.37
IoT	#Internet of Things OR #IoT	44536299	30.11
AI	#Artificial Intelligence OR #AI	46190309	51.51

3.1   Data Acquisition and Pre-processing

The first tier of the proposed method is responsible for crawling data against some queries. Data is crawled from twitter Application-programming interfaces (API) [5]. Different queries are used to get maximum public tweets regarding the keywords. Table 1 shows the keywords besides the obtained tweets and their percentage.

The retrieved data is unstructured i.e. containing text, special characters, images, and videos. Therefore, it needs some pre-processing steps to get the required format for further processing. We have performed a set of activities in order to present data in structured form. The list of activities is as follows:

- Firstly, we have converted all tweets into lower case in order to make consistent text.
- Secondly, we have removed the stopwords; the, a and so on.
- Thirdly, we have applied word stemming using Poster's algorithms for obtaining stem-word to reduce text noising.
- Fourthly, we have also removed emotions and special symbols used in tweets.
- Fifthly, we have removed the URLs as they were also not helpful for getting the required results.
- Sixthly, we have extracted the hashtag '#' data, as this is a method to spread opinion and information all over the network publicly.

Finally, all processed data is stored in the database.

3.2   Sentiment Analysis, Topic Detection and Data Distribution Analysis

The second tier of the proposed method is responsible for performing key roles. In this phase, firstly, we have used the pre-processed data for sentiment analysis to get the overall public perception about the keywords. Secondly, we have also analyzed the data distribution and investigated the nature of data using some statistical analysis. Thirdly, the topic modeling has been applied to identify the most significant words from the corpus. Three main steps of this tier are as follows:

A. Sentiment Analysis

For sentimental analysis of VR/AR, IoT, and AI, the cleansed data are retrieved from the database. Tags are assigned to each tweet based on their polarity. Tags can be positive, negative, or neutral. Positive tag means that the tweet is holding a positive opinion about the searched trend. Neutral tag means that the tweet is not either expressing positive or negative opinion about any trend. Similarly, negative tag means that the tweet contains negative words or expression. Some of the tweets are considered as garbage because these are not related to the context.

B. Data Distribution Analysis

Data is characterized by a distribution. Data would either be normally or abnormally distributed. We have identified the data distribution for IoT, AI, and VR/AR topics:

C. Topic modeling

We have applied the topic-modeling technique to extract the most significant words from the dataset. These terms or topics have been further analyzed to address the research questions. There exist various techniques, which are used for topic modeling such as LDA [16]. Different variants of LDA have been proposed [7], such as LSA is also used for the same purpose. Different studies have explained all the state-of-the-art topic-modeling techniques and their variants. Among all these models, LDA is the most widely used technique and an effective model. Hong and Davison [7] have performed the empirical analysis of various techniques and hence proved LDA as an effective computation model for identifying the hidden patterns in the corpus. In this technique, fuzzy have clustered of similar words such as VR/AR have been grouped together and extracted a significant topic from it.

3.3 Topics Analysis and Visualization

The aim of the third tier is to present the obtained and analyzed results. Sentimental analysis results have been visualized using Pie-graphs and bar-graphs to show the overall public perception in some particular keyword. The graphs show various aspects of public opinion. The data distribution analysis is represented in the form of a table to provide the comparison of retrieved data for all the keywords. Subsequently, the cloud graphs have been used for the presentation of extracted topics. Using cloud graph it is easy to show the most significant terms of the corpus along with the corresponding frequencies. Finally, we have calculated the correlation between each possible pair of keywords. For example, the correlation between V/AR to IoT, V/AR to AI, and AI to IoT.

4 Experiment Setup

In this section, we provide the experimental elaboration of our proposed study to perform certain experiments with some common operations as follows.

4.1 Data Collection

Distributed crawler has been designed to retrieve data from multiple clients at a time. It has been implemented in python. We have gathered data for V/AR, IoT, and AI. This data was in an unstructured format. The pre-processing steps (Sect. 3) are have performed all experiments using python Pyscripter IDE. We have converted all text into the lower case by using the Lower() function. Similarly, replace and re. sub () functions are used to remove all punctuation and subsequent URLs. After preprocessing, the structured data is stored in the NoSQL database. MongoDB NoSQL database stored the data in JSON format and provides its fast processing.

4.2 Sentiment Analysis

Sentiment analysis has been performed by identifying the polarity of the tweet. We have used a lexicon based sentimental approach to investigate the polarity of tweets. WordNet has been used for scoring the opinion words that how was the tweet positive, negative or neutral about the topic. Matplotlib package has been used for visualizing all the graphs in python.

4.3 Topic Detection

Topic detection has been done by LDA. LDA is a text-mining algorithm. It is a generative model based on the statistical Bayesian model. The main idea of this algorithm was that it considers a document which has been comprised of different topics. Where each topic was the probability distribution of words. Consequently, each word would appear on the topic depending on its occurrence probability. So mainly it models D documents consider as comprised of K latent topic. Each word presented the multinomial distribution over a W word vocabulary. The algorithm of LDA [17] is as follows:

1. Choose a topic $z_{ij} \sim Mult(\theta_j)$
2. Choose a word $x_{ij} \sim Mult(\phi_{z_{ij}})$.

Where the parameters of the multinomial for topics in a document θ_j and words in a topic ϕ_k have Dirichlet priors

4.4 Data Distribution Analysis

Since we have retrieved topics of three keywords, namely IoT, V/AR, and AI with different frequency distribution. Consequently, we have recommended Mann Whitney U test to analyze the significant relationship between their frequency distribution. In order to determine the closeness, we have considered three following null hypothesis:

HP-1. Frequency distribution of IoT and V/AR are significantly different.
HP-2. Frequency distribution of V/AR and AI are significantly different.
HP-3. Frequency distribution of AI and IoT are significantly different.

The p-value of Mann Whitney U test help to determine the acceptance or rejection of the above-mentioned hypothesis. For example, HP-1 will be accepted if its p-value <0.05 otherwise its alternative hypothesis (i.e. Frequency distribution of IoT and V/AR is same) will be accepted.

5 Results and Discussion

5.1 Response to RQ-1

In order to respond RQ-1, descriptive statistics regarding the frequency distribution of V/AR, IoT, and AI trends are reported in Table 2. The highest mean of the V/AR distribution depicts its familiarity and high user perception as compared to IoT and AI.

5.2 Response to RQ2

In order to respond RQ-2, the public perception and responses about V/AR, IoT, and AI are retrieved and showed in Fig. 2. It explains the public emotions regarding these trends. There are two main consequences retrieved from the results of Fig. 2.

1. IoT got significant positive responses with a minor difference from the positive responses of V/AR.
2. AI got significant negative responses as compared to the negative response of IoT and V/AR.
3. IoT has less negative responses as compared to V/AR and AI.
4. Neutral responses of IoT are approximately equal to neutral responses of V/AR.

Table 2. Data distribution

Statistic	V/AR	IoT	AI
Sample size	34	34	32
Range	117	120	120
Mean	46.529	16.735	23.563
Variance	1358	1227.7	1326.9
Std. deviation	36.85	35.038	36.426
Coef. of variation	0.79198	2.0937	1.5459
Std. error	9.8487	10.115	9.4052
Skewness	0.00568	2.1166	1.7123

a) IoT   b) V/AR   c) AI

Fig. 2. Sentimental analysis of (a) IoT, (b) V/AR and (c) AI

Subsequently, we depict the cloud graph in Fig. 3 which described the significant topics extracted for each trend. There are some subtopics with the least significance are found in AI, such as Job demand, Job loss, inequalities, health risk are some negative polarity tweets. However, combining all the comments made the 71.3% negative sentiments about AI.

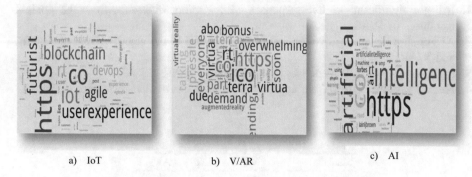

a) IoT   b) V/AR   c) AI

Fig. 3. Extracted topics for (a) IoT, (b) V/AR and (c) AI

5.3   Response to RQ3

In order to response RQ-3, we have performed Mann Whitney U test to assess the significant relationship between the frequency distribution of IoT, V/AR, and AI trends by considering three null hypothesis (Sect. 4.4). The list of the alternative hypothesis is as follows:

Alternative Hypothesis for HP-1. Frequency distribution of IoT and V/AR is the same or closely related.
Alternative Hypothesis for HP-2. Frequency distribution of V/AR and AI is the same or closely related.
Alternative Hypothesis for HP-3. Frequency distribution of AI and IoT is the same or closely related.

The p-value for IoT and V/AR (i.e. $p = 0.138 > 0.05$), V/AR and AI(i.e. $p = 0.227 > 0.05$), and IoT and AI(i.e. $p = 0.348 > 0.05$) indicate the rejection of the null hypothesis (Sect. 4.4) and describe the correlation between frequency distribution of topics of these trends, which describe the acceptance of above-mentioned alternative hypothesis. The p-value with IoT, V/AR, and AI are shown in Fig. 4.

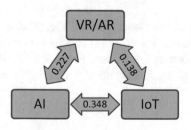

Fig. 4. Correlation between AI, IoT and V/AR

6   Implication to Researchers

The aim of the proposed study is to aid IT expert and data scientist to provide a systematic way to assess and target the evolution, sustainability, and forecasting of business progress. The main implications of the proposed study to the research community are as follows. Data scientists can understand the impact of any new technology on their data centers.

1. A businessman can make effective decisions through public perception regarding any target topic.
2. An economist can predict the correlation between the domain-specific terms and improve business planning.
3. The sentimental analysis results can also be used for emerging application development.
4. The flow of the proposed study can be replicated in certain domains, such as health, education, and transportation.

7   Conclusion

The public perception regarding new technologies can aid a data scientist, economist, IT leader, and a businessman in their adaptation and analyzing its impact on the business progress and forecasting. Recently, in terms of green computing, researchers have reported the implications of three technologies, namely Internet of Thing (IoT), Virtual/Augmented Reality (V/AR), and Artificial Intelligence (AI). In this paper, we have proposed a method comprised of three tiers, namely data acquisition and preprocessing, topic detection, and analysis of results. The aim of the first tier is to retrieved data using a crawler and applies several preprocessing techniques. We retrieved 147 million total public tweets from Twitter regarding IoT, V/AR, and AI. Subsequently, the aim of the second tier is to apply Linear Discriminant Analysis (LDA) to identify a topic for each trend and apply sentimental analysis. Finally, the aim of the third tier is to visualize the interpreted results. The main consequences of the proposed study are. (1) In terms of frequency distribution, we observed more tweets on AI (51.51%) as compared to V/AR (18.37%) and IoT (30.11%), (2) More positive comments for IoT as compared to V/AR and AI, (3) identification of few terms such as block-chain, futurist, user-experience, users-demand, bonus, and presale as subtopic for each trend. Finally, the proposed study will help to IT leaders, economist, and data scientist to describe the public perception and the impact of any new technology on the evolution, sustainability and progress of their systems.

References

1. Karamı, A., Dahl, A.A., Turner-McGrievy, G., Kharrazi, H., Shaw, J.G.: Characterizing Diabetes, Diet, Exercise, and Obesity Comments on Twitter, September 2017
2. Karami, A., Bennett, L.S., He, X.: Mining Public Opinion about Economic Issues: Twitter and the U.S. Presidential Election, February 2018
3. Yeruva, V.K., Junaid, S., Lee, Y.: Exploring social contextual influences on healthy eating using big data analytics. In: 2017 IEEE International Conference on Bioinformatics and Biomedicine (BIBM), pp. 1507–1514 (2017)
4. Virtual Reality Revenue to Hit $7.2B This Year, $75B By 2021 – Variety. http://variety.com/2017/digital/news/virtual-reality-industry-revenue-2017-1202027920/. Accessed 21 Apr 2018
5. Ramage, D., Dumais, S., Liebling, D.: Characterizing microblogs with topic models. In: ICWSM, and undefined. aaai.org (2010)
6. Text Mining: Classification, Clustering, and Applications - Google Books
7. Hong, L., Davison, B.D.: Empirical study of topic modeling in Twitter. In: Proceedings of the First Workshop on Social Media Analytics, SOMA 2010, pp. 80–88 (2010)
8. Java, A., Song, X., Finin, T., Tseng, B.: Why we Twitter. In: Proceedings of the 9th WebKDD and 1st SNA-KDD 2007 Workshop on Web Mining and Social Network Analysis, WebKDD/SNA-KDD 2007, pp. 56–65 (2007)
9. Krishnamurthy, B., Gill, P., Arlitt, M.: A few chirps about Twitter. In: Proceedings of the First Workshop on Online Social Networks, WOSP 2008, p. 19 (2008)
10. Arnold, C., Speier, W.: A topic model of clinical reports. In: Proceedings of the 35th International ACM SIGIR Conference on Research and Development in Information Retrieval, SIGIR 2012, p. 1031 (2012)
11. Myslín, M., Zhu, S.-H., Chapman, W., Conway, M.: Using Twitter to examine smoking behavior and perceptions of emerging tobacco products. J. Med. Internet Res. **15**(8), e174 (2013)
12. Bollen, J., Mao, H., Zeng, X.: Twitter mood predicts the stock market. J. Comput. Sci. **2**(1), 1–8 (2011)
13. Hussain, S., Keung, J., Sohail, M.K., Ilahi, M., Khan, A.A.: Automated framework for classification and selection of software design patterns. Appl. Soft Comput. **75**, 1–20 (2019)
14. Hussain, S., Keung, J., Khan, A.A.: Software design patterns classification and selection using text categorization approach. Appl. Soft Comput. **58**, 225–244 (2017)
15. Hussain, S., et al.: Implications of deep learning for the automation of design patterns organization. J. Parallel Distrib. Comput. **117**, 256–266 (2018)
16. Kolda, T.G., Sun, J.: Scalable tensor decompositions for multi-aspect data mining. In: 2008 Eighth IEEE International Conference on Data Mining, pp. 363–372 (2008)
17. Rubayyi Alghamdi, K.A.: A survey of topic modeling in text mining. Int. J. Adv. Comput. Sci. Appl. **6**(1) (2015)
18. Hussain, S.: A methodology to predict the instable classes. In: 32nd ACM Symposium on Applied Computing (SAC), Morocco, 4th to 6th April 2017. 13
19. Hussain, S.: Threshold analysis of design metrics to detect design flaws. In: ACM Symposium on Applied Computing (SRC) 2016, Pisa Italy, 4–8 April 2016 (2017)

Health Technology Innovation:
Emerging Trends and Future Challenges

The Challenges of European Public Health Surveillance Systems - An Overview of the HIV-AIDS Surveillance

Alexandra Oliveira[1,2(✉)], Luís Paulo Reis[2,3], and Rita Gaio[4,5]

[1] School of Health, Polytechnic Institute of Porto, Porto, Portugal
aao@ess.ipp.pt
[2] LIACC-Artificial Intelligence and Computer Science Laboratory, Porto, Portugal
[3] DEI-FEUP-Department of Informatics Engineering,
Faculty of Engineering of the University of Porto, Porto, Portugal
lpreis@fe.up.pt
[4] DM-FCUP-Department of Mathematics,
Faculty of Sciences of the University of Porto, Porto, Portugal
argaio@fc.up.pt
[5] CMUP-Center of Mathematics of the University of Porto, Porto, Portugal

Abstract. Surveillance has been defined as the continual scrutiny of all aspects in emerging and the spread of a disease that is pertinent to effective control, involving a systematic collection, analysis, interpretation, and dissemination of health data. Given their fragmentation several problems inherent to data must be recognized. This paper aims to provide an overview of European Public Health Surveillance Systems emphasizing their structure and main challenges. The HIV-AIDS surveillance is overview as a particular case.

The most common issues are unrepresentativeness, changes in the implementation through time, inconsistent use of case definitions, miss diagnoses, miss or fail to report a case, reporting delay, and errors during completion of the form or data entry. The HIV - AIDS surveillance is one of the most complex mainly due to the special epidemiology of the disease surrounding the transmission modes and the lack of treatment and all the socio-ecological framework involved.

Keywords: Surveillance system · Challenges · Overview ·
Data quality · HIV-AIDS

1 Introduction

Surveillance has been defined as the continual scrutiny of all aspects in emerging and the spread of a disease that is pertinent to effective control, involving a systematic collection, analysis, interpretation, and dissemination of health data [1,2]. Are implemented for detecting signs of a disease outbreak, rapid recognition of its presence and diagnosis of the microbial cause. Also, is used to identify

© Springer Nature Switzerland AG 2019
Á. Rocha et al. (Eds.): WorldCIST'19 2019, AISC 932, pp. 513–523, 2019.
https://doi.org/10.1007/978-3-030-16187-3_50

modes of transmission and at-risk population groups, and to define public measures for target prevention [1,3]. The collection of adequate data is vital to evaluate the burden of the disease and the impact of prevention and control programmes, aiming at effective and efficient responses.

Typically, surveillance systems are very complex and rely on processes and individuals, thereby can be found to differ substantially according to the disease, health condition, and country. Nevertheless, all traditional public health surveillance approaches use pre-specified case definitions, employ manual data collection and human decision-making [3,5].

From Population under Surveillance or Coverage until Central Government, it must be recognized that the surveillance is a fragmented system and that the data collection process is composed by several critical levels with unique mechanisms. It has a strong human decision-making component firmly imprinted in data which may affect the information quality registered on the system. From representativeness until simple data entry errors several issues may occur and affect negatively the aim for which the data was collected [2,3,6–8].

A particular case, and among the most complex, of a surveillance system is that for the HIV - AIDS mainly due to the special epidemiology of the disease surrounding the transmission modes and the lack of treatment and all the socio-ecological framework involved.

This paper aims to provide an overview of the typical European Public Health Surveillance Systems emphasizing their structure and main challenges in each step. The HIV-AIDS surveillance system is overview as a particular case. So in Sect. 2 it is provided a description of the organization of a traditional European public health surveillance systems, the data collection mechanisms the data flow throughout the system and the main challenges they face. In Sect. 3 it is presented the case study of the European Surveillance System of HIV-AIDS. Finally, in Sect. 4 are present the main conclusions.

2   European Public Health Surveillance Systems

For surveillance purposes, before counting cases it is necessary to decide what a "case" is. Case definitions are a set of standard clinical and laboratory criteria that unequivocally classify whether a person has a particular disease, syndrome or other health condition developed by epidemiologists [3,9]. These case definitions remove the potential bias and make the comparison between populations possible in different geographical locations and times. So, they are a fundamental cornerstone for standardizing the collection of data [3].

Typically, epidemiologists in national public health authorities collect confirmed cases from laboratories, general practitioners and hospitals which are common detection and primary infection diagnosis data sources [3]. Each case is record, processed and compiled in a national database. Then the information is analysed and made available to public health professionals, the general public and public health authorities for supporting decision-making in public health practices [3,6].

In Europe and, in accordance with the European Centre of Disease Control and Prevention (ECDC) founding regulation (Regulation (EC) 851/2004), all European Union (EU) Member States (28) have to provide to this centre in a timely manner, the available scientific and technical data on 52 communicable diseases and related special health conditions [10,11]. In the case of Human Immunodeficiency Virus, Tuberculosis and Influenza, the surveillance is conducted with the collaboration of the World Health Organization - Regional Office for Europe (Fig. 1). Considering the attributes of the disease and objectives for which they were implemented, surveillance systems may be classified as passive or active, as compulsory or voluntary and as comprehensive or sentinel. In a passive system, the data providers take the initiative to report the case to the public health agency which do not stimulate the reporting nor give feedback. By the contrary, an active surveillance relies on the initiative of public health official that contact relevant sources of data, stimulating them to report, send agencies alert or remind and give feedback of the results [3,12]. Some systems make data submission mandatory by law, professional edict, policy or guidance. Comprehensive systems include reports of cases that occur within the whole population of a geographical area covered by surveillance system while sentinel rely on notifications from a selected group of reporting sources [3,12]. Active and mandatory systems typically generates high-quality data, that is data with high levels of completeness, validity and timeliness.

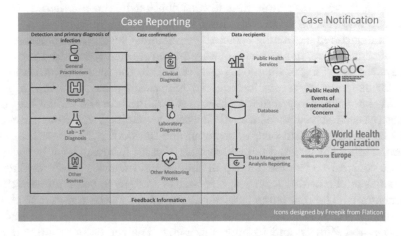

Fig. 1. Simplified flowchart of a generic surveillance system based on [3].

The format in which data is collected may be paper or electronic through an information system. The paper format has the advantage of requiring no special technical skill and it entirely circumvents the difficulties of interfacing between information systems and the disadvantage of needing to undergo a transcription step increasing the risks of introducing additional errors and do not enforce the completion of mandatory fields or other validation checks [3].

The most common problems in a surveillance system are unrepresentative or selection bias, changes in the system implementation through time, inconsistent use of case definitions, availability of cases, reporting delay, and processing errors such as errors during completion of the reporting form or data entry.

Changes in the system implementation through time are inevitable. As epidemiologists' knowledge about the natural history of the disease being surveyed and diagnosis techniques evolve its natural that the case definition change. Likewise, the evolution and adopting of different information technologies used for reporting a case may affect the behavior of system main stakeholders. Other important factors affecting data quality are changes in government funding, legislation or reporting sources. It must be recognized that any change in the reporting system may interfere with the number of cases recorded and thus with the statistical analysis, especially in time trend monitoring. Moreover, time changes in surveillance infrastructure, clinical practices and reimbursement policies may smite representativeness [3].

Generally, health conditions are not reported randomly. Diseases handled in a public health facility are reported disproportionately more frequently than those diagnosed by private practitioners; conditions that lead to hospitalization, are more likely to be reported than problems dealt with on an outpatient basis; diseases with testing practices implemented by the central government are more likely to be diagnosed and reported [2,3,9]. So, the information collected may not be representative of the affected population and thus the occurrence of the health-related event over time and its distribution in the population by place and person may not be accurately described [3].

The unregistered cases may occur at a community level, when patients do not seek professional care, in a surveillance system level when the system fails/misses or delays the reporting of diagnosed case or/and in the health department itself when, for example, cases are lost due to misclassification [3,13,14]. A general morbidity surveillance pyramid is often used to illustrate the availability of disease data at each surveillance level. "With each ascending level (from the community to healthcare institutions, to regional and national public health agencies) data availability shrinks and only a fraction of cases from the level below is captured. In contrast to the narrow tip of the pyramid which represents data held by national public health agencies, the base is wide as it holds all infections in the community. The difference between the numbers at the top and at the base can be considered cases lost to underestimation" which is the sum of cases lost to under-ascertainment and to underreporting. The typical morbidity pyramid is presented in Fig. 2 [13]. Depending on the disease, 5% to 80% of the cases that actually occur will be reported [2,3]. Under-ascertainment or under-diagnose of cases correspond to patients that are not diagnosed and hence not identified by the healthcare systems and it may occur when [3]:

1. patients do not visit healthcare services because they do not feel symptoms or feel only mild symptoms;
2. patients have low health literacy and do not recognize symptoms nor perceive the need to seek healthcare;

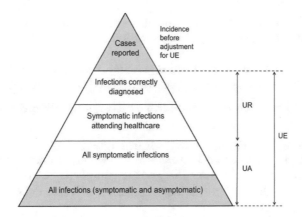

Fig. 2. Typical morbidity surveillance system [13] UR - Underreporting, UA- Underascertainment, UE - Underestimation

3. unequal geographical distribution of healthcare services exists;
4. diagnostic tools are unavailable;
5. routine surveillance does not capture marginalized high-risk groups;
6. physicians and the general population are unaware of the disease.

Reporting delay can be defined as the time between the diagnosis of a case (by the physician and/or the laboratory) and its report to the national surveillance system. Timely report may enable public health authorities to take fast and effective actions to prevent outbreaks by reducing disease transmission in a population [15–17]. Underreporting is often viewed as a reporting delay with infinity length. Reporting delays and underreporting may depend on a number of factors such as [3,4,18]:

1. the use of confirmatory laboratory testing;
2. the volume of cases identified in the reporting source;
3. the reporting process by the health care provider or the laboratory to the local, region, or state public health authority;
4. the case follow-up investigations to verify the case report or to collect additional case information;
5. periods of decreased surveillance system activity due to reduction of financial resources;
6. computer system down-time for maintenance, upgrades, or new application development;
7. data processing routines, such as data validation or error checking;
8. physicians workload;
9. different geographical location;
10. reporting format;
11. patients during visiting the doctor do not showing symptoms of the notifiable disease;

12. a patient has more than one reportable condition and only one is reported;
13. the patient is in a serious clinical condition and the physician focuses on the patient;
14. the registered case is present and registered in the public health surveillance database, but has been misdiagnosed, misclassified, or miscoded.

During completion of the reporting form or data entry, different types of errors may occur:

1. Errors in interpretation or coding, for example [2,3,19]:
   (a) Unclear case definitions or that are not widely known may lead to variation on the criteria use. The greater complexity of the diagnosis, "the greater the difficulty in reaching consensus on a case definition" and properly register the case.
   (b) frequently adjustments to the disease case definition.
   (c) emerging of new diseases or the definition of parameters that the surveillance system has not taken into account cannot be reported.
2. Errors of intention, for example [3,20,21]:
   (a) selecting only cases with good outcomes to report ("cherry-picking"). Avoidance or detection of intentional error can be challenging. Some approaches include checking for consistency of data between sites, assessing screening log information against other sources (e.g., billing data), and performing onsite audits (including monitoring of source records) either at random or on purpose.
   (b) when some of the information of the reporting form is self-reported; for example, in diseases like HIV, individuals may not report embarrassing conditions, socially discriminated or criminalizing activities.
   (c) stakeholders perceive negative consequences to their interests by reporting high infection rates, so intentionally report lower ones.
3. Errors when data is entered into the registry inaccurately, such as incompletion of the reporting form, data entry, transfer, or transformation accuracy. These errors are due to insertion, deletion or substitution of characters or writing wrong dates, such as [3,19]:
   (a) a laboratory value of 2.0 entered as 20;
   (b) it has been documented that data entry operators show a systematic digit preference when filling in numeric fields for age (ages ending in 0 and 5 are overrepresented) or date (01, 10, 15 and 20 are overrepresented);
   (c) digit preference may be also combined with the tendency to avoid certain 'unpleasant' numbers, such as 13.

The most common errors are random ones, during completion of the reporting form or during data entry, namely insertion of characters, missclassification of the disease, changing code of observations and writing wrong dates [3].

The main issues described in the previous sections are summarized in Fig. 3 and grouped by the key levels of the system. A particular case of a surveillance system is that for the HIV - AIDS. It differs from others in many ways, namely by reflecting the special transmission patterns, the long asymptomatic or with mild symptoms latency period of the infection, the lack of affordable treatment and cure, high case fatality rates, and the social stigma associated with it [22].

Fig. 3. Case reporting main issues from a surveillance system

3 European Surveillance of HIV-AIDS

Surveillance for HIV - AIDS is among the most complex systems posing a number of special ethical concerns due to the inclusion behavioural data and to the stigma and discrimination attached to the infection [22, 29]. Its main purposes are to determine the extent of the epidemic and to track eventual changes or trends overtime [23]. Are implemented in all countries in the World Health Organization - Regional Office for Europe, except Monaco and Liechtenstein, have established systems for monitoring the number of new HIV diagnoses [24].

In addition to the surveillance data common quality issues discussed in previous sections, one must recognize that countries vary in data collection methods and testing policies that may have an impact on the results and introduce bias in comparisons between countries [24–26]. Due to these variations, national data need to be compared carefully since countries with the largest number of diagnosed cases may reflect not the true scale of the epidemic but the efficacy in case finding.

It is widely recognized that reporting delay is an ever present issue that downward biases the HIV trend estimates. This bias is greater in most recent years and, to a lesser extent, in the 2 to 3 years prior to the reporting period [26]. In 2006, a European-wide survey of HIV surveillance systems in 44 countries found that, among the 16 countries that had examined reporting delay, the majority stated that 90% of the HIV diagnoses were reported within 6 months and only 3 countries stated that 75% of cases were reported in the same time period [24, 27]. In present days, it was estimated that the longest delays occurred in 9 countries: Greece, France, Italy, Luxembourg, the Netherlands, Poland, Portugal, Sweden, and the UK [25].

The duplication of HIV-AIDS reported cases occurs in several European surveillance systems [25, 26] and may have an origin in the following settings:

1. an individual may have more than one positive HIV test as a result of receiving health care in different settings or using both anonymous and named testing services;
2. a single positive test could be reported more than once, for example, by both the laboratory undertaking the testing and the clinician [24];
3. a single case reported by two physicians in two different settings; for example, a positive test on an IDU that is engaged in a drug treatment program may be reported, simultaneously by the clinic when the treatment occurs and by the general practitioner.

The completeness of key epidemiological variables such as age, sex, transmission category, CD4 count at diagnosis and migration status (either of the following: country of birth, the region of origin, country of nationality) vary over time and between countries [24–26]. In 2015, ECDC and WHOe reported the lowest completeness rate being observed on CD4 count [26]. Migration status could be determined for 88% of the diagnosed cases. On the other hand, the proportion of individuals with known transmission category was around 81.7% for those diagnosed in the period 2012–2014. It is important to notice that this is the probable transmission mode and that individuals may have more than one risk behaviour. Age and gender were consistently reported in proportions greater than 98% [25].

HIV testing practices have a direct effect on the extent to which HIV infections are diagnosed and reported and hence also on the data representativeness. Approaches to HIV testing vary widely in the European region, but most countries have a policy or strategy to offer HIV testing and counselling to most-at-risk groups and pregnant women [24, 28]. In 2015, a survey to a Primary Target Group Member States showed that, considering their national guidelines, 78% of the countries had post-test access to treatment, care and prevention services, 70% of the countries had voluntary, confidential testing with informed consent and post-test counselling, and 57% reported having a pre-test counselling or pre-test discussion. Dedicated HIV testing centres (e.g., for people at a high risk such as IDUs) are present in 65% of the countries. Testing of all pregnant women for HIV is recommended in 70% of the countries and this practice is offered to all most-at-risk population groups in 57% of the countries [28].

4   Conclusions

Surveillance systems have been established to accomplish the critical mission of identify and track epidemics. Typically, it depends on several processes and stakeholders, such as the population under surveillance, health care providers, laboratories and the Public Health Services, which challenges an efficient and effective detection and reporting of the diagnosed cases. This compartmentalized system may imprint several problems in the data which in turns may affect negatively the aim for which the data were collected. The most common problems are unrepresentativeness or selection bias of the population, changes in the system implementation through time, inconsistent case definitions, miss diagnoses, miss or fail to report a case, reporting delay, and errors during completion of the reporting form or data entry. This paper describes a typical European Surveillance system identifying the main issues and mechanisms that may be in their origin so specific content operations can be implemented.

The HIV-AIDS surveillance system is a specific case of surveillance systems due to the infection characteristics and its social repercussions. It must reflect the special transmission patterns, the long asymptomatic or with mild symptoms' latency period, the lack of affordable treatment and cure, high case fatality rates, and the social stigma associated with it [22]. More over, because of the long asymptomatic latency period and the effect of treatment, current HIV (asymptomatic) cases are hard to track and the reported AIDS cases reflect past infections that have occurred many years ago.

National surveillance systems are not identical across Europe. In addition to the common surveillance data quality issues, there are variations in data collection methods and testing policies which can introduce bias in comparisons between countries. Such as, the comparison of HIV-AIDS case numbers between countries must be done carefully.

Acknowledgements. Luís Paulo Reis and Alexandra Oliveira were partially founded by the European Regional Development Fund through the programme COMPETE by FCT (Portugal) in the scope of the project PEst-UID/CEC/ 00027/2015 and QVida+: Estimação Contínua de Qualidade de Vida para Auxílio Eficaz à Decisão Clínica, NORTE010247FEDER003446, supported by Norte Portugal Regional Operational Programme (NORTE 2020), under the PORTUGAL 2020 Partnership Agreement. Rita Gaio was partially supported by CMUP (UID/MAT/00144/2019), which is funded by FCT with national (MCTES) and European structural funds through the programs FEDER, under the partnership agreement PT2020.

References

1. Choffnes, E., Sparling, P., Hamburg, M., Lemon, S., Mack, A., et al.: Global Infectious Disease Surveillance and Detection: Assessing the Challenges-Finding Solutions, Workshop Summary. National Academies Press, Washington, DC (2007)
2. Teutsch, S., Churchill, R.: Principles and Practice of Public Health Surveillance. Oxford University Press, Oxford (2000)
3. European Centre for Disease Prevention and Control: Data quality monitoring and surveillance system evaluation. ECDC, Stockholm (2014)
4. Jajosky, R., Groseclose, S.: Evaluation of reporting timeliness of public health surveillance systems for infectious diseases. BMC Public Health **4**, 29 (2004)
5. Doyle, T., Glynn, M., Groseclose, S.: Completeness of notifiable infectious disease reporting in the United States: an analytical literature review. Am. J. Epidemiol. **155**, 866–874 (2002)
6. Chen, H., Hailey, D., Wang, N., Yu, P.: A review of data quality assessment methods for public health information systems. Int. J. Environ. Res. Public Health **11**, 5170–5207 (2014)

7. Waller, L., Gotway, C.: Applied Spatial Statistics for Public Health Data. Wiley, Hoboken (2004)
8. Karr, A., Sanil, A., Banks, D.: Data quality: a statistical perspective. Stat. Methodol. **3**, 137–173 (2006)
9. Centre for Disease Prevention and Control and others: Principles of epidemiology in public health practice: an introduction to applied epidemiology and biostatistics (2006)
10. European Centre for Disease Prevention and Control: Policy on data submission, access, and use of data within TESSy - 2015 revision. ECDC, Stockholm (2015)
11. European Centre for Disease Prevention and Control: Indicator-based surveillance. http://ecdc.europa.eu/en/activities/surveillance/Pages/index.aspx
12. MacDonald, P.: Methods in Field Epidemiology. Jones & Bartlett Learning, Burlington (2011)
13. Gibbons, C., Mangen, M., Plass, D., Havelaar, A., Brooke, R., Kramarz, P., Peterson, K., Stuurman, A., Cassini, A., Fevre, E., Kretzschmar, M.: Measuring under-reporting and under-ascertainment in infectious disease datasets: a comparison of methods. BMC Public Health **14**, 1–17 (2014)
14. Medicine, I., Prevention, B., Evaluation, P., Allocation, C.: Measuring What Matters: Allocation, Planning, and Quality Assessment for the Ryan White CARE Act. National Academies Press, Washington DC (2004)
15. Marinovic, A., Swaan, C., van Steenbergen, J., Kretzschmar, M.: Quantifying reporting timeliness to improve outbreak control. Emerg. Infect. Dis. **21**, 209–216 (2015)
16. Reijn, E., Swaan, C., Kretzschmar, M., van Steenbergen, J.: Analysis of timeliness of infectious disease reporting in the Netherlands. BMC Public Health **11**, 409 (2011)
17. Mauch, S.: Situational Assessment of the HIV/AIDS Notification System - A Portuguese Experience. National Coordination For HIV Infection (2009)
18. Pagano, M., Tu, X., De Gruttola, V., MaWhinney, S.: Regression analysis of censored and truncated data: estimating reporting-delay distributions and AIDS incidence from surveillance data. Biometrics **50**, 1203–1214 (1994)
19. Gliklich, R., Dreyer, N., Leavy, M.: Registries for Evaluating Patient Outcomes: A User's Guide [Internet], 3rd edn. Data Collection and Quality Assurance, 11 April 2014. https://www.ncbi.nlm.nih.gov/books/NBK208601/
20. Anthamatten, P., Hazen, H.: An Introduction to the Geography of Health. Routledge, London (2012)
21. Trick, W.: Decision making during healthcare-associated infection surveillance: a rationale for automation. Clin. Infect. Dis. **57**, 434 (2013)
22. World Health Organization and others: WHO report on global surveillance of epidemic-prone infectious diseases (2000)
23. Walker, N., Garcia-Calleja, J., Heaton, L., Asamoah-Odei, E., Poumerol, G., Lazzari, S., Ghys, P., Schwartlander, B., Stanecki, K.: Epidemiological analysis of the quality of HIV sero-surveillance in the world: how well do we track the epidemic? AIDS **15**, 1545–1554 (2001)
24. Platt, L., Jolley, E., Hope, V., Latypov, A., Vickerman, P., Hickson, F., Reynolds, L., Rhodes, T.: HIV epidemics in the European region: vulnerability and response. Directions in Development, International Bank for Reconstruction and Development/The World Bank (2015)
25. Rosinska, M., Pantazis, N., Janiec, J., Pharris, A., Amato-Gauci, A., Quinten, C., Network, E.H.S.: Missing data and reporting delay in the European HIV Surveillance System: exploration of potential adjustment methodology (2017)

26. European Centre for Disease Prevention and Control, C.R.O.: HIV/AIDS surveillance in Europe 2015. ECDC, Stockholm (2016)
27. EuroHIV: Report on the EuroHIV 2006 survey on HIV and AIDS surveillance in the WHO European Region. Institut de veille sanitaire, Saint-Maurice (2007)
28. European Centre for Disease Prevention and Control: HIV testing in Europe. Evaluation of the impact of the ECDC guidance on HIV testing: increasing uptake and effectiveness in the European Union. ECDC, Stockholm (2016)
29. Stoto, M.: Syndromic surveillance in public health practice. In: Institute of Medicine (ed.) Infectious Disease Surveillance and Detection (Workshop Report), pp. 63–72 (2007)

Healthcare Information Systems
Interoperability, Security and Efficiency

Steps Towards Online Monitoring Systems and Interoperability

Diana Ferreira[1]([⊠]), Cristiana Neto[1],
José Machado[2], and António Abelha[2]

[1] Informatics Department, University of Minho, Braga, Portugal
{a72226,a72064}@alunos.uminho.pt
[2] Algoritmi Research Center, University of Minho, Braga, Portugal
{jmac,abelha}@di.uminho.pt

Abstract. In the health area, there is, on a daily basis, an enormous amount of data being produced and disseminated. The fast-growing amount of collected data and the rich knowledge, possibly life-saving, that could be extracted from these data has demanded the search of new ways to ensure the reliability and availability of the information with an emphasis on the efficient use of information technology tools. Although the main focus of the information systems is the health professionals who contact directly with the patient, it is also imperative to have tools for the background of the health units (information services, managers of systems, etc.). The main purpose of this work is the development of an innovative and interactive web platform for the daily monitoring of the web services activities of a Portuguese hospital, Centro Hospitalar do Porto (CHP). This platform is a web application developed in React that aims to ensure the correct functioning of the web services, that are responsible for numerous tasks within the hospital environment, and which failure could result in disastrous consequences, both for the health institution and for the patients. The development of the web application followed the six stages of the Design Science Research (DSR) methodology and was submitted to the Strengths Weaknesses Opportunities and Threats (SWOT) analysis, which results were considered optimistic.

Keywords: Information technology · Monitoring · Web services ·
Web application

1 Introduction

The last years have been critical for the foundation of a new world-wide era, being particularly remarkable the influence of a diverse set of forces and tendencies associated with the acceleration of the scientific and technological progress in the information domain. The "Information Era" promotes, in a global context of increasing dynamism, a permanent revolution of knowledge and perceptions in practically all areas of human knowledge. Information is a decisive factor in the survival of any organization. An information system (IS) is a socio-technical subsystem of an organization that collects,

© Springer Nature Switzerland AG 2019
Á. Rocha et al. (Eds.): WorldCIST'19 2019, AISC 932, pp. 527–536, 2019.
https://doi.org/10.1007/978-3-030-16187-3_51

stores, processes, transmits and displays data, information, and knowledge relevant to it [1–5]. This meets Davis and Olson's (1985) vision of an IS, which is an integrated man/machine system that provides information to support operations and decision-making in organizations [6].

Information is especially crucial to the health organizations, either for its action, i.e., its decision-making process, or for its reaction, i.e., the control and correction of deviations from its action. The information technologies have the potential to ensure the efficient delivery of health care and improve the quality of services provided by health professionals as they provide comprehensive and credible information and support the decision-making process, both clinically and administratively, thereby reducing the incidence of adverse events and clinical errors [4, 7–9].

There is a whole set of economic, technological, social and political factors that impel health organizations to search for new ways to improve the quality of the services provided. Faced with this reality, these organizations should be prepared to make more informed and less intuitive decisions in the shortest time possible. Accordingly, health institutions have been adopting health information systems (HISs). HISs are complex socio-technical subsystems processing data, information, and knowledge to ease the management of clinical and administrative information and the planning, refinement and decision-making process of the different professionals of the health system [1, 4, 5, 10]. Since the main focus of a hospital is the patients' health status, most of these systems are dedicated to the diagnosis, treatment, and follow-up of patients. However, it is necessary to make decisions related to the computer processes of the hospital, such as web services, which also, indirectly, influence the quality of care. Therefore, the present study focuses on the development of an innovative and interactive web platform for the daily monitoring of the web services activities of the CHP.

The presented paper is organized as follows: Sect. 2 depicts the different stages of the DSR methodology that led to the development of the web platform, Sect. 3 refers to the discussion of the results achieved, and Sect. 4 discloses the main conclusions outlined with this work.

2 Monitoring Platform for Web Services Activity

A health institution must continuously monitor the technological solutions implemented and used in its hospital units. Monitoring means overseeing ongoing activities to ensure that they meet the standards and purposes established in the initial projection. It is a continuous process, consisting of three components: data collection, regular analysis of data and dissemination of data to all stakeholders. This definition can be adapted to the computing world by using the available electronic means for such monitoring. Regarding the health units' computer systems, monitoring involves evaluating the performance of services provided by computers or communication systems.

The main purpose of this project was to monitor the activities of the web services of the CHP through a monitoring platform, that is, a web application using Business Intelligence (BI) technology, in order to better control the aspects that condition the web services performance, using the periodic collection of data and the analysis of indicators. These indicators are parameters that should help to summarize and understand important factors related to the data, such as the number of errors reported in a certain period of time. In this way, it is possible to discover a potential common cause for the occurrence of errors and to plan and implement specific and efficient measures to reduce the incidence of errors in these critical operations, increasing the quality of health care delivery.

The BI technology is used to generate and present these parameters, as it allows the processing of data and the analysis of the information extracted from these data, in an easy and efficient way.

The web application was developed based on the DSR methodology (see Fig. 1). The first steps of this methodology concern the identification of a problem that needs to be solved and the definition of the objectives of the possible solution. All the subsequent phases aim to develop and evaluate the proposed solution. Accordingly, the following subsections contain the description of the web application development process and are organized according to the six phases of the DSR methodology.

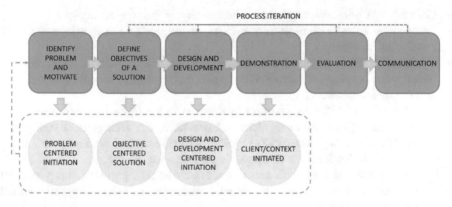

Fig. 1. Model of the DSR methodology. Adapted from [11].

2.1 Problem Identification and Motivation

In the first stage it was made an overview of the current state of the information systems of the CHP, in order to find problems and possible areas of intervention. There are two needs at the CHP that should be eliminated. Firstly, the need of a system to visualize various aspects of the activities of the web services was identified. Secondly, it is necessary to manage this activity and evaluate its behavior over time allowing a quick perception of performance problems and thus decreasing the time of its resolution. Hence, the present work arises from the need to continuously monitor the hospital environment as well as the operations performed in it and the results of its activities in order to understand if the web services are functional and if its performance meets the defined standards, in addition to timely notice the occurrence of errors to apply specific measures in the moment.

The core motivation is related to the need for managers to make fast and informed decisions in order to improve the productivity and efficiency of the organization and, at the same time, the quality of the care provided. There is an imperative need to simplify the presentation and analysis of information regarding the activity of the web services, in order to assist professionals in making decisions related to the control of these activities and the prevention of the occurrence of errors in the web services.

2.2   Objectives Definition

After the definition of the problem and the explanation of the surrounding motivations, all the objectives to be fulfilled were outlined. The main purpose of this project is to provide a dynamic, interactive, and user-friendly platform to allow key management staff to consult and manage the web services' performance in order to keep track of the hospital's operations and study the processes behavior optimizing them and improving its operational performance. In addition, some adjacent objectives were outlined:

- Collection of information related to the activity of the web services;
- Study of technical requirements necessary in the design and development of the monitoring platform;
- Decide what the best type of web service is to work as a basis for interoperability between the hospital databases and the platform developed, according to the needs and the means available;
- Creation of a user-friendly platform in React with the following purposes:
  - Gathering reliable summary data and providing key management staff greater visibility into the information required to make critical decisions and collect greater insights for the web service's performance indicators;
  - Identification of a set of features concerning targets, purposes and goals, data sources, indicators and their graphical representation, analytics support, and data processing approaches;
  - Creation of visual interacting live dashboards that enable the monitorization of hospital's operations and the study of processes behavior in order to optimize them and improve its operational performance.
- Maintenance of the proper functioning of the web services in order to ensure that the solutions satisfy the requirements for which they were developed.

2.3   Design and Development

Frameworks and Languages. The information generated and presented to the user in the form of a monitoring system contains BI modules, and basic information presentation structures (tables). To collect this data, a visual basic .NET function was created to obtain information about the web services' activities. This function is called in several phases of the web service execution, to ensure that the information regarding the occurrence of errors is obtained in all possible places for these errors to occur. It also allows to get the time when the web service was first called and the end time of its execution in order to be able to obtain the duration of execution of the web service and consequently to control and optimize it whenever necessary. This information was stored in a new table that was added to the hospital's SIL database.

This monitoring platform is, in fact, a web application that was developed using ReactJS, a JavaScript library for building user interfaces. PHP and SQL programming languages were also used, but to develop RESTful PHP Web services and query requests for Oracle databases, respectively. JavaScript executes HTTP requests to a server of the type REST implemented in the PHP language which constitutes the backend of the application, to make the data available. The data is selected through SQL queries and sent in the JSON format. Finally, Apache corresponds to the Web server that supports the Web application.

REST api is a need because it is the lightest way to create, read, update or delete information between different applications over the internet or HTTP protocol. PHP was the programming language used to build the RESTFUL server because of its easy implementation and interpretation using Apache.

It is extremely important to take into consideration the visual aspect of the application, as in certain cases it is directly related to the non-adoption of those applications. This is the main reason why this web application was developed in React, as it is a modern library that is taking over the front-end development.

Figure 2 shows the entire architecture of this web application, including its components and their interactions.

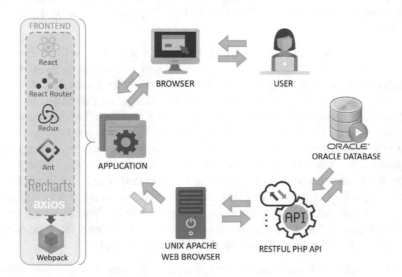

Fig. 2. Web application architecture.

Main Modules. Regarding the design of the interface, the monitoring platform has a side menu which grants access to three distinct modules, namely the Dashboards Module, Indicators Module, and Tables Module. The information provided could have been displayed in just one single page. However, it was unclear, overwhelming and confusing to the user. Accordingly, this decision was made to organize and present information in a way that is easy to read. Figure 3 shows the visual aspect of the web application interface.

Fig. 3. Overview of the web application interface.

It is important to note that the charts presented in the web application were built using the Recharts library that facilitates the construction of charts in React since it results from the combination of React and D3. Next, each module of the web application will be described.

Dashboards Module. Presents interactive dashboards, according to the selected filters. These filters concern the choice of the web service on which it is intended to analyze information about its activity and the choice of the time interval that limits this information. When clicking the dates filter, a calendar appears that allows the selection of any range of dates, and below, easy access options are available with some common date range options, like today or last week. This page presents some relevant statistical values, like the number of calls made to the web service and the average time of execution of that web service. Immediately next to the above statistics, two graphs are presented with information about the activity of the web services, one corresponding to the accesses to the web service and the other one related to the errors occurred in it. It is important to note that each of these graphs presents the possibility of filtering according to the methods of the web service, i.e., it allows to separate the data visualized according to the different methods that constitute the web service.

Indicators Module. Contains all the performance indicators of the web services considered relevant and important when monitoring its activities. These indicators were defined according to the manager's needs. After being gathered and developed, these indicators were analyzed, refined and checked for completeness. The final indicators were:

- Number of accesses to the web service in specific dates, like, today, last week and last year;
- Number of errors occurred in the web service in specific dates, like, today, last week and last year;
- Number of accesses to the web service per year;

- Number of accesses to the web service per month;
- Number of errors occurred in the web service per year;
- Number of errors occurred in the web service per month;
- Number of accesses to the web service according to its different methods;
- Number of errors occurred in the web service according to its different methods.

To ease the visualization and understanding of the indicators related to the number of accesses and errors by a specific date, it was decided to group them all on a chart (see Fig. 4). These indicators allow to make direct comparisons, enabling progress to be monitored, i.e., the advances and setbacks of the performance of these web services.

Fig. 4. Indicators related to the number of accesses to a web service by specific dates.

Tables Module. As the name suggests, displays information about the web services activities in the form of tables. The information displayed concern the accesses to the web service and the errors occurred. These tables are very interactive and allow not only to filter the table entries, either by typing in the free search filter or by selecting the interest values, as well as sorting those entries in ascending or descending order according to the fields that the user desires.

2.4 Demonstration and Evaluation

The demonstration stage aims to demonstrate the use of the artifact to solve one or more instances of the problem. At this point, the application must be submitted to specific tests in the hospital environment, using different web services and different contexts. The demonstration is only possible, after the successful implementation of the web application in the hospital environment.

In Information Technology (IT), any type of development project should be submitted to a set of tests and evaluations before being made available. The evaluation stage corresponds to the process of evaluating whether the developed solution solves the problem at hand and meets the requirements and specifications initially defined. Accordingly, after developing the web application, several tests were performed with information from different web services to verify and demonstrate its accuracy and completeness, allowing the identification and correction of errors. After considering the application stable and reliable, it is essential to test the solution in the hospital environment in order to correct eventual errors or inconsistencies and, consequently, improve the overall performance of the web application. This is the next step to perform in the development process of the platform, since the web application must be fully integrated in the hospital environment.

2.5   Communication

The present work was the object of study of a master thesis in the area of medical informatics and culminated with the publication of the respective dissertation. Furthermore, the contributions of this effort were disclosed in several academic lectures at the University of Minho.

3   Discussion of the Results

The solution proposed in this study provides a dynamic, interactive, and user-friendly platform which allows key management staff to consult and manage the web services performance in order to keep track of the hospital's operations and study the processes behavior, optimizing them and improving its operational performance. The main benefits of the adoption of this monitoring platform were:

- Display of key information in a summarized and objective way through a simple and intuitive interface;
- Provision of instant access to relevant information about the web services activity and greater visibility into the information required to make critical decisions;
- Monitorization of hospital's web services and the study of processes behavior to optimize them and improve its operational performance;
- Maintenance of the web services behavior and prevention of unnecessary failures;
- The monitoring platform is scalable and easy to maintain, as its architecture is of easy maintenance and the expansion of its functionalities is very simple;
- By using modern, innovative and efficient technologies, the CHP is able to follow the technological advances, which offers a competitive advantage and brings benefits in terms of system usability. For example, the use of ReactJS brings several rewards in terms of system performance, of which stands out a fast rendering due to the existence of a virtual DOM and the ability to reuse and combine components.

3.1   Proof of Concept

After completing the technical development of the solution proposed as a resolution to the problems identified in the scope of this study, a Proof of Concept (PoC) was carried out to prove the viability and usefulness of the developed solution. In addition to identifying potential failures or errors in the developed solution, the use of a PoC allows to verify if the solution satisfies the requirements and objectives for which it was initially designed. Thus, a SWOT analysis was carried out to briefly define the strengths and weaknesses (internal factors) of the resulting product, as well as the opportunities and threats that it is exposed (external factors). The results of this analysis are shown in Table 1.

Table 1. SWOT analysis

Strengths	Weaknesses
• High usability: intuitive and user-friendly tool; • High scalability; • Centralization of different information and functionalities concerning the web services performance in a summarized and objective way. • Web application easily adaptable to different health institutions that may need this type of clinical tool.	• Requires constant connection to the hospital internal network; • Runs slower, when the requests for information involve a high amount of data;
Opportunities	*Threats*
• Growing demand of health units to control the performance of services that occur in background that are not directly related with patient care; • Need to control the occurrence of errors and enable a faster response action; • Increase the quality and efficiency of healthcare using new technologies.	• Structural failures or changes in the databases that "feed" the platform may jeopardize its main functions; • Problems with network connectivity to the hospital intranet; • Lack of interest or commitment in the project.

4 Conclusions

An effective monitoring process allows to characterize a system and recognize its vulnerabilities and threats in order to determine the likelihood of a failure. The solution presented addresses this need by monitoring key information concerning the activity of numerous web services of a Portuguese Hospital. Modern and up to date technologies were adopted to create an intuitive and easy to interact with platform. The use of contemporary technologies, like React, helped in building a user interface with an attractive and pleasant design that makes the user experience simpler.

Accordingly, the execution of this case study culminated in the elaboration of an area restricted to the hospital's key management staff with live and interactive dashboards, including charts, tables, statistics, and, also, performance indicators, so that they can consult and manage the web services performance allowing greater control of its operation. Consequently, and since it also allows to observe in a timely manner the occurrence of errors or inconsistencies, it helps to identify and eliminate root causes of inefficiency and poor performance. This verification aims to optimize various functions and procedures as well as to ensure the web services functioning and efficiency. Ultimately, this streamlined approach enables and empowers executives and managers to check performance quickly, for an up-to-date glance at where things are going well and where improvements need to be made thus improving administrative and, consequently, clinical decision making which generates positive repercussions in the quality of the offered services and more so maximizes the financial returns. Accordingly, the use of innovative technologies and the new functionalities offered by the monitoring platform have enhanced the hospital modernization and improved the health care delivery. As future work, new forms to increase the loading of the application can be sought and studied. Furthermore, adapting the platform to be used in mobile devices could be highly beneficial. Last but not least, it would be significant to find new ways to raise the level of data security and confidentiality.

Acknowledgements. This work has been supported by FCT – Fundação para a Ciência e Tecnologia within the Project Scope: UID/CEC/00319/2019.

References

1. Haux, R., Ammenwerth, E., Winter, A., Brigl, B.: Strategic Information Management in Hospitals: An Introduction to Hospital Information Systems. Springer, New York (2004). https://doi.org/10.1007/978-1-4757-4298-5
2. Buckingham, R.A., Hirschheim, R.A., Land, F.F., Tully, C.J.: Information Systems Education: Recommendations and Implementation. Cambridge University Press, Cambridge (1986)
3. Avison, D.E., Wood-Harper, A.T.: Multiview - an exploration in information systems development. Aust. Comput. J. **18**(4), 174–179 (1986)
4. Miranda, M., Pontes, G., Gonçalves, P., Peixoto, H., Santos, M., Abelha, A., Machado, J.: Modelling intelligent behaviours in multi-agent based HL7 services. In: Computer and Information Science, pp. 95–106. Springer, Heidelberg (2010)
5. Miranda, M., Pontes, G., Abelha, A., Neves, J., Machado, J.: Agent based interoperability in hospital information systems. In: 2012 5th International Conference on BioMedical Engineering and Informatics, pp. 949–953. IEEE, October 2012
6. Davis, G.B., Olson, M.H.: Management Information Systems: Conceptual Foundations, Structure, and Development. McGraw-Hill Inc, New York City (1984)
7. Bonney, W.: Applicability of business intelligence in electronic health record. Procedia-Soc. Behav. Sci. **73**, 257–262 (2013)
8. Buntin, M.B., Burke, M.F., Hoaglin, M.C., Blumenthal, D.: The benefits of health information technology: a review of the recent literature shows predominantly positive results. Health Aff. **30**(3), 464–471 (2011)
9. Lee, J., McCullough, J.S., Town, R.J.: The impact of health information technology on hospital productivity. RAND J. Econ. **44**(3), 545–568 (2013)
10. Chen, R.F., Hsiao, J.L.: An investigation on physicians' acceptance of hospital information systems: a case study. Int. J. Med. Inf. **81**(12), 810–820 (2012)
11. Peffers, K., Tuunanen, T., Rothenberger, M.A., Chatterjee, S.: A design science research methodology for information systems research. J. Manag. Inf. Syst. **24**(3), 45–77 (2007)

Improving Healthcare Delivery with New Interactive Visualization Methods

Cristiana Neto[1(✉)], Diana Ferreira[1], António Abelha[2], and José Machado[2]

[1] Informatics Department, University of Minho, Braga, Portugal
{a72064, a72226}@alunos.uminho.pt
[2] Algoritmi Research Center, University of Minho, Guimarães, Braga, Portugal
{abelha, jmac}@di.uminho.pt

Abstract. Over the last years, the implementation and evolution of computer resources in hospital institutions has been improving both the financial and temporal efficiency of clinical processes, as well as the security in the transmission and maintenance of their data, also ensuring the reduction of clinical risk. Diagnosis, treatment and prevention of human illness are some of the most information-intensive of all intellectual tasks. Health providers often do not have or cannot find the information they need to respond quickly and appropriately to patient's medical problems. Failure to review and follow up on patient's test results in a timely manner, for example, represents a patient's safety and malpractice concern. Therefore, it was sought to identify problems in a medical exams results management system and possible ways to improve this system in order to reduce both clinical risks and hospital costs. In this sense, a new medical exams visualization platform (AIDA-MCDT) was developed, specifically in the Hospital Center of Porto (CHP), with several new functionalities in order to make this process faster, intuitive and efficient, always guaranteeing the confidentiality and protection of patients' personal data and significantly improving the usability of the system, leading to a better health care delivery.

Keywords: Information technology · Health information system · Electronic Health Records · Medical exams

1 Introduction

The progress of Information Technologies (IT) is an observable and unavoidable fact and currently plays a very important role in health services with the primary goal of contributing to a more efficient and high-quality health care delivery [1].

The implementation of Information Systems (IS) in hospital environments dates back to mid-1960s, given the high increase of clinical information over the years. By this time, its main functions were limited to administrative management. After 1970, larger hospitals gradually established internal information sectors [2]. Thus, the implementation of Hospital Information Systems (HIS) has improved the organization of the hospitals' large amount of information, with the purpose of not only automating, collecting and analyzing it, but also helping to support decision making.

© Springer Nature Switzerland AG 2019
Á. Rocha et al. (Eds.): WorldCIST'19 2019, AISC 932, pp. 537–546, 2019.
https://doi.org/10.1007/978-3-030-16187-3_52

A hospital is an institution with multiple resources that must be managed in the best way with the ultimate purposes of offering the patient a good service and optimizing profitability. To achieve these objectives, it is essential for each institution to have a good HIS for its current and integral management [3, 4]. The most common application of those types of systems is the Electronic Health Records (EHR) system.

EHR is a computerized health system where professionals record patients' clinical information. It aims to bring together all the health care provided to a particular patient and provide a cross-sectional analysis of the patient's medical history in different services and medical units. In addition to biometric information, current prescriptions and results from imaging and laboratory tests, new and more advanced mechanisms that already integrate EHR with decision support systems begin to emerge [5–7].

These systems are normally distributed and heterogeneous, but the interaction among them is a crucial demand these days. In this way, the interoperability among the HIS becomes an indispensable feature in health organizations. Interoperability is the capacity of two systems to interact with each other, ensuring the understanding of the process and the exchange of data on both sides [8].

In order to solve this problem, the Agency for Integration, Diffusion and Archive (AIDA) was created and implemented in some Portuguese hospitals. AIDA is a platform developed to allow the dissemination and integration of information generated in a healthcare environment, including information on Complementary Means of Diagnosis and Therapy (MCDT) (that includes medical exams information) presented by the AIDA-VIEW platform [9, 10].

Previous research has shown that health professionals often fail to review and act on test results accurately and properly. Although the reasons for the exams requests vary, accurate reporting of results is always crucial to ensure that appropriate action is taken. Several studies have also identified some practices that could improve the presentation of test results, such as highlighting the exams that still have to be visualized and the existence of groupers so related exams could be seen together. The EHR represents a significant step in the communication improvement and in the increasing of the relevant data availability and can potentially reduce communication problems associated with paper-based transmission of exams results [11, 12].

Thus, to address the issue of patients' safety and quality of care, the project described in this paper emerged to help health professionals analyze and act on test results in a safe, reliable, and efficient way. The next section of this paper presents a contextualization of hospital information systems, followed by the development section where the phases of this project are explained. The section four presents the discussion of results followed by the conclusions and future work.

2 Hospital Information Systems

A HIS can be defined as a subsystem of a hospital with a socio-technological development, which covers all hospital information processing [9]. Its main purpose is to contribute to the quality and efficiency of healthcare. This objective is primarily oriented to the patient after being directed to health professionals as well as the functions of management and administration [9, 10]. The projection and implementation of a HIS should focus on ensuring the efficient production of information in order to provide clinical decision-making resources. Thus, this implementation requires the existence of a management structure whose specific function focuses on the adequate allocation of resources and the definition of organizational rules [13, 14]. In order to provide complete and useful resources, a HIS should also allow the extraction of clinical and management indicators as a way to improve not only decision-making, but also planning and logistics processes. EHR can be assumed as a HIS for excellence and has replaced the traditional manual recording in Paper Clinical Process (PCP). EHR may include all hospital areas with a need for registration information. Some of the advantages of EHR are to provide accurate, up-to-date and complete information about patients at the time of care and to allow quick access to patient records for more coordinated and efficient care.

2.1 AIDA

In order to aggregate and consolidate all the information generated by a health unit, a solid and efficient process of integration and interoperability must be developed. The main goal of interoperability in healthcare is to connect applications so data can be shared, exchanged across the healthcare environment and distributed to medical staff or patients whenever and wherever they need it [15]. This process should take into account scalability, flexibility, portability and security (confidentiality, integrity and availability) when applied in a hospital environment.

With these goals in mind, the AIDA platform (Agency for Integration, Diffusion and Archive of Medical Information) has been established in some Portuguese hospitals, including the *Centro Hospitalar do Porto* (CHP), which in turn connects with several systems, with the main objective of integrating, disseminating and archiving large volumes of information from several HISs [8–10].

AIDA is a complex system composed by simple and specialized subsystems defined as intelligent agents responsible for tasks such as the communication between heterogeneous systems, the sending and receiving of information (e.g. clinical reports, images, a set of data, prescriptions, etc.), managing and archiving of information and responding to requests properly [16, 17]. AIDA's architecture is presented on Fig. 1,

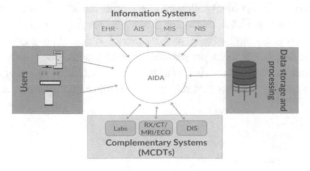

Fig. 1. AIDA platform.

where it is shown the information systems integrated by AIDA: EHR; Administrative Information System (AIS); Medical Information System (MIS); and Nursing Information System (NIS) [10].

AIDA-VIEW. One other way that EHRs improve the quality of care and patient outcomes, contributing to the health of the population, is through the continuous improvement of the clinical decision making, by conducting more easily clinical trials and other studies, managing clinical knowledge, and disseminating more quickly research results to providers and patients, incorporating them rapidly into decision-support technology, and tracking the resultant changes in patient outcomes [18].

Thus, there is a general interest in minimizing the time spent in the request and consultation of MCDTs, since there is an urgent need for a more rational and efficient use of available resources, in order to minimize the time spent by the physician in these tasks and in order to ensure the provision of health care with the maximum efficiency and quality. The AIDA platform has a built-in web-based MCDTs visualization tool, the AIDA-VIEW. However, this tool is quite basic and not very efficient, presenting several negative points such as:

- Poor usability since users only have two chances of consulting the MCDTs: seeing the last 10 presented by default or consulting by specialty;
- In the visualization of the MCDTs, one cannot have the perception of exams that have already been requested, which may lead to the unnecessary realization of multiple identical exams;
- The presentation of the MCDTs is the same regardless of the user that is accessing the application and the context, which means that, in most cases, users first see information that is not relevant to them;
- There is too much information presented in the initial page that may lead to difficulties in identifying the relevant information.

Consequently, the project to reformulate this platform arises in order to solve these and other issues.

3   Development

The development of this project followed one of the models of software development methodology, known as SDLC, short for Software Development Life Cycle, which is widely used in several engineering and industrial fields. The model used was the waterfall model that has 5 main phases: requirements definition, design, implementation, testing, and maintenance. This was the adopted model in order to lead to the attainment of a quality product that meets the original intentions of the client [19].

3.1   Requirements Definition

At this stage it was made an overview of the application in order to establish a basic project structure, to evaluate its feasibility and to describe the appropriate technical approaches. Here are edged the two main objectives of the software:

- Create a front-end user interface to replace the current MCDTs visualization platform;
- Create the backend required for the front-end operation.

The identification process of the functional and non-functional requirements took several forms. Initially, the immediate and more obvious requirements to potentiate the tool were identified. Subsequently, other requirements were collected from some doctors and nurses at the CHP during numerous meetings of the PCE working group.

Thus, the main functional requirements in terms of the tool interface were:

- Creation of a hierarchical classification structure of the MCDTs, aggregating them into several groups;
- Presentation of non-visualized MCDTs and MCDTs not visualized by their specialty;
- Free text search for MCDTs;
- Filtration by module and anatomical structure;
- Presentation of the MCDTs requests already done for that patient;
- Inclusion of the MCDTs by context, that is, MCDTs related to the context at the time of access to the platform;
- Possibility to change the display settings according to user preferences;
- Preview of the report's pdf;
- Possibility of adding notes to the MCDT, working as a sort of comments section.

The requirements of the backend were limited to the ability to connect the interface to the database to manage the information necessary to operate the platform.

The main non-functional requirements collected were:

- Usability of the platform, that is, any healthcare professional using the interface must be able to easily and intuitively access the information he/she needs;
- Interoperability, since there are already several systems implemented in the hospital, this platform must be compatible with them;
- Speed, since it is a hospital environment where the less time wasted in the search for information, the better the provision of services will be;
- Security, since the platform is dealing with some confidential information.

3.2   Design

This stage corresponds to the process of planning the problem-solving software solution. Regarding the design of the interface, the platform was designed to have 3 modules:

- an initial module where the MCDTs are separated by the stipulated aggregators (See Fig. 2). In addition to these aggregators, it is also possible to navigate through previous structure so the transition to this new platform is made smoothly. Half of the first page is filled with the information of the patient's MCDTs requests and if an MCDT is selected, that half of the page will present the pdf preview of the MCDT report.

- a second advanced filtering module where it is possible to perform a free text search or filter by module and anatomical location. On this page there is initially a complete list of all the MCDTs that would be filtered as the filters were selected.
- a third module with some indicators.

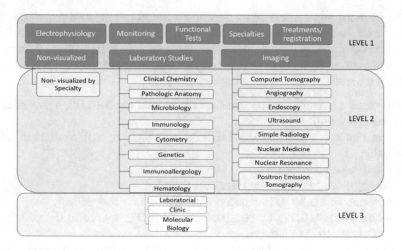

Fig. 2. Defined grouping for the exams.

In terms of the backend, it was necessary to decide which hospital databases to use, what tables of those databases were needed, and which tables to add. The hospital's databases used were SIL and PCE. Within these two databases the tables to be used were summarized tables that presented information about episodes, exams, exam requests, access logs to MCDTs, user documents, among others. The tables added to the database include: a table with the groupers (which maps between the MCTD's codes and the respective grouper) and a table with the notes added to the exams.

3.3   Implementation

This phase is the process of converting the whole requirements and blueprints into a production environment, that is where the real code is written and compiled into an operational application [19].

In order to meet the requirements presented above, more specifically the non-functional requirements, it is very important to choose the proper technologies to be used. For the construction of the desired web application it was decided to use a JavaScript library called ReactJS. ReactJS is a JavaScript library created by Facebook and launched in 2013. This library guarantees high performance in content rendering, is easy to learn and use since it is based on building small blocks of reusable code. At the server level, ReactJS contains very fast rendering of information, which makes it very useful for quick and constant queries of the content of interest. In the backend was used PHP to make the connection between the Oracle database and the interface.

In this sense, the MCDTs visualization platform developed has an architecture based essentially on 3 components (See Fig. 3): the database where the information is stored, a CRUD RESTful API Web service programmed in PHP and an interface in the part of the client accessible via Web browser developed in ReactJS and using related libraries.

Fig. 3. Representation of the AIDA-MCDT platform's architecture.

3.4 Testing and Maintenance

The testing step is also known as verification and validation and is a process to check that a software solution meets the original requirements and specifications and that it accomplishes its intended purposes [19]. So, after the construction of the application itself, several tests were performed with information from different patients in different contexts. The application was also tested by several health professionals at the CHP (doctors and nurses). This process allowed errors to be corrected in order to improve the application's performance. Initially, the testing process focus the successful integration of the platform in only one service. Then, this process is applied to all the services.

Maintenance is the process of modifying a software solution after delivery and deployment to refine output, correct errors, and improve performance and quality. Additional maintenance activities can be performed in this phase including adapting software to its environment, accommodating new user requirements, and increasing software reliability [19]. This phase is now in progress since the application is being fully integrated in the hospital.

4 Discussion of Results

The realization of this project enabled the development of a tool to support the decision and clinical practice, namely a user-friendly computer tool for visualization of MCDTs to be implemented in the CHP named AIDA-MCDT. The main advantages of updating the architecture and the functionalities of this system were:

- Use of a modern and powerful technology, namely the ReactJS library, which presents many advantageous characteristics for system performance, namely the ability to create, reuse and combine components, a virtual DOM that results in a faster performance and a simple integration process. The use of this technology contributes to the modernization of the HIS in the CHP.

- Implementation of new modules, components and functionalities, with emphasis on the advanced filtering that ease the search for relevant information more quickly, the display of not seen exams, the presentation of MCDTs requests, the organization of MCDTs by aggregators among others.
- Introduction of context-awareness in the application.
- Introduction of a small Business Intelligence section with only a few indicators (for example, number of patient's MCDTs per year, MCDTs performed by module and specialty and waiting time by specialty), opening doors for the development and deepening of this useful area.
- The web application is much more scalable and easier to maintain, since it presents a simple architecture and a simple process of expansion of its functionalities.

4.1   Proof of Concept

One of the most important steps in the design, planning, development, implementation and presentation of a prototype in IT is to perform a proof of concept since it can establish if the solution found fulfills the requirements and objectives initially defined. It also allows the identification of potential failures or errors in the proposed solution [20].

In this sense, a SWOT analysis was performed which briefly tries to define the strengths, weaknesses, opportunities and threats of the solution. Table 1 shows the results of this analysis, that is, the weaknesses, strengths, threats, and opportunities raised.

Table 1. SWOT analysis.

Strengths	Weaknesses
• High scalability; • High usability; • Innovation: the implementation of this platform enhances the modernization of the current system; • Improvement in the decision-making process	• Access to the platform is only possible through the internal network connection of the CHP; • Presence of inconsistent and unnecessary information in the tables that feed the platform; • Due to the amount of data and the complexity of some application features, it may become slower
Opportunities	Threats
• Reduction of medical errors; • Achievement of better quality and greater efficiency in the organization through the use of new technologies; • Decrease of paper usage and the consequent increase of computerization of the MCDTs	• Competition from other applications in the market; • Lack of interest on the part of health professionals to use a new computational tool; • Problems with network connectivity

5   Conclusion and Future Work

This project proved to be quite viable considering the current investment in the new information technologies to improve and streamline the decision-making process, which is much easier when the health professionals have easy access to the information needed at the moment. It is well known that the access to the MCDTs is a crucial step in the decision-making process, and the visualization of this information as quickly as possible, as well as the reduction of information that is not visualized and left to be forgotten can lead to a substantial decrease in the occurrence of clinical errors. This project is also very useful in terms of financial management since it avoids performing repeated exams and duplicated costs, showing the user which exams have already been requested.

The work done also proved to be very challenging since the application developed can be used by several health professionals with a wide range of academic backgrounds, from doctors to nurses, to health workers and technicians in all the specialties and modules of the hospital. In this way, several requirements were raised in order to satisfy the various needs that the hospital environment generates. Accordingly, the implementation of new ITs in the hospital institution, not only meets the needs of the healthcare professionals and improves their professional activity but also improves patients' experience as they would be less exposed to eventual clinical errors.

Thus, both the new functionalities present in the application and the technologies used have contributed to the modernization of the system and to the improvement of the health care delivery. This application also potentiates the reduction of hospital costs since it prevents the execution of duplicated exams when presenting the requested tests.

Despite all of this, some improvements can still be made like, for example, the inclusion of the request for new exams in this platform and its adaptation to mobile format. Finally, the implementation and use of this application imply several security measures, so, as future work, those security measures could be enhanced to ensure the maximum security of the data used in the platform.

Acknowledgments. This work has been supported by FCT – Fundação para a Ciência e Tecnologia within the Project Scope: UID/CEC/00319/2019.

References

1. Jardim, S.V.: The electronic health record and its contribution to healthcare information systems interoperability. Procedia Technol. **9**, 940–948 (2013)
2. Pai, F.Y., Huang, K.I.: Applying the technology acceptance model to the introduction of healthcare information systems. Technol. Forecast. Soc. Chang. **78**(4), 650–660 (2011)
3. Haux, R., Ammenwerth, E., Winter, A., Brigl, B.: Strategic Information Management in Hospitals: An Introduction to Hospital Information Systems. Springer, New York (2004)
4. Bertolini, M., Bevilacqua, M., Ciarapica, F.E., Giacchetta, G.: Business process reengineering in healthcare management: a case study. Bus. Process Manag. J. **17**(1), 42–66 (2011)

5. Neto, C., Dias, I., Santos, M., Peixoto, H., Machado, J.: Applied business intelligence in surgery waiting lists management. In: Healthcare Policy and Reform: Concepts, Methodologies, Tools, and Applications, pp. 1580–1594 (2018)
6. Carter, J.H.: Electronic Health Records: A Guide for Clinicians and Administrators. ACP Press, Philadelphia (2008)
7. Salazar, M., Duarte, J., Pereira, R., Portela, F., Santos, M.F., Abelha, A., Machado, J.: Step towards paper free hospital through electronic health record. In: Rocha, Á., Correia, A., Wilson, T., Stroetmann, K. (eds.) Advances in Information Systems and Technologies, pp. 685–694. Springer, Heidelberg (2013)
8. Oliveira, D., Duarte, J., Abelha, A., Machado, J.: Step towards interoperability in nursing practice. Int. J. Public Health Manag. Ethics (IJPHME) 3(1), 26–37 (2018)
9. Duarte, J., Salazar, M., Quintas, C., Santos, M., Neves, J., Abelha, A., Machado, J.: Data quality evaluation of electronic health records in the hospital admission process. In: 2010 IEEE/ACIS 9th International Conference on Computer and Information Science (ICIS), pp. 201–206. IEEE (2010)
10. Duarte, J., Portela, C.F., Abelha, A., Machado, J., Santos, M.F.: Electronic health record in dermatology service. In: Cunha, M.M., Varajão, J., Powell, P., Martinho, R. (eds.) ENTERprise Information Systems, pp. 156–164. Springer, Heidelberg (2011)
11. Poon, E.G., Gandhi, T.K., Sequist, T.D., Murff, H.J., Karson, A.S., Bates, D.W.: "I wish I had seen this test result earlier!": dissatisfaction with test result management systems in primary care. Arch. Intern. Med. 164(20), 2223–2228 (2004)
12. Singh, H., Naik, A.D., Rao, R., Petersen, L.A.: Reducing diagnostic errors through effective communication: harnessing the power of information technology. J. Gen. Intern. Med. 23(4), 489–494 (2008)
13. Duarte, J., Castro, S., Santos, M., Abelha, A., Machado, J.M.: Improving quality of electronic health records with SNOMED. Procedia Technol. 16, 1342–1350 (2014)
14. Lippeveld, T., Sauerborn, R., Bodart, C. (eds.): Design and Implementation of Health Information Systems, vol. 281. World Health Organization, Geneva (2000)
15. Miranda, M., Duarte, J., Abelha, A., Machado, J.M., Neves, J.: Interoperability and healthcare (2009)
16. Pontes, G., Portela, C.F., Rodrigues, R., Santos, M.F., Neves, J., Abelha, A., Machado, J.: Modeling intelligent agents to integrate a patient monitoring system. In: Pérez, J., et al. (eds.) Trends in Practical Applications of Agents and Multiagent Systems, pp. 139–146. Springer, Cham (2013)
17. Cardoso, L., Marins, F., Portela, F., Santos, M., Abelha, A., Machado, J.: A multi-agent platform for hospital interoperability. In: Ramos, C., Novais, P., Nihan, C., Corchado Rodríguez, J. (eds.) Ambient Intelligence - Software and Applications, pp. 127–134. Springer, Cham (2014)
18. Goldschmidt, P.G.: HIT and MIS: implications of health information technology and medical information systems. Commun. ACM 48, 68–74 (2005)
19. Bassil, Y.: A simulation model for the waterfall software development life cycle (2012)
20. Pereira, R., Salazar, M., Abelha, A., Machado, J.: SWOT analysis of a Portuguese electronic health record. In: Conference on e-Business, e-Services and e-Society, pp. 169–177. Springer, Heidelberg (2013)

A Multi-level Data Sensitivity Model for Mobile Health Data Collection Systems

Marriette Katarahweire[1(✉)], Engineer Bainomugisha[1], and Khalid A. Mughal[2]

[1] School of Computing and Informatics Technology, Makerere University,
P.O. Box 7062, Kampala, Uganda
{kmarriette,ibaino}@cis.mak.ac.ug
[2] Department of Informatics, University of Bergen,
P.O. Box 7800, 5020 Bergen, Norway
Khalid.Mughal@ii.uib.no

Abstract. Mobile Health Data Collection Systems (MHDCS) use a combination of encryption and user access control to grant or deny permissions to data collectors. However, the data in MHDCS is of diverse value and types which calls for different security measures. The level of sensitivity of data in electronic health is a function of the context characterised by the social environment, patient and content among others. When mobile devices are used for data collection and tracking participants, there is need for a more refined security system that allows finer controlled access to specific data elements. In this paper, we provide a conceptual design and prototype implementation for a data sensitivity model that enables attribute-based data access control, based on the level of sensitivity of the data involved. By allowing specific form data elements to have different security levels, we enhance the security of MHDCS and allow more use cases including the use of a single form to collect data for different stakeholders with diverse data needs and concerns.

Keywords: Mobile health · Data collection systems · Security · Data classification · Data sensitivity

1 Introduction

Mobile Health Data Collection Systems (MHDCS) enable extension of health services to communities without the need to physically visit a health facility. MHDCS allow data collection as electronic forms for surveys, clinical trials and interventions, immunisation campaigns, community-extension health visits, maternal and child health tracking and monitoring, and other purposes and services in the health sector. Several categories of data collected by MHDCS include meta data, patient data, identifiers, demographics and project related data among others. Therefore, the type of data elements collected vary in sensitivity, usage and intended audience for accessibility.

© Springer Nature Switzerland AG 2019
Á. Rocha et al. (Eds.): WorldCIST'19 2019, AISC 932, pp. 547–556, 2019.
https://doi.org/10.1007/978-3-030-16187-3_53

Like all mobile applications, MHDCS can be secured using native security mechanisms provided by the device operating system and those within the programming environment. Security and data access is controlled by the individual applications. Access control provides a means to limit access to the MHDCS data to authorised users [1]. As noted by [2], the security model of Android is "system centric" in the sense that applications statically identify the permissions that govern the rights to their data and interface instances at installation time. With this approach, the applications/developer has limited ability to whom the rights are given or how they are later exercised. However, the current data security system that is implemented by common MHDCS in low-resource settings such as DHIS 2 Tracker Capture app [3], Open Data Kit (ODK) [4] and mUzima [5] uses a combination of user access controls, encryption and other measures. This kind of security approach is binary in that you either see the entire form contents or nothing. To thwart other forms of attack such as over-the-shoulder eavesdropping, there is need provide additional security even when the user is authorized to access the data.

The diversity of data in MHDCS requires a closer look at the actual data collected and the context in which it is collected in order to provide appropriate security and privacy controls. As an example, we consider Fig. 1 which shows *Form 071* [6], designed and used by the Ministry of Health (MoH) in Uganda for antenatal visits. An expectant mother's form includes demographic data, medical history and medical findings, whose sensitivity, if disclosed, vary from no effect to potentially severe stigmatization.

(1) Serial No.	(2) Client No.	(3) Name of client	(4) Village + Parish	(5) Phone Number	(6) Age	(7) ANC Visit	(8) Gravida / Parity	(9) Gestational Age	(10) Expected Date of Delivery (EDD)	(11) Weight & Mid Upper Arm Circumference (WT & MUAC)	(12) Blood Pressure
						1 2 3 4+					
						1 2 3 4+					

(13) PMTCT codes		(14) Diagnosis	(15) ARV drugs	(16) Infant Feeding Option	(17) TB Status	(18) Haemoglobin	(19) Syphilis Test Results	(20) Family Planning Counseling	(21) TT	(22) IPT	(23) Free ITN	(24) Mebendazole
w	p											

Fig. 1. A data collection form used to track antenatal visits. In Uganda, an expectant mother is expected to visit the health facility at least 4 times. Adopted from Uganda HMIS [6].

Use cases for Form 071 enable different stakeholders that include community health works (CHEWS) and medical officers of varying qualifications and roles to use the same form for different purposes. For instance, the same form is used to track HIV/AIDS in pregnant mothers including prevention of mother to child transmission (PMTCT). Not every health worker that is accessing this form requires all the data elements and therefore the binary authentication model severely fails to enable a more refined access to data, both at the server or during collection and tracking. Sensitive data elements should be secured in terms of who, when, where and why (4W). We refer to the need to secure various levels of sensitive data in respect to 4W as the data sensitivity problem.

We argue that, to make it harder for attackers while easing accessibility for specific data elements, one needs to answer several questions in relation to MHDCS: (i) Why is the data being collected? (ii) What will the data be used for? (iii) How will the data be used? (iv) Who will use that data? (v) Are all data elements of same value? (vi) Is the value of the data elements the same in all contexts? These security needs are not adequately addressed by the current MHDCS. Responses to these questions from the security perspective should provide for more refined security considerations that can allow data elements to be secured based on varying considerations that include data sensitivity among others. Efforts to address the issue of finer-grained access control is noticeable in the cloud computing domain [7] where data in the cloud is shared with different audiences.

The contribution of this paper is a conceptual framework that aims to capture the data sensitivity problem and solution with related components. We show how the data sensitivity concepts fit into existing implementations, specifically mUzima and ODK, and provide detailed description for components introduced to support a multi-level data sensitivity security model. The rest of the paper is arranged as follows: Sect. 3 gives a brief description of the data sensitivity model. Section 2 is an overview of related work while Sect. 4 describes the conceptual model and its implementation. In Sect. 5 we give an illustrative example and Sect. 6 concludes the paper and discusses future work.

2   Related Work

In [8], a patient centric framework and a suite of mechanisms for data access control in Personal Health Records (PHR) stored in semi-trusted servers is proposed. The main concept behind the framework is to divide the system into two types of domains, namely public and personal domains (PSDs). The public domains consist of users who have access to the data based on their professional role such as doctors while personal domains contain users that are associated with the data owner such as family and these gain access rights assigned by the data owner. The access control is managed using attribute based encryption where the encrypted data can only be decrypted if the opponent and target share common attributes.

Sticky policies [9] are machine-readable policies that stick to data to define allowed usage and obligations as it travels across multiple parties, enabling users to improve control over their personal information. The data is originally encrypted and the sticky policy is read to provide access and data handling preferences of the user.

The authors in [10] develop a security framework for Android-based applications that dynamically enforces security policies based on the application's sensitive data and privileges. The security polices can be programmatically added in the medical application or be enforced dynamically by the server. Our work also attaches sensitivity levels to data and defines how different contexts can play a role in determining the sensitivity levels.

ODK, one of the widely used MDHCS platform provides optional, entire form encryption. It uses a random 256-bit symmetric key when a form is finalized [4]. We seek to provide form encryption on a per-attribute basis and depending on the sensitivity of the data.

3   Data Sensitivity Model

The data sensitivity model identifies three sensitivity levels $L = \{public, confidential, critical\}$ into which data may lie. Different security mechanisms can be applied to data in MHDCS depending on its sensitivity level. The Data Sensitivity Model also defines the appropriate security control mechanism listed in Table 1. Some controls are applicable to data on the mobile device while others apply to data stored on the server. The data sensitivity model is defined based on the data collection form that comprises of attributes in which a range of data values are captured. The model is summarised using definitions 1 and 2 below.

Definition 1. *A Data Sensitivity Model for a secure form is a tuple $F = (A, C, P, ps, L)$, where: – A is a set of attributes, C is a set of context variables. P is a set of sensitive parameters or simply parameters, $ps : C \cup P \times A \rightarrow L$ is a function that assigns a sensitivity value and L is a set of sensitivity levels. The function ps depends on the sensitive context variables and attribute parameters with thresholds that can be defined by domain experts or data collectors.*

Definition 2. *For a Data Sensitivity Model $F = (A, C, P, ps, L)$, the effective sensitivity level of an attribute a, is function $sens : L \times L \ldots L \rightarrow L$ such that given an attribute $a \in A$; $sens(a) = \max_{1 \leq i \leq n} (ps(x_i))$; $x \in C \cup P$.*

4   Conceptual Framework

During data collection, each data element should be transparently secured without user intervention. On retrieval of filled forms for patient tracking, the form should ensure that appropriate security measures are applied in a way to only facilitate the intended purpose of the user. We present the conceptual framework using Fig. 2 that comprises of part (a) for the static aspects and (b) the run-time interaction between the data collectors and the data collection forms. The conceptual framework consists of two main components, the form designer and the mobile device. Given that the form design is the first executable artifact in the development of MHDCS, it is important to embed security requirements at this stage.

During form design, the application context defines contextual information that relates to the MHDCS being designed. The application context information may warrant stronger or lower security measures. This information is used to

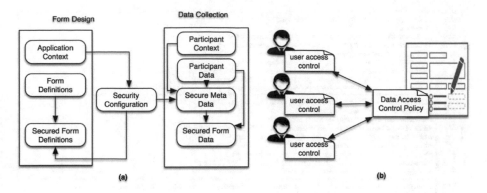

Fig. 2. Conceptual framework

generate the appropriate *security configuration file*. The *security configuration file* is attached to the form definition to create the *secure form definitions*, that are stored on the server for subsequent download by the data collectors.

When the data collector is interacting with the data collection form as indicated in Fig. 2(b), we have the data collectors with appropriate access controls on one hand, while on the other hand, we have the data collection form with a data access control policy that must be satisfied before a given data element can be accessed. Therefore, we use participant data, participant context and security configuration file to generate a *secure meta data file* that describes the security measures already enforced within the data collection form and those to be enforced at the server.

4.1   Specifications

The descriptive specifications for the *security configuration file* and *secure meta data file* are given in Fig. 3(a) using the Extended Backus-Naur Form (EBNF) notation [11]. The symbols `parameter` and `value` in Fig. 3(a) depend on the set of context parameters and corresponding values respectively. The symbol `string` has the same definition as used in general programming languages such as Java. The data sensitivity meta file is generated partly at form design time and the rest of details are automatically filled during data collection time. Each form instance is identified by the `instanceId` (line 3) and has different security controls depending on the sensitivity of the actual data captured. In line 5 of Fig. 3(b) each attribute has a sensitivity-level whose value can be `public`, `confidential` or `critical` as specified in line 8.

The security mechanisms to apply during form design and data collection are defined using the data sensitivity model whose details are specified in the data sensitivity configuration file. The specification of the data sensitivity configuration file is given in Fig. 3. The meta data file also uses EBNF [11] for its specification. Each form element has an `attributeDef` (line 3, Fig. 3(b)) that specifies the sensitivity definition, `senseDef` - a criteria for deciding the

(a)
```
1  DSConfig := "{" formName, formSensDef "}"
2  formName   := "form": string
3  formSensDef := "attributes": "[" attributeDef* "]"
4  attributeDef := "{" attributeName, sensDef "}"
5  attributeName := "name":string
6  sensDef := "sens" :  "{" <attributeSens>* "}"
7  <attributeSens> := "public":paraSens* , "confidential":paraSens* , "critical":paraSens*
8  paraSens   := "{" "param": parameter, "cond":condExp "}"
```

(b)
```
1  DSMetaDoc := "{" formName, instanceId, <attributeSensDef> "}"
2  formName   := "form": string
3  instanceId := "instanceId": string
4  <attributeSensDef> := "attributes": "[" attributeSensVal* "]"
5  attributeSensVal := "{" attributeName, sensVal "}"
6  attributeName := "name":string
7  sensVal := "sensitivity-level" :  sensLevel
8  sensLevel := "public" | "confidential" | "critical"
```

(c)
```
1  condExp := <relCond>{<op> <condExp>} | <parenthCond> {<op> <condExp>}
2  <relCond> :=<pValue><relOp><value>
3  <relOp>:= "==" | "<" | "<=" | ">" | ">=" | "!="
4  <parenthCond>:="(" <condExp> ")"
5  <op>:= AND | OR
6  <pValue>:=<char>*
7  <value>:=<quote><char>*<quote>
8  <quote>:="
```

Fig. 3. (a) is the EBNF syntax for data sensitivity meta file, (b) is the EBNF syntax for data sensitivity configuration file (c) is the EBNF syntax for condition expressions

sensitivity level of the data value during data collection. The `attributeSens` (line 7, Fig. 3(b)) partitions the data sets of each attribute into one of the sensitivity levels of `public`, `confidential` or `critical` which is then attached to the actual data in the meta data file. The level specifies the access controls and security measures as defined in Table 1. The evaluation of the appropriate level is based on the `condExp` (line 8, Fig. 3(b)) whose specification is given in Fig. 3(c).

4.2 System Components

Two main components are introduced (Fig. 4); the data classification engine and the security controller. The data classification engine maps the sensitivity levels to data elements and form attributes.

The security controller uses the security meta data file to apply and enforce security for access control and key management. The security controller is deployed on both the mobile device and server, while the data classification engine is deployed on the mobile device.

A data collection system has a single data-sensitivity configuration file

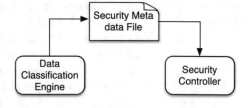

Fig. 4. Plugin component diagram

while each filled-in form has an automatically generated Secure Meta Data File which is also encrypted and stored with data on the server to facilitate enforcement when downloaded for future use. To detect and apply the data sensitivity security measures, we use a tag within the form definition to link to the corresponding configuration file. In this way, we avoid mixing security meta information with patient data in the same way external CSS are externally linked to HTML to separate styling from content. The security controller component ensures that data elements are secured as soon as they are filled such that security attacks including over the shoulder eavesdropping are mitigated immediately.

4.3   Security Mechanisms

In this section, we highlight how and when the different security mechanisms for each sensitivity level as defined in Table 1 are applied.

Table 1. Security mechanisms to apply to different sensitivity levels in MHDCS

Security mechanism	Public	Confidential	Critical
Encryption [12–14]	No	Yes	Stronger algorithm
Access Control [15]	No	Yes	According to need
Masking [16, 17]	No	Yes	Yes
Hashing [18]	No	Yes	Yes
Password Protect [19, 20]	No	Yes	Yes
Anonymise [16, 20, 21]	No	Yes	Yes
Storage Area [22]	No	Yes	Yes
De-identification [16, 20]	No	Yes	Yes

Encryption is used to encrypt the data values and requires that access be granted to a user whose credentials are signed by the same encryption key. The data remains encrypted both at the server, during transit and on the mobile device. Public data is not encrypted. Confidential and critical data is encrypted using Advanced Encryption Standard (AES) but with key length of 256 and 128 for critical and confidential data respectively. *Access Control* is assigned according to the roles of the users [15]. Each user can be associated with different access rights which will make the scheme more reliable and efficient.

Masking is the practice of hiding the password characters as entered by the user behind bullets ($\bullet$), asterisks ($*$), or similar camouflaging characters [16, 17]. Masking is now a de facto standard for avoiding over the shoulder eavesdropping of passwords and therefore its success can be extended to highly sensitive data. We apply masking in realtime as the data is being captured in the MHDCS form.

Hashing uses a hash function to emphasize on the integrity of the data and verify that the data elements are not tampered with. We use SHA-1 for its high security. *Password Protect* requires that before adjusting a data value, the user provides a password. Furthermore, sensitive files stored on the external storage such as medical images are password protected to prevent unauthorised access.

With *Anonymise* data is anonymised before access to ensure no tracking to a specific user especially in medical trials.

In relation to *storage area*, public data is stored on external storage of the mobile device while passwords and encryption keys are stored in the secure key store or shared preferences. The secure key storage offers a secure environment that protects the integrity and confidentiality of the cryptographic keys stored within the environment [22]. Confidential data is stored on internal storage but with other security mechanisms in place to grant or deny access depending on the data policy. Despite the availability of the secure key store on most mobile devices, recent studies found out that some developers do not use this environment [1]. With *de-identification*, personal identifiers are removed to make it hard for users to identify the owner of the data. We use K-anonymity [23] that masks the values of some potentially identifying attributes, called *quasi-identifiers* (QIs), such that the values of the QI attributes for any individual matches those of at least $k - 1$ other individuals in the same dataset.

5   Illustrative Example

For illustration, we consider form element 19 (Syphilis Test Results) in Fig. 1 which is a highly contagious sexually transmitted disease. As soon as a value is entered into the form element, the data classification engine component (Fig. 4) automatically generates the meta-data file whose fragment is indicated in Fig. 5(b).

In line 5 of Fig. 5(b), the overall sensitivity for attribute *disease* is set to "confidential". This is computed based on the parameter sensitivity function *ps* in

```
    1   {"form":"patient-monitoring",
    2    "attributes":
    3        [
    4          {"name":"disease",
    5           "sens":{
    6                 "public":[
    7                       {"param":"severity","cond":"pValue > 75%"},
(a) 8                     {"param":"mortality","cond":"pValue < 40%"},
    9                       {"param":"incidences","cond":"pValue == 'low'"},
   10                       {"param":"location","cond":"pValue == 'central'"}
   11                     ],
   12                "confidential":[
   13                       {"param":"severity","cond":"50% <= pValue <= 75%"},
   14                       {"param":"mortality","cond":"40% <= pValue <= 75%"},
                           {"par      inci                              oderat
```

```
    1   {"form":"patient-monitoring",
    2    "instanceId":"09090909",
    3    "attributes":
(b) 4        [
    5            {name:"disease","sensitivity-level":"confidential"}
    6            {name":"arv drugs","sensitivity-level":"critical"}
```

Fig. 5. (a) is the Data sensitivity configuration file showing the sensitivity definition of each attribute, (b) is the Secure data meta file annotated with sensitivity levels. Each form element is annotated with its corresponding sensitivity level

Table 2 defined by domain experts in which the *severity, mortality, incidences* and *location* are key factors in determining the sensitivity level of the Syphilis Test Results data value. The *ps* function is applied during the form design stage as indicated in Fig. 5(a) using the EBNF specification in Fig. 2.

Table 2. Sample definition for the parameter sensitivity function *ps* for attribute 'Syphilis Test Results'

	Severity	Mortality	Incidences	Location
Public	>75%	<40%	Low	Central
Confidential	>=50% & <=75%	>=40% & <=75%	Moderate	Western
Critical	<50%	>75%	High	Northern

The security mechanisms to be applied are Encryption, Access Control, Masking, Hashing, Password Protect, Anonymise, Storage Area and De-identification as specified in Table 1 in the "confidential" column.

6   Conclusion and Future Work

The proposed setup and implementation allow MHDCS to provide two levels of security. One is provided at the MHDCS application level and the other at the form level with finer-grained description that allows control on specific data elements. The form designers have the power to decide on who can access the data and how it can be accessed. By separating security data from participant data, the implementation provides an extensible mechanism in terms of security mechanisms that can be defined to control data access in MHDCS.

Unlike other security systems where access control schemes deny or grant access, an authenticated user with the right privileges will still have to satisfy the data control polices. Future work will involve testing the performance of the data sensitivity model on the mobile device in terms of time, processing power and energy among other parameters.

References

1. Katarahweire, M., Bainomugisha, E., Mughal, K.A.: Authentication in selected mobile data collection systems: current state, challenges, solutions and gaps. In: 2017 IEEE/ACM 4th International Conference on Mobile Software Engineering and Systems (MOBILESoft), pp. 177–178 (2017)
2. Ongtang, M., McLaughlin, S., Enck, W., McDaniel, P.: Semantically rich application-centric security in Android. Secur. Commun. Netw. **5**(6), 658–673 (2012)
3. DHIS2: Android tracker capture app (2016). https://docs.dhis2.org/2.25/en/android/html/android_tracker_capture.html. Accessed 1 Feb 2018
4. Open Data Kit: Open Data Kit Documentation (2017). https://docs.opendatakit.org/. Accessed 1 Feb 2018

5. mUzima: muzima (2016). http://muzima.org. Accessed 1 Feb 2018
6. Ministry of Health (Uganda): The health management information system (2014). http://www.gou.go.ug. Accessed 19 Nov 2018
7. Wang, G., Liu, Q., Wu, J.: Achieving fine-grained access control for secure data sharing on cloud servers. Concurrency Comput. Pract. Experience **23**(12), 1443–1464 (2011)
8. Li, M., Yu, S., Zheng, Y., Ren, K., Lou, W.: Scalable and secure sharing of personal health records in cloud computing using attribute-based encryption. IEEE Trans. Parallel Distrib. Syst. **24**(1), 131–143 (2013)
9. Pearson, S., Casassa-Mont, M.: Sticky policies: an approach for managing privacy across multiple parties. Computer **44**(9), 60–68 (2011)
10. Andow, B., Wang, H.: A distributed Android security framework. In: 2015 IEEE International Conference on Smart City/SocialCom/SustainCom (SmartCity), pp. 1045–1052 (2015)
11. Wang, Y.: A formal syntax of natural languages and the deductive grammar. Fundamenta Informaticae **90**(4), 353–368 (2009)
12. Ding, Y., Klein, K.: Model-driven application-level encryption for the privacy of e-health data. In: 2010 International Conference on Availability, Reliability and Security, pp. 341–346 (2010)
13. Mancini, F., Gejibo, S., Mughal, K.A., Valvik, R.A.B., Klungsøyr, J.: Secure mobile data collection systems for low-budget settings. In: 2012 Seventh International Conference on Availability, Reliability and Security, pp. 196–205 (2012)
14. Tawalbeh, L., Darwazeh, N.S., Al-Qassas, R.S., AlDosari, F.: A secure cloud computing model based on data classification. Procedia Comput. Sci. **52**(Supplement C), 1153–1158 (2015)
15. Boukayoua, F., Lapon, J., De Decker, B., Naessens, V.: Secure storage on Android with context-aware access control. In: Proceedings of the Communications and Multimedia Security: 15th IFIP TC 6/TC 11 International Conference, CMS 2014, Aveiro, Portugal, 25–26 September 2014, pp. 46–59 (2014)
16. Keerie, C., Tuck, C., Milne, G., Eldridge, S., Wright, N., Lewis, S.C.: Data sharing in clinical trials - practical guidance on anonymising trial datasets. Trials **19**(1), 25 (2018)
17. Tudur Smith, C., Hopkins, C., Sydes, M.R., Woolfall, K., Clarke, M., Murray, G., Williamson, P.: How should individual participant data (IPD) from publicly funded clinical trials be shared? BMC Med. **13**(1), 298 (2015)
18. Chen, B.C., Kifer, D., LeFevre, K., Machanavajjhala, A.: Privacy-preserving data publishing. Found. Trends Databases **2**(12), 1–167 (2009)
19. International Organisation for Standardisation: ISO 27799:2016 health informatics – information security management in health using ISO/IEC 27002 (2016). https://www.iso.org/standard/62777.html. Accessed 19 Jan 2018
20. Jones, E.: HIPAA 'Protected Health Information': What does PHI include? (2009). https://www.hipaa.com. Accessed 19 Dec 2017
21. Hrynaszkiewicz, I., Norton, M.L., Vickers, A.J., Altman, D.G.: Preparing raw clinical data for publication: guidance for journal editors, authors, and peer reviewers. BMJ **340**(7741), 304–307 (2010)
22. Cooijmans, T., de Ruiter, J., Poll, E.: Analysis of secure key storage solutions on Android. In: Proceedings of the 4th ACM Workshop on Security and Privacy in Smartphones & Mobile Devices, pp. 11–20 (2014)
23. Sweeney, L.: K-anonymity: a model for protecting privacy. Int. J. Uncertain. Fuzziness Knowl.-Based Syst. **10**(5), 557–570 (2002)

Application of Data Mining for the Prediction of Prophylactic Measures in Patients at Risk of Deep Vein Thrombosis

Manuela Cruz[1], Marisa Esteves[2(✉)], Hugo Peixoto[2],
António Abelha[2], and José Machado[2]

[1] Department of Informatics, University of Minho, Braga, Portugal
a79392@alunos.uminho.pt
[2] Algoritmi Research Center, University of Minho, Braga, Portugal
{marisa, hpeixoto, abelha, jmac}@di.uminho.pt

Abstract. In the last decades, with the increase in the amount of data stored in the healthcare industry, it is also extended the possibility of obtaining important information to support the decision-making process of health professionals. This article has as evidence to apply Data Mining (DM) techniques to health databases of patients with medical Deep Vein Thrombosis (DVT) risk, with the objective of classifying, based on different attributes obtained in medical discharge reports, the main prophylactic measures taken. Therefore, to achieve this goal, the free software Weka was used aiming to facilitate the process of DM, along with the algorithms chosen. In view of this, it was concluded that the service to which each patient is associated is the most relevant factor for prophylactic measures followed by the age range to which the patient belongs. This study also deduces that it can be possible to obtain classifiers capable of predicting the best prophylactic measures with a qualitative level similar as one of a health professional and, thereafter, it can be possible to obtain the classification.

Keywords: Deep vein thrombosis · Prophylactic measures · Data Mining · Classification · Prediction · Weka

1 Introduction

Deep vein thrombosis is a disease characterized by acute formation of thrombi in deep veins, being one of the main causes of hospital death [1]. Although prophylaxis for DVT is accepted as a well-established and effective strategy, with detailed recommendations that should be employed in all classes of hospitalized patients and disease prevention protocols available to all health professionals, many do not make use of them. The methods of prophylaxis, pharmacological or non-pharmacological, should be applied according to the degree of risk of DVT. Even after an hospital discharge, it should be maintained among those who still have some risk of DVT [2].

Computerized media has become more and more a potential for the prevention of DVT in health environments with programs that automatically calculate the need or not of prophylaxis in patients. So, it is intended to construct a model that can obtain better

Á. Rocha et al. (Eds.): WorldCIST'19 2019, AISC 932, pp. 557–567, 2019.
https://doi.org/10.1007/978-3-030-16187-3_54

results in order to contribute to the evolution of this type of diagnosis. The constructed model must be able to correctly classify this datatype.

Therefore, this study intends to present, explain, and prove how attributes obtained from discharge reports can be related to the prophylactic measures taken. The analysis of these results is accomplished through the response to two main research questions:

- Are the sex, the age group or the service to which the patient belongs factors relevant to the prophylaxis measures of deep vein thrombosis?;
- Is it possible to obtain classifiers capable of predicting the best prophylactic measures with a qualitative level similar as one of a health professional?

Thus, the purpose of this work was to find relationships between attributes by applying automatic learning techniques to data and defining models capable of assisting physicians in their decision-making process of prophylactic measures.

This article is divided into six different sections. After a first introduction to the manuscript (Sect. 1), all concepts and related works are presented in Sect. 2. Thereafter, Sect. 3 presents the materials and methods used in this study. Section 4 describes the implementation of the methodology (KDD) used in automatic learning in classification tasks. Section 5 presents the results obtained as well as their discussion. Finally, Sect. 6 includes conclusions and future work.

2   Background

2.1   Deep Vein Thrombosis

Deep vein thrombosis is characterized by the formation of a blood clot (thrombus) inside deep veins of the legs or the pelvic region. The blood clot will block the blood flow and cause the pressure inside the vein to increase. This clot can release and reach the lungs, with DVT being the main cause of pulmonary embolism. Deep vein thrombosis is a very frequent problem, mainly as a complication of other surgical and clinical infections [3].

Because it is the main cause of thromboembolism, it is extremely important that prophylactic measures are applied. Adequate prevention reduces DVT cases by two thirds, which makes them extremely important knowledge of different risk groups and their means of prevention [2].

There are two prophylactic measures that can be taken to prevent thrombosis: primary prophylaxis and secondary prophylaxis. The primary corresponds to the use of physical and/or pharmacological methods to prevent or minimize the chances of a patient developing the disease. This is the measure taken in most cases in clinical circumstances. The secondary concerns the early detection and treatment of DVT with the aim of preventing a possible thromboembolism: it should be used in patients in whom primary prophylaxis is contraindicated [2].

2.2   Data Mining

Data Mining is the process of using Artificial Intelligence (AI) techniques and statistical and mathematical functions that arises with the need to automatically extract potentially useful information from data, typically of high size and/or complexity, in a way that is understandable to users. The knowledge achieved can adopt various forms of representation, such as equations, trees or graphs, patterns or correlations [4–7].

The acquired knowledge is used to support the decision-making process in various processes, e.g., in medicine – in the diagnosis phase, a correct and rapid analysis of this large volume of data is important for the identification of pathologies [8].

In this study, classification techniques were used since the main objective was to find a model that could correctly identify which prophylactic measures a health professional should have towards a patient. The classification aims to evaluate the processed data and classifying them according to their characteristics. By using classification algorithms, one can define the relationships that exist between the attributes of processed data, which can determine a prediction [9].

2.3   Related Work

The healthcare industry has various information that causes the need for the application of Data Mining. There are some studies in the area of Deep Venous Thrombosis, but none uses DM to predict the prophylactic measures that should be applied to patients at risk of medical DVT.

Clemetente, Rocha, Rosewarne, Vaz, and Espíndola aimed to identify the best way to understand and analyze the data resulting from the monitoring of Deep Vein Thrombosis in an effective way. So, they used DM to produce models of representation or data prediction in order to positively support the activities of professionals involved with DVT. As results, they expected to identify the main causes of the occurrence of DVT or changes in its manifestation, in addition to its associated factors, to estimate the risk of DVT for a patient, as well as to simulate and describe the impact of adopting the prophylactic measures. They identified similar pictures of DVT that had different or similar evolution, as well as factors that influenced this difference or similarity [10].

Aljumah, Ahamad, and Siddiqui focused on predictive analysis of diabetic treatment using a regression-based DM technique. The dataset was studied and analyzed to identify the efficacy of different types of treatment for different age groups [11].

Bhatla and Jyoti aimed to provide a study of different DM techniques that can be employed in automated systems for predicting heart diseases using classification algorithms. As results, they concluded that the neural networks exceeded the remaining techniques of DM. Another conclusion to their analysis is that the decision trees also showed good accuracy with the help of genetic algorithms and the selection of subsets of resources [12].

3 Materials and Methods

3.1 Materials

The data used in this study are composed of 17942 patients belonging to the category of patients receiving medical treatment collected from November 3rd, 2015 to March 6th, 2017 of *Centro Hospitalar do Porto* (CHP). The patients' ages ranged from 1 to 107 years old with 9746 males and 8196 females. Later, the patients were divided by age groups, the ages being grouped from 0–15 years old, 15–29 years old, 30–59 years old, and more than 60 years old. Hospitalized patients were studied in 61 different services from the different hospital units of CHP. At the time of discharge, each patient was subjected to a clinical evaluation by his/her doctor, in which he/she answered a small questionnaire that, thereafter, led to identify which prophylactic measures the doctor should take with the patient. Each patient was assigned a last evaluation, in which it was concluded that approximately 11042 patients do not need any prophylactic measures.

3.2 Methods

The methodology developed in this study aims to treat information from a database. The Knowledge Discovery in Databases (KDD) aims to extract knowledge from databases. The KDD process tracks the entire course of knowledge discovery path in databases, from the way data is stored and accessed, to analyzing the performance of datasets because of their size and the way the results are interpreted and visualized.

There are five different phases that constitute the process of knowledge discovery: Selection, Preprocessing, Transformation, Data Mining, and Interpretation and Evaluation. In the next section, i.e. Sect. 4, each phase of the case study process will be described [11].

4 Implementation

4.1 Selection Phase

Throughout the data selection task, the attribute selection strategy was used. This technique reduces the number of redundant or irrelevant attributes of the data universe in question.

Thus, for the realization of this research, a database was used with patients referring to the risk of medical DVT, constituted by several attributes that could not be relevant to the study. Figure 1 shows the initial database. The first attribute that appears is the "ORDER" in which each patient was admitted; then the "EPISODE" that characterizes the admissions made by the patients; another attribute is the "ANSWERTIME" that refers to the response time that each doctor took to respond to the questionnaire; then the patients are identified by their process number "PROCESSNUMB"; the "AGE" that refers to the age that each patient has at the time of his/her admission, as well as the "SEX" that tells the sex of each one; "UTILCRIA" refers to a number created at the time of admission of each patient; the "SERVICE", i.e. the medical specialty to which

ORDER		EPISODE	ANSWERTIME	PROCESSNUMB	AGE	UTILCRIA	SERVICE	SEX	IDDATA	ANSWER
	15325215	17006543	19	535676	95	12111	31403	1	1	Q6
	15324945	17006833	7	578540	45	11144	31700	1	2	Q10
	15324572	17006707	17	976551	84	12268	31800	2	3	Q3

Fig. 1. Original dataset.

each patient is associated; and the "ANSWER" that indicates the prophylactic measure that was obtained at the end of each questionnaire.

Since it is intended to create a classification model for prophylactic measures, the attributes sex, age, service, and the prophylactic measures taken (answer) were chosen since it was concluded by the research team that they are the ones that can most influence the classification. In Fig. 2, the dataset with the selected attributes can be observed.

SEX	AGE	SERVICE	ANSWER
1	95	31403	6
1	45	31700	10
2	84	31800	3

Fig. 2. Dataset with the selected attributes.

Although the dataset can suffer a considerable reduction in volume, the integrity of the original data remains, that is, the mining of a reduced data group should be more efficient and should simultaneously produce the same analytical results [8].

4.2 Preprocessing Phase

Data preprocessing is thus an extremely important step throughout the entire process of knowledge discovery since it enables quality decisions to be based on quality data [8].

In this phase, methods were used to improve the performance of the algorithms since the larger the database to be analyzed, the more complex and time consuming the mining work will be, requiring a longer time in processing and obtaining results [13].

4.3 Transformation Phase

The transformation phase precedes the Data Mining phase. In this phase, data must be modified or consolidated in formats appropriate for the mining process.

Thus, in this phase a series of operations were involved, such as the construction of new attributes as the age group, in which the ages are grouped into five different bands, thus improving the mining process. Another operation performed was the normalization in which data were dimensioned in order to be inserted in a relatively short reference range, in this case between zero and one. In Fig. 3, the transformed data can be observed.

SEX	AGE GROUP	SERVICE	ANSWER
0	1	0,786825206	six
0	0,666666667	0,794275537	ten
1	1	0,796784066	three

Fig. 3. Knowledge Discovery in Databases (KDD) transformation phase dataset.

4.4 Data Mining Phase

This next phase is probably the most important in the whole process of KDD: the mining phase, where there is the application of specific algorithms for the extraction of data patterns [8]. For this phase, the Weka software, which has a vast set of automatic learning algorithms, was used. Thus, using this tool, six of the most important classification algorithms – NaiveBayes, J48, REPTree, Sequential Minimal Optimization (SMO), Logistic, and IBK – were applied to the study data target universe. The choice of these algorithms was since they are all methods of automatic learning, i.e., ranging from decision trees, classification rules, support vector machines, lazy learning, and bayesian networks. Thus, with this choice it is intended to study the behavior of the different algorithms in the classification of the data concerned and, thereafter, extract the classifiers that are proven to be more accurate.

4.5 Interpretation and Evaluation Phase

In this study, performance metrics were used to evaluate and validate the knowledge extracted, which also guarantees the reliability of the results achieved. These metrics are numerical measures that quantify the performance of a classifier. Thus, to certify the quality of the results obtained, the following metrics were used:

- Confusion matrix that allows the unambiguous visualization of the results of a given model;
- Accuracy, that is, the percentage of correctly classified instances, that is, the percentage of instances that the classifier correctly predicted;
- The percentage of instances incorrectly classified, that is, the number of instances that the classifier predicted incorrectly;
- Relative Absolute Error gives the discrepancy between an exact value of a classification and some approximation to it.

Finally, as stated previously, the experiments were performed using the Weka Data Mining tool. The files used during the experiences had the *arff* format. After loading the training set, it is necessary to define the algorithm that for the same set (when learning with *10-fold cross validation*) presented the best results.

5   Results and Discussion

In this subsection, the main results obtained are presented in order to be subsequently able to answer the two research questions specified in the Introduction section.

To select the best Data Mining algorithms for classification, initially, in a first experiment (E1), several tests were performed using the original dataset. Thus, at this stage, all the algorithms chosen were applied.

When comparing the results obtained, it was observed that globally the algorithm with the best performance was IBK, secondly Logistic, and the third best was J48. In Table 1, it is possible to observe the accuracy, as well as incorrectly classified instances, and relative absolute error. In the confusion matrix of the application result of the IBK algorithm, it is verified that the attributes that contain few instances will be classified incorrectly.

Table 1. Percentage of accuracy, incorrectly classified, and relative absolute error metrics for each algorithm in experiment E1

Algorithm	Experiment	Data approach	Incorrectly classified (%)	Accuracy (%)	Relative absolute error (%)
J48	E1	Without under or oversampling	27,42	72,58	69,28
NaiveBayes	E1	Without under or oversampling	27,47	72,53	68,96
IBK	E1	Without under or oversampling	27,35	72,65	68,50
Logistic	E1	Without under or oversampling	27,38	72,62	69,12
REPTree	E1	Without under or oversampling	27,44	72,56	69,31
SMO	E1	Without under or oversampling	27,29	72,71	139,09

Another experience (E2) that was realized was to change the age group by the age in order to determine which one would interfere more with the results. As it can be observed in Table 2, it was verified that the correctly classified instances go from 13034 to 12409, and that there is a slight decrease of the absolute error from 68.50% to 67.61%, so it was concluded that the use of age as an attribute was not the best option.

Table 2. Classification performance

	Experiment	Incorrectly classified	Correctly classified	Relative absolute error (%)
Age group	E2	4903	1309	67,61
Age	E2	5533	12409	68,50

Thereafter, another task that was accomplished was to verify in which way the attributes related to each other and, on the other hand, what was the weight that each attribute has in the classification of the answer. Consequently, it was verified that the attribute that relates more to the response is the age group, and the attribute that has the most weight in the classification was the service. In Table 3, it is possible to observe these values.

Table 3. Correlation of attributes with the response and attribute with more weight

	Correlation ranking filter		Information gain ranking filter	
	Rank attributes		Rank attributes	
Age group	0,2129	2	0,05899	2
Service	0,1144	3	0,42355	3
Sex	0,0637	1	0,00427	1
Selected attributes	2,3,1:3		3,2,1:3	

Thus, a third experiment (E3) was performed, in which the attributes were removed one by one, and the results of the classification without each one was verified, when applying the algorithm that had better classification, i.e. the IBK.

When analyzing Table 4, it is possible to conclude that by removing the service, the number of correctly classified instances decreases considerably, because, as already determined, this is the attribute that has the more weight in the classification process, it is also possible to observe that as the attributes that have greater weight are eliminated, the absolute relative error increases.

Table 4. Percentage of accuracy, incorrectly classified, and relative absolute error metrics for each scenario in experiment E3

Scenario	Experiment	Data approach	Incorrectly classified (%)	Accuracy (%)	Relative absolute error (%)
Without removing	E3	Without under or oversampling	27,35	72,65	68,50
Sex removed	E3	Without under or oversampling	27,35	72,64	68,50
Age group removed	E3	Without under or oversampling	28,56	71,44	70,91
Service removed	E3	Without under or oversampling	38,45	61,54	94,44

Finally, in the first experiment performed with the original dataset, it was concluded that a fairly high absolute error was obtained. This is due to the fact that, in this particular study, there are unbalanced classes. A method to avoid this problem was to manipulate the training samples so that the classes are no longer unbalanced, applying oversampling to the zero, two, three, four, five, and seven responses. The reduction of the training set, so that the classes become balanced, contributed to the resolution of this problem, by applying the undersampling to the six and ten responses, a total of

10698 cases remained in the sample. Thus, to perform these two tasks, the results obtained in the correlation of the attributes, as well as in the weight, were considered. After the classes were all balanced, a fourth experiment (E4) was carried out where the classification algorithms were applied.

Therefore, by comparing the results obtained with the different algorithms, the algorithm that obtained the best results was IBK. Nonetheless, although the J48 proved to have better classification results, it presents a relative absolute error superior to IBK, the worst performing algorithm was REPTree. That is, the first one presented a relative absolute error value of 22.99%, while the second one presented a value of 22.26%. Table 5 shows the relative absolute errors for each algorithm. By analyzing the confusion matrix of the classification for the IBK algorithm, it was verified that the classification was, thereafter, made more uniformly, and few false positives were obtained from each question, that is, an improvement in the results.

Table 5. Percentage of accuracy, incorrectly classified, and relative absolute error metrics for each algorithm in experiment E4

Algorithm	Experiment	Data approach	Incorrectly classified (%)	Accuracy (%)	Relative absolute error (%)
J48	E4	With under and oversampling	13,94	86,06	22,99
NaiveBayes	E4	With under and oversampling	21,88	78,12	35,53
IBK	E4	With under and oversampling	13,99	86,00	22,26
Logistic	E4	With under and oversampling	18,45	81,55	28,76
REPTree	E4	With under and oversampling	14,19	85,81	23,20
SMO	E4	With under and oversampling	17,37	82,63	86,81

In summary, by analyzing the results achieved, the research questions raised in the introduction section were answered. First, it was necessary to remove the attributes one by one and to study how they influence the classification of prophylactic measures. Through the IBK classification algorithm, it was discovered that the service to which each patient is associated is the most relevant factor for prophylactic measures, following by the age group to which the patient belongs. Through the second research question, it was possible to obtain classifiers capable of predicting what is the best prophylactic measure with a qualitative level similar to one of a health professional. When analyzing the results obtained from the experiences, it was also verified that the percentage of accuracy is high and the percentage of classification error is a low value. Therefore, it can be possible to obtain an accurate classification if, in a test phase, health professionals are advised to supervise the classification process.

6 Conclusion and Future Work

Finally, this study focused on the creation of a model, using Data Mining techniques, to classify from a health database the prophylactic measures to be applied to patients at risk of Deep Vein Thrombosis.

Initially, the DM techniques that could be used in order to achieve the objectives were studied, and a sequence of experiments was implemented to evaluate and compare the results obtained. The first experiment was performed on the original dataset, and the results were taken as reference. The best algorithm obtained to classify was the IBK. Then, a new balanced dataset was built based on the original dataset since the original dataset was not balanced in terms of positive and negative cases. After those modifications, significant improvements in the results were observed. Initially the percentage of accuracy was 72.65%, which was improved to 86%. The relative absolute error percentage had a positive sharp decrease from 68.50% to 22.26%.

In order to implement perfect classification systems, it would be necessary to obtain totally correct classifications on the datasets. Based on this study, future work is to continue to apply techniques of DM that allow to improve the results obtained. For example, build new datasets and, consequently, apply other DM algorithms, or even perform new post-processing techniques. Based on the results achieved, post-processing techniques that might be interesting to use to improve the classification would be oversampling and undersampling in order to standardize more the dataset. The main goal would be to reduce the percentage of absolute relative error to an insignificant value and to increase the percentage of accuracy until, in the future, the supervision of a health professional is not necessary at all.

Acknowledgments. This work has been supported by FCT – Fundação para a Ciência e Tecnologia within the Project Scope: UID/CEC/00319/2019.

References

1. Pitta, G.B.B., Gomes, R.R.: A frequência da utilização de profilaxia para trombose venosa profunda em pacientes clínicos hospitalizados. J. Vasc. Bras. **9**(4), 220–228 (2010)
2. Machado, N.L.B., Leite, T.L., Pitta, G.B.B.: Frequência da profilaxia mecânica para trombose venosa profunda em pacientes internados em uma unidade de emergência de Maceió. J. Vasc. Bras. **7**(4), 333–340 (2008)
3. Correia, A., Winck, J.C.: Trombose venosa profunda e Embolismo Pulmonar | Programa Harvard Medical School Portugal. Programa Harvard Medical School Portugal (2011). https://hmsportugal.wordpress.com/2011/03/28/trombose-venosa-profunda-e-embolismo-pulmonar/. Accessed 6 Nov 2018
4. Ramalho, V.V.: Modelo de data mining para deteção de embolias pulmonares. Instituto Superior de Engenharia de Lisboa (2013)
5. Oliveira, S., et al.: Clustering Data Mining models to identify patterns in weaning patient failures. Int. J. Biol. Biomed. Eng. **10**, 183–190 (2016)
6. Neves, J., et al.: A deep-big data approach to health care in the AI age. Mob. Networks Appl. **23**(4), 1123–1128 (2018)

7. Rodrigues, M., Peixoto, H., Esteves, M., Machado, J., Abelha, A.: Understanding stroke in dialysis and chronic kidney disease. Procedia Comput. Sci. **113**(4), 591–596 (2017)
8. Miguel da Silva Ferreira, P.: Aplicação de Algoritmos de Aprendizagem Automática para a Previsão de Cancro de Mama. Faculdade de Ciências da Universidade do Porto (2010)
9. dos S. Lima, T.: Estudo Comparativo dos Algoritmos de Classificação da Ferramenta WEKA. Centro Universitário Luterano de Palmas (2005)
10. Clemente, V., de Noronha Rocha, T.H., Gargano Lemos Rosewarne, T., Rocha, R., César Lopes Vaz, J., Pinto Espíndola, R.: Proposta de Gestão da Trombose Venosa Profunda Através de Mineração de Dados. Revista Acreditação: ACRED. (Ejemplar dedicado a: Revista Acreditação), vol. 3, no. 5, [s.n.], pp. 34–38 (2013). ISSN-e 2237-5643
11. Aljumah, A.A., Ahamad, M.G., Siddiqui, M.K.: Application of data mining: diabetes health care in young and old patients. J. King Saud Univ. Comput. Inf. Sci. **25**(2), 127–136 (2013)
12. Bhatla, N., Jyoti, K.: An analysis of heart disease prediction using different data mining techniques. Int. J. Eng. Res. Tecnol. **1**(8), 1–4 (2012)
13. de M. Nogueira, R.: Análise dos Impactos Harmônicos em uma Indústria de Manufatura de Eletroeletrônicos utilizando Árvores de Decisão. Universidade Federal do Pará, Instituto de Tecnologia (2015)

Predicting Low Birth Weight Babies Through Data Mining

Patrícia Loreto[1], Hugo Peixoto[2($\boxtimes$)], António Abelha[2], and José Machado[2]

[1] University of Minho, Braga, Portugal
a71934@alunos.uminho.pt
[2] Algoritmi Research Center, University of Minho, Braga, Portugal
{hpeixoto,abelha,jmac}@di.uminho.pt

Abstract. Low Birth Weight (LBW) babies have a high risk of developing certain health conditions throughout their lives that affect negatively their quality of life. Therefore, a Decision Support System (DSS) that predicts whether a baby will be born with LBW would be of great interest. In this study, six different Data Mining (DM) algorithms are tested for five different scenarios. The scenarios combine information about the mother's physical characteristics and habits, and the gestation. Results are promising and the best model achieved a sensitivity of 91,4% and a specificity of 99%. Good results were also achieved without considering the gestational age, which showed that the use of DM might be a good alternative to the traditional medical imaging exams in the prediction of LBW early in the pregnancy.

Keywords: Knowledge Discovery in Databases · Data Mining · Classification · Decision Support Systems · Low Birth Weight · CRISP-DM

1 Introduction

A newborn is a Low Birth Weight (LBW) baby when is born with a weight below 2500 grams. LBW babies have a greater risk of developing a variety of health conditions during the neonatal period and throughout their lives [1,2].

People born with LBW are also more likely to have a lower Health Related Quality of Life (HRQL). This measure evaluates the consequences that health conditions have in a person's psychological, social and physical welfare [3]. LBW newborns have, for example, a bigger chance of neonatal death and longer hospitalization [4,5]. In children, consequences include visual perceptual and visual motor deficit, that may have as result difficulties in reading and in mathematics, and higher risk of motor disabilities [1,6]. In adults, LBW is reflected in higher chances of being hypertensive and developing Chronic Kidney Disease (CKD) due to the fact that LBW is associated to a lower number of renal glomeruli [2].

It is possible to predict a newborn's weight through medical imaging exams. However, results may not be accurate, depend on several factors, such as the

© Springer Nature Switzerland AG 2019
Á. Rocha et al. (Eds.): WorldCIST'19 2019, AISC 932, pp. 568–577, 2019.
https://doi.org/10.1007/978-3-030-16187-3_55

multiplicity of the gestation and the mother's Body Mass Index (BMI), and good results are usually only achieved if the exam is performed closer to the delivery moment [7,8]. As for the factors that influence birth weight and help in predicting if a baby will be born with LBW, they include maternal factors, such as the mother's age, BMI and risk factors (e.g. hypertension), the multiplicity of the gestation and social and economic factors [9].

In this paper, it is studied the development of a Decision Support System (DSS) that uses Data Mining (DM) to predict whether a baby will be born with LBW, taking into account maternal and gestational factors.

2 Background

Hospitals usually collect patient data, analyze them at the moment of care and store them in databases. These data have great potential to help in decision-making and in the last few years this potential has been recognized and is now being applied in the development of DSS that use DM [10].

Knowledge Discovery in Databases (KDD) is a process of extracting useful knowledge from data. Its main goal is to find new, useful and valid knowledge through the application of algorithms. This process has five steps:

1. *Selection* - choice of the dataset;
2. *Pre-processing* - data cleaning and processing;
3. *Transformation* - data processing so that they become uniform (e.g. normalization) and suitable for the DM models;
4. *Data Mining* - choice of the task to be performed and the techniques to be used;
5. *Interpretation/Evaluation* - interpretation and evaluation of the patterns obtained in DM [11,12].

Since 1990, DM has been the most important tool to discover knowledge from large databases. DM is a process of discovering patterns and relationships within large amounts of data and of constructing models using the collected information that are later applied in decision-making [13]. In healthcare, DM can be useful in the prediction of future events, whether they are clinical, management or financial, improving the health care provided [10].

DM algorithms can be of two types: predictive and descriptive. Predictive algorithms are used in prediction and classification. Whether descriptive algorithms find associations, clusters and subgroups in the given data. In this paper, a data classification problem is studied. In these problems, a set of pre-classified instances is used to develop a model that can classify new instances correctly. This process has two stages: learning and classification. In the learning stage, a pre-classified training dataset is analyzed by a classification algorithm. Then, in the classification stage, pre-classified test data are applied to the algorithm and are used to estimate the accuracy of the algorithm. If results are good, the classification rules are applied to new data instances [14].

The main goal of Intelligent DSS is the supporting of decision-making processes. These systems use intelligent agents to perform tasks automatically and DM techniques to make predictions [10]. Clinical DSS assist health care providers not only in the diagnosis moment, but also in the initial consultation and through the follow-up. They have the advantage of making the decision-making process faster and more accurate, leading to less waiting times and fewer mistakes [15]. Several DM studies have been conducted to find out ways of supporting health professionals in decision-making and in the prediction of health conditions. In this section, two studies that apply DM tools in Obstetrics are presented.

Pereira et al. (2015) studied the influence of obstetric risk factors in predicting the type of delivery through DM. To guide the DM process the *Cross Industry Standard Process for Data Mining* (CRISP-DM) methodology was used and good results were acquired. Four different algorithms were implemented to induce the DM and Cross Validation and Holdout Sampling were the chosen sampling methods. A total of 26 variables were used and five scenarios were developed. The variables included data about the mother's age and physical information, a set of medical exams, information about the delivery and gestational age, among others. As for the target variable, four different approaches were evaluated and the most successful one was the one that divided deliveries in caesarean sections and vaginal deliveries. This approach achieved sensitivity and specificity values of 90,11% and 80,05%, respectively [16].

Brandão et al. (2015) conducted a study related to the Voluntary Interruption of Pregnancy (VIP) that had as main goal the evaluation of this process and the identification of the associated risk. The dataset used in the DM process contains information about the patient's age, whether she had previous VIP or pregnancies, professional status and gestational age, among others. In the DM process, three algorithms were chosen and they were applied to ten different scenarios with two approaches: without oversampling and with oversampling. The best result was achieved a sensitivity of approximately 93% for the scenario that did not considered the woman's marital status and gestational age [12].

3   Materials and Methods

The chosen software tool was *Orange*. *Orange* is an open-source software for data mining, machine learning and data analysis, developed by the Bioinformatics Laboratory of the Faculty of Computer and Information Science at University of Ljubljana and that is written in Python [17]. Orange was the chosen due to its intuitive interface and because, compared to other similar tools, it makes the development of algorithms and the comparison of different approaches easier [18].

As for the dataset, the data used in this study was extracted from an Obstetrics service from a Portuguese hospital. The dataset has 3163 lines and each line has information about a newborn and the respective birth and mother.

KDD was applied in this study and, for the DM stage, CRISP-DM was chosen, since it is related to the models in the KDD process [12]. It is a standard model for DM processes and has six steps: Business Understanding, Data Understanding, Data Preparation, Modelling, Evaluation and Deployment [19]. The classification techniques chosen to induce the DM models were AdaBoost (AB), Classification Tree (CT), K-Nearest Neighbours (KNN), Naïve Bayes (NB), Random Forest (RF) and Support Vector Machine (SVM).

4   Data Mining Process

As previously stated, the CRISP-DM methodology was followed to develop this study. In this section, all phases of the process are described.

4.1   Business Understanding

This study has as main goal the prediction of whether a baby will be born with low-weight given the mother's health information and habits, the multiplicity of the gestation and the gestational age. Therefore, the DM goal is to develop models that can predict if a baby will have LBW when is born.

4.2   Data Understanding

First, the necessary data was extracted. This data includes information about the mother's admission data and labour. Afterwards, the variables considered relevant were chosen and a dataset was constructed. The dataset is composed by data instances and each data instance is composed by nine variables:

- *Fetus's sex*;
- *Multiplicity* – whether the gestation is multiple;
- *Smoker* – whether the mother is a smoker during pregnancy;
- *Hypertension* – whether the mother is hypertensive;
- *Diabetes* – whether the mother has diabetes during the pregnancy;
- *Age* - the mother's age;
- *BMI* - the mother's body mass index in the moment of admission;
- *Gestational age* – gestational age in weeks when labour happened;
- *Fetus's weight*.

It is important to mention that the information for the *Smoker*, *Hypertension* and *Diabetes* variables were extracted from free-text fields and, therefore, in some cases this information might not have been registered. In Table 1 are some statistical measures of the numeric variables used in the dataset and in Table 2 are the percentages of occurrence of the remaining variables.

Table 1. Statistical measures of the numerical variables used in this study.

Variable	Minimum	Maximum	Mean	Standard deviation
BMI	14,18	52,6	29,75	4,78
Age	14	46	31,07	5,57
Gestational age	22	43	37,75	3,36

Table 2. Percentage of occurrence of non-numerical variables.

Variable	Class	Percentage
Sex	Feminine	48,75%
	Masculine	51,25%
Multiplicity	Simple	97,29%
	Multiple	2,71%
Smoker	Smoker	1,72%
	Non-smoking	98,28%
Hypertension	Hypertensive	0,13%
	Non-hypertensive	99,87%
Diabetes	Diabetic	3,39%
	Non-diabetic	96,61%

4.3   Data Preparation

After selecting the attributes and the instances that compose the dataset, it was necessary to pre-process the data. The instances that had null values or noise values were removed, leaving 2328 instances. These instances were the ones that were used in the DM process. Afterwards, the data was normalized so that all instances were represented by values between 0 and 1, except for the target value. The values for the target variable *Low Birth Weight* were obtained by defining that for babies with birth weight bellow 2500 grams the target value is *Yes* and for babies with weight equal or above that value the target value is *No*. For 86,55% of the instances the target variable was *No*.

After some DM models induction and since the number of LBW babies is considerably inferior to the number of babies who are not low-weight, it was applied oversampling so that algorithms can train with more instances. Oversampling was achieved by duplicating all data instances.

4.4   Modelling

After selecting and processing the data, the data mining models were induced using the chosen DM techniques with the default definitions that *Orange* applies. The sampling method for all cases was Cross Validation with 10 folds. The DM models were induced for the initial and for the oversampled dataset. Lastly, different scenarios were applied to identify which ones affect more the prediction of low-weight. The considered combinations were:

- S1: {All variables}
- S2: {Sex, Multiplicity, Age, BMI, Gestational Age}
- S3: {Sex, Multiplicity, Smoker, Hypertension, Diabetes, Gestational Age}
- S4: {Multiplicity, Age, BMI, Gestational Age}
- S5: {Sex, Multiplicity, Age, BMI}.

The data mining model is composed by the approaches (A), the scenarios (S), data mining techniques (DMT), sampling models (SM), data approaches (DA), and target (TG). In this case represented by the following equation:

$$DMM = \{A_f, S_i, DMT_y, SM_c, DA_b, TG_i\}$$

- A = {Classification}
- S = {S1, S2, S3, S4, S5}
- DMT = {SVM, CT, RF, NB, KNN, AB}
- SM = {Cross Validation}
- A = {without oversampling, with oversampling}
- TG = {low-weight}.

A total of 60 models were induced:

$$DMM = \{1A, 5S, 6DMT, 1SM, 2DA, 1TG\}$$

In Fig. 1 is the *Orange* workflow that was developed for this study.

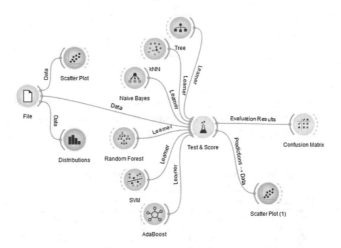

Fig. 1. Project workflow.

4.5   Evaluation

The chosen DM software provides a set of tools to help in this stage, such as the Confusion Matrix. The Confusion Matrix allows the acquisition of the number of True Positives (TP), False Positives (FP), True Negatives (TN) and False Negatives (FN) that can then be used to calculate the models' sensitivity and specificity. For this study, the chosen evaluation parameters were sensitivity, sensibility and accuracy. Sensitivity is the most important parameter, because in this case study the main goal is to correctly identify the LBW newborns and, since the number of LBW newborns is much lower than the number of newborns that are not LBW, the accuracy and specificity values by themselves may induce in error.

The best results for each algorithm are presented in Table 3. For the AB algorithm, for example, the best sensitivity, specificity and accuracy were achieved for the same scenario and approach (S1 with oversampling).

Table 3. Best model for each algorithm.

Algorithm	Sensitivity			Specificity			Accuracy		
	Scenario	Approach	Value	Scenario	Approach	Value	Scenario	Approach	Value
AB	S1	With oversampling	91,4%	S1	With oversampling	99,0%	S1	With oversampling	98,0%
CT	S1	With oversampling	61,8%	S2	With oversampling	98,5%	S2	With oversampling	93,5%
KNN	S2	With oversampling	64,1%	S2	With oversampling	97,6%	S2	With oversampling	93,1%
NB	S1	With oversampling	34,8%	S5	With and without oversampling	99,6%	S1, S3	With (S1, S3) and without (S3) oversampling	87,6%
RF	S2	With oversampling	77,8%	S2	With oversampling	98,8%	S2	With oversampling	96,0%
SVM	S3	Without oversampling	64,5%	S3	Without oversampling	90,0%	S3	Without oversampling	86,6%

To find out the best model, threshold values were introduced. For sensitivity the threshold value was 80%, for specificity was 97% and for accuracy was 95%. The models that achieved these three threshold values are presented in Table 4.

Table 4. Models that achieved the threshold.

Algorithm	Scenario	Approach	Sensitivity	Specificity	Accuracy
AB	S1	With oversampling	91,4%	99,0%	98,0%
AB	S2	With oversampling	91,1%	98,8%	97,8%
AB	S4	With oversampling	89,3%	99,0%	97,7%
AB	S5	With oversampling	82,4%	97,9%	95,9%

5 Discussion

After analyzing the obtained results, it is possible to conclude that the algorithm that best fits the given data is AB and the worst one is NB. AB performed better than any other algorithm, this possibly happened because it works well for a large number of variables and gives less importance to outliers. As for NB, the obtained results show that overfitting probably happened and that the variables are correlated, thus the results are not good. On the opposite side is the RF models that did not have the best results, but still performed better than the remaining algorithms except for AB. Another conclusion is oversampling significantly improved results, specially the sensitivity values, which are the most important ones in this case study. This possibly happened because with oversampling there are more instances to train with.

As for the best model, the defining criteria was the sensitivity value, since the specificity values were all close to each other. The best model uses the scenario S1 (all variables), with a sensitivity of 91,4%. However, it is also important to remember that this scenario uses the variable *Gestational Age*, that has a great influence in the results. S1 results are not far ahead from the scenario S2 results, even tough S2 does not include the variables *Smoker*, *Hypertension* and *Diabetes*. This is probably because there were women that had these conditions but their information was not recorded. Because of that, only a small part of the pregnant women were considered to have these conditions and, therefore, these variables did not have much influence. This can also explained why no model with the scenario S3 achieved the threshold. Comparing the values obtained for S2 and S4, whose only difference is the presence or absence of the variable *Sex*, respectively, it is possible to conclude that the variable *Sex* influenced the results, since S2's sensitivity (91,1%) was higher than S4's sensitivity (89,3%). It is also important to mention that despite having a lower sensitivity value (82,4%), the S5 scenario still is one of the best results, even though it does not use the variable *Gestational Age*. The non-use of this variable is a good component of a system that gives the probability of a fetus being born with LBW early in the pregnancy, so that further measures can be taken in that moment.

6 Conclusions and Future Work

The obtained results are promising and show that the prediction of LBW through DM may be a good and viable alternative to the traditional medical imaging exams. Furthermore, the results obtained for this dataset suggest that the DM process presented hereby has the potential to be used in real-life scenarios.

The factors that influence the most these predictions are the mother's age and BMI, the multiplicity of the gestation and gestational age. However, good results were also achieved without the gestational age as a variable. This is important for the development of a DSS that predicts if it is likely that a fetus will have LBW early in the pregnancy, allowing the application of measures to correct that and avoiding the consequences that this condition may cause.

In the future, it would of great interest to improve the results without using the gestational age. To do so, more data is needed to train and to test algorithms. It would also be interesting to study the influence of risk factors (smoker, hypertension and diabetes) if there was more information recorded about them. One of the struggles of this study was the lack of information about them, because these data are in free-text fields. Thus, the introduction of fields that assess this information independently would be a good addition to the records.

Acknowledgments. This work has been supported by FCT – Fundação para a Ciência e Tecnologia within the Project Scope: UID/CEC/00319/2019.

References

1. Poole, K.L., Schmidt, L.A., Missiuna, C., Saigal, S., Boyle, M.H., Van Lieshout, R.J.: Childhood motor coordination and adult psychopathology in extremely low birth weight survivors. J. Affect. Disord. **190**, 294–299 (2016). https://doi.org/10.1016/j.jad.2015.10.031
2. Mañalich, R., Reyes, L., Herrera, M., Melendi, C., Fundora, I.: Relationship between weight at birth and the number and size of renal glomeruli in humans: a histomorphometric study. Kidney Int. **58**(2), 770–773 (2000). https://doi.org/10.1046/j.1523-1755.2000.00225.x
3. Wolke, D.: Born extremely low birth weight and health related quality of life into adulthood. J. Pediatr. **179**, 11–12 (2016). https://doi.org/10.1016/j.jpeds.2016.09.012
4. de Castro, E.C.M., Leite, Á.J.M., de Almeida, M.F.B., Guinsburg, R.: Perinatal factors associated with early neonatal deaths in very low birth weight preterm infants in northeast brazil. BMC Pediatr. **14**(1), 312 (2014)
5. Bahado-Singh, R.O., Dashe, J., Deren, O., Daftary, G., Copel, J.A., Ehrenkranz, R.A.: Prenatal prediction of neonatal outcome in the extremely low-birth-weight infant. Am. J. Obstet. Gynecol. **178**(3), 462–468 (1998). https://doi.org/10.1016/S0002-9378(98)70421-1
6. Perez-Roche, T., Altemir, I., Giménez, G., Prieto, E., González, I., Peña-Segura, J.L., Castillo, O., Pueyo, V.: Effect of prematurity and low birth weight in visual abilities and school performance. Res. Dev. Disabil. **59**, 451–457 (2016). https://doi.org/10.1016/j.ridd.2016.10.002
7. Dimassi, K., Douik, F., Ajroudi, M., Triki, A., Gara, M.F.: Ultrasound fetal weight estimation: how accurate are we now under emergency conditions? Ultrasound Med. Biol. **41**(10), 2562–2566 (2015). https://doi.org/10.1016/j.ultrasmedbio.2015.05.020
8. Khalil, A., D'antonio, F., Dias, T., Cooper, D., Thilaganathan, B.: Ultrasound estimation of birth weight in twin pregnancy: comparison of biometry algorithms in the stork multiple pregnancy cohort. Ultrasound Obstet. Gynecol. **44**(2), 210–220 (2014). https://doi.org/10.1002/uog.13253
9. Yadav, H., Lee, N.: Maternal factors in predicting low birth weight babies. Med. J. Malays. **68**(1), 44–47 (2012)
10. Portela, F., Santos, M.F., Silva, Á., Rua, F., Abelha, A., Machado, J.: Preventing patient cardiac arrhythmias by using data mining techniques. In: 2014 IEEE Conference on Biomedical Engineering and Sciences (IECBES), pp. 165–170. IEEE (2014). https://doi.org/10.1109/IECBES.2014.7047478

11. Fayyad, U., Piatetsky-Shapiro, G., Smyth, P.: From data mining to knowledge discovery in databases. AI Mag. **17**(3), 37 (1996)
12. Brandao, A., Pereira, E., Portela, F., Santos, M.F., Abelha, A., Machado, J.: Predicting the risk associated to pregnancy using data mining. In: Proceedings of the International Conference on Agents and Artificial Intelligence, ICAART 2015, vol. 2, Lisbon, Portugal. SciTePress (2015)
13. Khademolqorani, S., Hamadani, A.Z.: An adjusted decision support system through data mining and multiple criteria decision making. Procedia Soc. Behav. Sci. **73**, 388–395 (2013). https://doi.org/10.1016/j.sbspro.2013.02.066
14. Han, J., Pei, J., Kamber, M.: Data Mining: Concepts and Techniques. Elsevier, Amsterdam (2011). https://doi.org/10.1007/978-1-4899-7993-3_104-2
15. Castaneda, C., Nalley, K., Mannion, C., Bhattacharyya, P., Blake, P., Pecora, A., Goy, A., Suh, K.S.: Clinical decision support systems for improving diagnostic accuracy and achieving precision medicine. J. Clin. Bioinform. **5**(1), 4 (2015). https://doi.org/10.1186/s13336-015-0019-3
16. Pereira, S., Portela, F., Santos, M.F., Machado, J., Abelha, A.: Predicting type of delivery by identification of obstetric risk factors through data mining. Procedia Comput. Sci. **64**, 601–609 (2015)
17. Naik, A., Samant, L.: Correlation review of classification algorithm using data mining tool: WEKA, Rapidminer, Tanagra, Orange and Knime. Procedia Comput. Sci. **85**, 662–668 (2016)
18. Yadav, S.K., Bharadwaj, B., Pal, S.: Data mining applications: a comparative study for predicting student's performance. arXiv preprint arXiv:1202.4815 (2012)
19. Chapman, P., Clinton, J., Kerber, R., Khabaza, T., Reinartz, T., Shearer, C., Wirth, R.: Crisp-dm 1.0 step-by-step data mining guide (2000)

New Pedagogical Approaches with Technologies

Using Virtual Reality Tools for Teaching Foreign Languages

Bruno Peixoto[1(✉)], Darque Pinto[1], Aliane Krassmann[2], Miguel Melo[3],
Luciana Cabral[4], and Maximino Bessa[1,3]

[1] Universidade de Trás-os-Montes e Alto Douro, Vila Real, Portugal
`brunomepeixoto@gmail.com`
[2] Universidade Federal do Rio Grande do Sul, Porto Alegre, Brazil
[3] INESC TEC, Porto, Portugal
[4] CITCEM and Instituto Politécnico de Bragança, Bragança, Portugal

Abstract. Among the wide application areas that Virtual Reality (VR)
can have a major impact, one is Education. However, this potential is
still unexplored, and one of these gaps has to do with language learning. Listening activities, which are often only supported by audio, are
thought of to be demanding area when it comes to learning a second or
foreign language and so therefore an interesting area for VR to take place.
This pilot study therefore presents the perceptions of foreign language
teachers regarding a novel medium for delivering listening activities to
their students: Virtual Reality technology. The results show that foreign
language teachers are of the opinion that this technology can help motivate students and potentiate the student's learning curve regarding the
listening of a foreign or second language.

Keywords: Virtual Reality · Media in education ·
Teaching/learning strategies

1 Introduction

Nowadays in students' lives technology is everywhere, from interactive games to
the growing plethora of digital devices. All this exposure to technology means
that their learning expectations are likely to be very different from previous
generations where technology was not widespread [1]. This scenario makes the
traditional teaching techniques sometimes fail when it comes to engaging the
contemporary students and, consecutively, creates an inability to motivate them
to learn [2].

One of the most important factors regarding foreign or second language learning is to be exposed to it, by talking or listening to it. Listening sessions can be
defined as when listeners are able to convert visual and auditory clues into information about what is going on in any given situation [3]. They are part of most
language courses and international proficiency tests as the Test of English as
a Foreign Language (TOEFL) TOEFL and the International English Language

© Springer Nature Switzerland AG 2019
Á. Rocha et al. (Eds.): WorldCIST'19 2019, AISC 932, pp. 581–588, 2019.
https://doi.org/10.1007/978-3-030-16187-3_56

Testing System (IELTS), in the case of English language. In addition, they are the hardest ones, as it requires the student the ability to carefully pay attention to understand in real time what is being said.

The use of multimedia content in education is considered to be quite productive. Videos, images, sounds, animations or simulations are just a few types of multimedia that can be used in a meaningful way due to its interactivity and flexibility [4]. Virtual Reality (VR) technology, described as the use of an immersive computer-simulated environment in which people can interact with [5], incorporates these traditional multimedia and significantly elevates the level of user immersion, particularly at the level of visual perception. It replaces interaction with immersion; it replaces the desktop metaphor with a world metaphor; and it replaces direct manipulation with symbiosis [6]. The high levels of immersion, authenticity and interaction provided by VR allow the user to believe that he is within the computer-simulated environment [7], which is called the sense of presence. This aspect gives the opportunity to help students comprehend complex or abstract concepts in situations that have cost or security constrains or even which may no longer exist [8]. Studies by the VIRART group [9] from (PAÍS) concluded that VR applications are an asset in teaching, more specifically in especial communication languages.

From 2012 onwards, VR technology started to attract the public insterest with the success of Oculus Rift, a wearable and affordable Head Mounted Display (HMD), with stereoscopic displays that is considered to be comfortable and lightweighted. In 2014 Google launched Google Cardboard, which could turn any smartphone in an HMD. Due to this recent expansion and cost reduction, few studies have already investigated the very benefits of VR in foreign or second language teaching and learning, being mostly restricted to theoretical ones. Schwienhorst [10], for example, reports that the investigation of VR field in foreign language teaching is still largely unexplored, although incredibly helpful not only to bring language learners closer to the language culture but also to create realistic simulations that wouldn't even exist in the real world. Lin and Lan [11], by their turn, suggest that foreign language teachers are now expected to employ VR technologies as a way to help their students, which changes the teacher's role from the person with all the answers to facilitators who support students in the virtual world path [9]. In contrast, most of today's teachers still lack the knowledge and skills regarding this use, making them reluctant to incorporate new media into their teaching [12,13]. Thus, according to Youngblut [9], VR developers and teachers should work together to overcome this impasse.

Bearing in mind the above, we developed an immersive VR experience aimed at English language teaching, focusing in the listening activities. In order to verify if teachers can see the benefits of VR to improve the classic listening exercises or not, we conducted a pilot study having a teacher-based sample to investigate the experience. The research was therefore carried out together in a private school, *the English Institute of Vila Real*, Portugal, with seven foreign language teachers familiarised with the traditional listening exercises and for whom the VR technology and educational tools were yet to be experimented and applied to language teaching and learning.

2   Methods

Listening comprehension, especially in foreign language learners, requires the listener to observe attentively as much visual and auditory clues as possible inserted in an audio, in order to have a real idea of what is going on in the given narrative. Since in most basic and language schools the listening exercises are mostly supported by audio streaming, an immersive VR application was developed to help teachers explain a particular scenario in a listening session. Four from the seven teachers who participated in the pilot study were not only responsible for the dialogue script for the VR application but also gave voice to Non-Player Characters (NPCs). The methodology adopted consists of a quasi-experimental design, cross-sectional study with a quantitative focus. The sampling technique used was the nonprobabilistic convenience sampling procedure.

2.1   Sample

The sample consisted of teachers. Namely, seven participants (six female and one male), between 35 and 53 years old ($M = 44,50$, $SD = 6,02$) who were foreign language teachers from *the English Institute of Vila Real*, in Portugal. The teachers were accustomed to deal with listening exercises by means of audio in their classes and had no previous VR experience.

2.2   Materials

A laptop computer was used for the experiments. It had Intel Core i7-7700HQ processor and NVIDIA GeForce 1070 graphics card; it was responsible for running the game engine and all the input and output devices required for the experiment were connected to it. For the visual stimulus, the HTC Vive HMD was used. This HMD features a 110° viewing angle and a resolution of 1080×1200 pixels per eye. The audio stimulus was delivered with a noise cancellation headphones (Bose QuietComfort 25).

In order to support the study, a realistic virtual environment (VE) was created by using the Unity game engine and it consist of two scenarios (Fig. 1, one representing a formal language dialogue, consisting of an office, and the other an informal one, consisting of an English pub.

The scenarios were filled with props that characterize an office room and a pub, respectively, and NPCs to meet the dialogue script. In both scenarios, the user would be placed at a table surrounded by the NPCs, as if he/she is part of the plot. The movement of the characters was produced with motion capture animation to ensure a realistic simulation, using an OptiTrack tracking system with 8 Prime 13 cameras.

Fig. 1. View of the VE during the formal scenario (top) and informal scenario (bottom)

Each virtual scenario was developed having into account the study by Melo et al. [14] to ensure that Presence was not affected by the exposure time. Thus, the formal scenario had a length of 2:36 minutes and the informal scenario a length of 1:29 min.

2.3   Instruments

Concerning the questionnaires as presented: (1) a simple sociodemographic questionnaire to determine the sample characteristics and teaching experiences regarding multimedia content in class; (2) a 6 item 7-point Likert scale questionnaire based in After-Scenario Questionnaire (ASQ) to assess the teachers' satisfaction with the virtual experience [15]. (3) a 4 item 7-point Likert scale created by the authors to understand the teachers' satisfaction with the immersive VR application as an educational tool. The questionnaires were presented in printed format in the teachers' first language, Portuguese. A rough translation of the questionaries (2) and (3) is presented below.

(2) Questionnaire based in After-Scenario Questionnaire (ASQ):

In general...
... I liked the experience I just had
... The experiment discouraged me

... I felt discomfort throughout the experience
... I'm satisfied with the ease with which the tasks were completed
... I'm satisfied with the time it took to complete the tasks
... I'm satisfied with the information given by the application.

(3) Questionnaire regarding the immersive VR application as an educational tool:

In general...
... I am satisfied with this application as a teaching method
... I believe this experiment complements in a positive way the traditional listening exercise
...I believe this experiment facilitates teaching techniques
...I would use this technology as a method of teaching in my classes.

2.4   Procedure

All experiments were conducted in a classroom at *the English Institute of Vila Real*, where external variables were controlled. Each teacher participated individually. It started with instructions about the experiment and with the participants filling in the sociodemographic questionnaire in the sequence. Afterwards, they were asked to seat, and the HMD and the headphones were properly placed with the help of the researchers.

The experiment was preceded by a pre-exposure in the VE, where the participants were given a period of around one minute to get acquainted with the VE. In this phase, the participant was immersed in the respective scenario, but without the NPCs nor the dialogue. This phase happened before each scenario. After that period, the NPCs appeared and the dialogue started right away.

After ending the two VR scenarios, the participants got help with the removal of all the equipment and were guided to a table so that they could complete the remaining questionnaires.

Overall, every participant took between 15 to 20 min to finish the whole procedures, including the questionnaires.

3   Results

The sociodemographic data shows that 85,7% of the participants use multimedia content in class at least occasionally ($M = 1,14$, $SD = 0,53$). When inquired the types of multimedia they use, 100% answered sounds (music, recordings of voice, etc.), 85,7% images (photographs, drawings, maps, figures, etc.), 57,1% videos (entertainment, educational, animations, etc.), and only 14,3% reported the use of simulations and others. It was questioned what was their perceived level of knowledge on VR, in which 71,4% of the participants said that they only have basic knowledge about this technology and how it works.

The scores for user satisfaction regarding the virtual experience (ASQ) were very positive with an overall mean of 6,81 and a standard deviation of 0,78 in

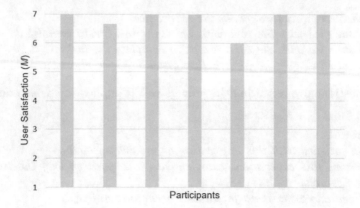

Fig. 2. User satisfaction regarding the virtual experience

a scale from 1 to 7, where a higher value represents a higher user satisfaction (Fig. 2).

Regarding to the teachers' satisfaction with the immersive VR application as an educational tool, the results were even more positive, with M = 6,86 and SD = 0,45 (Fig. 3). A closer analysis revealed that the participants strongly agree that: (1) this experience complements in a positive way the classic listening exercise; (2) this experiment facilitates and helps with the language teaching methods; (3) they would like to use this technology as a way of teaching in their classes.

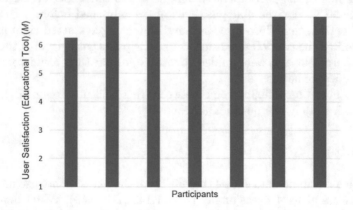

Fig. 3. User satisfaction regarding the VR application as an educational tool

4   Discussion

Since its emergence, digitally generation has been losing interest in the traditional teaching techniques [1,2]. So nowadays, the teacher's role is to try and convert the established learning theories into new practices [1] while guiding, supporting and facilitating their students' learning [4]. Since foreign language learners should convert both visual and auditory clues into information [3] and most schools only use audio in listening exercises, an immersive VR application regarding this topic seems ideal to help the student, while motivating them by providing a new way of learning.

Rickel and Johnson [16] point out that VR makes simulations closer to real-life experiences, because instead of observing the simulated world through a workspace window, students are immersed in a 3D environment where they can improve their skills through more realistic practices. Thus, VR adds another layer of interactivity which could substantially improve the students' concentration during listening exercises, and as a consequence, help them to understand an memorize what is being said. In addition, it allows to place the student in a foreign experience, in a simulated environment that resembles and evokes very culture, society and country of the language being learned, that is, the real context of the language.

Even though teachers seem to be hesitant to incorporate new digital media into their teaching [2], our results show that they are willing to use VR as a teaching technique and take advantage of the mentioned benefits. However, since the students will be unaware of their surroundings while using HMD, constant supervision is needed to assure their safety, especially with the minors. As the results also show, teachers are unfamiliar with this technology, so convincing them of its benefits is the first step towards this incorporation, in which this study made a contribution.

5   Conclusion

The main objective of this pilot study was to evaluate the perceptions of foreign language teachers regarding the benefits of a VR application as a tool to help their students with listening sessions, rather than the classic listening exercises where the student only has access to audio. To this end, a VR application was developed which allowed the user to be immersed in the virtual world representing formal and informal scenarios. After testing it with foreign language teachers, we concluded that there is a unanimous opinion that such technology is not only attractive but can also help to motivate students and potentiate the student's learning curve regarding the listening of a foreign language.

The main limitation of the pilot study is the sample size. Future work intends to broaden the scope of the study to both get, a larger sample, with more teachers and schools participating and, extend this VR-based learning tool to incorporate more learning activities. Afterwards, with the lessons learned we plan to design a pedagogic framework that could help teachers in making successful use of VR, towards more digitally enhanced classes.

Acknowledgments. This work is financed by the ERDF – European Regional Development Fund through the Operational Programme for Competitiveness and Internationalisation - COMPETE 2020 Programme and by National Funds through the Portuguese funding agency, FCT - Fundação para a Ciência e a Tecnologia within project POCI-01-0145-FEDER-028618 entitled PERFECT - Perceptual Equivalence in virtual Reality For authEntiC Training.

We would like to thank Cláudia Peixoto, Marta Peixoto and the other foreign languages teachers from *the English Institute of Vila Real* for their collaboration in this experience.

References

1. Wang, R., Newton, S., Lowe, R.: Experiential learning styles in the age of a virtual surrogate. Int. J. Arch. Res. ArchNet-IJAR **9**, 93–110 (2015). https://doi.org/10.26687/archnet-ijar.v9i3.715
2. Prensky, M.: Digital Game-Based Learning. Paragon House, New York (2007)
3. Alatis, J.: Georgetown University Round Table on Languages and Linguistics (GURT) 1990: Linguistics, Language Teaching and Language Acquisition: The Interdependece of Theory, Practice and Research, Georgetown University Round Table on Languages and Linguistics series. Georgetown University Press (1990)
4. Andresen, B., van den Brink, K., Unesco Institute for Information Technologies in Education: Multimedia in Education: Curriculum. Unesco Institute for Information Technologies in Education (2013)
5. Machover, C., Tice, S.E.: Virtual reality. IEEE Comput. Graph. Appl. **14**(1), 15–16 (1994). https://doi.org/10.1109/38.250913
6. Psotka, J.: Immersive training systems: virtual reality and education and training. Instr. Sci. **23**(5), 405–431 (1995). https://doi.org/10.1007/BF00896880
7. Alqahtani, A.S., Daghestani, L.F., Ibrahim, L.F.: Environments and system types of virtual reality technology in stem: a survey. Int. J. Adv. Comput. Sci. Appl. **8**(6). https://doi.org/10.14569/IJACSA.2017.080610
8. Christou, C.: Virtual Reality in Education, Ch. 12, pp. 228–243. IGI Global (2010). https://doi.org/10.4018/978-1-60566-940-3.ch012
9. Youngblut, C.: Educational Uses of Virtual Reality Technology. Institute for Defense Dnalyses (1998)
10. Schwienhorst, K.: The state of VR: a meta-analysis of virtual reality tools in second language acquisition. Comput. Assist. Lang. Learn. **15**(3), 221–239 (2002). https://doi.org/10.1076/call.15.3.221.8186
11. Lin, T.-J., Lan, K.: Language learning in virtual reality environments: past, present, and future. Educ. Technol. Soc. **18**, 486–497 (2015)
12. Prensky, M.: Don't Bother Me Mom-I'm Learning!, 1st edn. Paragon House, New York (2006)
13. Albion, P.: Self-efficacy beliefs as an indicator of teachers' preparedness for teaching with technology. Creative Education
14. Melo, M., Vasconcelos-Raposo, J., Bessa, M.: Presence and cybersickness in immersive content: effects of content type, exposure time and gender. Comput. Graph. **71**, 159–165 (2018)
15. Lewis, J.R.: IBM computer usability satisfaction questionnaires: psychometric evaluation and instructions for use. Int. J. Hum. Comput. Interact. **7**(1), 57–78 (1995). https://doi.org/10.1080/10447319509526110
16. Johnson, W., Rickel, J.: Steve: an animated pedagogical agent for procedural training in virtual environments. SIGART Bull. **8**, 16–21 (1997)

Virtual Reality in Education: Learning a Foreign Language

Darque Pinto[1]([⊠]), Bruno Peixoto[1], Aliane Krassmann[2], Miguel Melo[3], Luciana Cabral[4,5], and Maximino Bessa[1,3]

[1] Universidade de Trás-os-Montes e Alto Douro, Vila Real, Portugal
Darquepinto@gmail.com
[2] Universidade Federal do Rio Grande do Sul, Porto Alegre, Brazil
[3] INESC TEC, Porto, Portugal
[4] CITCEM, Porto, Portugal
[5] Instituto Politécnico de Bragança, Bragança, Portugal

Abstract. There are still open questions about the effectiveness of Virtual Reality (VR) in Education when compared to conventional learning methods. This paper studies the feasibility of a VR-based learning tool and the possible differences in knowledge retention across a VR learning method and a conventional audio method, when it comes to learning a foreign language. Also, the students' sense of presence and satisfaction were studied. For such purpose, a user study was conducted and results revealed that while presence and satisfaction were higher in Virtual Reality, the knowledge retention score remains the same across both experimental conditions.

Keywords: Virtual reality · Language learning · Media in education · Simulations

1 Introduction

With the widespread of social networks, virtual communities, 3D virtual worlds, and human avatars, education has been challenged to develop new didactic approaches and strategic learning methods, integrating technology to enhance learning performance and motivation. The attention of learners is no longer attracted to the same resources that were used in the past, it is necessary to innovate to make students feel engaged in the learning process.

Among the different technologies available, Virtual Reality (VR) has been proven to be valuable for educational purposes [1,2]. VR can be described as a technology that provides a sensation of being immersed in a digital environment [3]. The main objective of VR is to make the user feel close to another reality, using the human's five senses. This technology integrates a diversity of devices that may be used to help create a realistic and multisensory experience (e.g., Head-Mounted Displays (HMD), motion tracking, etc.) [4]. This enables us to replicate or to create

© Springer Nature Switzerland AG 2019
Á. Rocha et al. (Eds.): WorldCIST'19 2019, AISC 932, pp. 589–597, 2019.
https://doi.org/10.1007/978-3-030-16187-3_57

different convincing realities and simulations, allowing to change the users' perception of the world and their capacity to store, share and transmit information, a fact that increased the interest of educational researchers.

As a person experiences the computer generated environment rather than the actual physical location, the computer world becomes the user's world, which is called the sense of presence [5]. For Slater and Wilbur [6], the fundamental idea is that individuals with a high sense of presence experience the virtual environment as the reality more involving than the surrounding physical world, to the point of considering it as a place visited. That is why attention and involvement (or engagement) are common responses from users associated with the sense of presence [7].

In this sense, North and North [2] and Makransky, Terkildsena and Mayer [8] emphasize that sense of presence in VR can contribute to learning. This is because, according to Hassell et al. [9], when experiencing high levels of presence concentration, which is a key factor for learning, will be focused on the activity that occurs in the virtual environment.

Felix [10] explains that "skills and knowledge are best acquired within realistic contexts[...]" suggesting that learning can be enhanced by the stimulus of interaction with environments and people [11]. That is, placing the student in the real context where the knowledge is applied, which is called situated learning [12]. In the case, of foreign language, it is common to observe that people decide to study abroad to fast acquire listening and speaking skills, while inserted in its real context of the language existence, where they will literally be surrounded by it. Thus, it is highlighted the benefits of this pedagogic approach for foreign language learning.

Emerging VR technology offers an authentic hyper-immersive experience at an affordable price and in a technically feasible format for large cohort teaching for the first time. It is therefore absolutely relevant and convenient to investigate how VR might impact the student learning [13]. Thus, we hypothesize that (1) is it possible to use VR-based learning tools to learn a foreign language, and (2) that such VR-based learning tools are more effective regarding the learning process. Thus, we propose such a virtual environment and conduct a user study to clarify the tangible results of VR for education, focusing on the knowledge retention of listening exercises of foreign language learning, in comparison with the Audio format.

The remainder of the paper is structured as follows: we start reviewing articles from the same area, followed by a description of the used methods and their results. We finally our study by adding some discussions comparing our results with the literature knowledge answering the researches' objective.

2   Related Work

The study by Ijaz, Bodgdanovych and Trescak [14] compared the outcomes of unsupervised learning from a text document, an educational video, and a 3D virtual world, all with the same content of historical facts from the ancient city of Uruk. From the experiment, a better performance was observed in the virtual world group over the text and video groups. Because of that, they reached the conclusion that studying in the virtual world would result in better academic performance, although the text group was the fastest mode.

Makransky, Terkildsen and Mayer [8] is one of the few studies that actually investigates immersive VR, comparing the learning outcomes with a desktop display version of the same simulation, in the context of Science. They found out that the students felt more presence with the VR but they actually learned less, compared with the low-immersive version. On their perspective, the lower results are justified by the excitement of using VR for the first time, as well as the overwhelming aspect of the more cognitive load inherent with the use of HMD.

Christopoulos, Conrad and Shukla [15] conducted a research in which they identified that the interactivity of virtual objects, combined with the interactions between the peers, helped the students feel a sense of presence in the virtual world and, consequently, experience the learning material more intensively; something that, according to the authors, turned out to be more effective than "just studying".

It is reasonable to assume that with increased realism the student's engagement with a virtual environment will be improved and learning outcomes will as well outdo as a consequence. However, insufficient evidence is available in the existing literature to confirm those links [13].

In this research, it is intended to endorse the efficiency of Virtual Reality aiming the learning of a foreign language, compared with a conventional use. Our goal is to analyze the sense of presence, satisfaction, and knowledge retention between VR and audio.

3   Methods

3.1   Participants and Design

The experiment was performed by 12 participants, ages between 12 and 15 years old (M = 13.67, SD = 0.745), attending classes of J5 level of English (being J0 the basic level and J8 the advanced), at a private English school in Portugal, therefore guaranteeing the sample consistency. Students were contacted and invited to participate in the experiment. Participation was entirely voluntary on the part of the students and there was no grading associated with the exercise. As volunteers were underage, written consent was provided by their legal guardians prior to the implementation of the study.

3.2   Variables

The independent variable was the learning mode (VR vs. Audio). The dependent variables were the sense of presence, satisfaction and knowledge retention, which were measured by a multiple-choice post-test questionnaire.

3.3   Materials

The materials used in the study included a simulation delivered in two different versions (VR and Audio), a knowledge retention objective test, and a sense of presence and a satisfaction questionnaire. The knowledge retention test was written in English, and the questionnaires (presence and satisfaction) were written in the user's own language: Portuguese.

The simulation contains a dialogue similar to the ones commonly used in the English listening exercises. It was created by using the Unity game engine and it consisted of two scenarios. The first scenario present a formal English dialogue (in an office), while the second one presents an informal conversation (in a pub). The scenarios were filled with props to characterize it, and Non-Player Characters (NPCs) to meet the dialog script. In both scenarios, the user would be placed at a table surrounded by the NPCs engaged in a narrative fitting each scenario. The formal scenario takes 02:36 min while the informal takes 01:29 min. For the two modes of instruction, the simulation was the same, just varying the ways of the stimulus: Audio and VR.

Two laptop computers were used for the experiment: an ASUS ROG G752 (Intel Core i7-7700HQ processor and NVIDIA GeForce 1070 graphics card) for the VR experience, and an HP OMEN 17-an (Intel Core i7-7700HQ processor and NVIDIA GeForce 1050 graphics card) for the audio experience.

Concerning the visual VR stimulus, the HTC Vive HMD was used featuring a 110° viewing angle and a resolution of 1080 × 1200 pixels per eye. The audio stimulus in VR was delivered with a noise cancellation headphones (Bose QuietComfort 25). The same headphones were used on the audio version. The VR stimulus was developed having into account the study by Melo et al. [16] to ensure that presence was not affected by the exposure time.

3.4   Instruments

Three instruments were used in this research, each one to evaluate on of the three aspects investigated: sense of presence, satisfaction and knowledge retention. To measure the sense of presence, the validated Portuguese version of Igroup Presence Questionnaire (IPQ) [17] was selected, because it gives insights values three dimensions: spatial presence, involvement, and experienced realism. It contains fourteen likert scale affirmations, where the users answered between one (totally disagree) and five (totally agree).

The After-Scenario Questionnaire (ASQ) [18] was used to evaluate the satisfaction of the participant with the system usability, namely the following: ease of task completion, time to complete and adequacy of support information. It is composed with six likert scale affirmations where the users answered between one (totally disagree) and seven (totally agree).

An eight questions multiple-choice test was created by the English teachers of the school, based on the story that is presented in both scenarios, thought of to evaluate students' knowledge retention after the experiment. Since the test has a full value of 100%, each answer has a value of 12.5%. The questions of this test are presented below.

Questions for the formal scenario:
1. *Why is Paul Jones not at the meeting? Because he went to:*
2. *What was the Director worried about?*
3. *When was the meeting held?*
4. *What was the main topic of the discussion at the meeting?*

Questions for the informal scenario:
5. *At the pub, how are they feeling at first?*
6. *Was Anna able to contact her colleague who was abroad?*
7. *Which of the countries listed below did Paul not go to?*
8. *Were the colleagues happy at the end of the meeting at the pub?*

The instruments were answered in a pen-and-paper based format, immediately after the student undertook the experiment, in order to ensure high response rates and the currency of responses.

3.5 Procedure

Participants were tested individually in a room, with a controlled environment, at the English school. A sociodemographic questionnaire was given before the beginning of the experience. To avoid confusing the results with a specific type of content, the two different modes of instruction (audio and VR) were tested for both the scenarios, in a random order. Each participant first tested one scenario in one mode and the other using the other mode of instruction, alternatively. In this way, after each scenario, the user had to answer a knowledge test, a presence, and a satisfaction questionnaire.

Prior to the experiment itself, participants in the VR condition entered in a training room for one minute, which was the same scenario of the simulation, but without the NPCs, just the scene. The aim was to allow the subjects, who could not be familiar with virtual environments, to become accustomed to maneuvering the HMD. Overall, each participation took around 20 min, including the instruments filling.

4 Results

Regarding the presence and satisfaction, VR condition achieved better results than audio. In Fig. 1 we have a graph that shows the results of the questionnaires and in Table 1 we have the same more detailed results.

In Fig. 2 it is represented the percentage of the knowledge retention test, separated by scenario, being 100% all the questions answered correctly and 0% all wrong answers. In none of the conditions it was reached the minimum to consider as a positive result (above 50%). Apart from this low outcome, some differences were observable between both scenarios. In the formal scene, virtual reality had 30% of correct answers and the audio got a better result with more than 30%. In the informal scene, we observed the opposite. As regards, the final results virtual reality had more than 40% and audio less than that value.

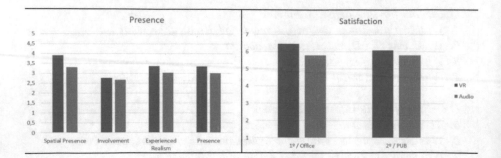

Fig. 1. IPQ and ASQ graph

Table 1. M and S.D Table

	Experiment			
	VR		Audio	
Presence Scale	**M**	**S.D.**	**M**	**S.D.**
Spatial Presence	3.907	.394	3.310	.412
Involvement	2.778	.845	2.679	1.083
Experienced Realism	3.361	.458	3.036	.891
Presence	3.349	.362	3.008	.579

	Experiment			
	VR		Audio	
Satisfaction Scale	**M**	**S.D.**	**M**	**S.D.**
Satisfaction	6.3	.48	5.8	.77

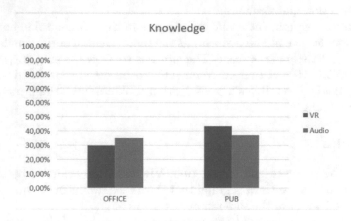

Fig. 2. Knowledge test graph

5   Discussion

After reviewing the knowledge retention test and the IPQ and ASQ questionnaires answered by the participants, some interesting results were collected.

In terms of the sense of presence, we could have expected that Virtual Reality would have better results than audio and in Fig. 1 and Table 1 we can see the difference between them happening on the values. Thus, our results corroborate the Makransky, Terkildsen, and Mayer [8] study where they found out that the presence is bigger in VR condition when comparing learning between VR and a desktop display, when testing the learning of science.

The same contrast happened with the measure of satisfaction. VR is a preference among participants. We theorize that this happens due to the fact that it was the very first time trying Virtual Reality for the most of them, causing therefore enthusiasm and worship for this kind of learning.

In Fig. 2 we can also see that the results of the knowledge retention test are lower than we could expect but they are very similar with the audio version.

Since VR was better than the conventional use in one of both scenario, and vice versa, the results about knowledge retention were inconclusive. Thus, we could not corroborate the Ijaz, Bodgdanovych, and Trescak [14] study, where they reached the conclusion that VR was more efficient for a better academic performance, comparing to a conventional method (text version, in their case). Regarding this difference between results, we speculate that this happens due to the same fact as the satisfaction results, it was the first time experimenting VR for most of the students, which can lead to some natural distraction. While some of the participants were enjoying the environment itself, they did not pay attention to the dialogue. This cannot justify the difference between scenarios and that is why the knowledge retention results are inconclusive.

Since, according to Christopoulos, Conrad, and Shukla [15] study, interactivity with virtual objects make the participants learn more effectively than just "studying", this kind of tool in VR could well be a solution for better results in the knowledge retention. This conclusion may well lead to another investigation including interactivity in the scenarios, comparing results with and without it. However, at this stage, we aimed comparing VR with "listening". The mentioned interaction is thought of to be a huge advantage, comparing to audio alone.

6   Conclusions

According to our investigation, we can conclude that the VR results of the knowledge retention test are very similar to the audio version, validating our hypothesize (1) where we state that it is possible to use VR-based learning tools to learn a foreign language.

However, also because the learning outcomes were very similar in both conditions, the results are inconclusive for the assumption (2) that VR-based learning tools are more effective regarding the learning process.

The investigation of the sense of presence and the satisfaction aspects increased the analysis, showing that students actually felt more present and satisfied in the VR condition, which points out to the potential benefits of integrating this technology in teaching, to improve the students' positive feelings, besides capturing and maintaining their attention, especially in disciplines considered more "boring". Slater [19] have demonstrated that users whose primary representation systems are visual are more likely to experience presence in a visual virtual environment than users whose primary representation systems are auditory or kinaesthetic. This claims for further investigation on whether students learning preferences (learning styles) could influence or not in the outcomes of the treatments provided in this research.

Within the limitations of the study, it is certainly the novelty factor of VR technology, which may have influenced the learning outcomes. Nevertheless, it is expected that VR will become more popular as the devices become cheaper, so it is necessary to study its impact on education to be prepared for this integration. As a new medium, VR offers something that other media used in education cannot, that is, an immersive experience that can place the student in a specific learning context, evoking similar reactions and emotions [2]. Thus, the potential of situated learning [12] for foreign language must be explored, which can be a good cost-effective approach.

In addition, the small sample size is clearly a limitation of the study, avoiding to extrapolate the results to other contexts. Also, the age group of the subjects may have distort the measure of the sense of presence, because as they were under age they probably do not have the experience of being in an office or in a pub (formal and informal scenarios, respectively), and the instrument had assertions as "the experience in the virtual environment seemed to me as real as my daily experiences" and "the virtual environment seemed to me as real as the world I know" [7]. Thus, future endeavours are planned to extend this study to more participants, in a context of adults and, if possible, with subjects familiar with VR technology to mitigate the novelty effects.

In conclusion, this research makes a contribution to understanding the potential of VR technology to support and enhance situated learning, elucidating the main advantages and drawbacks in the context of listening exercises for foreign language learning.

Acknowledgments. This work is financed by the ERDF – European Regional Development Fund through the Operational Programme for Competitiveness and Internationalisation - COMPETE 2020 Programme and by National Funds through the Portuguese funding agency, FCT - Fundação para a Ciência e a Tecnologia within project POCI-01-0145-FEDER-028618 entitled PERFECT - Perceptual Equivalence in virtual Reality For authEntiC Training.

We want to thank Marta Peixoto, Cláudia Peixoto and the others teachers from *the English Institute of Vila Real* for their support and collaboration in this study.

References

1. Mikropoulos, T.: The unique features of educational virtual environments. In: Proceedings E-society 2006, International Association for Development of the Information Society, vol. 1 (2006)
2. North, M., North, S.: The sense of presence exploration in virtual reality therapy. J. Univers. Comput. Sci. **24**, 72–84 (2018)

3. Góomez-García, M., Trujillo-Torres, J.M., Aznar-Díaz, I., Cáceres-Reche, M.P.: Augment reality and virtual reality for the improvement of spatial competences in physical education (2018)

4. Chesham, R.K., Malouff, J.M., Schutte, N.S.: Meta-analysis of the efficacy of virtual reality exposure therapy for social anxiety. Behav. Chang. **35**(3), 152–166 (2018). https://doi.org/10.1017/bec.2018.15

5. Witmer, B.G., Singer, M.J.: Measuring presence in virtual environments: a presence questionnaire. Presence Teleoperators Virtual Environ. **7**. https://doi.org/10.1162/105474698565686.

6. Slater, M., Wilbur, S.: A framework for immersive virtual environments (FIVE): speculations on the role of presence in virtual environments. Presence Teleoperators Virtual Environ. **6**, 603 (1997). https://doi.org/10.1162/pres.1997.6.6.603

7. Lessiter, J., Freeman, J., Keogh, E., Davidoff, J.: A cross-media presence questionnaire: the ITC-sense of presence inventory. Presence **10**, 282–297 (2001). https://doi.org/10.1162/105474601300343612

8. Makransky, G., Terkildsen, T.S., Mayer, R.E.: Adding immersive virtual reality to a science lab simulation causes more presence but less learning. Learn. Instr. https://doi.org/10.1016/j.learninstruc.2017.12.007

9. Hassell, K., Coutin, P., Nugegoda, D.: A novel approach to controlling dissolved oxygen levels in laboratory experiments. J. Exp. Mar. Biol. Ecol. **371**, 147–154 (2009). https://doi.org/10.1016/j.jembe.2009.01.013

10. Felix, U.: The web as a vehicle for constructivist approaches in language teaching. ReCALL **14**(1), 2–15 (2002). https://doi.org/10.1017/S0958344002000216

11. Yeh, Y.-L., Lan, Y.-J., Lin, Y.-T.R.: Gender-related differences in collaborative learning in a 3D virtual reality environment by elementary school students. J. Educ. Technol. Soc. **21**(4), 204–216 (2018)

12. Matusov, E., Bell, N., Rogoff, B.: Situated learning: legitimate peripheral participation. Am. Ethnol. **21**, 918–919 (1994). https://doi.org/10.1525/ae.1994.21.4.02a00340. jean lave, etienne wenger

13. Wang, R., Newton, S., Lowe, R.: Experiential learning styles in the age of a virtual surrogate. Int. J. Arch. Res. ArchNet-IJAR **9**, 93–110 (2015). https://doi.org/10.26687/archnet-ijar.v9i3.715

14. Ijaz, K., Bogdanovych, A., Trescak, T.: Virtual worlds vs books and videos in history education. Interact. Learn. Environ. **25**(7), 904–929 (2017). https://doi.org/10.1080/10494820.2016.1225099

15. Christopoulos, A., Conrad, M., Shukla, M.: Interaction with educational games in hybrid virtual worlds. J. Educ. Technol. Syst. **46**(4), 385–413 (2018). https://doi.org/10.1177/0047239518757986

16. Melo, M., Vasconcelos-Raposo, J., Bessa, M.: Presence and cybersickness in immersive content: effects of content type, exposure time and gender. Comput. Graph. **71**, 159–165 (2018). https://doi.org/10.1016/j.cag.2017.11.007

17. Schubert, T., Friedmann, F., Regenbrecht, H.: The experience of presence: factor analytic insights. Presence Teleoperators Virtual Environ. **10**(3), 266–281 (2001). https://doi.org/10.1162/105474601300343603

18. Lewis, J.R.: IBM computer usability satisfaction questionnaires: psychometric evaluation and instructions for use. Int. J. Hum. Comput. Interact. **7**(1), 57–78 (1995). https://doi.org/10.1080/10447319509526110

19. Slater, M., Steed, A., McCarthy, J., Maringelli, F.: The influence of body movement on subjective presence in virtual environments. Hum. Factors **40**, 469–477 (1998). https://doi.org/10.1518/001872098779591368

Pervasive Information Systems

Overpass Information System

Evaluation Model for Big Data Integration Tools

Ângela Alpoim, Tiago Guimarães, Filipe Portela$^{(\boxtimes)}$,
and Manuel Filipe Santos

Algoritmi Research Center, University of Minho, Guimarães, Portugal
{tsg,cfp,mfs}@dsi.uminho.pt

Abstract. Given the growing demand and need by enterprises for data and information to positively support the decision-making process, there is no doubt about the importance of selecting the correct and appropriate integration tool for the different types of business. For this reason, the essential objective of this study is to create a model that will serve as a basis to evaluate the different alternatives and solutions that exist in the market able to overcome the big data integration challenges. The evaluation process of data integration product begins with the definition and prioritisation of critical requirements and criteria. In this evaluation model, the characteristics evaluated are categorised into three main groups: ease of integration and implementation, quality of service and support, and costs. After identifying the essential criteria and characteristics, it is necessary to determine the weights that these criteria should have in the evaluation. Then, it needs to verify which solutions existing in the market best fit the needs of the business and can satisfy them more effectively. And lastly, compare those solutions adopting this framework. It is essential to carry out a weighted evaluation, based on well-defined criteria like ease of use, quality of technical support, data privacy and security. This process is fundamental to verify if the solution offers what the organization needs if it meets the business requirements and their integration needs.

Keywords: Big data fabric · Evaluation model · Big data integration tools

1 Introduction

Connectivity is one of the most remarkable features of today's society. The way that the Internet promotes the connection between people, businesses and even objects is one of the factors that originated the latest perspective of the term "intelligence".

In an increasingly competitive world, it is essential for organizations' success to "be intelligent". It can be translated into making quick and agile decisions and transforming so fundamental resource, that it is information, in knowledge capable of helping to take the best decisions [1]. The truth is that we are facing the era of big data. Data is generated, analysed and used on an unprecedented scale, and decision-making based on this data is being applied to all aspects of society [2]. Never have so many records been generated about what people do, think, feel or desire like today. Therefore, the daily interactions of people with pervasive systems, create traits that capture various aspects

© Springer Nature Switzerland AG 2019
Á. Rocha et al. (Eds.): WorldCIST'19 2019, AISC 932, pp. 601–610, 2019.
https://doi.org/10.1007/978-3-030-16187-3_58

of human behaviours and allow machine learning algorithms to extract valuable information about users and their actions as well as the need of evaluating and analysing data accordingly. For instance, a bank's transaction logs can be used to advert unusual activity in client's accounts [3].

Managing and analysing this massive amount of data, from the perspective of information sharing in social networks, through the analysis of banking transactions, online surveys, access to websites or even with the emergence of connected devices (such as smartphones or smartwatches) can be seen, simultaneously, as one of the most significant benefits and challenges of organizations. It is as important to obtain and generate information as to be able to process it quickly [4]. And this is one of the major challenges: to organise and model the data to facilitate the process of association, transformation, processing and analysis of the collected data in order to make the best decisions promptly [5]. This case requires the exploration of appropriate resources, new methods as well as detaining the appropriated technology [6]. The process of selecting the correct and appropriate integration tool for the different types of business is crucial, given the growing demand and need by companies for data and information to positively support the decision-making process of their business. In this sense, an evaluation model was developed, which is a guide that can be adapted taking into account the different requirements, needs and evaluation criteria of different users, in order to select the big data integration solution that best suits the needs of each business.

This document is structured in 5 sections. The first one advances a brief introduction in order to contextualise this study. Section two describes and characterises the concept of big data, big data fabric and presents three examples of big data fabric tools. In section three is presented examples of three areas of application of big data. Section four is represented by a framework to evaluate big data integration tools, describing the main criteria that can be used to make this evaluation. Section five describes the results of this study and lastly, section six, advances the main conclusions regarding this paper.

2   Background

2.1   Big Data

Innovation and technological development combined with greater accessibility to digital devices have led to the emergence of what is considered, by many, to be the Era of big data. As a result of this impact that the role of technologies is currently taking on people and their lives, there is an "explosion" in the quantity, diversity and availability of digital data in real time [7]. According to Gupta [8], big data can be defined as "high-volume, -velocity and -variety information assets that demand cost-effective, innovative forms of information processing for enhanced insight and decision making". Hashem [9] complements this definition referring big data as "a set of techniques and technologies that require new forms of integration to uncover large hidden values from large datasets that are diverse, complex, and of a massive scale".

The intelligent reading of this information and data is essential, because, according to some studies, the appropriate use of big data can play a very useful economic role for organizations, promoting innovation, competitiveness and productivity in all segments [10]. Organizations are hoping to gain benefits in many areas, such as e-commerce, e-government, health and safety. These benefits and values that organizations expect to create will also depend on the strategies adopted and the goals they intend to achieve [11].

2.2 Big Data Fabric

The concept "data fabric" has emerged as an approach to help organizations dealing better with the rapid growth of data. This term refers to the technology that creates a convergent platform that supports the storage, processing, analysis and management of the enormous diversity of data that exists today, such as text, images or sensor data [12]. According to Forrester's study [13] this concept can be defined as: "Bringing together disparate big data sources automatically, intelligently, and securely, and processing them in a big data platform technology, such as Hadoop and Apache Spark, to deliver a unified, trusted, and comprehensive view of customer and business data" [12]. Big data fabric helps companies in this procedure of process, transform, integrate and secure rapidly large amounts of data on big data platforms to support a strong view of the customer and the business [13, 14].

2.3 Examples of Big Data Fabric Tools

According to Forrester's studies [12, 13], most companies that have a big data fabric platform have been integrating several open source technologies such as Apache Flume, Spark, Hadoop, and have been supporting the platform with commercial products for data integration, security, governance, machine learning and data preparation technologies. However, organizations have realised that customising a big data fabric implementation in this way and still meeting business requirements requires significant time and effort. It is the reason why vendor solutions such as Talend, IBM and Informatica (described in the following points) are beginning to emerge, providing some or all the layers of the big data fabric architecture [14].

Talend. Talend Data Fabric is a complete solution that provides all integration needs (batch, streaming, real-time and cloud) all in one platform in a standard set of easy-to-use solutions. This tool was built on the data integration solution and allows users to access, transform, move and synchronise big data, taking advantage of the Apache Hadoop and making the platform very easy to use. With this solution, it is possible to work with large volumes of data at high speed to perform the necessary analyses, perform real-time integrations, share information promptly and to clean and organise the data to create a single view of the client [15].

IBM. The InfoSphere Information Server Enterprise Edition version of IBM is a data integration platform that includes a family of products that enable users to understand, monitor, clean, transform and provide data. It also collaborates to bridge the gap between business and Information Technology. InfoSphere Information Server provides capabilities to provide a highly scalable and flexible integration platform that handles all data volumes, from the smallest to the largest [16].

Informatica. Informatica Platform offers a range of data integration products as part of its Intelligent Data Platform. This platform performs the collection of any type of data (structured, semi-structured and unstructured), through any integration standard (either real-time or streaming, for example), from any source (database, data warehouses, big data, social networks), from any location (local data, cloud, hybrids). And it transforms this data into reliable, secure, accessible, timely and actionable intelligence [17].

3   Evaluation Model

Lima [18] presents a proposal for a "Big Data Complexity Framework", which aims to help find and identify a big data problem, according to the dimensions of the 3 V's of big data (volume, variety, velocity) and the corresponding level of complexity. It is essential to assess the organization's needs and understand whether these needs are met with big data resources and whether these organizations really need to make big data investments [18]. This initial approach can be adapted to detect whether a project involves big data and then use the model and framework proposed in this article to evaluate the big data integration tools available in the market. The practical application of these frameworks could simplify and support organizations in decision-making processes. The data integration product evaluation process begins with the definition and prioritization of key requirements and criteria. There will be a set of variables that will vary according to the different business contexts and the needs of the organizations, such as different types of data that will be needed to process, the source and destination systems involved, and the forms of integration that will be needed. For this reason, it is vital to carry out a weighted evaluation, based on well-defined criteria, mainly to understand if the solution offers what the organization needs if it meets the business requirements and their integration needs.

In order to carry out this study and the creation of this model, a set of criteria must be defined, and different weights must be assigned to them because it is considered that there are criteria with different weights and levels of importance. The characteristics evaluated are categorised into three main groups: ease of integration and implementation; quality of service and support; and costs.

The criteria defined were based on the literature review of studies such as Marakas & O'Brien [19], Lněnička [20] and Altalhi [21], in addition to Forrester's [13] and Gartner's [14] reports. Also the websites from G2 Crowd[1] and Gartner peer insights[2] contributed to the perception of the most important criteria when evaluating a big data fabric tool.

The evaluation of each of the parameters can be performed using a scale of 1 to 10. In which 1 represents the lowest possible score and 10 the maximum.

---

[1] https://www.g2crowd.com/.

[2] https://www.gartner.com/reviews/home.

Before assigning the weights to the criteria, two important factors must be considered:

- The AHP (Analytic Hierarchy Process) method. This method was developed by Thomas Saaty in 1980 and there are a diversity of articles that explore and explain the AHP method, as is the case of Saaty [22] and Lněnička [20];
- And the sensitivity analysis, in order to support the decision-making process. After scoring all the criteria, the percentages relative to the weights are applied, and the total evaluation is presented at the end.

Table 1. Evaluation model for big data integration solutions

Metrics	Features	Description	Weight	Alternatives		
				1	2	3
Ease of integration and implementation	Connecting to data sources and destination support	Ability to interact with a variety of different types of data structures, including relational and non-relational databases, XML, different data types and multiple file formats				
	Data security and privacy	Whether the solution can overcome the challenges of privacy and data security effectively				
	Simple and complex transformations	Integrated capabilities for achieving data transformation operations, including fundamental transformations (such as data type conversions, string manipulations and simple calculations) and complex transformations (such as sophisticated large-scale analysis operations)				
	Ease of implementation and integration	A width of support for hardware and operating systems on which data integration processes can be implemented. Also, it is essential to have a set of features in this type of solution to facilitate the integration process, such as diversity of pre-built connectors (time-saving) and to guarantee portability of the solution				
	Scalability and adaptability	Whether the solution can quickly expand to meet business needs				
Quality of service and support	Usability	The ease of use of the tool, associated with the fact that it is intuitive, easy to handle and easy to learn. This perspective will vary according to the skills of the professionals involved				

(*continued*)

Table 1. (*continued*)

Metrics	Features	Description	Weight	Alternatives		
				1	2	3
	Quality of technical support and documentation available	The existence of efficient and timely technical support with high availability as well as adequate and quality documentation that responds promptly and effectively to the technical obstacles that may arise to users during the exploration of the tool. A wide range of options regarding customer support programs (such as forums) is also considered a key factor				
Costs	Free trial	If the solution presents a free trial, in order to understand if it meets the needs of the business				
	Professionals with the right skills	The existence in the organization of experts in data integration or have enough budget to hire professionals who have experience and knowledge in handling the chosen integration tool				
	Price flexibility	The licensing and pricing methods are easy to understand, and the costs are attractive.				
	Return on investment	If the solution has a significant impact on the business (high profits) about the investments that were made				
Evaluation			**100%**			

This process will be performed for all the alternatives that the user pretends to compare.

This model can be seen as a framework that users can follow to evaluate their several options when intending to opt for a big data fabric solution. Firstly, it is important to define the weight that each criterion must have and then evaluate how each of the alternatives meet each of the requirements. It is also essential and recommended to conduct a sensitivity analysis, since it helps and supports the decision-making process considerably.

4   Results

This study attends to provide a holistic view of those that are considered the main requirements that must be taken into account when choosing one of the many solutions available in the market. Some examples of these solutions have also been provided in chapter 3 of this article. The model produced in this study can be seen as a framework that users can adapt taking into account the different requirements and needs of the several areas and business. With this framework, it is intended to pass the idea that before assessing and selecting a data integration solution, it is essential to evaluate what are the primary and "mandatory" features and functionalities to the business. After identifying the essential criteria and characteristics, it is necessary to carry out a weighting, to verify which solutions in the market better suit the needs of the business and that can more effectively satisfy them. It is important to realise that different businesses represent different needs. However, the criteria that were described in this model was broad and could, in fact, be adapted to any industry. Organizations must indeed understand which are the "must have", the different use cases of the business and then adapt this model accordingly based on the strengths and weaknesses of the existing data fabric solutions in the market.

The three solutions described in chapter 3 were evaluated based on this framework A sensitivity analysis was led, in order to understand how the ideal solution is sensitive for any changes in the input values of one or more parameters. The three solutions were ranked by a decision support software (Decision Pad[3]) (Fig. 1), showing that Talend had the highest score (7,8), followed by IBM (6,2) and lastly Informatica (5,4). The weight sensitivity analysis gives an idea of how the rankings respond to changes in weights, a useful way of seeing which aspects to take more into account and that are important. As it is possible to see in Fig. 2, the vertical line "At" represents the current weight, and the lines intersect when a change in weight of the criterion occurs causing a change of classification.

It is possible to conclude that, in this case, the assigned weights are evenly distributed, since the line "At" is distanced from the crossing of the three lines.

Matrix	Notes	Scores	Weights	Weight Sensitivity				
				Rank in 3	1	2	3	
				(Previous Rank)	1	2	3	
				Score	7,8	6,2	5,4	
				(Previous Scores)	7,8	6,2	5,4	
				Alternatives	Talend	IBM	Informatica	
Criteria				Weights				
Ease of Integration and Implementation				60%	Very Good	Good	Good	
Usability				19%	Very Good	Good	Fair	
Quality of technical support and documentation				9%	Very Good	Very Good	Good	
Return on investement and flexibility of prices				12%	Good	Good	Fair	

Fig. 1. Evaluation matrix

---

[3] http://www.apian.com/downloads/proceed-dp.php.

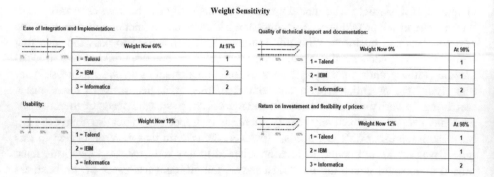

Fig. 2. Weight sensitivity analysis

5 Conclusion and Future Work

Having large amounts of data may have no value if there is no way of visualising, interacting and getting the right insights into it. An intelligent data integration strategy can effectively help companies leverage information from a variety of sources and use them to support business goals and objectives. After all, that's the big goal when implementing a big data fabric solution.

However, choosing the right big data integration solution for a business is not that simple. Organizations should not seek and purchase data integration software that has the highest level of features and functionality they can find on the market. In fact, they should acquire the data integration tool that best fit their needs.

The "must have" list of each company differs based on different requirements and different needs. However, in general, there are a set of essential data integration features for most organizations that can be pointed in three big categories: ease of integration and implementation, quality of service and support and costs. These are described and represented as a framework in Table 1.

It is essential to address a set of evaluation criteria that best reflect organizational needs by selecting and adopting a big data fabric platform, allowing organizations to assess, choose, and take the most appropriate solution that ensures that all stakeholders' perceptions are considered in the evaluation process. And that is the main contribution of this study: a framework that is a guide that can be adapted taking into account the different requirements, needs and evaluation criteria of different users to select the big data integration solution that best suits the needs of each business.

Although this model can be broadly adopted across all sectors and areas of application, it would be interesting, as future work, to apply it to specific areas such as medicine or the financial area, for example. For this reason, as future work, it is intended to demonstrate the practical applicability of the framework presented in this study, highlighting some areas of the success of the application of big data fabric. For that, it is essential to adjust the criteria according to the needs and contexts of the business and use cases.

Acknowledges. This work has been supported by FCT – Fundação para a Ciência e Tecnologia within the Project Scope: UID/CEC/00319/2019.

References

1. Zikopoulos, P., Eaton, C.: Understanding Big Data: Analytics for Enterprise Class Hadoop and Streaming Data. McGraw-Hill Osborne Media, New York (2011)
2. Srivastava, D.: Big data integration. In: Proceedings of the 19th International Conference on Management of Data, p. 3. Computer Society of India, December 2013
3. Baron, B., Musolesi, M.: Interpretable Machine Learning for Privacy-Preserving IoT and Pervasive Systems. arXiv preprint arXiv:1710.08464 (2017)
4. Volpato, T., Rufino, R.R., Dias, J.W.: Big Data – Transformando Dados em Decisões. University of Paranaense, Paranavaí (2014)
5. Cassavia, N., Dicosta, P., Masciari, E., Saccà, D.: Data preparation for tourist data big data warehousing. In: Proceedings of 3rd International Conference on Data Management Technologies and Applications, pp. 419–426. SCITEPRESS-Science and Technology Publications, Lda, August 2014
6. Oussous, A., Benjelloun, F.Z., Lahcen, A.A., Belfkih, S.: Big data technologies: a survey. J. King Saud Univ. Comput. Inf. Sci. **30**(4), 431–448 (2017). https://www.sciencedirect.com/science/article/pii/S1319157817300034
7. Pulse, U.G.: Big data for development: Challenges & opportunities. Naciones Unidas, Nueva York, mayo 2012
8. Gupta, S., Chaudari, M.S.: Big data issues and challenges. Int. J. Recent. Innov. Trends Comput. Commun. **3**(2), 062–066 (2015)
9. Hashem, I.A.T., Yaqoob, I., Anuar, N.B., Mokhtar, S., Gani, A., Khan, S.U.: The rise of "big data" on cloud computing: review and open research issues. Inf. Syst. **47**, 98–115 (2015)
10. Lima, C.A.R., Calazans, J.D.H.C.: Pegadas Digitais: "Big Data" E Informação Estratégica Sobre O Consumidor. NT – Sociabilidade, novas tecnologias, consumo e estratégias de mercado do SIMSOCIAL, 2013 (2013)
11. Günther, W.A., Mehrizi, M.H.R., Huysman, M., Feldberg, F.: Debating big data: a literature review on realizing value from big data. J. Strateg. Inf. Syst. **26**, 191–209 (2017)
12. Izzi, M., Warrier, S., Leganza, G., Yuhanna, N.: Big Data Fabric Drives Innovation and Growth. Next-Generation Big Data Management Enables Self-Service and Agility (2016)
13. Hoberman, E., Leganza, G., Yuhanna, N.: The Forrester Wave™: Big Data Fabric, Q2 2018. Tools and Technology: The Data Management Playbook (2018)
14. Beyer, M., Thoo, E., Zaidi, E.: Gartner Magic Quadrant for Data Integration Tools (2018)
15. Talend Data Fabric. A single, unified platform for modern data integration and management. https://www.talend.com/products/data-fabric/. Accessed 21 Oct 2018
16. IBM InfoSphere Information Server. Flexibly meet your unique requirements — from data integration to data quality and data governance. https://www.ibm.com/analytics/information-server. Accessed 1 Nov 2018
17. Informatica Intelligent Data Platform - Powered by CLAIRE. www.informatica.com/nl/products/informatica-platform.html. Accessed 2 Nov 2018
18. Lima, L.: Big data for data analysis in financial industry (Master dissertation). University of Minho, Guimarães, Portugal (2014)
19. Marakas, G.M., O'Brien, J.A.: Introduction to Information Systems. McGraw-Hill/Irwin, New York (2013)

20. Lněnička, M.: AHP model for the big data analytics platform selection. Acta Inform. Prag. **4**(2), 108–121 (2015)
21. Altalhi, A.H., Luna, J.M., Vallejo, M.A., Ventura, S.: Evaluation and comparison of open source software suites for data mining and knowledge discovery. Wiley Interdiscip. Rev. Data Min. Knowl. Discov. **7**(3), e1204 (2017)
22. Saaty, T.L.: Decision making with the analytic hierarchy process. Int. J. Serv. Sci. **1**(1), 83–98 (2008)

Adaptive Business Intelligence in Healthcare - A Platform for Optimising Surgeries

José Ferreira[1], Filipe Portela[1], José Machado[2],
and Manuel Filipe Santos[1(✉)]

[1] Department of Information Systems, University of Minho,
Guimarães, Portugal
mfs@dsi.uminho.pt
[2] Department of Informatics, University of Minho, Braga, Portugal

Abstract. Adaptive Business Intelligence (ABI) combines predictive with prospective analytics in order to give support to the decision making process. Surgery scheduling in hospital operating rooms is a high complex task due to huge volume of surgeries and the variety of combinations and constraints. This type of activity is critical and is often associated to constant delays and significant rescheduling. The main task of this work is to provide an ABI based platform capable of estimating the time of the surgeries and then optimising the scheduling (minimizing the waste of resources). Combining operational data with analytical tools this platform is able to present complex and competitive information to streamline surgery scheduling. A case study was explored using data from a portuguese hospital. The best achieved relative absolute error attained was 6.22%. The paper also shows that the approach can be used in more general applications.

Keywords: Decision support systems · Adaptive Business Intelligence

1 Introduction

The idea that moves this work is the absence of Adaptive Business Intelligence (ABI) approaches in hospitals and other health institutions to improve the quality of service through efficient scheduling of surgeries in the decision process. Nowadays, decision support systems integrate prediction and optimization capabilities making it possible to minimise or maximise a specific goal. Complex processes to be solved by humans, can be addressed by computer based decision support systems, in a more straightforwardly way, allowing for a better efficiency and effectiveness of the decisions, saving human and financial resources. ABI systems are decision support systems, capable of predicting, optimising and adapting to external changes.

Work has been conducted in order to extend the capabilities of an already existing hospital business intelligence platform with ABI features. Real data collected from the hospital were used as a proof of concept of the platform developed. The hospital will benefit from an improvement in the quality of the information they have. The platform

Á. Rocha et al. (Eds.): WorldCIST'19 2019, AISC 932, pp. 611–620, 2019.
https://doi.org/10.1007/978-3-030-16187-3_59

can help predicting the surgery time and optimisating surgeries scheduling along the shifts vailable. This solution allows the minimisation of time wastes of the total duration of the necessary shifts and avoid the delays that are generated by the lack of use of shifts. This paper is divided into six sections, beginning with an introductory section and finishing with a section dedicated to conclusions. The related concepts and existing ABI solutions are presented in second and third sections respectively. The case study and their results are detailed in the remaining sections.

2 Background

2.1 Adaptive Business Intelligence and Data Mining

Adaptive Business Intelligence is the combination of a business intelligence system, data mining, prediction methods and optimisation techniques, that is, it combines prediction, optimisation and adaptability into a system capable of answering two fundamental questions [1]: What is likely to happen in the future? And what is the best decision right now? This system is used to solve many business problems in the real world, ranging from demand prediction, scheduling, fraud detection and investment strategies to significant benefits and savings [2].

Adaptive Business Intelligence brings together the techniques and tools of the database, data warehouse, data mining, prediction, optimisation and adaptability to increase versatility. Furthermore, it allows business managers to make better decisions, thus improving efficiency, productivity and competitiveness. The Adaptive Business Intelligence System is divided into three components [2]: 1. Prediction (e.g. projections of the standard time of the event); 2. Decision making - almost perfect (e.g. scheduling of surgeries) and 3. Adapting predictions and optimisation of external changes. The optimisation is a technique of searching for better parameter values. The purpose of this technique is to find the parameter values that minimise the prediction error, based on the prediction model data. This technique maximizes the number and duration of surgeries allocated to the available shifts based on the predicted duration of the surgeries [2].

Data mining is a science of exploring large amounts of data, which aims to find consistent standards. With Data Mining it is possible to extract implicit and unknown information before exploitation, making it useful to solve problems of various applications in domains like business, health, science and engineering [4, 7]. Daryl Pregibons described the concept of data mining as a "blend of statistics, artificial intelligence, and database research" [3], is still currently the affirmation. The Exploratory Data Analysis (EDA) techniques used in data mining are divided into two parts, which are [4–7]: Computational Methods (e.g. statistical, classification, regression, others) and Data Visualisation.

2.2 Modern Optimisation

Modern optimisation is the name given to the methods known as meta-heuristics, and, as mentioned above, they are applied to minimise or maximise a solution to obtain a satisfactory result that solves a problem. The problems that modern optimisation proposes to address are complex and do not have any specialized optimisation algorithm. They include problems with discontinuities, dynamic changes, multiple objectives or hard and soft constraints that are difficult to manipulate [8] is because the hard constraint cannot be violated because of factors involving laws and physical, and soft constraint can only be adjusted by multi-objective optimisation [8], so that optimisation is always useful in several business areas. During the implementation of a modern optimisation method the user should ever consider the following aspects [2]: representation of the solution; objective function; evaluation function.

3   Scheduling Optimization Platform

3.1   Method and Tools

The platform was built using Weka tool for the prediction block and RStudio tool for the optimisation block. The Weka tool is an open source software issued under the GNU General Public License, which aims to add algorithms for machine learning for data mining tasks, which brings together various techniques, particularly for this case is the regression being through this technique that is done to predict the duration (default time) of the event. RStudio is a more user-friendly tool for the R language, which allows the use of modern optimisation methods as a solution to the problem, with the purpose of local search optimisation being hill climbing and simulated annealing.

3.2   ABI Architecture

This platform was developed based on the Adaptive Business Intelligence architecture. This platform consists of two blocks, namely a prediction block, which predicts the duration of each surgery or more generally an event, and an optimisation block, which aims to minimise waste of shifts, preventing delays and increasing efficiency and effectiveness of schedule. Figure 1 shows the full flow of the scheduling optimisation platform wherein the next Subsects. 3.3 and 3.4 are explained in detail.

3.3   Predicting Block

The first step is to predict the standard time of each event (surgery), through the training of models in data mining, in the Weka tool. In the ABI platform prediction block are used algorithms for machine learning regression, predictive modelling in Weka tool. The techniques used are as follows: Linear Regression (LinearRegression); k-Nearest Neighbours (IBk); Decision Tree (REPTree); Support Vector Regression (SMOreg); Multi-Layer Perceptron (MultilayerPerceptron).

Before training these models, it is necessary to identify which variables are more correlated to the target variable. The target corresponds to the event that is intended to be predicted, in this case the duration of the surgeries. Feature selection is obtained through the algorithm "CorrelationAttributeEval", which makes a ranking of the occurrences of those, and then used the 10 to 15 most relevant characteristics. After this phase, the estimators are trained, to determine which one of them has a lower error rate.

After choosing the model with the lowest error, Weka tool was used to fill the surgery duration for the data set that will be used for optimization.

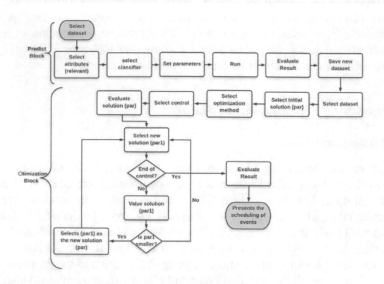

Fig. 1. Platform architecture flow

3.4   Optimisation Block

The result of the prediction block is a new dataset containing events with the duration time predicted in the previous block. The methods used in this block are hill climbing and simulated annealing, which require the use of a script file containing the functions called "hclimbing" and "optim" to obtain the best result, which in this case is the less waste between the two methods. The functions of the algorithms and the input parameters for hill climbing method and the simulated annealing method were set as [8]: hclimbing (par, fn, change, lower, upper, control, type) and optim (par, fn, method, gr, control). The initial solution (par) was obtained randomly or sequentially shifts through the list of surgeries of specialities matching the shift, assigning each dimension index a turn. The lower and upper parameters represent the lowest and highest values for each dimension, varying from 1 to the number of shifts of a dimension. The evaluation function calculates the wastes of each shift by subtracting the total standard time from the assigned shift and returns the total sum of shift wastage. It is also in this part of the function that the constraints are defined and the unwanted paths are penalized by assigning a high value to the waste so that this solution is not chosen. The constraints found for the proper functioning of the optimisation were: Date of the event <Date of shift and Event deadline> Date of shift and Waste >= 0.

The method is consists of a finite number of iterations through the control function, which contains a list of braking and controls the maximum number of iterations and the frequency control information. The simulated annealing method has only the difference on the trace parameter (set to true), which displays the trace information about the optimisation progress only if they are positive. The iteration limit will be the number that will go through the optimisation and the control and allows for the presentation of a report containing the solution of each one of the iterations, where each following solution is changed through the change function.

The change function aims to generate the next solution, which will undergo minor changes to the previously generated solution, and so on. This function has the following input parameters [8]: hchange (par, lower, upper, dist, and round).

To obtain the optimum value in the hill climbing method is used the minimization option since the goal is to minimize the wastage of the shifts of each surgical speciality. In order to understand the best model to be used as an optimisation method, it will be necessary to compare the two solutions obtained, choosing what for each surgical speciality presents the smallest waste, avoiding delays in the surgeries and, consequently, expenses in unnecessary shifts.

4 Case Study

A case study will be presented making use of CRISP-DM methodology [9] and a dataset of a Portuguese health institution.

4.1 Business Understanding

The first phase of CRISP-DM is a kind of problem statment. Previous works in this area [9–11] only addressed the time duration prediction. Did not address the optimisation problem and did not solve the wasting of time of the shifts. This can result in a great inefficiency. In this work, the immediate goal is to minimise the waste of each shift, for each one of the surgical specialities, using only the necessary shifts. The estimation of the value for the duration of the surgical specialities is used in order to proceed with the optimisation based on this standard time. For this, regression machine learning algorithms were used to predict the standard time of each surgery, and the satisfactory solution will be achieved using a modern optimisation method. This method of optimisation finds the solution through the neighbouring settlements, that is, it goes through several iterations, calculating the waste of each shift until finding the smallest value for this problem.

This way, it is avoided that future surgeries are delayed due to the unavailability of operating rooms and shifts, since this platform, through the study of the various possible combinations, will select the most satisfactory solution, in the sense of maximising efficiency and effectiveness.

4.2 Data Understanding and Preparation

Due to data set limitations, the tests of each block were performed with different datasets. A dataset was used to predict the duration time of the surgeries and a different dataset was used for the optimisation. This division happens because the table of surgeries available to being used in the optimisation block does not have sufficient surgeries for the forecasting process. Nevertheless, and despite the data used for the tests being different, the cohesion of the platform is not impaired, since the end user has enough data for the forecast result to be used as the standard time for the surgeries in the optimisation. Table 1 presents some statistics of the dataset.

Table 1. Sample of the statistical analysis of some specialities (time in seconds)

Speciality	Number of shifts	Quantity of surgeries	Standard time shifts	Standard time surgeries	Total waste
OPHTHALMOLOGY	1078	3162	34905600	5307459	29598141
PEDIATRIC OPHTHALMOLOGY	23	30	909900	116817,8	793082,2
VASCULAR SURGERY	569	796	19389000	3456938	15932062
ORTHOPEDICS	875	1440	26650800	9509328	17141472
PEDIATRIC ORTHOPEDICS	94	163	3187800	848623,6	2339176,4

4.3   Modelling

Regression models (REPTree, LinearRegression, IBk, MultilayerPerceptron and SMOreg) were induced using the Weka tool to predict the operating time in the operating room. At this stage, for testing the prediction block, was used data from the surgical specialities that included the most significant number of surgeries in the entire dataset, namely the orthopaedic speciality. Tables 2 and 3 identify the variables that appear to have a more descriptive power by the occurrence of only a few values, selecting the 15 attributes most relevant for the predicting task.

Table 2.   Attributes with more descriptive power

Attributes	Punctuation
TEMPO6	0.969341
TEMPO7	0.942850
TEMPO2	0.677621
TEMPO8	0.376831
SERVICE	0.345452
GDH	0.299350
MCONBASE	0.296299
COD_ANESTESIA	0.233695
PROCS1	0.233320
INTERVENCAO1	0.232584
COD_INTERV_CIRURGICA	0.230257
COD_SALA	0.144060
PROCS8	0.117009
COD_PATOLOGIA	0.092911

Table 3. Summary of the regression with 15 attributes

REPTree	IBk	SMOreg	LinearRegression	MultilayerPerceptron
Correlation coefficient	0.9958	0.813	0.9972	0.9955
Mean absolute error	269.7897	2107.5949	214.699	313.8986
Root mean squared error	405.5077	2850.6335	332.11	425.2081
Relative absolute error	7.8189%	61.0813%	6.2223%	9.0973%
Root relative squared error	9.2617%	65.1079%	7.5853%	9.7117%
Total number of instances	237	237	237	237

For the optimisation task, modern optimisation methods were implemented, as mentioned above, to obtain the minimum value of the total wastes of the shifts:

- hclimbing (par, fn, change, lower, upper, control, type);
- optim (par fn, method, gr, control).

These methods were used for 32 surgical specialities, allowing annual and monthly optimisation, to find the best solution. The first step is to obtain the initial solution, given randomly or sequentially, through the necessary shifts and the size of the surgeries. Next, the evaluation function is applied, which aims to go through the constraints and then add up the total wastes of each shift. Each waste value is obtained by subtracting the standard time from the surgeries, considering a turnover time of 17 min (except for the last surgery), at the time of the assigned shift, and at the end, the total waste is returned to the solution obtained. For each one of the iterations a complete set of waste values is obtained to be compared among alternative solutions.

The iterations end in the limit assigned in the optimisation method, which will return the best value of the solutions obtained. In the end, the satisfactory solution is chosen from the two optimisation methods tested.

4.4 Evaluation

The evaluation phase intends to promote a benchmarking among the obtained models. SMOreg and REPTree models, generated with the top 15 attributes, presented the lowest relative absolute error, i.e. 6.22% and 7.82% of the occurrences, respectively. The relative root quadratic error percentages are 7.59% and 9.26%, respectively.

Initially, it becomes evident that a large amount of data made it difficult to reach the business objective, since the models obtained were complex and large, despite high accuracy values. In this way, two iterations were performed in which only the TOP 10 and TOP 15 of the most relevant attributes for the analysis of the Orthopedics speciality were tested, with satisfactory results. In these two methods, only the characteristics that were documented for the assignment of TEMPO5 (room time) were checked, and no significant changes were obtained. Finally, it was possible to get a model with the lowest percentage of error, enabling the prediction of the duration of each surgery.

In this phase, results of the optimisation methods used to solve the problem were also evaluated. The steps performed in the optimisation block were satisfactory, since there was a significant reduction of the shifts, allowing the placement of all the surgeries without the occurrence of problems, from the preparation of the data until the modelling, being able to minimise the waste of the shift effectively and efficiently. The platform is validated by comparing hill climbing and simulated annealing methods, always choosing the best solution between the two solutions found, namely the solution that presents the least total waste of the shifts for each speciality.

5   Results

5.1   Analysis of the Results Obtained in the Prediction Block

Prediction Block includes the trained models as described above. Analysing the error percentages of each model pemits to conclude that the algorithm of Support Vector Regression (SMOreg) is more efficient (because it is the one that presents the lowest percentage about 6% relative absolute error). Table 4 presents an example containing ten surgeries of the orthopaedic speciality, the expected duration time, predicted duration and error.

Table 4. Prediction sample of the duration of orthopaedic surgeries (time in seconds)

Surgery	Actual	Predicted	Error
1	8100	8049.027	−50.973
2	7020	7127.73	107.73
3	15300	15304.194	4.194
4	5100	5165.485	65.485
5	9000	9026.354	26.354
6	4200	4267.64	67.64
7	9300	9331.48	31.48
8	10560	10761.003	201.003
9	6960	7042.839	82.839
10	7860	7721.515	−138.485

5.2   Analysis of the Results Obtained in the Optimisation Block

For the dataset of surgeries and shifts, only the results of two specialities, namely the one with the smallest and the one with the highest number of surgeries, will be discussed, because it is enough to understand the minimisation of the wastes. The speciality with higher duration times of surgery is vascular surgery, which contains 796 surgeries (3456938 s) to be distributed for a total of 19389000 s, corresponding to 569 possible shifts. Two types of optimisation have been applied to solve this problem: global optimisation for one year; and distributed optimisation for each month of a year. In this method, the optimisation was performed for each of the 12 months that have pendent surgeries, using as a dimension the capacity of surgeries of a respective speciality for the respective month. For orthopaedic a satisfactory result has been attained

Table 5. Monthly sequential optimisation of waste for vascular surgery

	Total waste
January	127089.205125
February	134182.871451
March	171413.244737
April	95961.169928
June	5401.727598
July	23751.340097
August	9194.226731
September	9194.226731
October	29114.226731
Total	605302.200000

using of the initial solution generated sequentially, yielding a total loss smaller. The results for each month are as in Table 5.

For this speciality, it can be concluded that the goal was achieved through monthly optimisation. It was possible to distribute all surgeries by the available shifts, even if they were of a large number and having reduced their shifts by more than half, which means that is possible to introduce more surgeries even though within these restrictions. After analysing the results, the method of optimisation process would be the most profitable for the scheduling of surgeries in the orthopaedic speciality. Nevertheless, for a smaller number of surgeries the optimisation method that offers a better result is the annual optimisation method, precisely because of the ability to make greater management of all possible combinations along a year (which could never happen with large datasets). Despite of these results, it is strongly recommended to use monthly optimisation method, precisely because it is the one that would best fit any dataset.

6   Conclusions

The use of an ABI approach combining predictive capacities with modern optimisation methods reduces the unforeseen delays in the events of a given organization or institution. To increase the effectiveness of the service, an ABI platform was created using the R environment to facilitate the surgery scheduling task. A dataset of a hospital containing data on surgeries and shifts was used to test the platform. The results are promising and make this approach an efficient and effective ABI event scheduling platform, adaptable to any organisation or institution that needs to schedule large lists of events. It can result in a reduction in unforeseen delays and an increase in the effectiveness of that service, by minimising the waste of shifts. However, and considering the temporal limits and the lack of attributes of the dataset it was not able to address some relevant points that were left for future work, namely: the prioritisation of the events; the implementation of a graphical interface; the Implementation of a system of analysis of the results through tables, to facilitate the visualisation of the distribution of the events. The approach should be tested considering more complex and abundant data.

Acknowledgments. This work has been supported by FCT – Fundação para a Ciência e Tecnologia within the Project Scope: UID/CEC/00319/2019.

References

1. Michalewicz, Z., Schmidt, M., Michalewicz, M., Chiriac, C.: Adaptive business intelligence: three case studies. Stud. Comput. Intell. **51**, 179–196 (2007)
2. Michalewicz, Z., Schmidt, M., Michalewicz, M., Chiriac, C.: Adaptive Business Intelligence (2006). https://doi.org/10.1007/978-3-540-32929-9
3. Pregibon, D.: Data mining. Stat. Comput. Graph. Newsl. **7**, 8 (1996)
4. Gorunescu, F.: Data Mining: Concepts, Models and Techniques. Intelligent Systems Reference Library, vol. 12 (2011) https://doi.org/10.1007/978-3-642-19721-5
5. Hastie, T., Tibshirani, R., Friedman, J.: The Elements of Statistical Learning. Elements, 1, pp. 337–387 (2009). https://doi.org/10.1007/b94608
6. Jiawei, H., Kamber, M., Han, J., Kamber, M., Pei, J.: Data Mining: Concepts and Techniques. Morgan Kaufmann, San Francisco (2012)
7. Witten, I.H., Frank, E., Hall, M.A.: Data Mining Practical Machine Learning Tools and Techniques (2005). ISBN: 9780120884070
8. Cortez, P.: Modern Optimisation with R. Springer, Guimarães (2014)
9. Azevedo, A., Santos, M.F.: KDD, SEMMA and CRISP-DM: a parallel overview. In: Proceedings of the IADIS European Conference on Data Mining 2008, pp. 182–185 (2008)
10. Peixoto, R., Ribeiro, L., Portela, F., Filipe Santos, M., Rua, F.: Predicting resurgery in intensive care - a data mining approach. Procedia Comput. Sci. **113**, 577–584 (2017). https://doi.org/10.1016/j.procs.2017.08.291
11. Coelho, D., Miranda, J., Portela, F., Machado, J., Santos, M.F., Abelha, A.: Towards of a business intelligence platform to portuguese misericórdias. Procedia Comput. Sci. **100**, 762–767 (2016). https://doi.org/10.1016/j.procs.2016.09.222

Mobile CrowdSensing Privacy

Teresa Guarda[1,2,3(✉)], Maria Fernanda Augusto[1,4],
and Isabel Lopes[3,5]

[1] Universidad Estatal Península de Santa Elena – UPSE, La Libertad, Ecuador
tguarda@gmail.com, mfg.augusto@gmail.com
[2] Universidad de las Fuerzas Armadas-ESPE, Sangolqui, Quito, Ecuador
[3] Algoritmi Centre, Minho University, Guimarães, Portugal
isalopes@ipb.pt
[4] BITrum-Research Group, C/ San Lorenzo 2, 24007 León, Spain
[5] UNIAG (Applied Management Research Unit),
Polytechnic Institute of Bragança, Bragança, Portugal

Abstract. Information and communication technologies have been evolving rapidly, providing more and more alternative ways of accessing information, accessible through multiple paths and devices, transforming IoT into a giant digital system. This heterogeneous diversity ecosystem allows travel to the CrowdSensing era. Information in its role of reducing uncertainty is responsible for finding ways and shortcuts that increase efficiency and reduce the waste of time in the decision-making process. The massive spread of IoT devices has led to CrowdSensing, a sensor-based ecosystem of many different formats and technologies, in which information created by sensors is essential for decision-making processes and for improving business process efficiency. Service providers are beginning to design new business models that consider the smart things scenario, providing sensing as a service (SaaS). This article aims to provide insight into the Mobile CrowdSensing application environment by focusing on issues related to privacy of participant.

Keywords: IoT · Sensing · Mobile CrowdSensing · Privacy

1 Introduction

Wireless technologies, sensor networks, smart, wearable networks, coupled with new IoT models and the diffusion of their uses, are creating new sources of business value, and bringing numerous challenges to privacy, security, and data integrity [1].

Gartner estimates that by 2020, 20 billion things will be connected to the Internet, including smartphones, computers, and dedicated-function objects. Mentioning that IoT will have a major impact on the economy, turning many businesses into digital businesses and facilitating new business models [2].

In 2020, Gartner estimates internet-connected things will outnumber humans 4-to-1, creating new dynamics for marketing, sales and customer service [2].

Cloud computing, and mobile devices are the primary assistant for the reception and use of sensors information. Currently in the Smart Cities, each citizen with a mobile device becomes a provider of relevant information to improve the public administration. Smartphones are equipped with a number of embedded sensors such as GPS, accelerometer, gyroscope, brightness, microphone, camera and others that allow the detection of environmental data, and so they can be interpreted and processed by various applications, creating this global scenario the opportunity for the development of applications that make use of the ability to detect mobile devices, transforming the collected data into useful information [3].

The scope of CrowdSensing is a newly application paradigm that enables ubiquitous mobile devices with enhanced sensor capabilities to collect and share local information toward a common goal [4]. Sensing devices potentially collect confidential data from individuals, being privacy a key problem [5]. The sensing data must be protected against unauthorized access, being CrowdSensing participant privacy an emerging challenge.

The aim of this article is to provide an insight into the Mobile Crowd Sensing application environment by focusing on issues related to privacy of participant in CrowdSensing systems.

In Sect. 2 we present the MCS concepts. After in Sect. 3, some MCS privacy threats are presented. Finally, in Sect. 4 we presents some final considerations.

2   Mobile CrowdSensing

The process of sensing is used in the management of Smart Cities through the monitoring of urban areas and observation of the dynamics of communities, aiming to provide managers with information essential for decision-making on a wide range of subjects, such as the sensing of environmental factors allow authorities or agencies to obtain data and inform the public about traffic conditions, noise pollution, air pollution, water quality, public safety, among other things, informing what happens, when and what happens when something happens [6].

In the context of smart cities, sensing combines the omnipresence of smartphones with the ability of sensors to collect data that depict different aspects of the city, being used to improve citizens' lives and to help decision makers in city management [3].

A new sensing paradigm called Mobile CrowdSensing (MCS) has been taking advantage of the wide range of capabilities to monitor and share common interest information collected through sensors embedded in mobile devices to support individuals and businesses in the decision-making process [7]. The collection can be carried out opportunistically or in a participatory way. In an opportunistic way the user initially has access to the application for data sensing, and later sends autonomously the detected data to a back-end server for processing; in turn, in the participatory way, users performs an action through the smartphone providing the sensed data [8–10].

MCS refers to the great diversity and heterogeneity of sensors through which individuals collectively share data and extract information to measure and map phenomena of common interest [11], and uses mobile devices equipped with sensors to collect data from the surrounding environment [10].

MCS provides a new way of seeing and perceiving the world, involving anyone in the process of sensing, allowing to increase the service of IoT, as well build smarter heterogeneous networks that interconnect things with things, things with people, and people with people [12]. Being MCS environments highly dynamic, where mobile devices, the data type of each sensor and the quality in terms of accuracy, latency and reliability can change randomly [3, 13].

MCS uses crowd sourcing for large-scale sensing, leveraging the mobility and sensing capabilities of participants' devices as well as the existing communication infrastructure, making deployment easier and reducing costs, since it is not necessary to build a specific infrastructure, as in conventional sensors networks [13].

Compared to other sensing approaches with user participation, in the case of MCS, all data collection and information actions (registration) are done from the mobile device of the participating user of the service, connected to any Internet access network [14], often without the need to register in real time.

The overall model of activities to be developed in MCS has three dimensions: Sensing activity, Data generation; and Data processing (Fig. 1). In the 1st dimension, activities are defined according to the objectives outlined, and an application is created \made available to the participants. The 2nd dimension, data generation can be performed in the individual context by mobile sensing, or in the social context in mobile social networks (MSN), including all data collection and registration. In the 3th dimension is carried out the processing of the data collected from all the participants, to post prior dissemination of results.

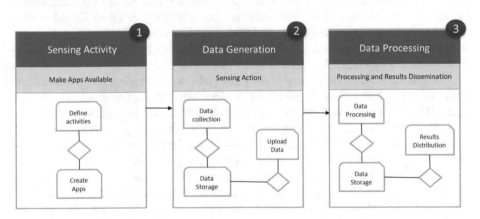

Fig. 1. Mobile CrowdSensing activities model dimensions.

Table 1. Mobile CrowdSensing categories.

Categories	Description
Environmental	*Monitoring natural phenomena such as levels of noise or pollution in a particular city. These applications allow the monitoring of several large-scale environmental phenomena*
Infrastructure	*Include large-scale phenomena related to public infrastructure. Examples of this type of application include the road conditions, availability of parking, traffic congestion measurement in real time, among others*
Social	*Participants share monitored information among themselves to collaborate for a common cause*

MCS applications can be classified into three different categories based on the type of phenomena to be monitored: environmental; infrastructure; and social [15] (Table 1).

3   Participant Privacy

WP29 (Article 29 Data Protection Working Party), European privacy and data protection adviser, established by Article 29 of Directive 95/46/EC, has decided to give a specific opinion on the consideration that IoT represents a large number of privacy and data protection challenges, some of which are new and other traditional ones, which will increase simultaneously with the exponential increase of data processing, resulting from the continuous evolution of IoT.

WP29 identifies the following privacy and data protection challenges in IoT: lack of control and asymmetry of information; quality of user consent; redefinition of the original data processing; identification of patterns and relationships; and limitations on the possibility of maintaining anonymity when using services (Table 2). And these are also the challenge of MCS.

Sensing devices potentially collect sensitive data from individuals [5], being privacy a key problem. Data captured by sensing must be protected against unauthorized (unauthorized) access, and be used only for the efficiency of some CrowdSensing services or activity to be performed. And that should be done with the knowledge and endorsement of who is making the information available (the participant), complying with the data protection laws in force in the country.

The guarantee of privacy is one of the pillars of modern society and the rule of law. It can be defined as a right of control by the individual about the circulation of their personal information, a right to not have their data registered or used by third parties.

GPS sensor readings usually record private information of participants, and when GPS sensing data are sharing participants' privacy can be compromised. Therefore, it is necessary to preserve the security and privacy of the participant.

On the other hand, in MCS systems, personal information may not be obtained directly, but inferred from aggregated data, as is the case of objects\things with RFID tags, which allow the user to be traced and identified and may create privacy problems. Sharing personal data on MCS systems can raise privacy concerns. it is essential that new techniques for protecting user privacy are developed, allowing their devices to contribute reliably. Then it's necessary to ensure that participants' data are not disclosed to unreliable third parties

Privacy is the right of each individual to maintain and control the set of information that surrounds him or her and may decide whether, when, why and by whom this information can be obtained and used. Due to MCS unique characteristics, privacy involves the right of the user/participant to remain intruder-free, and autonomous. The privacy in MCS has concerns with the direct disclosure of the identity of the participants as well as with the disclosure of sensitive attributes that allow to infer about the identity of the participants [13, 15].

From the participants' point of view, privacy threats can occur when the participant receives a specific task and shares their preferences during the assignment of this task

Table 2. WP29 privacy and data protection challenges in IoT.

Challenges	Description
Lack of control and asymmetry of information	The interaction between objects that communicate automatically, and between objects and back-end systems will result in the generation of data streams that can hardly be controlled with the traditional tools used to ensure the proper protection of the interests and rights of the data subjects. This issue of lack of control also concerns areas such as cloud computing or big data, and is even more challenging when it is thought that different emerging technologies can be used in combination, as is the case with MCS
Quality of user consent	*In many cases, the user may not be aware of the processing of data by certain devices. The possibility of rejecting certain services is not a viable alternative in IoT, and the classic mechanisms used to obtain consent are difficult to apply. Therefore, new ways of obtaining user consent for connected devices should be considered by their manufacturers*
Redefinition of the original data processing	*The increased amount of data generated by IoT in combination with modern data analysis and cross-matching techniques may give rise to secondary uses of the same data, whether or not related to the processing purpose initially assigned to the devices. That is, apparently insignificant data collected from devices can be used to infer information with a totally different purpose from the initial one*
Identification of patterns and relationships	*Although each device generates data streams in isolation, its collection and subsequent analysis can easily reveal individual patterns, behavior, preferences and habits. As seen in the redefinition of the original data processing, knowledge can be generated from trivial information, through profiling the sensor data*
Limitations on the possibility of maintaining anonymity when using services	*The full development of IoT capabilities can put pressure on the current possibilities of anonymous use of services and limit the possibility of remaining anonymous*

or notifies the server that accepted the task, in this case some attributes such as location, types of tasks in which the participant is interested, as well as some attributes of the sensor can be revealed [16]. In this case it is possible to argue that this information alone may not violate privacy, but may allow the "attacker" to track the tasks selected by the participant and thus reveal their identity or other sensitive attributes [17]. Among the attributes that can be used to track participants we can refer participant IDs and IP addresses, then participant's privacy must be protected at device level when communicating with the server, on server storage, and on processing.

As a way to maintain the privacy of participants various techniques are used: user preferences; anonymity; user preferences; anonymity in the distribution of tasks; Data Disturbance (see Table 3) [16, 17] [18].

Table 3. Techniques to maintain the privacy of participants.

Techniques	Description
Anonymity	*Although the use of anonymity techniques is very white, the intent is to remove any information that may identify participants or other entities during the distribution and performance of tasks*
User preferences	*Allow participants to configure their privacy preferences, thus enabling them to control the data collection process to be sensed*
Anonymity in the distribution of tasks	*Data collection from a sensor is usually triggered by tasks that specify the sensing modes based on, so tasks are only distributed to devices that meet the requirements of these*
Data disturbance	*Data disturbance adds noise to the sensor data before it is shared, and noise can be added to the data without compromising its accuracy*

4   Conclusions

At present cloud storage mechanisms are mature, sensor technologies and the evolution of IoT implementation models are quite accelerated, and the access devices are ready.

With the evolution of mobile computing, MCS emerges as a new term referring to the sharing of information collected from different mobile devices in order to measure and map phenomena of common interest. In order to perform this collection, there are two possible ways: in a participatory way, where the users exercise an action through the smartphone and make the data available; and opportunistically, where the user initially releases the application's access to the sensed data and the latter, in turn, almost autonomously, sends the data to a back-end server for processing.

Assuming that millions of individuals have at least one mobile device, CrowdSensing applications emerge as an inexpensive and time-consuming alternative, reducing efforts for the development of specialized sensor infrastructures.

Lacking the analytical mechanisms that allow efficiency gains from debugging the information generated by the CrowdSensing, the synchronization between the different technologies and IoT makes the MCS era a reality, which is possible with the sum of the technologies and tools of analysis, with the services of companies specialized in projects of crow sensing, thus making the sensing as a service.

MCS applications are gaining in popularity due to the creation of diverse systems and applications, conquering and involving more and more people, networks and group of collaborators. In the other hand, the use of MCS also has its risks, in this context privacy.

In spite of the privacy techniques, the participant's privacy must be one of the fundamental points in the construction of future comprehensive privacy-preserving architecture.

References

1. Guarda, T., Bustos, S., Torres, W., Villao, F.: Botnets the cat-mouse hunting. In: 2018 International Conference on Digital Science (2018)
2. Gartner: Leading the IoT. Gartner Insights on How to Lead. Mark Hung, Gartner Research Vice President (2017)
3. Guarda, T., Augusto, M.F., Díaz-Nafría, J.M.: Crowd sensing and delay tolerant networks to support decision making at the routing level. In: 2018 13th Iberian Conference on Information Systems and Technologies (CISTI) (2018)
4. Khan, W.Z., Xiang, Y., Aalsalem, M.Y., Arshad, Q.: Mobile phone sensing systems: a survey. IEEE Commun. Surv. Tutor. **15**(1), 402–427 (2013)
5. Chen, Y., Zhou, J., Guo, M.: A context-aware search system for internet of things based on hierarchical context. Telecommun. Syst. **62**(1), 77–91 (2016)
6. Theunis, J., Stevens, M., Botteldooren, D.: Sensing the environment. In: Participatory Sensing, Opinions and Collective Awareness, pp. 21–46 (2017)
7. Guo, B., Yu, Z., Zhou, X., Zhang, D.: From participatory sensing to mobile crowd sensing. In: International Conference on Pervasive Computing and Communications Workshops (PERCOM Workshops) (2014)
8. Miorandi, D., Carreras, I., Gregori, E., Graham, I., Stewart, J.: Measuring net neutrality in mobile Internet: towards a crowdsensing-based citizen observatory. In: IEEE International Conference on Communications Workshops (ICC), pp. 199–203 (2013)
9. Liu, J., Shen, H., Zhang, X.: A survey of mobile crowdsensing techniques: a critical component for the internet of things. In: 2016 25th International Conference on Computer Communication and Networks (ICCCN), pp. 1–6 (2016)
10. Guo, B., Wang, Z., Yu, Z., Wang, Y., Yen, N.Y., Huang, R., Zhou, X.: Mobile crowd sensing and computing: the review of an emerging human-powered sensing paradigm. In: International Conference on ACM Computing Surveys (CSUR) (2015)
11. Peng, D., Wu, F., Chen, G.: Pay as how well you do: a quality based incentive mechanism for crowdsensing. In: Proceedings of the 16th ACM International Symposium on Mobile Ad Hoc Networking and Computing (2015)
12. Jian, A., Xiaolin, G., Jianwei, Y., Yu, S., Xin, H.: Mobile crowd sensing for internet of things: a credible crowdsourcing model in mobile-sense service. In: International Conference on Multimedia Big Data (BigMM) (2015)
13. Han, G., Liu, L., Chan, S., Yu, R., Yang, Y.: HySense: a hybrid mobile crowdsensing framework for sensing opportunities compensation under dynamic coverage constraint. Commun. Mag. **55**(3), 93–99 (2017)
14. Bellavista, P., Corradi, A., Foschini, L., Ianniello, R.: Scalable and cost-effective assignment of mobile crowdsensing tasks based on profiling trends and prediction: the participact living lab experience. Sensors **15**(8), 18613–18640 (2015)
15. Biskup, J.: Security in Computing Systems: Challenges, Approaches and Solutions. Springer, Heidelberg (2008)

16. Chon, Y., Lane, N.D., Kim, Y., Zhao, F., Cha, H.: Understanding the coverage and scalability of place-centric crowdsensing. In: Proceedings of the 2013 ACM International Joint Conference on Pervasive and Ubiquitous Computing (2013)
17. Pournajaf, L., Garcia-Ulloa, D.A., Xiong, L., Sunderam, V.: Participant privacy in mobile crowd sensing task management: a survey of methods and challenges. ACM SIGMOD Rec. **44**(4), 23–34 (2016)

Author Index

© Springer Nature Switzerland AG 2019
Á. Rocha et al. (Eds.): WorldCIST'19 2019, AISC 932, pp. 629–631, 2019.
https://doi.org/10.1007/978-3-030-16187-3

Printed in the United States
By Bookmasters